The Origin and Early Evolution of Life: Prebiotic Systems Chemistry Perspective

The Origin and Early Evolution of Life: Prebiotic Systems Chemistry Perspective

Editors

Michele Fiore
Emiliano Altamura

MDPI • Basel • Beijing • Wuhan • Barcelona • Belgrade • Manchester • Tokyo • Cluj • Tianjin

Editors
Michele Fiore
Institut de Chimie et
Biochimie Moléculaires et
Supramoléculaires,
University of Lyon Claude
Bernard Lyon 1
Villeurbanne
Italy

Emiliano Altamura
Department of Chemistry
University of Bari Aldo Moro
Bari
Italy

Editorial Office
MDPI
St. Alban-Anlage 66
4052 Basel, Switzerland

This is a reprint of articles from the Special Issue published online in the open access journal *Life* (ISSN 2075-1729) (available at: www.mdpi.com/journal/life/special_issues/Prebiotic_Systems_Chemistry_Biomolecules).

For citation purposes, cite each article independently as indicated on the article page online and as indicated below:

LastName, A.A.; LastName, B.B.; LastName, C.C. Article Title. *Journal Name* **Year**, *Volume Number*, Page Range.

ISBN 978-3-0365-4470-0 (Hbk)
ISBN 978-3-0365-4469-4 (PDF)

© 2022 by the authors. Articles in this book are Open Access and distributed under the Creative Commons Attribution (CC BY) license, which allows users to download, copy and build upon published articles, as long as the author and publisher are properly credited, which ensures maximum dissemination and a wider impact of our publications.

The book as a whole is distributed by MDPI under the terms and conditions of the Creative Commons license CC BY-NC-ND.

Contents

About the Editors vii

Preface to "The Origin and Early Evolution of Life: Prebiotic Systems Chemistry Perspective" ix

Emiliano Altamura and Michele Fiore
The Origin and Early Evolution of Life: (Prebiotic) Systems Chemistry Perspective
Reprinted from: *Life* **2022**, *12*, 710, doi:10.3390/life12050710 1

Addy Pross
How Was Nature Able to Discover Its Own Laws—Twice?
Reprinted from: *Life* **2021**, *11*, 679, doi:10.3390/life11070679 5

Thomas Geisberger, Jessica Sobotta, Wolfgang Eisenreich and Claudia Huber
Formation of Thiophene under Simulated Volcanic Hydrothermal Conditions on Earth—Implications for Early Life on Extraterrestrial Planets?
Reprinted from: *Life* **2021**, *11*, 149, doi:10.3390/life11020149 13

Benton C. Clark, Vera M. Kolb, Andrew Steele, Christopher H. House, Nina L. Lanza and Patrick J. Gasda et al.
Origin of Life on Mars: Suitability and Opportunities
Reprinted from: *Life* **2021**, *11*, 539, doi:10.3390/life11060539 21

Daniele Fulvio, Alexey Potapov, Jiao He and Thomas Henning
Astrochemical Pathways to Complex Organic and Prebiotic Molecules: Experimental Perspectives for In Situ Solid-State Studies
Reprinted from: *Life* **2021**, *11*, 568, doi:10.3390/life11060568 67

İrep Gözen
Did Solid Surfaces Enable the Origin of Life?
Reprinted from: *Life* **2021**, *11*, 795, doi:10.3390/life11080795 89

Ylenia Miele, Gábor Holló, István Lagzi and Federico Rossi
Effect of the Membrane Composition of Giant Unilamellar Vesicles on Their Budding Probability: A Trade-Off between Elasticity and Preferred Area Difference
Reprinted from: *Life* **2021**, *11*, 634, doi:10.3390/life11070634 99

Adrien D. Garcia, Cornelia Meinert, Friedrich Finger, Uwe J. Meierhenrich and Ewald Hejl
Racemate Resolution of Alanine and Leucine on Homochiral Quartz, and Its Alteration by Strong Radiation Damage
Reprinted from: *Life* **2021**, *11*, 1222, doi:10.3390/life11111222 111

María J. Dávila and Christian Mayer
Membrane Structure Obtained in an Experimental Evolution Process
Reprinted from: *Life* **2022**, *12*, 145, doi:10.3390/life12020145 127

Manesh Prakash Joshi, Luke Steller, Martin J. Van Kranendonk and Sudha Rajamani
Influence of Metal Ions on Model Protoamphiphilic Vesicular Systems: Insights from Laboratory and Analogue Studies
Reprinted from: *Life* **2021**, *11*, 1413, doi:10.3390/life11121413 141

Paul G. Higgs
When Is a Reaction Network a Metabolism? Criteria for Simple Metabolisms That Support Growth and Division of Protocells
Reprinted from: *Life* **2021**, *11*, 966, doi:10.3390/life11090966 157

Xin-Yi Chu, Si-Ming Chen, Ke-Wei Zhao, Tian Tian, Jun Gao and Hong-Yu Zhang
Plausibility of Early Life in a Relatively Wide Temperature Range: Clues from Simulated Metabolic Network Expansion
Reprinted from: *Life* **2021**, *11*, 738, doi:10.3390/life11080738 183

Jeremy Kua, Alexandra L. Hernandez and Danielle N. Velasquez
Thermodynamics of Potential CHO Metabolites in a Reducing Environment
Reprinted from: *Life* **2021**, *11*, 1025, doi:10.3390/life11101025 193

Arthur Omran, Josh Abbatiello, Tian Feng and Matthew A. Pasek
Oxidative Phosphorus Chemistry Perturbed by Minerals
Reprinted from: *Life* **2022**, *12*, 198, doi:10.3390/life12020198 217

Alexander Nesterov-Mueller and Roman Popov
The Combinatorial Fusion Cascade to Generate the Standard Genetic Code
Reprinted from: *Life* **2021**, *11*, 975, doi:10.3390/life11090975 229

Judit E. Šponer, Jiří Šponer, Aleš Kovařík, Ondrej Šedo, Zbyněk Zdráhal and Giovanna Costanzo et al.
Questions and Answers Related to the Prebiotic Production of Oligonucleotide Sequences from 3′,5′ Cyclic Nucleotide Precursors
Reprinted from: *Life* **2021**, *11*, 800, doi:10.3390/life11080800 241

Jozef Nahalka and Eva Hrabarova
Prebiotic Peptides Based on the Glycocodon Theory Analyzed with FRET
Reprinted from: *Life* **2021**, *11*, 380, doi:10.3390/life11050380 255

Alexander Nesterov-Mueller, Roman Popov and Hervé Seligmann
Combinatorial Fusion Rules to Describe Codon Assignment in the Standard Genetic Code
Reprinted from: *Life* **2020**, *11*, 4, doi:10.3390/life11010004 265

Mario Rivas and George E. Fox
Further Characterization of the Pseudo-Symmetrical Ribosomal Region
Reprinted from: *Life* **2020**, *10*, 201, doi:10.3390/life10090201 283

Benjamin D. Lee and Eugene V. Koonin
Viroids and Viroid-like Circular RNAs: Do They Descend from Primordial Replicators?
Reprinted from: *Life* **2022**, *12*, 103, doi:10.3390/life12010103 297

About the Editors

Michele Fiore

Michele Fiore obtained a PhD in Agrobiology and Agrochemistry in 2007 at the University of Naples, Federico II, Italy. From 2008 to 2011, he was a postdoc under the guidance of Professors A. Dondoni and A. Marra at the University of Ferrara, Italy, working on radical reactions. From 2012 to 2014, he was a postdoc, and then became a research associate at the I2BM (Prof. O Renaudet) on vaccine development. Since 2014, he has held the position of Associate Professor and obtained the title of HDR, discussing a thesis on prebiotic synthesis and the role of phospholipids and other amphiphiles at the University of Lyon. His research interests include ürebiotic chemistry, symmetry breaking under prebiotic conditions, carbohydrate chemistry, vaccines, systems chemistry, and fluorophores.

Emiliano Altamura

Emiliano Altamura was awarded a PhD in Chemical and Molecular Science (Innovative Materials Chemistry), discussing the thesis "Bio-mimetic Cell-like Soft Matter Systems" at the Chemistry Department of the University of Bari Aldo Moro in Italy, in 2014. After six years as a postdoc working on the project "Giant vesicles as minimal cell model systems", he received a position as assistant professor at the same university (December 2020). Research interests include systems chemistry, biophysics, synthetic biology, origin of life, giant lipid vesicles, enzymatic reaction in compartmentalized systems, artificial photosynthetic cells, and lipo-thermogels.

Preface to "The Origin and Early Evolution of Life: Prebiotic Systems Chemistry Perspective"

What is life? How, where, and when did life arise? These questions have remained most fascinating over the last hundred years. Aristoteles postulated 2400 years ago that "nature does not do anything endless" and many years later the biologist Carl Richard Woese (1928–2012) emphasized the urgency of conducting in-depth studies in search of what we call life with the following words: "Biology today is no more fully understood in principle than physics was a century or so ago. In both cases the guiding vision has (or had) reached its end, and in both, a new, deeper, more invigorating representation of reality is (or was) called for". Systems chemistry is the way to go to better understand this problem and to try and answer the unsolved question regarding the origin of Life. Self-organization, thanks to the role of lipid boundaries, made possible the rise of protocells. The role of these boundaries is to separate and co-locate micro-environments, and make them spatially distinct; to protect and keep them at defined concentrations; and to enable a multitude of often competing and interfering biochemical reactions to occur simultaneously.

The aim of this Special Issue is to summarize the latest discoveries in the field of the prebiotic chemistry of biomolecules, self-organization, protocells and the origin of life. In recent years, thousands of excellent reviews and articles have appeared in the literature and some breakthroughs have already been achieved. However, a great deal of work remains to be carried out. Beyond the borders of the traditional domains of scientific activity, the multidisciplinary character of the present Special Issue leaves space for anyone to creatively contribute to any aspect of these and related relevant topics. We hope that the presented works will be stimulating for a new generation of scientists that are taking their first steps in this fascinating field.

Michele Fiore and Emiliano Altamura
Editors

Editorial

The Origin and Early Evolution of Life: (Prebiotic) Systems Chemistry Perspective

Emiliano Altamura [1] and Michele Fiore [2,*]

1. Chemistry Department, University of Bari Aldo Moro, Via Orabona 4, 70125 Bari, Italy; emiliano.altamura@uniba.it
2. Institut de Chimie et Biochimie Moléculaires et Supramoléculaires, Université de Lyon, Claude Bernard Lyon 1, Batiment Lederer, Bureau 11.002, 1 Rue Victor Grignard, F-69622 Villeurbanne, France
* Correspondence: michele.fiore@univ-lyon1.fr

Citation: Altamura, E.; Fiore, M. The Origin and Early Evolution of Life: (Prebiotic) Systems Chemistry Perspective. *Life* **2022**, *12*, 710. https://doi.org/10.3390/life12050710

Received: 25 March 2022
Accepted: 7 May 2022
Published: 10 May 2022

Publisher's Note: MDPI stays neutral with regard to jurisdictional claims in published maps and institutional affiliations.

Copyright: © 2022 by the authors. Licensee MDPI, Basel, Switzerland. This article is an open access article distributed under the terms and conditions of the Creative Commons Attribution (CC BY) license (https://creativecommons.org/licenses/by/4.0/).

1. Systems Chemistry

Aristotle considered that "nature does not do anything endless". Today, despite the enormous scientific achievements reached in the field of the applied science, such as medicine, biology, engineering, physics and chemistry, many question marks accompany the life of scientists. Mankind has always asked one question "where do we come from? How could life have emerged from an inanimate collection of organic molecules, minerals and water?" International scientific figures, often chemists, who believe that life emerged spontaneously from a mixture of molecules in a prebiotic land agree that studies on the spontaneous origin of life are needed [1–3].

Molecular biology is undergoing a transformation, moving from studying cellular processes as they currently are; both basic understanding (analytical approach) and hypothesis testing by a constructive procedure (synthetic approach), using artificial models. This second route, which complements the first one, requires a degree of understanding that is ultimately based on chemistry and physics—this is why we say, "from the bottom". For example, building artificial analogues of biochemical processes reveals whether we have really understood how a system works or whether we fail because something is still missing (according to the motto: "What I cannot create, I do not understand", by the Nobel Prize winning physicist Richard Feynman).

"*Biology today is no more fully understood in principle than physics was a century or so ago. In both cases the guiding vision has (or had) reached its end, and in both, a new, deeper, more invigorating representation of reality is (or was) called for*". The late biologist Carl Richard Woese (1928–2012) emphasized the urgency of conducting in-depth studies in search of what in the early days of the formation of the universe and then of our planet gave rise to what is called life [4]. In 2005, the term "**Systems Chemistry**" appeared in a conference on Prebiotic Chemistry and Early Evolution (Chembiogenesis 2005 in Venice, Italy). This field of chemical research has its roots in a number of different areas such as dynamic combinatorial chemistry, self-assembly and self-organization, prebiotic chemistry, minimal self-replicating molecules, metabolic and non-metabolic networks and autocatalytic systems. In few words, systems chemistry is the science of studying the networks of interacting molecules to form new functions from a set of molecular components at different hierarchical levels with emergent properties. The goal now is to understand, perhaps from experimental reproduction, the formation of a compartmentalized chemical system far from thermodynamic equilibrium, a system which is kinetically stable and can self-maintain (metabolism), dynamically evolve and is capable of dividing/self-reproducing, which we then could call "*alive*" or "*animate*", according to one of the definition of life proposed in the literature in 2005 [5].

2. The Seminal Idea of Systems Chemistry and the Bottom-Up Approach

The main purpose of synthetic chemistry is the preparation of pure, designed compounds by using well-defined synthetic pathways and multi-step reactions. The preparation of a complex mixture of chemical compounds can be highly interesting since simple mixtures of nonreacting molecules can trigger chemical reaction networks, including feedback loops and elements of nonlinearity form. From such systems, new, unexpected and unpredicted emergent properties could arise, while none of the components alone have these properties. Systems chemistry is therefore an extremely interesting and very new way to do and to think and to rethink chemistry beyond the fundamental insight that we can obtain.

From the origin-of-life point of view, the top-down approach consists of simplifying the machinery of contemporary cells in order to obtain the minimal but efficient system which would resemble primitive cells and which could explain how they worked. Systems chemistry though can be interpreted as a **bottom-up approach**, which aims at piecing together the parts of a whole to give rise to a new and more complex system with emerging proprieties [6–9]. Concerning the study of protocellular entities, artificial lipid vesicles can be used as biomimetic systems to study and gain insights into the physiological functions of biological cells. These are cell-sized, artificial compartments consisting of natural or synthetic amphiphilic molecules, creating an artificial entity to mimic the structure and some essential properties of a natural cell, and artificial reaction networks are used to program the functions of protocells conferring specific tasks [10]. Indeed, the reduced complexity of the system allows the investigation of the dynamics of the cell behavior, minimizing the interference from the cellular complexity [11]. Since the conditions to observe the transition from inanimate to living matter are still missing, the bottom-up approach can help to clarify some aspects of this transition by building models with increasing complexity, from very early cells, not yet alive, to the first artificial living cell. Protocell systems can help us think about the minimal complexity required for the "emergence" of properties and/or the "emergence" of life. This method of proceeding tends to reproduce the process, which led to the outbreak of life and biology in chemistry.

3. The Aim of Prebiotic Systems Chemistry of Biomolecules

The aim of the systems chemistry of biomolecules is to combine chemistry, physics and biology to give plausible answers, supported by rigorous scientific experiments, to some questions concerning the origin of life and the evolution of synthetic chemical systems complexity. Kuhn, Pross, Pascal, Szostak and Luisi, among others, highlighted the problems of understanding the transition from chemical to biological processes and the abiotic synthesis of biomolecules. With this aim, we have asked more than fifty selected scientists to participate in this Special Issue: nineteen original articles or reviews have been published so far. We wish to thank them all for their efforts and their scientific commitment.

Acknowledgments: Michele Fiore and Emiliano Altamura wish to thank all the contributors of the Special Issue of *Life* (ISSN 2075-1729): "The Origin and Early Evolution of Life: Prebiotic Systems Chemistry Perspective". MF's daily work is dedicated to the memory of his beloved daughter, Océane (2015–2017).

Conflicts of Interest: The authors declares no conflict of interest.

References

1. Pross, A. Seeking the Chemical Roots of Darwinism: Bridging between Chemistry and Biology. *Chem.-A Eur. J.* **2009**, *15*, 8374–8381. [CrossRef] [PubMed]
2. Kauffman, A. *Investigations*; Oxford University Press: Oxford, UK, 2000.
3. Whitesides, G.M.; Boncheva, M. Beyond molecules: Self-assembly of mesoscopic and macroscopic components. *Proc. Natl. Acad. Sci. USA* **2002**, *99*, 4769–4774. [CrossRef] [PubMed]
4. Woese, C.R. A New Biology for a New Century. *Microbiol. Mol. Biol. Rev.* **2004**, *68*, 173–186. [CrossRef] [PubMed]
5. Pross, A. Toward a general theory of evolution: Extending Darwinian theory to inanimate matter. *J. Syst. Chem.* **2011**, *2*, 1. [CrossRef]

6. Fiore, M.; Madanamoothoo, W.; Berlioz-Barbier, A.; Maniti, O.; Girard-Egrot, A.; Buchet, R.; Strazewski, P. Giant vesicles from rehydrated crude mixtures containing unexpected mixtures of amphiphiles formed under plausibly prebiotic conditions. *Org. Biomol. Chem.* **2017**, *15*, 4231–4240. [CrossRef] [PubMed]
7. Fayolle, D.; Altamura, E.; D'Onofrio, A.; Madanamothoo, W.; Fenet, B.; Mavelli, F.; Buchet, R.; Stano, P.; Fiore, M.; Strazewski, P. Crude phosphorylation mixtures containing racemic lipid amphiphiles self-assemble to give stable primitive compartments. *Sci. Rep.* **2017**, *7*, 18106. [CrossRef] [PubMed]
8. Altamura, E.; Comte, A.; D'Onofrio, A.; Roussillon, C.; Fayolle, D.; Buchet, R.; Mavelli, F.; Stano, P.; Fiore, M.; Strazewski, P. Racemic Phospholipids for Origin of Life Studies. *Symmetry* **2020**, *12*, 1108. [CrossRef]
9. Fiore, M.; Buchet, R. Symmetry breaking of phospholipids. *Symmetry* **2020**, *12*, 1488. [CrossRef]
10. Altamura, E.; Albanese, P.; Marotta, R.; Milano, F.; Fiore, M.; Trotta, M.; Stano, P.; Mavelli, F. Chromatophores efficiently promote light-driven ATP synthesis and DNA transcription inside hybrid multicompartment artificial cells. *Proc. Natl. Acad. Sci. USA* **2021**, *118*, e2012170118. [CrossRef]
11. Lopez, A.; Fiore, M. Investigating Prebiotic Protocells for a Comprehensive Understanding of the Origins of Life: A Prebiotic Systems Chemistry Perspective. *Life* **2019**, *9*, 49. [CrossRef] [PubMed]

Review

How Was Nature Able to Discover Its Own Laws—Twice?

Addy Pross

Department of Chemistry, Ben Gurion University of the Negev, Be'er Sheva 8410501, Israel; pross@bgu.ac.il

Abstract: The central thesis of the modern scientific revolution is that nature is objective. Yet, somehow, out of that objective reality, projective systems emerged—cognitive and purposeful. More remarkably, through nature's objective laws, chemical systems emerged and evolved to take advantage of those laws. Even more inexplicably, nature uncovered those laws twice—once unconsciously, once consciously. Accordingly, one could rephrase the origin of life question as follows: how was nature able to become self-aware and discover its own laws? What is the law of nature that enabled nature to discover its own laws? Addressing these challenging questions in physical-chemical terms may be possible through the newly emergent field of systems chemistry.

Keywords: origin of life; dynamic kinetic stability; cognition; chemical evolution; systems chemistry

1. Introduction

The origin of life problem is striking in that it can be posed in several seemingly unrelated ways. Beyond the direct question: 'how did life emerge?', one could ask how it was possible for function and purpose—key life characteristics—to have emerged from an inanimate universe in which such characteristics would presumably have been non-existent? [1,2]. Functional biology, a central sub-discipline within biology [3], is rooted in life's functional character, so that character must be inexorably linked to the origins question. Or, to pose the origin question in yet another way, how was an information processing system able to arise? Information transmission requires both a sender and a receiver—but how did senders and receivers come to be? How did matter become cognitive thereby enabling information transmission and processing to take place?

Yet another facet of the life problem remains no less perplexing. In exploring the nature of the universe, we must begin with certain axiomatic assumptions, primarily, the physical existence of the universe, together with laws that govern its behavior. That is the starting point from which all scientific endeavor commences. But a deep puzzle then arises. As will now be discussed, examination of living things reveals that through the life process, nature has acquired knowledge of its own laws, and developed a means of exploiting those laws. What then is the law of nature that enables nature to discover itself? Addressing that question, and how it relates to the origin of life problem, is the topic of this essay.

2. Discussion

2.1. Nature, the Technologist

We are all familiar with the concept of technology, the application of scientific knowledge for practical purposes. It is ubiquitous. Starting from simple applications, such as stone tools, human ingenuity has taken us far, and today's range of technological applications is mesmerizingly large. Just think of modern communication, transport, health management, manufacturing. Michael Faraday and James Clerk Maxwell could not have imagined to what extent their pioneering work on the laws governing electromagnetism and electromagnetic radiation could change the world. However, while the idea that humankind has been able to exploit nature's laws for practical purposes is obvious to all, there is a truly mysterious feature of technological development, one generally overlooked, that needs to be mentioned. Technological development and innovation did not start with us humans, but began literally billions of years earlier, before man ever set foot on the earth, in fact, at a time when all life on earth was still microbial. The striking reality is that nature's technological achievements, even for simplest life, have far exceeded human technological achievement that we, in our anthropocentric arrogance, take so much credit for. A comparison of human versus natural technological capability is revealing.

Take flying for example. The Swiss mathematician, Daniel Bernoulli, uncovered the principles governing flight in 1738 when he discovered the relationship between pressure, density and velocity in a flowing fluid. That understanding eventually led to the Wright brothers taking their first flight almost two centuries later. Nature, however, has been exploiting those same aeronautical principles for millions of years, well before humans existed. Insects, for example, already took to the air some 320 million years ago, and flight would not have been possible without the insect's ability to accommodate the specific structural and dynamic requirements that derive from those aeronautical principles. More remarkably, nature's 'understanding' of aeronautical principles extends far beyond the ones laid down by Bernoulli. In bees, for example, the principles of flight are particularly complex and go beyond the thrust and lift concepts governing avian and aircraft flight. Careful study of how bees fly has revealed additional forces that come into being through wing rotation and vortex formation, as the short and stubby bee body would not support flight through just wing beating. The fluid dynamics enabling bee flight turn out to be different to those that enable a plane or bird to fly [4].

In the common house fly, the above mentioned technological prowess goes even further, demonstrating capabilities currently quite out of reach of human engineers. Flies have mastered the knack of routinely landing on a ceiling upside down, a remarkable aerodynamic feat. The successful inverted landing of a fly depends on a series of several perfectly timed and highly coordinated actions. First, the fly increases its speed, then it undertakes a rapid body rotation, much like a cartwheel, then it proceeds to extend its legs, before finally landing by a body swing that pivots around the legs that have attached to the ceiling, and all of this coordinated with the fly's visual and other sensory capabilities. Such capability greatly exceeds current robot technology resulting in engineers utilizing high-speed videography of the fly's inverted landing to learn how the fly carries out such a sophisticated maneuver [5]. Human engineers are not proud—they are more than happy to consult with common house flies to further their technological capabilities!

The aeronautical examples described above are just arbitrary ones from an almost infinite list of exceptional natural technological capabilities, and from many fields of human endeavor. For example, you might think that the chemical giant Dupont was the first to make polymeric materials, starting with nylon, the first synthetic fiber [6]. Nature, however, has been routinely manufacturing a range of natural polymeric materials for eons, whether silk, wool, or cotton. However, it is when one examines nature's technological capabilities at the molecular level, that a truly astounding picture comes to light. No matter where one looks within the biological cell, one sees extraordinary technological capabilities, all quite staggering from a human technological perspective. Consider the capabilities of the ribosome, that microscopic entity located in the thousands in every living cell, and able to synthesize proteins from a supply of amino acids in an assembly-line type process, based on information coded into the cell's DNA sequence. That molecular machine is able to churn out required proteins in the space of a few seconds [7]. The synthetic chemist, while able to combine amino acids to form peptides and simple proteins, can only gaze in wonder at the staggering efficiency and specificity of that ribosomal system.

Discussing the ribosomal protein generation system brings to the fore the issue of information processing technology. That leads directly to Claude Shannon's landmark theory of information published in 1948 [8]. However, a moment's thought indicates that nature has been applying information theory well before Shannon published his ideas and, together with Alan Turing and John von Neumann, initiated the information age. The role of nucleic acid sequences in governing life processes reveals that nature was into digital information transmission and storage billions of years ahead of those legendary human information pioneers.

My point is simple. Through an evolutionary process, nature has succeeded in uncovering and exploiting technological principles far beyond what we humans, even today, are able to achieve. Without doubt, nature is the ultimate technologist—the supreme information theorist, polymer scientist, energy engineer, electrical engineer, operations manager, architect, systems chemist, photochemist, synthetic chemist, molecular biologist, to name just a few of nature's wide-ranging capabilities. It is staggering to realize that every living thing is continually exploiting nature's laws, and able to do so far more creatively than any human technologist, past or present, and most remarkably, in the case of simple life, they are doing so without those life forms even knowing that such laws exist!

2.2. The Origin of Technology

The question now arises: how could such cognitive capabilities have come into being? How was nature able to become the ultimate technologist? Darwinian natural selection is clearly a significant part of the answer to that question. Darwinian selection is traditionally viewed as a blind algorithmic process, akin to what takes place during the execution of a computer program [9]. Such algorithmic processes have turned out to be enormously powerful, as clearly demonstrated by the exceptional technological innovations that computer programs themselves have brought into our lives.

However, pointing to the existence of such an algorithmic process is not sufficient. Something more is required. To better understand that additional requirement, consider a computer programmed to play chess. Thanks to the algorithmic process taking place within that computer, a computer may well outplay any human opponent. Interestingly, however, that chess-playing computer does not know that a human game called chess even exists, let alone that it is playing the game at any given moment. A computer only does what all computers do—it reduces any problem to a set of 'dumb' calculations. In similar fashion, the simplest living entity, a bacterium, is totally unaware of nature's laws, but, again, through a step-by-step algorithmic process—in this case, that of natural selection—extraordinary technological advances came to be. Just as a chess computer does not know it is playing chess, a bacterium does not know that it is utilizing information theory to produce proteins, that it is actioning its immune system to protect itself from viral attack, or that it is propelling itself toward a food source.

However, the evolutionary process then takes a staggering, almost incomprehensible turn. Through that supposedly blind algorithmic process, an entirely new and implausible technological innovation emerged—mind. Nature appeared to have stumbled upon a new dimension of reality, one that seemingly extends beyond the physical. However, though that new dimension might appear separate from the physical, it is important to recognize that the non-physical dimension of mind can only be accessed through the physical dimension. Or, put differently, all indications are that the non-physical world is an evolutionary outcome rooted within that physical world, and therefore necessarily part of it.

The significance of that development is hard to overstate. It means that nature began to see, not just physically, as in sight, but conceptually, as in mind. Mind—the ultimate natural innovation, one we humans continually use, but have barely begun to understand. This remarkable innovation, in addition to its many practical benefits, enabled us humans to begin to discover natural law *consciously*, in stark contrast to the seemingly undirected algorithmic approach that nature followed for billions of years prior to that evolutionary development. The implications of that new pathway to scientific discovery cannot be overstated—*nature managed to uncover and exploit its own laws, not once, but twice!* Once, mindlessly—algorithmically; once, mindfully—consciously. Moreover, most wonderfully, the *mindful* way allowed us to begin to investigate, to understand, and to marvel at the *mindless* way. In the ultimate irony, through the technological innovation of mind, nature discovered a means of exploring itself. Through what started off as a seemingly blind algorithmic process, the cosmos became able to explore itself, to become self-aware.

The above discussion now enables us to rephrase the life question in an unconventional manner: how was a material system, capable of undergoing such an extraordinarily innovative algorithmic process, able to come about? Just as you cannot program a computer to play chess, if you do not have a computer on which that program can be installed, you cannot have technological advance through an algorithmic evolutionary process without there being a suitable material substrate—in this case, a simple life form—on which the blind algorithmic process can operate. How did that initial life form, the equivalent of the computer in the chess analogy, come to be? Hidden within natural law *there must exist the primal means by which nature was able to create such a system, one with the extraordinary potential to discover itself!* Agency, mind, technological innovation, scientific discovery—whether natural, whether human—all arose out of that primal system. However, it means that hidden away within the natural order, lie the physical means for exploring the natural order. It also suggests that the non-material aspects of reality—mind and consciousness—which accompanied the emergence of advanced life, can be physicalized, though the term is used here in a broader sense to the one usually employed. To support that less traditional view, I would argue that any manifestation of a physical system, such as mind and consciousness, even if it is not within traditional physical bounds, should be viewed as an extension of the physical world, and therefore subject to physical consideration. In any case, where natural selection was able to lead, human thought, like a good tracker, can surely follow. The path from mindless to mindful was presumably stepwise and rooted in material change, just like all evolutionary processes, and therefore, a legitimate subject for physical study, though one still to be explored. However, that is for the future. The current and primary challenge is more immediate: to discover how the first step along that extraordinary emergence process—from inanimate beginnings to simplest life—was at all possible. How could it have begun?

2.3. Kinetic and Thermodynamic Stabilities

Change characterizes the world, and the fundamental law that describes the direction of change in nature is the Second Law of Thermodynamics [10,11]. However, as we are all aware, that Second Law directive leads toward death, not life. The equilibrium state, the state that all physical-chemical systems are directed toward, is the antithesis of life. So how could the central principle governing natural change, one antithetical to life, lead to the emergence of life, to a system able to become cognitive and explore the universe?

One way of expressing the Second Law is to state that material systems are driven toward stability for thermodynamic reasons—to maximize the systems' entropy [10,11]. However, through research in a relatively new area of chemistry termed systems chemistry [12], it has become increasingly clear that certain physical-chemical systems can be stable for *kinetic*, rather than thermodynamic reasons [13,14]. The persistent state that is achieved can be classified as one of *dynamic kinetic stability (DKS)* [15–18], as opposed to the more common stability kind, thermodynamic stability. Physical examples of dynamic kinetic stability are familiar—a water fountain or a waterfall. The dynamic, non-equilibrium steady state is maintained, as long as energy and material resources are continually provided.

However, the discovery by van Esch, Eelkema and co-workers in 2010 that a regular chemical system can be induced into such a persistent dynamic, energy-fuelled kinetic state [19,20] was a game changer. It made clear that there exists a new dimension of chemical possibility, a kinetic dimension. In that kinetic dimension, change derives primarily from kinetic factors, rather than thermodynamic ones, and revolves around the possible existence of thermodynamically unstable chemical systems maintained in a stable (persistent) dynamic kinetic state. What is particularly striking about this chemical state is that it can lead to unusual material characteristics, characteristics we normally associate with biological entities—self-healing, adaptive, even communicative [21]. Indeed, the DKS state depends on an energy-fuelled, irreversible, cyclical process, very much akin to life's metabolic processes, which are also energy-fuelled, cyclical, and irreversible. The discovery of DKS systems is therefore able to contribute to the establishment of a more coherent physicochemical framework for accommodating living systems [22]. Given the evident generality of this new kind of chemistry, we have proposed a descriptive term for it, *dynamic kinetic chemistry* [22]. With the discovery of dynamic kinetic chemistry, life chemistry may have finally found its physical-chemical home!

A further striking feature associated with this newly discovered kinetic domain is that for any given chemical system in a particular thermodynamic state, there exist, potentially at least, an extensive array of structurally distinct energized kinetic states associated with that single thermodynamic state, and all accessible through the tuning of the system's kinetic parameters [19,20,22]. In fact, biological systems have taken full advantage of the inherent multiplicity of kinetic states, as it is through the interplay of those multiple states that living cells are able to carry out key biological functions, for example, modulating the structural characteristics of the cell's cytoskeleton, thereby enabling cell motility and material transport to take place [23]. Thus, the discovery of this new chemical domain appears to be a significant step toward the longstanding goal of better relating chemical and biological processes. Not surprisingly, great interest has been directed toward the preparation of such systems [21,24,25]. A detailed kinetic analysis of these systems has also been undertaken [26,27].

So where does replication come into the life picture? Just as a replicative capability in some chemical system on its own, is insufficient to explain the emergence of life, so a DKS state in isolation is also unable to explain the life phenomenon. However, when the DKS state is *combined* with a replicative capability, unexpected chemical possibilities become possible, in fact, the door to life's emergence appears to open up. The kinetics associated with replication systems in the DKS state predicts that increasingly kinetically stable systems will form [28,29], and that realization helps lead to the formulation of a Persistence Principle [30], which can offer certain material insights beyond those provided by the Second Law. The principle may be stated as follows: *all material systems are driven toward more persistent forms*. As Grand pointed out with incontestable logic: "things that persist, persist, things that don't, don't" [31]. The point, however, is that persistence can come about for either kinetic or thermodynamic reasons, and that duality reveals the dual character of the stability concept: stability has *two* discrete facets: *time* and *energy*. Whereas the Second Law addresses stability primarily through the energy facet, the Persistence Principle does so through the time facet [30]. In fact, it is the kinetic power of replication that enables the Persistence Principle to extend its scope beyond that of the Second Law, as it helps explain the existence of highly persistent, yet thermodynamically unstable, systems.

The bottom line: the existence of replicative systems in a DKS state opens up an evolutionary path toward kinetically stable systems of enhanced persistence, where kinetic considerations, not energetic ones, play the dominant role. As a result, novel material characteristics can arise, eventually leading to life. Indeed, as will now be discussed, it would have been through the emergence of replicative DKS systems that would have led to the emergence of one of life's most striking and central characteristics—cognition.

2.4. The Origin of Cognition

Cognition has been traditionally defined as "the mechanisms by which animals acquire, process, store, and act on information from the environment" [32], and is normally associated with neural organisms. More recently, however, cognition has been understood to manifest in aneural life forms [33,34]. In fact, it has recently been claimed that simplest life—bacteria—exhibits strikingly sophisticated cognitive capabilities, despite their aneural nature [33,34]. That realization leads to the obvious question: could a *chemical* system be able to exhibit rudimentary cognitive behavior, to perceive, process, and react to information? The answer: apparently, *yes*, and it is on this very point that the biological significance of the DKS state is further reinforced.

Recent work on chemical DKS systems suggests that such systems may have crossed the threshold for the emergence of a basic cognitive capability. Thus Merindol and Walther [21] have described how DKS systems are able to self-heal, sense, adapt, communicate—key cognitive characteristics—despite being unambiguously within the chemical domain. Moreover, theoretical considerations that have addressed the underlying basis for cognition, have concluded that the ontological essence of cognitive agents is that they are "dynamical systems" [35,36]. That view is satisfyingly consistent with DKS's dynamic character.

However, with the chemical emergence of simple cognitive systems, a profoundly significant evolutionary event would have taken place. A cognizant chemical system that emerged, and which was structurally able to evolve and adapt to better exploit its environment in the drive toward increasing persistence, would find new structural and organizational possibilities. In fact, that process over extended time led to the emergence of the bacterial cell, a major milestone along the evolutionary path toward increasingly persistent forms. Incredibly, several billion years along, it led to the emergence of animals with neural systems and to the ultimate evolutionary discovery of mind. Significantly, however, the entire evolutionary process was governed by nature's incessant drive toward increasingly persistent forms.

We are thus able to identify nature's central principle that enabled nature to discover itself—*it is nature's fundamental drive toward persistent forms.* In contrast to the common biological view that considers the evolutionary process to be a blind algorithmic process [9], there *is* a general directive for that evolutionary process. If, within a segment of physical matter, a replicative system is able to become activated into a persistent DKS state, then its natural drive toward increased persistence could lead that system to explore and exploit its world to further enhance its persistence, beginning with cognition, ultimately leading to consciousness, and all through an evolutionary process of continuing complexification. Indeed, it is nature's inherent drive toward persistent forms that is responsible for life's extraordinary complexity, and that tendency is already apparent at the chemical level. As noted by Otto and co-workers in their recent molecular Darwinian study, DKS stable replicators that are *more* complex, are selected over ones that are *less* complex, despite the intrinsically slower replication rate of those more complex ones [37].

Of course, once the energy source and/or essential replicative building blocks enabling the evolutionary process to continue, have been dissipated, cognition and the transient formation of information will also dissipate, just as all material information forms eventually dissipate. Ultimately, matter continues to barrel on toward its final destination of heat death.

What then was the precise chemical form of the system that led to the emergence of cognition, enabling nature to explore itself? We will likely never know, as the historic record has itself largely dissipated through the degrading effect of time. No matter, we are now in a position to characterize that chemical system in general terms. The chemical system that was potentially able to evolve into life, to become increasingly cognizant, to learn to explore the world outside of itself, was one that was *replicative, evolvable, and in the DKS state*, effectively the very same characteristics that define the living state. A physical-chemical description of what life is thus emerges. And that realization then leads us to the ultimate life challenge: can the human life form now proceed to synthesize unnatural life forms? Our technological experience suggests the answer is most likely *yes*. After all, much of human technological achievement has been to mimic nature's creative achievements. It seems only reasonable to conclude that what nature was able to achieve mindlessly, we humans can achieve mindfully. The synthesis of simple chemical proto-life would now appear possible. However, this means that after centuries of scientific division, biology may be closer to becoming conceptually integrated within the physical sciences. We may finally be nearer to understanding life's fundamental physical-chemical essence. The implications of such physical-chemical understanding would appear to be profound.

Funding: This research received no external funding.

Institutional Review Board Statement: Not applicable.

Informed Consent Statement: Not applicable.

Data Availability Statement: Not applicable.

Acknowledgments: I thank Sijbren Otto and Robert Pascal for their most helpful comments on an earlier draft of this paper.

Conflicts of Interest: The author declares no conflict of interest.

References

1. Kauffman, S.A. *A World beyond Physics*; Oxford University Press: Oxford, UK, 2019.
2. Ruiz-Mirazo, K.; Briones, C.; de la Escosura, A. Chemical roots of biological evolution: The origins of life as a process of development of autonomous functional systems. *Open Biol.* **2017**, *7*, 170050. [CrossRef]
3. Mayr, E. *Toward a New Philosophy of Biology*; Harvard University Press: Cambridge, MA, USA, 1988; pp. 24–25.
4. Altshuler, D.L.; Dickson, W.B.; Vance, J.T.; Roberts, S.P.; Dickinson, M.H. Short-amplitude high-frequency wing strokes determine the aerodynamics of honeybee flight. *Proc. Natl. Acad. Sci. USA* **2005**, *102*, 18213–18218. [CrossRef] [PubMed]
5. Liu, P.; Sane, S.P.; Mongeau, J.-M.; Zhao, J.; Cheng, B. Flies land upside down on a ceiling using rapid visually mediated rotational maneuvers. *Sci. Adv.* **2019**, *5*, eaax1877. [CrossRef]
6. DuPont. *Nylon: A DuPont Invention*; DuPont International, Public Affairs: Wilmington, NC, USA, 1988.
7. Rodnina, M.V. The ribosome in action: Tuning of translational efficiency and protein folding. *Protein Sci.* **2016**, *25*, 1390–1406. [CrossRef] [PubMed]
8. Shannon, C.E. A Mathematical Theory of Communication. *Bell Syst. Tech. J.* **1948**, *27*, 379–423.
9. Dennett, D.C. *Darwin's Dangerous Idea*; Penguin: London, UK, 2008.
10. Kondepudi, D.; Prigogine, I. *Modern Thermodynamics: From Heat Engines to Dissipative Structures*; Wiley: Chichester, UK, 1998.
11. Lineweaver, C.H.; Egan, C.A. Life, gravity and the second law of thermodynamics. *Phys. Life Rev.* **2008**, *5*, 225–242. [CrossRef]
12. Ashkenasy, G.; Hermans, T.M.; Otto, S.; Taylor, A.F. Systems Chemistry. *Chem. Soc. Rev.* **2017**, *46*, 2543–2554. [CrossRef]
13. Lotka, A.J. Natural selection as a physical principle. *Proc. Natl. Acad. Sci. USA* **1922**, *8*, 151–154. [CrossRef]
14. Whitesides, G.M.; Grzybowski, B. Self-assembly at all scales. *Science* **2002**, *295*, 2418–2421. [CrossRef]
15. Pross, A.; Khodorkovsky, V. Extending the concept of kinetic stability: Toward a paradigm for life. *J. Phys. Org. Chem.* **2004**, *17*, 312–316. [CrossRef]
16. Pross, A. Toward a general theory of evolution: Extending Darwinian theory to inanimate matter. *J. Syst. Chem.* **2011**, *2*, 1. [CrossRef]
17. Pross, A.; Pascal, R. How and why kinetics, thermodynamics, and chemistry induce the logic of biological evolution. *Beilstein J. Org. Chem.* **2017**, *13*, 665–674. [CrossRef]
18. Pross, A. *What Is Life? How Chemistry Becomes Biology*, 2nd ed.; Oxford University Press: Oxford, UK, 2016.
19. Boekhoven, J.; Brizard, A.M.; Kowlgi, K.N.K.; Koper, G.J.M.; Eelkema, R.; van Esch, J.H. Dissipative self-assembly of a molecular gelator by using a chemical fuel. *Angew. Chem. Int. Ed.* **2010**, *49*, 4825. [CrossRef]
20. Boekhoven, J.; Hendriksen, W.E.; Koper, G.J.M.; Eelkema, R.; van Esch, J.H. Transient assembly of active materials fueled by a chemical reaction. *Science* **2015**, *349*, 1075. [CrossRef] [PubMed]
21. Merindol, R.; Walther, A. Materials learning from life: Concepts for active, adaptive, and autonomous molecular systems. *Chem. Soc. Rev.* **2017**, *46*, 5588. [CrossRef]
22. Pascal, R.; Pross, A. Chemistry's kinetic dimension and the physical basis for life. *J. Syst. Chem.* **2019**, *7*, 1.
23. Fletcher, D.A.; Mullins, R.D. Cell mechanics and the cytoskeleton. *Nature* **2010**, *463*, 485–492. [CrossRef] [PubMed]
24. Tena-Solsona, M.; Wanzke, C.; Riess, B.; Bausch, A.R.; Boekhoven, J. Self-selection of dissipative assemblies driven by primitive chemical reaction networks. *Nat. Commun.* **2018**, *9*, 2044. [CrossRef] [PubMed]
25. Te Brinke, E.; Groen, J.; Herrmann, A.; Heus, H.A.; Rivas, G.; Spruijt, E.; Huck, W.T. Dissipative adaptation in driven self-assembly leading to self-dividing fibrils. *Nat. Nanotechnol.* **2018**, *13*, 849–855. [CrossRef] [PubMed]
26. Ragazzon, G.; Prins, L.J. Energy consumption in chemical fuel-driven self-assembly. *Nat. Nanotechnol.* **2018**, *13*, 882–889. [CrossRef]
27. Astumian, R.D. Stochastic pumping of non-equilibrium steady-states: How molecules adapt to a fluctuating environment. *Chem. Commun.* **2018**, *54*, 427–444. [CrossRef]
28. Szathmáry, E.; Gladkih, I. Sub-exponential growth and coexistence of non-enzymatically replicating templates. *J. Theor. Biol.* **1989**, *138*, 55–58. [CrossRef]
29. Lifson, S. On the crucial stages in the origin of animate matter. *J. Mol. Evol.* **1997**, *44*, 1. [CrossRef]
30. Pascal, R.; Pross, A. Stability and its manifestation in the chemical and biological worlds. *Chem. Commun.* **2015**, *5*, 16160. [CrossRef] [PubMed]
31. Grand, S. *Creation*; Harvard University Press: Cambridge, MA, USA, 2001.

32. Shettleworth, S.J. *Cognition, Evolution and Behavior*; Oxford University Press: New York, NY, USA, 1998.
33. Lyon, P. The cognitive cell: Bacterial behavior reconsidered. *Front. Microbiol.* **2015**, *6*, 264. [CrossRef]
34. Shapiro, J.A. All living cells are cognitive. *Biochem. Biophys. Res. Commun.* **2021**, *564*, 134–149. [CrossRef]
35. van Gelber, T. The dynamical hypothesis in cognitive science. *Behav. Brain Sci.* **1998**, *21*, 615–665. [CrossRef] [PubMed]
36. Smolensky, P. Information processing in dynamical systems: Foundations of harmony theory. In *Parallel Distributed Processing*; Rumelhart, D.E., McClelland, J.L., Eds.; MIT Press: Cambridge, MA, USA, 1986; Volume 1, pp. 194–281.
37. Yang, S.; Schaeffer, G.; Mattia, E.; Markovitch, O.; Liu, K.; Hussain, A.S.; Ottele, J.; Sood, A.; Otto, S. Chemical Fueling Enables Molecular Complexification of Self-Replicators. *Angew. Chem. Int. Ed.* **2021**, *60*, 11344–11349. [CrossRef]

Communication

Formation of Thiophene under Simulated Volcanic Hydrothermal Conditions on Earth—Implications for Early Life on Extraterrestrial Planets?

Thomas Geisberger, Jessica Sobotta, Wolfgang Eisenreich and Claudia Huber *

Lehrstuhl für Biochemie, Department Chemie, Technische Universität München, Lichtenbergstraße 4, 85748 Garching, Germany; thomas.geisberger@tum.de (T.G.); Jessy.Sobotta@web.de (J.S.); wolfgang.eisenreich@mytum.de (W.E.)
* Correspondence: claudia.huber@tum.de

Abstract: Thiophene was detected on Mars during the Curiosity mission in 2018. The compound was even suggested as a biomarker due to its possible origin from diagenesis or pyrolysis of biological material. In the laboratory, thiophene can be synthesized at 400 °C by reacting acetylene and hydrogen sulfide on alumina. We here show that thiophene and thiophene derivatives are also formed abiotically from acetylene and transition metal sulfides such as NiS, CoS and FeS under simulated volcanic, hydrothermal conditions on Early Earth. Exactly the same conditions were reported earlier to have yielded a plethora of organic molecules including fatty acids and other components of extant metabolism. It is therefore tempting to suggest that thiophenes from abiotic formation could indicate sites and conditions well-suited for the evolution of metabolism and potentially for the origin-of-life on extraterrestrial planets.

Keywords: thiophene; acetylene; transition metal sulfides; hydrothermal conditions; early metabolism; origin-of-life

Citation: Geisberger, T.; Sobotta, J.; Eisenreich, W.; Huber, C. Formation of Thiophene under Simulated Volcanic Hydrothermal Conditions on Earth—Implications for Early Life on Extraterrestrial Planets? *Life* **2021**, *1*, 149. https://doi.org/10.3390/life11020149

Received: 16 December 2020
Accepted: 12 February 2021
Published: 16 February 2021

Publisher's Note: MDPI stays neutral with regard to jurisdictional claims in published maps and institutional affiliations.

Copyright: © 2021 by the authors. Licensee MDPI, Basel, Switzerland. This article is an open access article distributed under the terms and conditions of the Creative Commons Attribution (CC BY) license (https://creativecommons.org/licenses/by/4.0/).

1. Introduction

Recent findings by the Curiosity mission have shown the existence of thiophene and some of its derivatives on Mars [1,2]. Within this context, different possibilities for their abiotic as well as biotic formation were discussed. Thiophene was even suggested as a biomarker in the search for life on Mars [2], whereas a hydrothermal abiotic origin was also considered [1]. Likewise, thiophenes are common pyrolysis products from meteoritic macromolecular materials. For example, these compounds are produced by aqueous and/or thermal alteration of carbonaceous chondrites like the Murchison meteorite [3,4]. On Earth, thiophenes can be detected in volcanic gas discharges and in fluid emissions related to hydrothermal systems [5]. In submarine basins, like the Guaymas basin, thiophene derivatives can be detected after hydrothermal pyrolysis of organic material [6,7]. Furthermore, thiophenes are suggested to be at least a part of the organic sulfur found in the globules of 2.72 Ga years old stromatolites of the Tumbiana Formation. In the Western Australian Dresser Formation, stromatolites are dated back to 3.5 Ga and are considered as one of the most ancient traces of life on Earth [8,9]. On an industrial scale, thiophene is produced from butane and sulfur at 560 °C, from sodium succinate and phosphorous trisulfide, and from acetylene and hydrogen sulfide at 400 °C on alumina [10].

We have shown earlier that acetylene is also an excellent source for primordial carbon fixation, especially in combination with carbon monoxide, e.g., for the synthesis of short chain fatty acids [11] and intermediates of extant carbon fixation cycles [12]. On Earth, acetylene is present in fumarolic exhalations [13]. It can also be found extra-terrestrially, for example, on Saturn's moon Titan [14]. It was also proposed that explosive volcanism may have injected ~6×10^{12} g/year of acetylene into the atmosphere of early Mars [15].

The importance of sulfides in an origin-of-life scenario is emphasized by Wächtershäuser's Iron-Sulfur-World hypothesis [16]. Based on this hypothesis, we here report the facile abiotic formation of thiophene and some of its derivatives from acetylene and metallo-sulfides, especially NiS, under aqueous conditions at 105 °C. In context with the formation of potential building units and reaction networks for the emergence of metabolism under the same conditions [11,12] and capitalizing on recent hypotheses [2], the detection of extraterrestrial or terrestrial thiophenes could therefore indeed be indicative of early metabolic evolution under chemoautotrophic conditions.

2. Materials and Methods

All chemicals were purchased from Sigma Aldrich GmbH (Steinheim, Germany) in the highest purity available. Acetylene 2.6 (acetone free) was purchased from Linde AG (Pullach, Germany), and CO 2.5 and argon 4.6 were purchased from Westfalen AG (Münster, Germany). In a typical run (run 1, Table 1), a 125 mL glass serum bottle was charged with 1.0 mmol $NiSO_4 \cdot 6H_2O$ and closed with a silicon stopper. The bottle was evacuated three times and filled with argon, finally resulting in a de-aerated state. Subsequently, the bottle was filled with 3.5 mL argon-saturated water (calculated for a final volume of 5 mL) to dissolve the $NiSO_4$ and with 1.0 mL argon-saturated 1 M Na_2S solution. In this mixture, a precipitate of black NiS is immediately formed due to its low solubility constant of 1×10^{-22} [17,18] in aqueous solution. Furthermore, the bottle was filled with 0.5 mL 1 M NaOH solution, and finally with 120 mL of acetylene gas using gas-tight syringes for injection. The freshly precipitated NiS acted as a putative transition metal catalyst for the reaction and the molar variations of Na_2S to $NiSO_4$ resulted in free sulfide ions in the solution. In runs 10–12, 17, and 20, a mixture of 60 mL CO and 60 mL acetylene was used as gaseous phase. Instead of $NiSO_4 \cdot 6H_2O$, runs 2, 11, and 14 were loaded with 1 mmol $FeSO_4 \cdot 7H_2O$ and runs 3, 12, and 15 were loaded with 1 mmol $CoSO_4 \cdot 7H_2O$. In run 9, $NiSO_4 \cdot 6H_2O$ and $FeSO_4 \cdot 7H_2O$ were combined. Otherwise, the settings were identical to the above described procedure. Reactions were carried out at 105 °C. pH-Variations were achieved through the addition of 0.1–1.0 mL 1M H_2SO_4 or NaOH. For safety reasons (danger of explosion) and for technical reasons, the reactions were carried out at low pressure (1 bar) of acetylene. After 1 day (24 h) or 7 days, the reaction mixture was allowed to cool down and, after vigorous shaking, 1 mL was taken out and centrifuged at 10,000 rpm for 10 min. For the isolation of thiophenes, the supernatant and the solid residue were extracted separately with 1 mL ethyl acetate. The organic phases were dried over Na_2SO_4 and directly analyzed with gas chromatography-mass spectrometry (GC-MS). GC-MS analysis was performed with a GC-2010, coupled with MS-QP2010 Ultra (Shimadzu GmbH, Duisburg, Germany) with a 30 m × 0.25 mm × 0.25 μm fused silica capillary column (Equity TM5, Supelco, PA-Bellefonte, USA) and an AOC-20i auto injector. Temperature program and settings: 0–6 min at 40 °C; 6–25 min at 40–280 °C, 10 °C/min; injector temperature: 260 °C; detector temperature: 260 °C; column flow rate: 1 mL/min; scan interval: 0.5 sec; and injection volume 0.1 μL. For detection of thiophene derivatives, a larger injection volume of 3 μL was used. Peak assignment was achieved by comparison with the retention times and mass spectra of purchased reference compounds, as well as with data from the National Institute of Standards and Technology (NIST) spectral library. Thiophene showed a retention time of 3.7 min. Retention times for derivatives are given in Table S2. Quantification was performed by external calibration using known concentrations of thiophene. Runs without a transition metal compound or without acetylene were performed for comparison.

Table 1. Transition metal catalyzed formation of thiophene. Reactions were performed with 120 mL (5.36 mmol) or 60 mL (2.68 mmol) acetylene and freshly precipitated sulfides under aqueous conditions at 105 °C. $NiSO_4$, $CoSO_4$ and $FeSO_4$ were used as hydrates (see method section). Reactions were performed for 24 h (run 1–12) or 7 days (run 13–18). pH-Values were measured at the end of the reaction time. Run 1 was performed three times showing a representative standard deviation of 14%. Thiophene concentrations were given in mM for the separated organic extracts of supernatants and solid sulfides as well as total concentration in the 5 mL setups. Yields are given in mol% conversion based on acetylene.

Run	$NiSO_4$ (mmol)	$FeSO_4$ (mmol)	$CoSO_4$ (mmol)	Na_2S (mmol)	NaOH (mmol)	CO (mL)	C_2H_2 (mL)	pH_{end}	Extract Supernatant (mM)	Extract Solid (mM)	Total Conc. (mM)	Total Yield (%)
1	1	-	-	1	0.5	-	120	9.7	0.379	0.709	2.175	0.406
2	-	1	-	1	0.5	-	120	9.0	<0.001	<0.001	<0.001	<0.001
3	-	-	1	1	0.5	-	120	9.0	0.047	0.046	0.185	0.035
4	1	-	-	1	-	-	120	6.5	0.337	1.243	3.160	0.590
5	1	-	-	1.5	-	-	120	11.0	0.393	0.740	2.268	0.423
6	1	-	-	2	-	-	120	13.5	0.333	0.129	0.925	0.173
7	-	1	-	1.5	-	-	120	12.0	0.023	0.066	0.177	0.033
8	-	1	-	2	-	-	120	13.5	0.002	0.002	0.008	0.001
9	0.5	0.5	-	1	0.5	-	120	11.0	0.377	0.503	1.761	0.329
10	1	-	-	1	0.5	60	60	9.5	0.220	0.067	0.574	0.214
11	-	1	-	1	0.5	60	60	9.0	0.002	<0.001	0.004	0.001
12	-	-	1	1	0.5	60	60	9.5	0.075	0.476	1.103	0.412
13	1	-	-	1	0.5	-	120	10.1	0.135	0.618	1.505	0.281
14	-	1	-	1	0.5	-	120	8.5	0.001	0.001	0.003	0.001
15	-	-	1	1	0.5	-	120	8.7	0.023	0.012	0.070	0.013
16	1	-	-	1	-	-	120	7.1	0.040	0.468	1.015	0.190
17	1	-	-	1	0.5	60	60	7.8	0.225	0.213	0.877	0.327
18	-	1	-	1	0.5	60	60	7.6	0.000	0.001	0.002	0.001

3. Results

We reacted acetylene at 105 °C for one day or seven days under strictly anoxic aqueous conditions, with freshly precipitated nickel sulfide, iron sulfide, cobalt sulfide, or mixtures thereof. In some runs, CO was added additionally as another putative reactant. The pH values were measured at the end of the reaction time (Table 1). After the indicated periods, the reaction mixtures were separated by centrifugation into a clear liquid supernatant and a black solid residue. Supernatants and solid residue were extracted separately using ethyl acetate. These extracts were finally analyzed by GC-MS. In a blank run without any addition of transition metal, under otherwise identical conditions to run 1, no formation of thiophene or thiophene derivatives was observed. Run 1, using NiS as the sulfide compound and catalyst at pH 9.7, is defined as standard run and was performed three times showing the formation of 2.2 mM thiophene as a mean concentration with a representative standard deviation of 14%. Under these conditions, thiophene formation was observed in a broad pH range from pH 5 to pH 11. However, yields were pH dependent as shown in Figure 1 and Table S1, with a pH optimum in the neutral range. Up to 3 mM thiophene were detected at pH 6.5 (run 4). Thiophene was also formed in comparable amounts in runs 3 and 10, using CoS or a mixed FeS/NiS catalyst, whereas only low amounts of thiophene were formed in the presence of FeS alone (Table 1, runs 2,7,8, and 14; Figure 2). Interestingly, in one third of the reactions, the amount of thiophene in the residue was up to two times higher than in the corresponding supernatants (Table 1) which reflects a strong binding of thiophene to the metal sulfide surfaces. This led us to the question as to whether the sulfur in thiophene derives from the solid NiS or, alternatively, from free sulfide in the solution. We therefore increased the amount of Na_2S in runs 5–8. In run 5 which contained 0.5 mmol additional free sulfide, the amount of thiophene was not significantly changed (Table 1, run 5 vs. run 1). In run 6 with 1 mmol additional sulfide, the yield was diminished to one half (run 6 vs. run 1). This could again indicate that solid nickel sulfide served as the reacting agent and not the free sulfide, with a possible blockage of catalytic sites through excess sulfide. Otherwise, free sulfide ions shifted the pH to a more alkaline value which is less suited for thiophene formation (Figure 1). In the presence of FeS alone, only traces of thiophene were detected. In the presence of 0.5 mmol additional sulfide (run 7 vs. run 2), the thiophene yield was significantly enhanced, but again lowered in the presence of 1 mmol free sulfide (run 7 vs. 8). This could indicate a different reaction mechanism of FeS catalysis involving free sulfide ions and, possibly, a different pH dependency compared to NiS catalysis. Next, we chose a longer reaction time of 7 days, in an attempt to estimate reaction kinetics. We observed that the elongation of the reaction time in the NiS/acetylene system from 24h to one week was not favorable for thiophene formation (run 1 vs. 13, 3 vs. 15, 4 vs. 16), whereas thiophene yields increased in the NiS/acetylene/CO system (run 10 vs. 13) and the FeS/acetylene system (run 2 vs. 14). This showed that the formation of thiophene is a complex process influenced by many parameters, involving also consecutive reactions to other products. In Table 1, yields in mol% conversion based on acetylene are given additionally to the measured concentrations. The maximum conversion rate of 0.59% is reached for the NiS experiment at a nearly neutral pH value (run 4). For safety reasons the reactions were carried out at low pressure (1 bar). At a high sub-seafloor pressure (maybe >1000 bar), yields would be increased because of negative volumes of reaction [19,20].

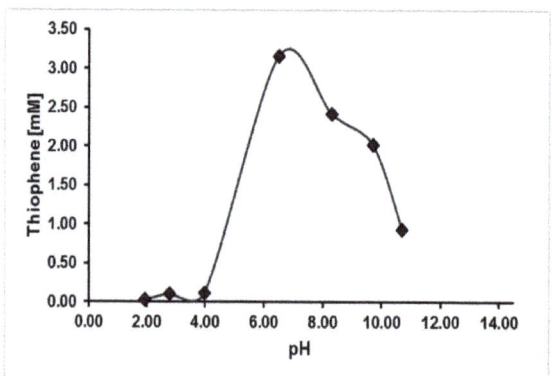

Figure 1. pH-Dependent formation of thiophene in the presence of NiS. Reactions were performed with 5.36 mmol acetylene and 1 mmol freshly precipitated nickel sulfide under aqueous conditions at 105 °C. Reactions were performed for 24 h and pH values were measured at the end of the reaction time.

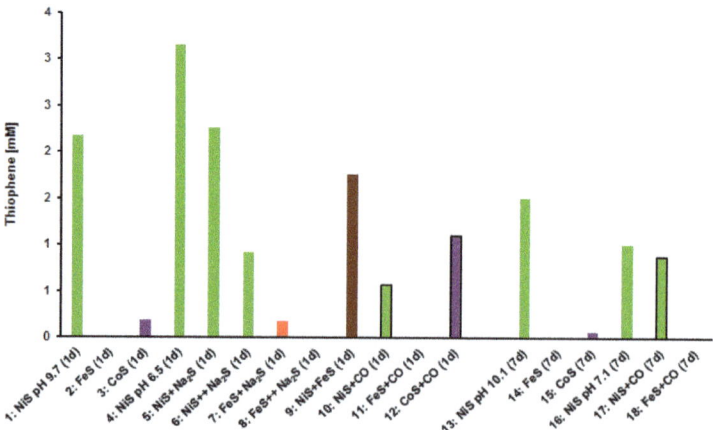

Figure 2. Total yield of thiophene formed in experiments as described in Table 1. Reactions were performed for 24 h (1 d) or 7 days (7 d). +Na$_2$S/++Na$_2$S$^-$ imply free sulfide ions because of a higher molar ratio of Na$_2$S to NiSO$_4$. Bars are colored according to the metal sulfide used: NiS: green; FeS: red; CoS: purple; NiS/FeS: brown; Runs with CO show black frames.

Further analysis by GC-MS led to the detection of several thiophene derivatives. Next to thiophene (1, Figure 3), 2-ethylthiophene (2), 3-ethylthiophene (3), 2,3-dimethylthiophene (4), ethyl-vinyl-sulfide (5), tetrahydrothiophene (6), 3-ethynylthiophene (7), 3-thiophenthiol (8), 5-methylthiophen-2-carboxaldehyde (9), 2[5H]-5-methylthiophenon (10), 2-vinylthiophene (11), 2-acetyl-5 methyl-thiophene (12), 2-acetylthiophene (13), thiophen-2-carboxaldehyde (14), cyclohex-2-enthion (15), cis-1,4-dithiapentalene (16), trans-1,4-dithiapentalene (17) and benzo[b]thiophene (18) were observed in the extracts of the reaction mixtures at estimated concentrations of 0.01–0.03 mM (Figure 3, Table S2). Product identification was performed by comparison with commercially available standards and/or comparison to the NIST14 database (Figure S1 and Figure S2, in the supplementary materials). Individual amounts of each derivative were not calculated, but ratios of thiophene to the total amount of thiophene derivatives are given in Table S2. Reactions performed in the presence of NiS showed ratios from 1.0–17.9, whereas the highest ratio of 25.4 was observed in the reaction setting using a mixed NiS/FeS catalyst. Reactions performed in the presence of CoS showed ratios in the range of 0.3 to 2.8, indicating higher amounts of derivatives in comparison to NiS. The low ratios in FeS settings (<0.8) were due to the low amounts of thiophene found in these settings. The decrease of the ratio thiophene to thiophene derivatives by addition of CO in the presence of NiS (run 1 vs. 10) could indicate follow up reactions initiated by CO. The fact, that no thiophene derivatives were detected after 7 days in the presence of CO could indicate reaction or degradation steps towards products, which are not covered by our experimental setup.

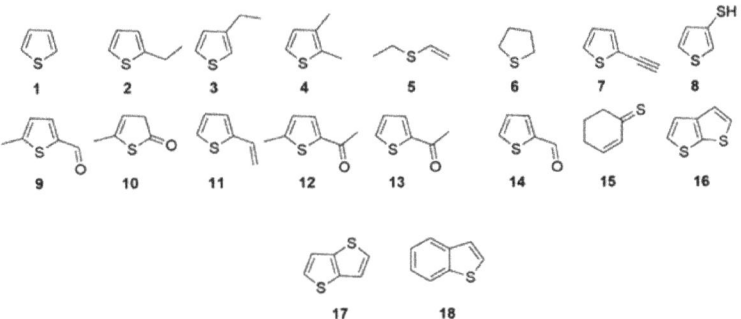

Figure 3. Thiophene and thiophene derivatives as formed from acetylene and nickel sulfide. Structures were verified by analytical standards and/or spectral libraries.

When thiophene was used as a starting material under conditions as described for run 11, tetrahydrothiophene (6), 2/3-ethylthiophene (1,2) and 5-methyl-thiophen-2-carboxaldehyde (9) were observed by GC-MS.

These findings demonstrate the formation of thiophene and its derivatives as products from acetylene and nickel sulfide under relatively mild hydrothermal conditions with the opportunity for further evolution. In Figure 4, a mechanism is proposed in analogy to the reaction of acetylene and hydrogen sulfide in super basic media [21]. In this scheme, the sulfur atom of NiS reacts with two molecules of acetylene in a concerted one-step mechanism. The so formed divinyl sulfide was not detected probably due to rapid conversion into by dehydrogenation to thiophene (1) or reduction to ethyl-vinyl-sulfide (5). Thiophene (1) could then react with further acetylene or sulfide to form 2-vinylthiophene (11), 2-ethylthiophene (2), 3-ethylthiophene (3) or 3-thiophenthiol (8). Further, it could be reduced to tetrahydrothiophene (6) or react with CO to form thiophen-2-carboxaldehyde (14). Additional experiments including stable isotope labelled precursors are required to unravel this mechanism in more detail. However, the various products observed under these conditions clearly imply that products downstream of thiophene are formed in a reaction network that could further evolve.

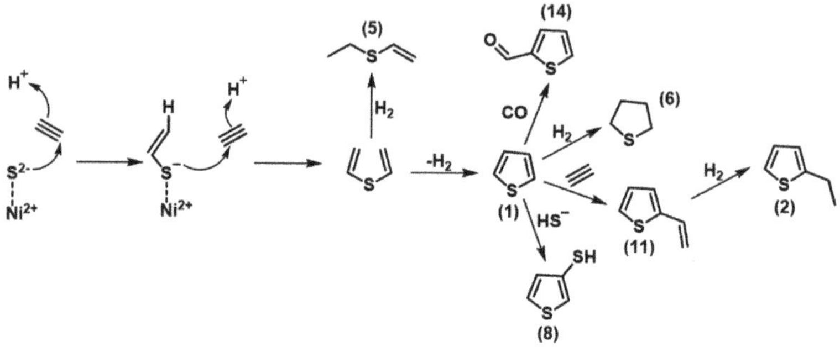

Figure 4. Hypothetical mechanism for the formation of thiophene and its derivatives from acetylene and nickel sulfide. ≡ signifies acetylene (C_2H_2).

4. Discussion

Chemical reactivity on Earth and Mars could be determined by metal sulfide catalysis, e.g., by FeS catalysis, since both planets contain high amounts of iron in their mantles [22]. Nickel and cobalt as members of the iron group are often found together with iron and are also of special interest as catalysts in the "iron–sulfur theory" for the origin-of-life [23]. Earth's core consists of 80–90% Fe-Ni alloys and 2.3wt% sulfur [24]. Fe-Ni sulfides are present on Earth as well as on Mars through ultramafic lava eruptions [25] and additional Ni is deposited on Mars through meteoritic impact [26]. On Earth, the formation of mixed NiFeS minerals, for example, of pentlandite $(FeNi)_9S_8$ and violarite $FeNi_2S_4$ is investigated in the context of serpentinization, a potential key process for metabolic evolution on early Earth [27].

In earlier work starting from acetylene, CO or cyanide under simulated hydrothermal conditions, we showed that nickel, especially NiS, is a potent catalyst for the formation of organic molecules, such as fatty acids, intermediates serving in biological carbon fixation and amino acids [11,12,28]. As we can now show, NiS catalyzes additionally the formation of thiophene from acetylene. The observed low thiophene formation with FeS underlines the importance of nickel minerals in this context.

Organic sulfur compounds in general are indeed essential for life and play an important role in the sulfur cycle on Earth [29]. In addition to thiophene, methanethiol and carbonyl sulfide can be found in terrestrial hydrothermal exhalations and were also used for simulated primordial synthesis of biomolecules [23,30]. In the biological context, thiophene and its derivatives (such as benzo-thiophenes) are sometimes considered as secondary biomarkers, preserving the original n-alkane chain or carbon skeleton of biomolecules in sulfur-bound forms at different lithofacies [31]. Organic sulfur compounds are more stable under sulfide rich geological conditions than their biological precursors (e.g., functionalized lipids), and can therefore be found in ancient sedimentary rocks [32].

In extant biochemistry, thiophene is still conserved as a structural part of biotin, which is a prosthetic group for carboxylase classed enzymes, like the pyruvate carboxylase [33]. Furthermore, thiophenes can be found as structural components in the quinone fractions (e.g., caldariellaquinone—benzo[b]thiophene-4,7-quinone) of extreme thermophilic and acidophilic archaeons, like *Caldariella acidiphila* and *Sulfolobus solfataricus* [34,35].

According to recent literature [7,36], thiophenes *per se* should not be named biomarkers, due to their possible abiotic origin, but the presence of thiophenes, as easily detectable molecules, could indicate samples or sites, which should be investigated in more detail for the presence of additional organic molecules like amino acids or fatty acids. It is also suggested that terrestrial origin-of-life conditions could be used as a guideline in the search for life on extraterrestrial planets [37].

5. Conclusions

We here could show the abiotic formation of thiophenes from acetylene and transition metal sulfides under aqueous conditions, which we consider as a valid simulation of volcanic hydrothermal settings on early Earth, but also on other planets. Under identical conditions, we could previously demonstrate the conversion of acetylene and CO into short chain fatty acids (C_3–C_9) and other C_2–C_4 compounds including metabolic intermediates of carbon fixation in extant life [9,10]. These compounds are also considered as important precursors for a potential chemo-autotrophic origin-of-life. It is therefore tempting to speculate that the detection of thiophenes on planets reflects possible habitats for the early emergence and evolution of metabolism and life.

Supplementary Materials: The following are available online at https://www.mdpi.com/2075-1729/11/2/149/s1, Table S1: pH dependent formation of thiophene in the presence of NiS. Table S2: Identified thiophene derivatives and their retention times. Figure S1: GC/MS chromatograms comparing thiophene and commercially available thiophene derivatives, Figure S2: GC/MS mass spectra comparing reaction products to commercially available thiophene standards and mass spectra from NIST14 library.

Author Contributions: Conceptualization, C.H., W.E; data curation, T.G.; funding acquisition, C.H.; methodology, J.S., T.G.; supervision, C.H., W.E.; visualization, T.G.; writing—original draft, C.H.; writing—review and editing, T.G., C.H. and W.E. All authors have read and agreed to the published version of the manuscript.

Funding: This research was funded by the Deutsche Forschungsgemeinschaft (DFG, German Research Foundation)—Project-ID 364653263—TRR 235 and the Hans-Fischer-Gesellschaft (Munich, Germany).

Acknowledgments: We thank Günter Wächtershäuser for continuous support and valuable discussions and we thank Felicia Achatz for practical assistance.

Conflicts of Interest: The authors declare no conflict of interest.

References

1. Eigenbrode, J.L.; Summons, R.E.; Steele, A.; Freissinet, C.; Millan, M.; Navarro-González, R.; Sutter, B.; McAdam, A.C.; Franz, H.B.; Glavin, D.P. Organic matter preserved in 3-billion-year-old mudstones at Gale crater, Mars. *Science* **2018**, *360*, 1096–1101. [CrossRef]
2. Heinz, J.; Schulze-Makuch, D. Thiophenes on Mars: Biotic or Abiotic Origin? *Astrobiology* **2020**, *20*, 552–561. [CrossRef]
3. Ehrenfreund, P.; Sephton, M.A. Carbon molecules in space: From astrochemistry to astrobiology. *Faraday Discuss.* **2006**, *133*, 277–288. [CrossRef]
4. Sephton, M.A. Organic compounds in carbonaceous meteorites. *Nat. Prod. Rep.* **2002**, *19*, 292–311. [CrossRef] [PubMed]
5. Tassi, F.; Montegrossi, G.; Capecchiacci, F.; Vaselli, O. Origin and distribution of thiophenes and furans in gas discharges from active volcanoes and geothermal systems. *Int. J. Mol. Sci.* **2010**, *11*, 1434–1457. [CrossRef] [PubMed]
6. Kawaka, O.E.; Simoneit, B.R.T. Hydrothermal pyrolysis of organic matter in Guaymas Basin: I. Comparison of hydrocarbon distributions in subsurface sediments and seabed petroleums. *Org. Geochem.* **1994**, *22*, 947–978. [CrossRef]
7. Simoneit, B.R.T. A review of current applications of mass spectrometry for biomarker/molecular tracer elucidation. *Mass Spectrom. Rev.* **2005**, *24*, 719–765. [CrossRef]
8. Lepot, K.; Benzerara, K.; Rividi, N.; Cotte, M.; Brown, G.E.; Philippot, P. Organic matter heterogeneities in 2.72 Ga stromatolites: Alteration versus preservation by sulfur incorporation. *Geochim. Cosmochim. Acta* **2009**, *73*, 6579–6599. [CrossRef]
9. Van Kranendonk, M.J.; Philippot, P.; Lepot, K.; Bodorkos, S.; Pirajno, F. Geological setting of Earth's oldest fossils in the ca. 3.5 Ga Dresser Formation, Pilbara Craton, Western Australia. *Precambrian Res.* **2008**, *167*, 93–124. [CrossRef]
10. Mishra, R.; Jha, K.K.; Kumar, S.; Tomer, I. Synthesis, properties and biological activity of thiophene: A review. *Der Pharma Chem.* **2011**, *3*, 17.
11. Scheidler, C.; Sobotta, J.; Eisenreich, W.; Wächtershäuser, G.; Huber, C. Unsaturated C-3,C-5,C-7,C-9-Monocarboxylic Acids by Aqueous, One-Pot Carbon Fixation: Possible Relevance for the Origin of Life. *Sci. Rep.* **2016**, *6*, 27595. [CrossRef] [PubMed]
12. Sobotta, J.; Geisberger, T.; Moosmann, C.; Scheidler, C.M.; Eisenreich, W.; Wächtershäuser, G.; Huber, C. A Possible Primordial Acetyleno/Carboxydotrophic Core Metabolism. *Life* **2020**, *10*, 35. [CrossRef]
13. Igari, S.; Maekawa, T.; Sakata, S. Light hydrocarbons in fumarolic gases: A case study in the Kakkonda geothermal area. *Chikyukagau* **2000**, *34*, 7.

14. Singh, S.; McCord, T.B.; Combe, J.P.; Rodriguez, S.; Cornet, T.; Mouélic, S.L.; Clark, R.N.; Maltagliati, L.; Chevrier, V.F. Acetylene on Titans surface. *Astrophys. J.* **2016**, *828*, 55. [CrossRef]
15. Segura, A.; Navarro-Gonzalez, R. Production of low molecular weight hydrocarbons by volcanic eruptions on early Mars. *Orig. Life Evol. Biosph.* **2005**, *35*, 477–487. [CrossRef] [PubMed]
16. Wächtershäuser, G. Groundworks for an evolutionary biochemistry: The iron-sulphur world. *Prog. Biophys. Mol. Biol.* **1992**, *58*, 85–201.
17. Hollemann, A.F.; Wiberg, N. *Lehrbuch der Anorganischen Chemie*, 101st ed.; Walter de Gruyter: Berlin, Germany, 1995; p. 1582.
18. Sillen, L.G.; Martell, A.E. Stability Constants of Metal-Ion Complexes. In *Lange's Handbook*; Special Publ. No. 17; The Chemical Society: London, UK, 1964; pp. 8-6–8-11.
19. Matsumoto, K.; Sera, A.; Uchida, T. Organic Synthesis under high pressure. *Synthesis* **1985**, *18*, 1–26. [CrossRef]
20. Klärner, F.-G.; Wurche, F. The effect of pressure on organic reactions. *J. Prakt. Chem.* **2000**, *342*, 609–636.
21. Trofimov, B.A. New Reactions and Chemicals Based on Sulfur and Acetylene. *Sulfur. Rep.* **1983**, *3*, 83–114. [CrossRef]
22. Halliday, A.N.; Wänke, H.; Birck, J.-L.; Clayton, R.N. The accretion, composition and early differentiation of Mars. *Space Sci. Rev.* **2001**, *96*, 197–230.
23. Huber, C.; Wächtershäuser, G. Activated Acetic Acid by Carbon Fixation on (Fe,Ni)S Under Primordial Conditions. *Science* **1997**, *276*, 245–247. [CrossRef]
24. Allegre, C.J.; Poirier, J.-P.; Humler, E.; Hofmann, A.W. The chemical composition of the Earth. *Earth Planet. Sci. Lett.* **1995**, *134*, 515–526.
25. Burns, R.G.; Fisher, D.S. Evolution of sulfide mineralization on Mars. *J. Geophys. Res. Solid Earth* **1990**, *95*, 14169–14173. [CrossRef]
26. Yen, A.S.; Mittlefehldt, D.W.; McLennan, S.M.; Gellert, R.; Bell, J.F.; McSween, H.Y.; Ming, D.W.; McCoy, T.J.; Morris, R.V.; Golombek, M. Nickel on Mars: Constraints on meteoritic material at the surface. *J. Geophys. Res.* **2006**, *111*, E12S11. [CrossRef]
27. Klein, F.; Bach, W. Fe–Ni–Co–O–S Phase Relations in Peridotite–Seawater Interactions. *J. Petrol.* **2009**, *50*, 37–59. [CrossRef]
28. Huber, C.; Kraus, F.; Hanzlik, M.; Eisenreich, W.; Wächtershäuser, G. Elements of metabolic evolution. *Chemistry* **2012**, *18*, 2063–2080. [CrossRef] [PubMed]
29. Kertesz, M.A. Riding the sulfur cycle–metabolism of sulfonates and sulfate esters in Gram-negative bacteria. *FEMS Microbiol. Rev.* **2000**, *24*, 135–175.
30. Leman, L.J.; Orgel, L.E.; Ghadiri, M.R. Carbonyl sulfide-mediated prebiotic formation of peptides. *Science* **2004**, *306*, 4. [CrossRef] [PubMed]
31. Van Kaam-Peters, H.M.; Rijpstra, W.I.C.; De Leeuw, J.W.; Damsté, J.S.S. A high resolution biomarker study of different lithofacies of organic sulfur-rich carbonate rocks of a Kimmeridgian lagoon (French southern Jura). *Org. Geochem.* **1998**, *28*, 151–177. [CrossRef]
32. Sinninghe Damste, J.S.; Rijpstra, W.I.C.; Kock-van Dalen, A.C.; de Leeuw, J.W.; Schenck, P.A. Quenching of labile functionalised lipids by inorganic sulphur species: Evidence for the formation of sedimentary organic sulphur compounds at the early stages of diagenesis. *Geochim. Cosmochim. Acta* **1989**, *53*, 13. [CrossRef]
33. Jitrapakdee, S.; Wallace, J.C. Structure, function and regulation of pyruvate carboxylase. *Biochem. J.* **1999**, *340*, 1–16. [CrossRef]
34. Lanzotti, V.; Trincone, A.; Gambacorta, A.; De Rosa, M.; Breitmaier, E. 1H and 13C NMR assignment of benzothiophenquinones from the sulfur-oxidizing archaebacterium Sulfolobus solfataricus. *FEBS J.* **1986**, *160*, 37–40. [CrossRef] [PubMed]
35. Zhou, D.; White, R. Biosynthesis of caldariellaquinone in Sulfolobus spp. *J. Bacteriol.* **1989**, *171*, 6610–6616. [CrossRef] [PubMed]
36. Simoneit, B.R.T. Prebiotic organic synthesis under hydrothermal conditions: An overview. *Adv. Space Res.* **2004**, *33*, 88–94. [CrossRef]
37. Longo, A.; Damer, B. Factoring Origin of Life Hypotheses into the Search for Life in the Solar System and Beyond. *Life* **2020**, *10*, 52. [CrossRef]

Article

Origin of Life on Mars: Suitability and Opportunities

Benton C. Clark [1,*], Vera M. Kolb [2], Andrew Steele [3], Christopher H. House [4], Nina L. Lanza [5], Patrick J. Gasda [5], Scott J. VanBommel [6], Horton E. Newsom [7] and Jesús Martínez-Frías [8]

1. Space Science Institute, Boulder, CO 80301, USA
2. Department of Chemistry, University of Wisconsin—Parkside, Kenosha, WI 53141, USA; kolb@uwp.edu
3. Earth and Planetary Laboratory, Carnegie Institution for Science, Washington, DC 20015, USA; asteele@carnegiescience.edu
4. Department of Biochemistry and Molecular Biology, Pennsylvania State University, State College, PA 16807, USA; chrishouse@psu.edu
5. Los Alamos National Laboratory, Los Alamos, NM 87545, USA; nlanza@lanl.gov (N.L.L.); gasda@lanl.gov (P.J.G.)
6. Department of Earth and Planetary Sciences, Washington University in St. Louis, St. Louis, MO 63130, USA; vanbommel@wunder.wustl.edu
7. Institute of Meteoritics, Department of Earth and Planetary Sciences, University of New Mexico, Albuquerque, NM 88033, USA; newsom@unm.edu
8. Institute of Geosciences (CSIC-UCM), 28040 Madrid, Spain; j.m.frias@igeo.ucm-csic.es
* Correspondence: bclark@spacescience.org

Citation: Clark, B.C.; Kolb, V.M.; Steele, A.; House, C.H.; Lanza, N.L.; Gasda, P.J.; VanBommel, S.J.; Newsom, H.E.; Martínez-Frías, J. Origin of Life on Mars: Suitability and Opportunities. *Life* **2021**, *11*, 539. https://doi.org/10.3390/life11060539

Academic Editors: Michele Fiore and Emiliano Altamura

Received: 22 April 2021
Accepted: 1 June 2021
Published: 9 June 2021

Publisher's Note: MDPI stays neutral with regard to jurisdictional claims in published maps and institutional affiliations.

Copyright: © 2021 by the authors. Licensee MDPI, Basel, Switzerland. This article is an open access article distributed under the terms and conditions of the Creative Commons Attribution (CC BY) license (https://creativecommons.org/licenses/by/4.0/).

Abstract: Although the habitability of early Mars is now well established, its suitability for conditions favorable to an independent origin of life (OoL) has been less certain. With continued exploration, evidence has mounted for a widespread diversity of physical and chemical conditions on Mars that mimic those variously hypothesized as settings in which life first arose on Earth. Mars has also provided water, energy sources, CHNOPS elements, critical catalytic transition metal elements, as well as B, Mg, Ca, Na and K, all of which are elements associated with life as we know it. With its highly favorable sulfur abundance and land/ocean ratio, early wet Mars remains a prime candidate for its own OoL, in many respects superior to Earth. The relatively well-preserved ancient surface of planet Mars helps inform the range of possible analogous conditions during the now-obliterated history of early Earth. Continued exploration of Mars also contributes to the understanding of the opportunities for settings enabling an OoL on exoplanets. Favoring geochemical sediment samples for eventual return to Earth will enhance assessments of the likelihood of a Martian OoL.

Keywords: origin of life; Mars; prebiotic chemical evolution; early Earth; astrobiology; CHNOPS; transition elements; sample return; exoplanets

1. Introduction

The history of the proposition whether there is life on Mars has been a roller-coaster of disjoint conclusions. Early conjectures by astronomers working at the limits of observability included technosignatures (canals) and seasonal vegetation patterns ("wave of darkening"). More modern studies cast doubt, and the initial space exploration of Mars with snapshot images during fast flybys left the impression that the red planet was far more Moon-like than Earth-like (Mariner 4, 6, 7). Although the residual water ice cap was observed in the south polar region and water vapor detected in the sparse atmosphere [1], the crater-littered surfaces and lack of evidence of a substantial hydrosphere cast further doubt as to whether Mars was ever habitable.

Despite its short useful life after the subsidence of a global dust storm, the Mariner 9 orbiter imaged numerous geological features, such as braided and networked channels, which attested to stages of fluid erosion, including flowing water and runoff [2], notwithstanding the confirmation that the contemporary climate of Mars is too cold and too dry for the significant persistence of liquid water. The Viking missions verified the cold, dry surface and discovered a global soil that was lacking in organics or any firm indications of metabolic activities [3]. Worse, the global soil was virtually identical in geochemistry and chemical reactivity on opposite sides of the planet. Mars was bland, and uninteresting. This idea prevailed for 20 years, during which no missions were sent to Mars.

Following this long hiatus, subsequent missions to Mars ranging from highly instrumented orbiters to long-lived rovers have radically changed that view of uniformity. Orbiters have detected mineralogical diversities across the planet, while the Mars Exploration Rover (MER) and Mars Science Laboratory (MSL) rovers have revealed even greater diversity at the local scale. These missions have broadened and deepened the understanding of the changes in climate and history of liquid water [4], while also vastly increasing the knowledge of the chemical diversity [5,6] and the sedimentary history [7] of Mars.

Although much has been discovered concerning the habitability of Mars (e.g., [4–8]), an equally important consideration is whether the environments on Mars were just as conducive to the abiotic origin of life (OoL) itself. Could life have arisen on Mars? [9–16]. A hallmark of a biosphere of Life is its ability to adapt its forms such that it has species and strains of organisms which variously are able to survive and even sometimes prosper in nearly all conceivable aqueous environments. Additionally, many hostile environments which are intermittent can be tolerated until more suitable conditions arise. In contrast, prebiotic chemical evolution (PCE), leading to the first life form and creation of a biosphere, is more likely restricted to not only certain limited conditions but may even require a specific sequence of special environmental changes for it to succeed [13].

In addition to liquid H_2O, for an OoL, the starting organics, essential elements, and access to energy sources are needed. We shall, therefore, examine the extent to which Mars as a planet may have been able to supply these specific conditions during its existence, including the past, the present, and the future. Although the OoL on Earth is itself not completely understood, we may use the current state of knowledge to constrain the possible precursor environments, chemistries, and materials that would have been necessary on Mars for a similar origin.

From homogeneous magmas, brines, and atmosphere come settings which host a wide diversity of mineral grains, evaporite salts in sequences, aqueously altered minerals, as well as raindrops, snowflakes, and micro-climates. Homogeneity can lead to heterogeneity, especially when driven by changes in temperature, pressure, chemical environment, and other factors. Life similarly arises, but a variety of hypotheses exist which envision different physical and chemical conditions under which it can arise or has arisen. The extent to which these variously conceived conditions were available on Mars is a topic worthy of attention because it affects the likelihood that life could arise ab initio, perhaps even more than once, in any given planetary system. By examining conditions on Mars over deep time, a better picture of the early environments on ancient Earth may become apparent and, therefore, the search for life or prebiotic evolution is inexorably linked to the search for our own origins on Earth. The rover missions, and the MSL Curiosity rover in particular, have made many discoveries which greatly increase the prospect that life could have arisen on Mars.

2. Materials and Methods

Some concepts for the origin of life are focused on the prebiotic organic chemistry which must take place to create the complex functions and structures that are associated with living entities. Other hypotheses focus on the environmental conditions that nurture such activities, including the sources of the ingredients and energy that are needed, and certain specific settings in which the OoL could occur.

We shall summarize the breadth of these concepts and their chosen settings, noting the common and unique aspects of each. In the subsequent sections, the current knowledge of the status and history of Mars will be assessed for its compatibility with these various scenarios.

The most fundamental recognized requirement for life is the medium of liquid H_2O. Liquid water provides multiple favorable physicochemical factors, including its solvation capabilities, wide temperature range, polar properties, and so forth, as well as enabling the mobility of its constituents while, overall, being contained. Furthermore, it is biochemically essential, with one-third to one-half of all metabolic reactions having H_2O molecules as either reactants or products [17]. All these properties of water are also at play in PCE scenarios. It is, therefore, the most fundamental component needed for life as we know it (LAWKI).

2.1. Organic Molecules and Elements

The standard compositional requirement for life is for the elements carbon, nitrogen, etc. making up the CHNOPS group. These include the primary formation elements (CHONS) which make up the amino acids, and hence proteins. In certain other critical biochemicals, sulfur does not play a principle role but phosphorus does: RNA, DNA, phospholipids, ATP, and other constituents. Unless a potentially habitable environment has these six elements available, LAWKI is not possible.

Laboratory studies of prebiotic chemical evolution have accelerated in recent decades, with many promising results. To reflect this progress, some of the groups spearheading various investigations are portrayed in the rough timeline of Figure 1. Their findings have identified several key pathways from simple organic compounds to the classes of complex molecules utilized by extant organisms. They also have evaluated many candidate reaction sequences which either do or do not produce significant yields, or have very specific requirements that may or may not be plausible in the early times of a lifeless planet.

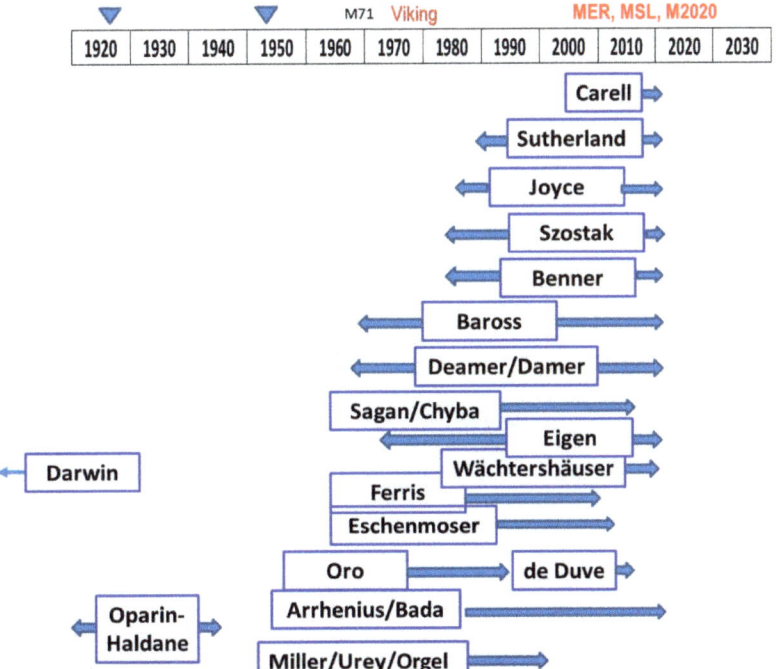

Figure 1. Some example groups that have spearheaded investigations into the origin of life, showing accelerated intensity in the study of prebiotic organic evolution pathways, subsequent to the findings of Miller–Urey experiment in 1953 [18] (Each block is one decade, e.g., 2010 = 2010 to 2019).

Following the discovery that amino acids could be formed by lightning-like electrical discharges in a mixture of simple gases [18], numerous laboratory investigations have sought to elucidate natural processes and reactions leading to PCE that could occur on early Earth.

Early work by de Duve [19] focused on the high-energy bond in thioesters formed by combining simple organic molecules, volcanic H_2S, and carboxylic acid molecules. His observation was that thioesters are deeply ingrained in many pathways of contemporary metabolism and could have been a precursor to the use of the high energy phosphate bonds. In biological systems on Earth, the energy-rich thioester bond provides the energy source for the phosphorylation of ADP to ATP, which makes sulfur-based energy management indispensable [20]. However, there are many metabolic pathways that operate completely independent of ATP, relying on acetyl-CoA to drive reactions, including the biosynthesis of fatty acids. The structure of acetyl CoA includes the relatively simple sulfur-containing molecule cystamine at the end where the thioester bond would be, as well as, at the other end, the relatively complex 3′-phosphorylated ADP, which acts as a handle for molecular recognition and attachment. This structure supports the notion that thioester sulfur chemistry might have progressed, independent of a phosphate-based system.

A one-pot synthesis of thioesters using FeS and peroxysulfate or UV stimulation has been demonstrated without the need for enzymes [21]. A combinatorial analysis of coupling of redox reactions to thioester PCE has led to an "organo-sulfur" protometabolism hypothesis [22]. Thioesters may have played a role in peptide ligation, as organic catalysts rather than requiring metal cofactors [23].

The "iron-sulfur world" analyzed by Wächtershäuser [24–26] proposes catalytic properties of Fe-S for forming organic compounds, and takes advantage of higher temperatures and sulfide availability at hydrothermal vents. Often using CO, thiols, and metal sulfide catalysts, these reactions can form activated acetic acid with steps similar to those of the Wood–Ljungdahl carbon fixation and acetogenesis pathways [27]. Thioesters formed abiotically can be partially protected from subsequent hydrolysis through incorporation into vesicle walls [28]. Abiotically formed thioesters might have been the foundation of the Earth's origin of life (e.g., [24,29]) and/or the basis of early cellular metabolism using a limited set of genes [30].

The Sutherland group has shown that a milieu of HCN (and some simple derivatives), plus sulfur species and a Cu catalyst can variously produce the precursor molecules for all three of the most fundamental classes of biomolecules: RNA, proteins, and lipids, with just simple variations in environmental conditions [12,31–35]. This set of reactions, referred to as a "cyanosulfidic" protometabolism, also requires UV light, as well as Fe, Ca, and P. This group of investigators has envisioned a set of bodies of water with interconnecting streams to produce serial syntheses of ever more complex assemblages of the precursor molecules needed for a protocell.

Cyanide destruction by reactions with H_2O and/or HCHO can be prevented through formation of ferrocyanide and salts thereof, with subsequent thermal decomposition to restore the availability of cyanide feedstock for PCE [12,36].

It has now also been shown that the S-containing amino acid, cysteine, can catalyze the elongation of peptides [37]. Incorporation of cysteine monomers at strategic locations in the amino acid chain of a protein allows the formation of disulfide covalent bonds which rigidize the protein in its optimum three-dimensional configuration for its powerful enzymatic activity.

The Carell group has demonstrated [38–41] the synthesis of all four nucleotides for RNA from just ribose (formed elsewhere) and other small molecules (HCN, NH_3, nitrate), using FeS, sulfite, Cu, Zn (or Co), B, and carbonate, with the help of wet–dry cycling and selective changes in physical conditions (temperature, pH). This can be done with the same simple ingredients ("one-pot") and is another breakthrough in finding likely PCE pathways [40].

Suggestions for boron participation in prebiotic chemistry has a significant history [42]. Benner and colleagues have investigated the use of boron minerals in the stabilization against rapid degradation of the key sugar, ribose [43–45], as well as pointing out other ways in which mineral versions of this element are useful for prebiotic evolution [42,44–46]. With Ni^{2+}, borate, urea, and a cyclic phosphate, they demonstrate the formation of ribonucleotides [47]. It is even proposed that perhaps Mars is a more likely locale for OoL than early Earth because of its arid environment, which is favorable to wet–dry episodes and concentrations of borate and other minerals [10]. They also invoke the protective effect of magmatically released SO_2 in producing a stable sulfonate derivative as a temporary storehouse of labile HCHO during PCE [48].

Earlier, the Ferris group had shown that oligomers of RNA could form under the catalytic action of montmorillonite clays on ribonucleotides [49,50].

The acetyl CoA (Wood–Ljungdahl) pathway is the most primitive and simplest sequence to formation of important organic molecules via CO_2 fixation with H_2, and can be catalyzed by native $Fe°$, $Ni°$, $Co°$ and also to some extent by $Mo°$ or $W°$ [51]. This overall exergonic reaction can also be catalyzed by Ni_3Fe (awaruite) [52]. In microbes, its enzymatic infrastructure includes cofactors of Co, Ni, Fe and S. This pathway is only one of at least six different pathways in the world of biology by which CO_2 fixation proceeds. It is the simplest and also the only one which can also be accomplished without enzymes, plus the only one that can have a net generation of ATP rather than a consumption of ATP [52]. Hence, it is often considered as one of the most rudimentary, fundamental, and earliest forms of metabolism [27,52].

Deamer and collaborators have championed the amphiphilic compounds extractable from carbonaceous meteorites for their ability to form vesicles, and the role these might play in the very earliest PCE by enabling the compartmentalization of key components that go into making up the system that the protocell needs for its primitive functions [53].

High concentrations of salts, including Mg^{2+}, are found to destabilize the formation of vesicles from amphiphilic precursor molecules [53]. It has, therefore, been proposed that subaerial hot springs may be more conducive to the OoL than deep oceanic hydrothermal vents (OHV) because typical spring waters have lower salt concentrations, especially for Mg^{2+}, than the seawater environment surrounding OHVs (with example spring waters concentrations of 0.03–3 mM Mg^{2+} compared to 53 mM for seawater and 9–19 mM in the Lost City white smoker [54]). However, it is now also found that certain biological amino acids can stabilize fatty acid membranes, even in the presence of high ionic strength, including solutions such as NaCl and Mg^{2+} salts [55]. Another substitute for Mg^{2+} is Fe^{2+} ions in solution [56], if they can be prevented from becoming oxidized. An organic chelator, citrate, still allows for RNA replication enhanced by Mg^{2+}, but prevents the Mg disruption of vesicle formation while enhancing their permeability to RNA tetramers [57]. Coacervates and other encapsulation pathways are also being investigated [58].

The Deamer and Damer group describes hot springs in a volcanic environment as the setting for the origin of life [15,54,59–62]. This group has also emphasized the wide-ranging benefits of wet–dry cycles to promote dehydration polymerizations of a wide variety of monomers [60], including RNA, DNA, and proteins.

Progress continues to be made, for example by the Joyce group [63,64], in test-tube evolution experiments of ribozymes in high levels of Mg^{2+} to evaluate the RNA World hypothesis for the origin of the genetic basis for life [62,63,65–67].

Transition Elements. For the management of the extraordinarily complex chemical pathways in a metabolic system, the organism must orchestrate reactions by means of catalysts. In the organisms which have emerged as the biosphere has proliferated and evolved over geologic time, exquisite protein-based enzymes have emerged, about 40% of which employ one or more of the transition metal elements or divalent cations from group 2 in the periodic table (Mg, Ca) [68]. The group 4 transition elements, Ti and Zr, are relatively abundant minor and trace elements, respectively, in basaltic materials, but have had little if any significant biological involvement. However, the first row transition elements in the Period Table subsequent to Ti, namely V, Cr, Mn, Fe, Co, Ni, Cu, and Zn, are known for their various catalytic properties and as essential elements in many if not most biological systems [68,69]. They have also been proposed as essential progenitors to an origin of life [70]. Additional key catalytic elements include Mo and W, which are lower-row transition elements in Group 6 along with Cr, which is not involved with living systems to any major degree. However, Cr^{3+} (in addition to Zn^{2+} and $Fe°$) has been shown to promote certain reactions of the reverse Krebs cycle, whose origins are also suggested to have been an early anabolic biochemical pathway for CO_2 fixation using H_2O [71].

Metalloenzymes presumably were gradually evolved to their current extraordinarily efficient form [72], but it is hypothesized that for some, or many, the metal ions themselves may have been the original primitive catalysts [69,71,73,74].

Although these trace elements are needed, it is at levels that are species-specific. Each element has a minimum concentration and a maximum acceptable level before toxicity or stress sets in. Any given species or strain is adapted to certain environments, and generally has sensing, as well as active transport systems which import or export elements as needed for optimum metabolic functionality [68,69,75].

Early peptides would have had important interaction with metal ions for various functions, including Mg, Zn, Fe-S, Cu and Mn [76]. It has been hypothesized that primitive oligopeptides were functional with only four amino acids (Gly, Ala, Val, Asp) which have specific domains, the binding metals of which range across Mg, Mn, Zn, and Ni [77].

One of the most important enzyme cofactors is a combination of ions, the [FeS] clusters which are fundamental to electron storage and transfer. Proteins such as the ferredoxins, which participate in photosynthesis, nitrogen fixation, and assimilation of hydrogen, nitrogen, and sulfur, rely on such clusters. There are [FeS(Ni)] clusters in hydrogenases, which promote oxidations of H_2, one of the most important catalytic functions in the biosphere.

Molybdenum plays several important roles in extant organisms, including the extraordinarily important nitrogenase enzyme ([MoFeS] cluster) in diazotrophic organisms for the fixation of atmospheric dinitrogen. Although Mo is the common cofactor, there are also vanadium-based nitrogenases, and even an Fe-only version. Several members of the Euryarchaeota phylum of the Archaea can fix nitrogen without Mo (possibly indicating a more rudimentary function), whereas members of the Crenarchaeota phyla utilize Mo and Cu for nitrification and denitrification, respectively [74].

Mo-containing enzymes are also involved in nitrate and sulfate metabolism. Because of a lack of the sufficient availability of soluble Mo prior to the Great Oxidation Event [78], other elements, such as tungsten and vanadium, may have provided the needed catalytic activity for these metabolic functions.

The composition of an ancient metallome from 3.33 Ga carbonaceous residue in the Barberton indicates the participation of Fe, V, Ni, As, and Co in that example of early organisms. This residue was also modestly enriched in Mn, Cu, and Zr, but with an absence of detectable Mo and Zn [79].

Mulkidjanian [80] has emphasized the role of zinc ions, among others, in the progress toward biological activity. Although zinc today is not essential in all procaryotes, it is prominent in many functions in eukaryotes and the hormones of their multicellular forms, with special use of the "zinc finger" proteins. Altogether, Zn in proteins may perform as many as six different general functions, but was presumably was only sparsely available on early Earth because of its insolubility as ZnS [69]. On a planet where Zn is readily available, one or more of its potential functions may be broadly utilized in early metabolism, such as the one-pot synthesis of nucleotides [38].

Another role of transition elements, especially for Zn and Ni, is their enhancement of adsorption of nucleotides onto montmorillonite and nontronite clays [73].

2.2. Energy Sources

Overcoming the entropic barriers to life's high degree of organization requires inputs of energy, as do many of the chemical reactions of its metabolism. In addition to the usual energy sources, such as heat, sunlight (including UV), ionizing radiation from galactic cosmic rays (GCR), and lightning, there are the chemical disequilibria that result when high-temperature magma releases volatiles or is quenched, as well as when air and water react with regolith, rocks, and each other.

Chemotrophs can live solely from the energy of chemical reactions. The variety and, in many cases, versatility of chemotrophs to obtain energy from many different redox couples that result from geochemical and atmospheric processes continues to amaze. For example, an evaluation of the Gibbs energy for 730 redox reactions among 23 inorganic reactants (from 8 elements: CHNOS plus Fe, Mn, As) possible in a shallow-sea hydrothermal vent found that almost one-half were exergonic reactions [81]. Intricate syntrophic relationships between communities also allows for the extensive exploitation of multiple sources of latent chemical energy in the environment. The extraordinarily high level of sophistication in metabolic processing of various redox couples by microbial species, especially in the archaea, would not be fully available to processes of prebiotic chemical evolution, so it is necessary to identify specific appropriate energy sources for the OoL.

2.3. Planetary Settings

Beginning with Darwin's musing that life may have started with some "warm, little pond" with ingredients of N, P, and energy, there have been pond scenarios for an OoL. These days, the feedstocks usually contemplated are molecules residing in a reducing atmosphere with CH_4, NH_3, and/or H_2. From the energetics of lightning and its chemical products, a variety of organics, including amino acids [18], are produced. Others invoke rare but fortuitous contributions of pre-formed organics directly into ponds from sources, such as comets [82] or carbonaceous meteorites [83]. From the interplay of several factors, ponds of intermediate size are the more likely settings than larger or smaller bodies of water [13].

Because of the multitude of biochemicals that make up a living system, it is envisioned that ponds and streams must undergo changes that bring together different ingredients. Although most PCE experiments are conducted at circumneutral pH, there are some which benefit or begin from a more acidic, or more basic, environment [14,33]. RNA stability increases with acidic pH and with Mg^{2+} rather than Fe^{2+} in solution [16].

As a setting or combination of settings with spatial and temporal variations progresses from prebiotic chemical evolution to replicating molecules, vesicles, and orchestration of catalyzed metabolic functions, it enables the protocell. Further development can lead to the colonizer cell to complete the transition from being a wholly inanimate locale to one which contains the earliest forms of life which can potentially establish a biosphere. This is termed the macrobiont [13] because this setting is macroscopic in scale yet contains molecular- and microbe-scale entities which have biological capabilities. Before this organizational transition to the birth of life, such a potentially appropriate setting for an origin of life is simply a proto-macrobiont.

A quite special type of setting is the oceanic hydrothermal vent (OHV), usually in a group comprising a vent field. These "smoker" chimneys exude a rich variety of constituents extracted by seawater circulating through hot basalt beneath the seafloor at magmatically active locations, especially at tectonic-plate boundaries. A variety of possibilities for an origin of life have been addressed for the vent chimneys which form from effusing streams [84,85], including porous cavities that trap components [86] for reactions [87] and can concentrate products by thermogravitational trapping [88]. Recent attention has moved from the acidic black-smokers to the much less abundant alkaline white-smoker chimneys because of the production of H_2 by the latter via the serpentinization reaction between H_2O and ultramafic minerals [84,85,87,89,90]. With the high pressures that the deep sea affords, H_2O can remain liquid at temperatures of several hundred °C, which shortens the time, increases the efficiency, and enhances the extractions of soluble ions into the stream. These effusions are rich in sulfides, especially those containing Fe, Cu, and Zn. Sulfide-rich bubbles with Fe and Ni, produced by seepage from vent fields, have been proposed as a micro-setting for PCE leading to life [29,91].

Smokers also include other bio-essential elements, such as Mo, Co, Ni, Se, as well as additional elements [92] not generally associated with the needs of living organisms or the OoL. The vent fluids' dissolved loads enable a diverse, but highly localized macrofauna by providing not only these essential nutrient elements, but also chemical energy from abundant and diverse redox couples.

In addition to sub-oceanic settings, there is strong interest in subaerial settings where hydrothermal activity is prominent [15,54,61,93]. These include hot springs, geysers, mud pots and other manifestations in geothermal areas, driven by shallow magmatic chambers and groundwater.

Although transient, the energy deposited during the impact cratering process into wet or permafrost-laden regolith to form warm crater lakes with subsurface hydrothermal activity has been a scenario of increasing interest as a macrobiont for the OoL [94–96]. The Chicxulub impact crater on Earth includes abundant evidence of creating a hydrothermal system and supporting colonies of sulfate-reducing organisms for ~3 million years [95]. Additionally, craters as small as ~25 km diameter may produce buried hydrothermal activity which persists for ~1 Myr, while ~5 km craters could result in hydrothermal systems that persist for thousands of years [96].

Another model ties the origin of life to a single cataclysmic event [97], the impact of a planetary-scale object which is large enough to have a native Fe core, which then reduces available H_2O to create a H_2-rich atmosphere. While causing a transient environment that would be sterilizing and also delivering the siderophile elements that are in the Earth's mantle, its aftermath is proposed to enable the origin of the RNA world.

Wet–dry and freeze–thaw cycles. Several decades ago, there began the search for physical processes which could promote certain chemical reactions which do not proceed well in water. These reactions include dehydrations to enable polymerization. One class is the formation of oligopeptides from amino acids and another is the synthesis of polynucleotides (RNA, DNA) from their constituent nucleotide monomers (ATGUC). Although modern cells routinely perform these functions with the help of enzymes and ATP energy, the challenge for PCE remains. There are a few paths to these results using "condensing agents", but these are generally not plausible natural chemicals. What was learned in these early studies was that thermal cycling can help promote polymerization reactions, but drying a prebiotic milieu and then re-wetting, followed by more cycles, is an even more effective way to promote the dehydration to form oligomers [40,46,60,98–101]. Similarly, freeze–thaw cycles can promote ligations for an RNA world [102–105]. Much additional study has demonstrated the general and powerful applicability of these plausible natural oscillations in environmental conditions to achieve highly relevant results for PCE [93].

River deltas are of considerable interest because they intrinsically include braided streams which enable ponds that are intermittently semi- or totally isolated, and then brought into communication and mixing as outflows change in volume or rate. The Gilbert-type delta in Garu crater on Mars is estimated to have developed over a period of as much as 10^5 years [106]. However, deltas can also form and desist rapidly, as in the <1 kyr estimate for the fluvial delta in Jezero crater [107], which may be too rapid for relevant chemistries to successfully proceed.

Subaerial semi-arid environments, including geothermal areas, are ideal for generating repetitious wetting and drying. Ponds in cool environments, especially along their foreshores, can experience both wet–dry and freeze–thaw events. Mudflat patterns can create further heterogeneity by trapping pond constituents temporarily in juxtaposed mini-environments [13] which could allow semi-independent progressions along the PCE arrow toward life with subsequent flooding to bring them together.

Porous sediments enable the phenomena of geochromatography to separate constituents. In this scenario, PCE components move along aqueous gradients containing heterogeneous concentrations of anions and cations, allowing for the separation of organic molecules based on size and charge [108]. Furthermore, in such a scenario, the concentration of anion and cations could also provide environments of differing water activity, effectively becoming a pseudo "wet and dry" environment. A feature is that such an environment could provide natural compartmentalization of molecules based on hydrophilic/hydrophobic reactions.

3. Results

Although Mars has many specific differences from Earth ($10\times$ smaller mass, $200\times$ less atmosphere, $6\times$ to $20\times$ less H_2O per unit surface area, $50\times$ less volcanism, and no plate tectonics or spreading centers), it nonetheless has many, perhaps all, of the ingredients needed for various scenarios proposed for the origin of life. The relevant comparison must be made with Earth itself, not the known or envisioned exoplanets, because Earth is demonstrably the only locale where life did actually once successfully originate (unless it came via lithopanspermia from Mars itself [10]).

The key properties of an incipient macrobiont include the availability of essential ingredients in terms of water, key elements and molecules, access to energy sources, and existence of the variety of local settings with physical and chemical properties which have been suggested to be adequate or essential for the processes leading to life. Sources that are intrinsically available on Mars as well as exogenous sources of ingredients (comets, asteroids) could each contribute.

3.1. Elements and Molecules on Mars

At its surface, current-day Mars offers an oxidizing environment [109–112]. This is contrary to what generally would be desired for most reactions in PCE [9]. However, both early Earth and early Mars are now thought to have had abundant atmospheric species in reduced form, such as H_2 and CH_4, in order to achieve a sufficient greenhouse effect to enable liquid H_2O at their surfaces [4,113–115].

The primary igneous minerals of Mars generally mimic those prominent on Earth, especially those in basalts, including olivines, pyroxenes, feldspars, apatites, etc. It is the mafic minerals, olivine and pyroxene, which can react with H_2O at moderate to high temperatures to produce H_2 in the reaction to form serpentines and magnetite [116] and, by additional reaction with CO_2 to produce methane [110]. These serpentinization reactions can also liberate important minor and trace elements from the crystal lattices of these minerals.

3.1.1. Organics and CHNOPS Elements

Not only are these CHNOPS and other elements essential to all life as we know it, but their molecular or mineral form is important because it affects reactivity and availability in aqueous media (e.g., valence state; solubility).

Carbon. From magmatic outgassing, the early Martian environment could have hosted significantly greater quantities of CO_2 in its atmosphere. The total inventory of CO_2 is estimated to have been in the range of 1 to 3 bar [117], based on loss rates observed by the MAVEN mission and orbital observations of carbonates on Mars, or less than 1–2 bar [118,119] when based on the size–frequency distribution of the ancient craters. The current nominal partial pressure of CO_2 on Mars is 6 mbar (plus traces of CO). Compared to Earth's 0.4 mbar of CO_2, this is more than adequate to support the biosynthesis of the array of organic molecules needed by living cells across a broad biosphere.

Both planets are thought to have had atmospheres with CO_2 more in the 1+ bar range, in order to have a sufficient greenhouse effect to prevent runaway freezing [115,120]. The Earth's higher rate of magmatic devolatilization has resulted in massive carbonate deposits, equivalent to tens of kilobars of CO_2 when integrated over geologic time [121].

Other sources of carbon on an emerging world can come from exogenous inputs, such as comets, carbonaceous meteorites, and interplanetary dust particles. Compared to carbonate and CO_2, an advantage of these sources is that the carbon will be in relatively reduced molecular forms important to PCE, such as hydrocarbons, N-rich organics (including cyanide), and even abiotically synthesized amino and carboxylic acids [122].

Although Mars has a 72% smaller cross-section for impact than Earth and a weaker gravity field, its cross-section for impacts scales the same as its surface area, such that the density of craters on land could be very similar. At the location of Mars farther out in the Solar System, it would be expected to experience a greater flux of C-bearing impactors by being nearer to the organically enriched bodies in the outer asteroid belt, as well as Kuiper belt objects and comets. Furthermore, from the relatively high Ni content of the Martian global soil (ranging from 400 to 600 ppm [123,124]) compared to the Ni in Martian basalts (165 ppm in Adirondack, 81 ppm in BounceRock, 79 ppm in Shergotty meteorite [125]), it can be inferred that the contribution of these exogenous delivery sources to the shallower regolith of Mars resulted in relatively higher concentrations of organics than was for Earth, which had more extensive regolith turnover and obscuration due to higher rates of extrusive volcanism and sedimentary processing. The expanse of Earth's global-scale ocean would have also lessened the relative importance of these endogenous contributions because its high dilution factor likely reduced their concentrations to be low, or negligible.

Organic compounds, such as HCN, may be produced in a reducing atmosphere by various processes [34]. These include production by photochemical or ionizing radiations reactions [126], by lightning [115], and even some by ionization during atmospheric passage of hypervelocity bolides and their impact ejecta [127]. Cometary delivery has long been considered a potential major source of HCN [128,129].

Hydrogen. This element mostly occurs in its oxidized form, as H_2O. Although water is the foremost use for H atoms to support both life and the origin of life, some must also be available to form organic molecules. For example, the dry mass of a typical microbe includes 1.77 H atoms per C atom [130]. Likely forms of H in the early atmospheres include not only H_2O vapor, but also reduced molecules such as H_2, CH_4, NH_3, HCN, and H_2CN_2. As noted above, water molecules can be decomposed to give off H_2 through the serpentinization reaction, but there are also other sources of H_2 early in Mars history [116].

Little direct evidence is available for the presence of significant H_2 on either early Mars or Earth. However, the extensive hydrologic activity now abundantly evident for the earliest history of Mars mandates the presence of a stronger atmospheric greenhouse effect than currently, and the best solution of that uncertainty is if Mars had an early ~1 or 2 bar CO_2 atmosphere but combined with 1% to 20% of H_2 gas [113,131], with similar scenarios for the early Earth [115]. Modern models which implicate reducing atmospheres for these planets negate the arguments at one time raised against the theories of Haldane and Oparin, as well as the pioneering experiments by Miller and Urey [18], all of which presumed a more reducing rather than oxidizing atmosphere early in our planet's history when the OoL was occurring.

Nitrogen. This element is primordially expected to be gaseous at the surface, in the form of the relatively inert N_2. The current Martian atmospheric level is only 0.16 mbar, about 5000× lower than on Earth. However, the majority of terrestrial organisms cannot metabolize dinitrogen gas, and instead must obtain this element from ammonium salts, nitrates, or nitrogenous organic molecules in the soil. A few microorganisms, the diazotrophs, are able to "fix" N_2 gas into one or more of these compounds. These organisms utilize Fe-S-Mo- or V-based nitrogenase enzymes to accomplish these conversions.

A Martian biosphere may have been limited in its vigor by this shortfall of N, just as many ecosystems on Earth are often growth-limited by the availability of nutrient N. However, nitrates have now been found in Martian soils [132,133]. On early Earth, prior to biological N_2 fixation, nitrates and nitrites formed by lightening may have been the source of N needed for prebiotic chemical evolution [134]. The nitrate ion was sought in the Wet Chemistry Laboratory (WCL) soil water experiment on the Phoenix mission [135,136] but was not found at a detection limit that would be equivalent to 25 mM (if at a 1:1 W/R ratio). However, nitrate has been detected by the Evolved Gas Analyzer (EGA) in certain Gale crater soils, albeit at only ~5 µM equivalent (300 ppm), and up to about 3× this amount in some samples [132]. Organic nitrogen compounds (pyrroles and imines) were found in the Tissint Martian meteorite, linked to electrochemical reduction on mineral surfaces [137]. The reduction of N_2 to NH_3 is a well described process in electrochemistry using $Fe°$/magnetite [138] or pyrite [139] as electrocatalysts.

Due to its rarity, the exogenous sources of useable N may be relatively much more important than for other elements, since some meteorites and the comets are known to have a significant content of N-rich organics [140], including cyanides [141]. Cometary delivery of HCN has been shown be a major potential source of cyanide during individual impacts [129]. It has also been hypothesized, however, that, as impacts proceed, they will re-liberate significant amounts of regolith nitrate and result in intermediate, quasi-steady state concentrations for N (soil and atmosphere) [142]

Early Mars had more N_2, since it can be lost by atmospheric escape and fixation into soil. Estimates based on isotopic composition and various loss mechanisms range from $13\times$ [143] to up to $200\times$ [144] higher concentrations than current levels. In comparison, early Earth's N_2 may have been only $2\times$ or perhaps even less than the modern value [145]. Taking these ranges into account allows for early Mars original pN_2 to have been as high as ~3% that of Earth's.

Oxygen. In most planetary atmospheres, O_2 is extremely low (with Earth as an exception because of its biosphere's abundance of oxygenic photosynthesizing organisms). Oxygen is a key atom in the biochemistry of life, with one O atom for every two C atoms in the organic makeup of the microbial cell, about twice as many as N atoms [130]. The O in carboxylic acids, esters, sugars, phosphates, and a myriad of other organic molecules needed for the OoL is readily derived by reaction with the hydroxyl radical from the water medium and from the photochemical processing of the relatively abundant CO_2 in the atmosphere and its soluble byproducts in the aqueous phase. Although the present-day Martian atmosphere has very low O_2 abundances, recent modeling suggests that Mars may have had multiple, cyclical episodes of higher atmospheric O_2 (e.g., ~10 mb) in the Noachian [146], which may have provided both challenges and opportunities for macrobiont chemical evolution. Indeed, carboxyl- and carbonyl-rich macromolecular organic carbon compounds have been described from three Martian meteorites so far [137]

Phosphorus. It was learned early that phosphorous could be extracted from Martian shergottite meteorites with mild acidification [147]. Detailed experiments on a variety of Mars-relevant P-containing minerals (merrillite, whitlockite, and chlorapatite) show significantly increased dissolution rates compared to the terrestrially more common fluorapatites, as well as the strong effect of promotion of solubilization by acidification [148].

From the Mars rover missions, mobile phosphorous was indicated at the Independence outcrop [149]. The numerous Wishstone and Watchtower rocks on Husband Hill have very high concentrations of phosphorus (>5 wt%, as P_2O_5-equivalent [150,151], compared to the 0.4–0.7 wt% in Mars igneous rocks, such as Adirondack-class [152] and shergottites [125]).

Abundant nodules, some of which are phosphorus-, Mn-, Ca- and S-rich, have been discovered in the Ayton sample near the Groken drill site in Gale crater, with up to 18 wt% equivalent P_2O_5 [153], and could be much higher for a Mn-P phase (without $CaSO_4$).

The Martian meteorite NWA 7034 is a basaltic breccia with clasts representing four distinct lithologies. One of these lithologies, termed "FTP" (enriched in Fe, Ti, and P), has a P_2O_5 abundance range of 6 to 12 wt% (1-sigma), as chlorapatite [154]. This "Black Beauty" meteorite is thought to be representative of average Martian crust and global soil because of its bulk element composition, although it does not contain the high S, Cl, and Zn of Martian soil. It also contains clasts enriched in Mn^{4+} oxides, which have 2.4 wt% P_2O_5, indicative of aqueous alteration [155].

The abundant iron-nickel meteorites on the surface of Mars [156–158] could also become a source of utilizable P from their content of the mineral schreibersite, $(Fe,Ni)_3P$ [159,160]. The minimum abundance from the Curiosity transects is $800/km^2$ [158], which implies $\sim 10^{11}$ irons on the surface of Mars. It also has been shown that, if ammonia solution is present, the readily soluble amidophosphate can be formed and would be an appropriate

P-source for PCE [160]. The higher Ni in Martian soil may be an indicator for microparticulates with this composition.

Given that early Mars and Earth could have had significant cyanide [12,36], as assumed by many hypotheses for the OoL, it has been shown that ferrocyanide or $MgSO_4$ plus Na cyanide can enable the formation of organophosphates from hydroxyapatite [161].

Sulfur. Mars is clearly a sulfur-rich planet, compared to Earth [162,163]. This may be especially important because, as pointed out above, so many concepts for the early prebiotic chemical pathways make significant and often essential use of organic molecules containing sulfur atoms and/or inorganic compounds or minerals involving sulfur. This is the case for the cyanosulfidic pathways of the Sutherland group [35,109,164–166], the thioesters of de Duve [19], the Fe-S World of Wächtershäuser [25,26], the SO_2 sequestration of HCHO of the Benner group [48], the cysteine primordial precursor of the Powner group [37], the sulfides for the hydrothermal vents [91,167], etc. On an early, wet, reducing and more sulfur-rich Mars, thiol-containing organics may have been far more prevalent than on early Earth, leading to sulfur organic chemistry in a variety of surface and subsurface environments and providing widespread cyanosulfidic chemistry and/or thioester chemical energy for a Martian origin of life.

Martian basalts, as evidenced by the shergottites and other meteorites, generally contain S (as sulfides) at a level [125] which is higher by about an order of magnitude than the composition of MORB basalts [168], and the global soils are enriched by a factor of more than an order of magnitude over the shergottites, with S equivalent to 4 to 8 wt% SO_3 [123,124,169,170].

The Martian soils and sediments often have S as sulfates. Magnesium-rich sulfates are found in duricrusts [169], the Burns formation [171], soil trenches [172], evaporites in Gale crater [173], and Murray formation [174]. Orbital observations of Mg sulfates in both monohydrated (kieserite) and polyhydrated states have been discovered in many deposits [5]. $MgSO_4$ is highly soluble, which can provide the Mg^{2+} cations used in many prebiotic chemical sequences, including the RNA world. Some microbes not only tolerate but grow well under 2 M $MgSO_4$ concentrations [175].

Extensive ferric sulfate soil horizons occur on Husband Hill and Home Plate [150,176]. Mg and Fe sulfates are also a common constituent of Murray formation rocks in Gale crater [150].

Less soluble than $MgSO_4$ are minerals composed primarily of $CaSO_4$, which were nevertheless once in solution, as evidenced by their widespread occurrences as veins in Gale Crater [6] and also in Columbia Hills at Gusev crater [177].

The source of the high S in Martian soil is generally posited to be via magmatic release into the atmosphere [162,178,179], variously as H_2S, SO_2, or SO_3, depending on the oxygen fugacity of the source at the time of release. Once in the atmosphere, the H_2S may be quickly oxidized, mediated by strong UV photochemistry [112,180]. The SO_x forms will yield the sulfurous and sulfuric acids once they interact with water (aerosol, or subaerial). Similarly, the HCl and Cl_2 released from magma can also impart acidity to the soil. In spite of the global soil's putative high content of these acidifying species, its pH was found by the Phoenix mission to be circumneutral, with a pH of 7.7 ± 0.1 [181], which indicates the alteration of mafic mineral grains or interaction with intrinsic carbonate [182] to provide neutralization by their mild alkalinity in solution.

Not only the widespread sulfates in soils and sediments, but also the presence of sulfites has been inferred for some samples, based on release of reduced S-containing volatile compounds during EGA analyses by the SAM instrument [183].

In view of the widespread incorporation of [FeS] clusters into various critical enzymes, a PCE path to their generation has been investigated, using environmental UV to photooxidize the Fe^{2+} that would be generally available in early planetary environments [165]. Early formation of protoferredoxins would be a major step toward establishing electron storage and transport chains for a variety of biochemical pathways. That Mars is more abundantly endowed with both Fe and S lends support for the potential rise and evolution of these Fe-S based functions.

Organic molecules. The search for organic molecules on Mars has been long and determined. Early exploration with the Viking mission concluded that organic compounds were at levels below about 1 ppb, and the detection of chlorobenzene was ascribed to contamination [184]. These low levels in spite of infall of carbonaceous meteorites and interplanetary dust particles, not to mention the demonstration of organic synthesis from atmospheric conditions under UV irradiation [185], were explained as being the result of the strong oxidation power of the Martian atmosphere, driven by photochemical reactions [109].

Organics discovered in numerous Martian meteorites are indicated to have been synthesized by electrochemical reduction in the amount of CO_2 by exposure of magnetites, Fe-sulfides and brines [137].

Subsequently, the MSL mission's organic analyzer SAM detected ~10 ppb of halogenated organic molecules in some samples [186]. Additionally, still later, aromatics, aliphatics, thiophenes and other S-C compounds were detected at up to ~20 ppm C in mudstone samples from the Murray formation in Gale crater [187]. The concentrations of organics (released at high temperatures >600 °C) in Martian meteorites are 8 to 14 ppm in Tissint [188], indicating that indigenous abiotically synthesized organic materials provide a pool of building blocks for prebiotic reactions on Mars. For our example of 1:1 W/R ratio in mud, these would be a concentration of a factor of three or so less than 1 nM, even if all organics were soluble. Various concentration mechanisms envisioned in macrobiont scenarios could enable PCE to proceed.

Although the Viking GCMS had the demonstrated capability for detecting thiophenes as well as other organics [184], it did not detect such levels in any of the samples at Utopia Planitia or Chryse Planitia. This was apparently due to several differences for Viking analyses: only global soil samples could be acquired, rather than lithified sediments which may have been much better protected against atmospheric oxidation and ionizing radiation (GCR); the maximum pyrolysis temperature was 500 °C, whereas the organics detected by SAM were only released above this temperature; the sample size was smaller in mass and flash heated then held at maximum temperature for only 30 s, compared to a 10× larger sample and much slower temperature ramp for SAM, which resulted in several hours at temperatures of 500 °C to 820 °C; and, finally, a cost descope eliminated the direct-inlet to the Viking mass spectrometer, requiring the evolved volatiles to pass through the GC column before injection into the MS.

The contribution of accreted meteoritic matter to the regolith places a floor on organic matter that would be expected, independent of abiotic or biotic synthesis, and assuming no subsequent oxidative destruction. The nickel contents of the global Martian soils cluster around 450 ppm Ni [123,124], but Ni in the source igneous rocks is much lower (average of ~100 ppm for SNC meteorites [125] and MER igneous rocks in Gusev crater). If this excess of 350 ppm is solely due to meteoritic contributions, it implies an upper limit of 1000 ppm for exogenous carbon because the ratio of C/Ni (wt/wt) in CI meteorites and Tagish Lake is 3.1 [189]. This is, of course, far higher than the levels of organics detected so far on Mars. Not all meteorites which can contribute Ni are carbonaceous, but it is expected that the organic-rich versions would be more plentiful at the orbit of Mars and also would be more readily incorporated into soil because of their relative fragility and susceptibility to disintegrative weathering than the irons or other non-carbonaceous meteorites.

Clay minerals. Alteration of feldspars can lead to montmorillonite minerals. These have been detected at many locations on Mars by orbital spectroscopy, including other smectite clays of Mg/Fe varieties [5] as well as kaolinite, chlorite and illite clays, plus hydrated silica [190]. However, orbital observations are limited in spatial resolution and discrimination ability, such that additional discoveries of montmorillonite geochemistry that could not be detected from orbit have also been made in the Independence outcrop on Husband Hill [149], in the Esperance boxwork [191] in the rim of Endeavour Crater [177], and in numerous mudstones and sandstones in Gale crater [6]. Although clays are generally not major hosts of bioavailable CHNOPS elements, they do have important capabilities for the physi- and chemisorption of elements and PCE molecules from solutions, as well as catalytic roles for oligomerization reactions [49,192].

3.1.2. Transition Elements

The significance of an enrichment of an element over its primary abundance in its igneous mineral precursor is not just that a 10-fold increase in boron, copper, or other trace element drives key reactions $10\times$ faster or more complete, or that some critical threshold has been crossed. Most importantly, any enrichment or any depletion in an element typically indicates that it has been subjected to aqueous dissolution and transport [193]. Hence, it becomes "bioavailable" at the location detected, or, if depleted, at potentially some other location where it has become even more concentrated by aqueous-mediated processes. Depletions as indicators of mobilization have been more difficult to ascertain, but many different cases of significantly reduced levels of Mg, Fe, Mn, Ni, or Cr, have been detected [6,7,149,150,170,171,177].

Less common non-aqueous processes of segregation, such as magmatic differentiation or eolian sorting, may be responsible for changes but, if not, then the departure from the normal range of concentrations from the primary sources is indicative of mobilization. Aqueous environments are what are needed for the OoL because they provide the means for key reactants to come together in the same medium, which is already a prerequisite for abiotic progression towards and sustenance of LAWKI

In Gale crater, Stimson formation sandstones have been extensively leached of Mg, Al, Mn, Fe, Ni, and Zn [194] from their original mineral phases, as well as probably many other elements not detectable by the rover-based analytical systems.

Iron is ubiquitous on Mars. It has also already been demonstrated to occur in much more than a dozen different mineral forms, thanks to the Mössbauer instruments on the MER rovers [176], as well as remote sensing of minerals ranging from hematite to Fe-smectites. Although the oxidized, Fe^{3+} forms are more common, there are several mixed $Fe^{2+}Fe^{3+}$ minerals, as well as the Fe^{2+} in the primary igneous minerals olivine and pyroxene. Iron and nickel can also be available in their native forms ($Fe°$ and $Ni°$) from the siderite meteorites which are surprisingly abundant at the surface of Mars [156,157].

Copper has been discovered at anomalously high concentrations at nearly a dozen locations in samples along the route of the Curiosity rover [195]. In the Kimberly formation, a concentration occurrence as high as 1100 ppm was found, some two orders of magnitude higher than for typical Martian meteorites of igneous composition [196] and much higher than typical crustal abundances on Earth; Figure 2. Copper enrichment at 580 ppm was also detected in target Liga at Gale [197], and a level of 230 ppm was discovered in the Independence outcrop [149] on Gusev crater's Husband Hill. These data indicate that, not only had Cu been rendered mobile, but enrichments were relatively common, lending credibility to the possible occurrence on Mars of the Cu-catalyzed cyanosulfidic metabolic pathways to precursors for the three fundamental classes of biochemicals. Tracking of Cu abundances has shown that enhanced levels are found over a wide range of occurrences in Gale crater, especially in areas of phyllosilicate abundances [195].

Nickel and zinc have been routinely detected at surprising levels by the APXS (Alpha Particle X-ray Spectrometer) for more than a thousand measurements of soils, rocks, and sediments on three rover missions, with generally much higher abundances than for terrestrially analogous materials. These enrichments can be one or two orders of magnitude over terrestrial averages, Figure 2, and are only those which have been discovered inadvertently and generally without opportunities to further trace their origins.

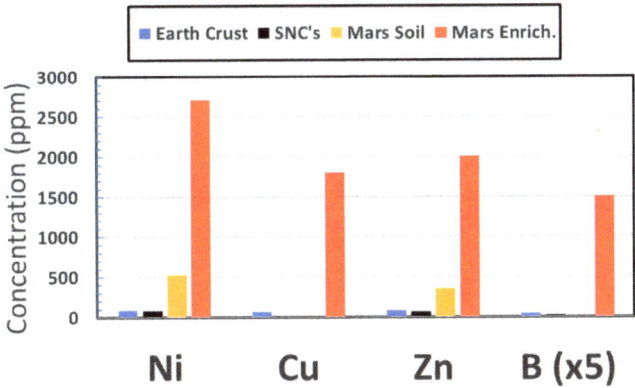

Figure 2. Examples of enriched occurrences of some key transition elements discovered during the MSL mission, compared to the Earth's average crustal concentrations, SNC meteorites [125], and a typical Martian global soil composition [123]. "Mars enrichments": nickel maxima (except for meteorites) for three rover missions (MER, MSL); Cu at Gale crater [196]; Zn also all three missions, plus up to 8000 ppm in Gale; boron at Gale [198].

Manganese. Enriched concentrations of Mn have been discovered repeatedly in Gale crater [199,200] and also in isolated occurrences at Endeavour crater [201]. These have been interpreted to implicate higher environmental oxidation potential and the presence of appreciable O_2 in the past to produce MnO_2 [200]. More recently, indigenous Mn oxides in the 4+ (oxidized) state have been identified in the Black Beauty meteorite pairs NWA 7034 and 7533, with the Mn-rich clasts containing up to 65 wt% MnO_2 [155]. Additionally, MnO_2 precipitation scavenges Zn and Ni, but not Cr [202], as seen in the Gale samples, and has led to the suggestion of the possible involvement of other oxidants, such as atmospheric agents (O_3), nitrates or perchlorates, and an Eh above +500 mV for a pH~8 [202], whereas an Eh of about +300 mV was measured in the soil by the Phoenix mission [181]. These trace element correlations do not occur, however, for the high Mn-Mg-sulfate rock coatings discovered at Endeavour crater, which may indicate different conditions or mechanism(s), such as alternatives that have also been suggested for concentrating Mn on Mars [203].

Cobalt is very difficult to detect by APXS, because of obscuration of its K_α and K_β X-ray emissions by K_α lines of the much more abundant Fe and Ni. However, target Stephen in Gale crater provided a special opportunity, for which a Co concentration of 300 ppm was detected [197], a nearly tenfold enrichment over Shergotty cobalt [125].

Vanadium can vary over a range of roughly 50 to 500 ppm for various basalts and meteorites, with Shergottites at ~300 ppm [125]. In contemporary terrestrial soils, it can range widely from a few ppm to ~500 ppm [204]. At the higher levels, it would be enough for detection on Mars by X-ray fluorescence spectroscopy, except that two other elements which are normally at higher concentrations overlap too closely in emission energies for the accuracy of non-laboratory measurements (Ti K_β overlaps V K_α, and Cr K_α overlaps V K_β X-ray emissions). However, if a Mars sample were enriched in V, while being lower in Ti than typical, a positive detection for V might be possible. The smaller analytical spot (~10^4 times smaller area) of the PIXL XRF instrument on the Perseverance rover could in principle make such a determination if a "reduction spot" precipitate enriched in V, a potential chemical biosignature [205,206], were detected.

Tungsten is typically found at 0.1 to a maximum of 1 ppm in terrestrial basalts [207] and shergottites [125]. However, detecting W by remote XRF would be by its L_α emission, which is unfortunately sandwiched between the more common Ni K_β and Zn K_α, both of which are typically at levels of hundreds of ppm in Mars soils.

Molybdenum is also present at ~1 ppm in basalts and meteorites, which is far too low for rover-based XRF detection. Additionally, its K_α emission will generally be obscured by the K_β from Zr, which typically occurs at one to two orders of magnitude higher concentrations. However, an oxide form of Mo, the molybdate ion MoO_4^{2-}, is soluble and could be mobilized and potentially detected if formed and sufficiently concentrated apart from minerals with nominal or lower concentrations of Zr.

For the V-W-Mo triumvirate, the most likely possible detection before samples are returned to Earth would be for V. These three elements often correlate in enrichments [204], such that detection of any one of these could be an indicator for the other two.

The elements As and Se can substitute for their corresponding higher row elements (P and S) in certain circumstances and seem to be essential elements for some organisms. However, they often are toxic and it is unknown if either may have had any important role in the early emergence of protometabolism.

3.1.3. Other Key Elements

In addition to fundamental feedstock elements and catalytic ions, there are several elements which seem necessarily attendant to the origin of life because of their special properties and/or abundances. These include electrolyte elements, such as Na and K, accompanied by Cl, as well as stabilization elements, such as boron and certain divalent cations, such as Mg and Ca.

Electrolytes. Several ions inside cells are typically at far different concentrations from the medium they are in, as modulated by various controlling factors: ion channels which can be gated open or closed; active pumps which utilize chemical energy to transport ions against their concentration gradients; uptake and sequestration by organic constituents. Typically, for LAWKI, the element potassium is brought inside the cell, while Na and Cl are reduced relative to their concentrations in sea water. Another key element, Mg^{2+}, is roughly at the same total concentration inside and outside, except that the large majority of the inside portion is bound up with ribosomes, ATP, proteins, and other macromolecules, such that its free concentration in the cytoplasm is greatly reduced [130].

Potassium. Wet–dry cycling yields for oligopeptide formation from glycine has been shown to be enhanced by as much as $10\times$ when deliquescent salts are present, especially those of potassium phosphates, which also invoke a possible direct relevance of K to the OoL [208]. In its native igneous form, it is predominantly found in feldspars, although there can be occurrences in certain other minerals, such as the micas. In addition, its high-temperature polymorph, sanidine, has also been discovered in some samples on Mars [209]. Once K is released as the result of weathering, its various forms are quite soluble, providing great mobility and accessibility. However, it is readily adsorbed by minerals and organic matter. It is also a principal component of illite clay, which has been detected on Mars from orbit [5] and by Curiosity-based measurements in the Gale crater [210], and is often associated with other phyllosilicates, such as montmorillonite.

On Earth, K-feldspar is common in continental rocks, whereas it is minor or lacking in oceanic basalts, and the types of mafic and ultramafic assemblages that were prevalent prior to the formation of the continents. On Mars, K is at relatively low concentration in the global soil (0.5 wt% K_2O), but occurs in several locations at higher concentrations (typically 0.25–0.5 wt%, rarely above 1.5%, but as high as 3.7 wt%, compared to Earth's crustal average of 2.8 wt% and local values often much higher). Three different igneous polymorphs of K-feldspar have been found in Gale samples [6]. However, various other samples, such as Oudan, have no detectable crystalline K-feldspar but K_2O is inferred at the level of 1.7 wt% in the amorphous material (which accounts for almost one-half of that sample) [6]. Similar amounts are inferred for the amorphous components of ordinary aeolian soils. K-bearing hydrated sulfate salts such as jarosite [5,176] and alunite [5] have also been discovered. At the Phoenix polar site, the measured concentration of K^+ in aqueous solution with Martian soil [136] would be equivalent to about 10 mM for a 1:1 water/soil ratio.

Sodium is sometimes correlated with potassium in Mars samples, as well as with aluminum, which implicates feldspars as the actual, or original source. In a few cases, with higher values of Cl, the Na is correlated with that element, further indicating that Na^+

ions were available in solution (although a major role in either biology or the PCE leading to an OoL is not typically attributed to chlorine). Positive ions are generally needed for charge balance, since many organics and the phosphates are negatively charged at neutral pH. There is also the need for osmotic balance of the intracellular fluid with respect to the extracellular medium.

Magnesium and calcium. The Group 2A elements of Mg and Ca are prominent bioinorganic chemicals [68], albeit with different functions. As widely acknowledged [69] and described previously, Mg^{2+} plays major roles in numerous biochemical and enzymatical processes of the key molecules of life. For example, at least five separate functions in ribosomal activity require Mg^{2+} [211] (although Fe^{2+} and Mn^{2+} can substitute for some of these functions [212]). Ca^{2+} also contributes too many important biological functions, although the ionic sizes and hence charge densities are quite different between these two cations [69] and lead to different utilizations. The salts of Mg are highly soluble, while Ca halides are also highly soluble but the sulfate is only sparingly soluble. Ca occurs in many primary minerals, especially the pyroxenes, plagioclase feldspars, and apatites, and is susceptible to release by aqueous alteration of these, with susceptibility generally in the order listed. Given the ubiquitous various occurrences of $CaSO_4$ on Mars, Ca^{2+} has clearly been an available ion.

The global soil of Mars has a spectral signature evidencing ~2% $MgCO_3$ [182]. Magnesite is poorly soluble in circumneutral H_2O, but is readily solubilized by mild acids. A low water/rock ratio for a mix of global soil and water (wt/wt) could produce a high concentration of Mg^{2+}, if all sulfate is present as soluble $MgSO_4$. Although there is not yet direct X-ray diffraction evidence for $MgSO_4$, such as kieserite or in higher hydrated states, strong correlations between Mg and S are observed in many locations, as cited above, while the S is known to be sulfate for a variety of reasons. For a typical SO_3 concentration of 6 wt% in Martian global soil [123] and a water/rock ratio of 1:1, the Mg^{2+} from the equivalent of 9 wt% $MgSO_4$ could reach as high as 750 mM, or about 15 times greater concentration than in Earth's ocean waters. However, it is not clear that this amount of Mg is available from soil. The Phoenix polar mission measured soluble Mg and SO_3 separately and concluded that a likely 2 wt% of soluble $MgSO_4$ was present in soil in that area [213], inferring a Mg^{2+} concentration of 166 mM for our example W/R = 1, which is within the range of 83 to 185 mM, as derived from the Mg^{2+} concentration (within error bars, across different samples) measured. It is unknown whether the Phoenix soil has as high total SO_3 as the typical global soil measured at six other mission locations (all in equatorial or mid-latitude locations). However, the early conclusion that the widespread Martian soil can supply large amounts of Mg^{2+}, sulfate anion, and chlorine and oxychlorine species is secure. Whether this global soil that is universally available in the present epoch also had this same composition in the Noachian, prior to the theiikian interval [214] when abundant bedded sulfates were deposited, is not yet determined but would seem less likely. Nonetheless, that the Martian lithosphere is sulfur-rich compared to terrestrial soils seems incontrovertible. Hydrothermal processing of mafic rocks on Earth, including at the suboceanic vents, typically result in high concentrations of S, chiefly in the reduced form of metal sulfides. Our example of a 1:1 ratio for W/R is a mud, whereas a pond will allow soil particles to settle. The saturation concentration of $MgSO_4$ is 2900 mM at +20 °C (2200 mM at 0 °C). Thus, if a pond leaches its bottoms and sides, as well as its foreshore and any airfall dust, the $MgSO_4$ could rise to very high levels. This salt is also very hygroscopic and forms several high-order hydrates (e.g., epsomite at 7 H_2O per $MgSO_4$).

Muddy water, as opposed to a wet mud, could reduce the Mg^{2+} concentration to a greater extent. This could still easily be adequate to foster RNAzyme activity since the test-tube evolution experiments are successful when conducted at high Mg^{2+} levels [63], typically 50 to 200 mM [56,62]. Although cations accelerate the natural degradation of RNA in solution by hydrolytic cleavage, the Mg^{2+} catalyzes this less severely than Fe^{2+} or Mn^{2+}, and also helps stabilize the three-dimensional conformations of RNA, while at pH < 5.4, increasing Mg^{2+} concentration to 50 mM actually slows down the degradative cleavage reaction [16].

Boron. At Gale crater up to 300 ppm boron has been detected [198,215], Figure 2, but the CCAM instrument uses laser ionization breakdown spectrometry (LIBS), which can detect B only in low-Fe samples, such as $CaSO_4$ veins, because of interfering emission lines from the otherwise ubiquitous Fe on Mars. Using different techniques in the laboratory to analyze the Nakhla Martian meteorite, boron has been found to be enriched to levels of 160 ppm in alteration zones associated with Fe-rich smectite clay [216]. On Earth, this element can also be found enriched in hot springs [217].

3.1.4. Elements Availability

Mars' endowment with the elements of life is adequate to supply not only the nutrients for microbial LAWKI (i.e., habitability), but also the feedstocks and catalysts needed for an origin of life. The enrichments noted above confirm the extraction and concentration of key ingredients. For these ingredients to be available, however, they must have adequate solubility in aqueous media. The solubility product (K_{sp}) for compounds of these elements is generally high, but depends on the valence state. Aside from the K_{sp} for pure H_2O, there can be dependencies on other components, but the most important mitigating factors can be the pH and oxidation potential (Eh) within the aqueous medium. Thus, because much of the Fe on Mars is now in the Fe^{3+} form, its K_{sp} is extremely low, except for conditions where the very low pH and Eh portion of the relevant Pourbaix diagram is realized. However, it is the ferrous form that is catalytic and involved in (FeS) clusters. At the low Eh for early Earth, the expected concentration of Fe^{2+} in the ocean would be as much as four orders of magnitude higher than the Fe concentration today, while Co and Mn would also be higher; in contrast, the Cu, Mo, Zn, and Ni concentrations would be much lower [218]. Cycling between more oxidizing and more reducing atmospheric states, as recently proposed [146], could induce significant variations in relative ionic concentrations among the various redox-sensitive elements. Combined with the discovery of ferrous smectite in the Gale crater and laboratory oxidation experiments lends credence to the hypothesis that the Fe^{3+} smectites observed by orbital spectroscopic mapping were originally in the Fe^{2+} form before being altered further [219].

Since Mars is a sulfur-rich world in comparison to the surface of the Earth, when volcanic emissions of H_2S and SO_2 are converted to SO_3 by photochemical byproducts [180,220,221], then Martian shallow aqueous reservoirs will have their pH lowered, resulting in the formation of sulfates [222] which are highly soluble and can provide high levels of availability for all relevant elements, except for Ca and Fe^{3+}. Jarosite in the Burns formation has been cited as a clear indicator of significant acidity (pH ~3) at the time and the location where it was formed [171]. As basic environmental minerals react and drive the pH toward neutrality, new emissions can reverse the process in shallow lakes and ponds, to restore the higher levels of needed elements. However, such reactions require time, depending on the minerals available as well their grain size and armoring effects [178], providing the macrobiont with slowly varying pH which may facilitate some steps of PCE. Volcanic emissions of Cl_2 and HCl can also produce chlorides that form highly soluble salts of these elements [162,223]. In the widespread Martian global soil, the S/Cl ratio (atom/atom) is ~4:1 [123,124].

From the X-ray diffractograms of the CheMin instrument, most samples of soils and sediments at Gale have a significant component of X-ray amorphous material (15 to 70 wt%) [6,224]. By assuming elemental compositions of the clays, igneous silicates, and other minerals exhibiting diffraction peaks, the net elemental composition of the amorphous components (AmC) can be inferred from the APXS measurements of the bulk sample. The resulting compositions of AmC are extraordinarily disparate among samples (e.g., SiO_2 at 29 to 75 wt%, FeO of 5 to 30 wt%, SO_3 of 1 to 22 wt%, and Cl as high as 6 wt%). Amorphous material can include $MgSO_4$, which has been widely detected by other means but not in crystalline form by CheMin. Given the non-consistent composition of this AmC material, and general lack of correlation between most elements, many elements must be individually mobilized, with relative concentrations resulting from various local conditions at the time of immobilization. This implies a wide range of element availability for Na, Mg, Si, P, S, Cl, K, Ca, and Fe. This phenomenon is not restricted to Gale crater. The dark coating of the Esperance montmorillonite-composition fracture fills [191] at Endeavour crater matches Gale's amorphous material in the JohnKlein sample (a mudstone in the Yellowknife Bay formation [225] with 19% AmC) [224] for all major elements analyzed if that sample would simply have more $MgSO_4$ and some $MgCl_2$, as shown in Figure 3. Finding such similar amorphous materials in both a mudstone at Gale crater and, some 17,000 km distant, as a coating along the rim of Endeavour crater, implies that amorphous materials may be ubiquitous on Mars, and hence its elements would be widely available. It is noteworthy that only the MSL mission, with its CheMin diffractometer, has had the capability to detect and infer the composition of amorphous materials.

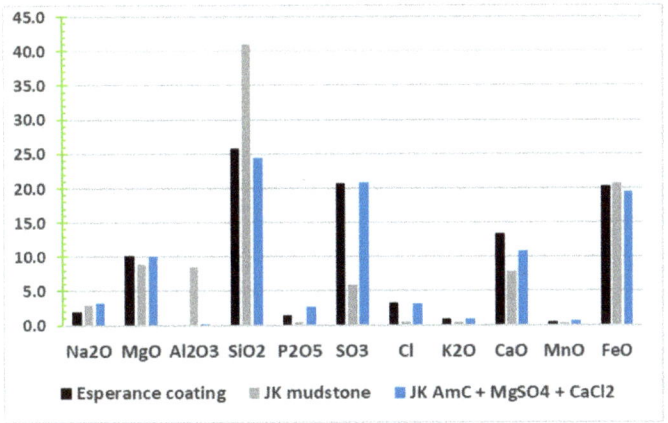

Figure 3. Evidence for similar amorphous material at widely separated sites: Esperance coating [191] at Endeavour crater compared to JohnKlein (JK) and its amorphous component [224] plus salts at Gale crater (85% JK AmC, 13% $MgSO_4$, 2% $CaCl_2$).

3.2. Energy Sources on Mars

At aphelion, Mars receives only a little more than one-third the solar flux of Earth, and at its perihelion still only one half, although these are more than adequate for photosynthesis. This factor of between $2\times$ and $3\times$ less sunlight, assuming a transparent atmosphere, is not a significant difference for the UV flux, and the diurnal cycle for the very early Earth was shorter from the closer proximity of the moon.

A variety of redox energy couples would have been available on early Mars [226,227], in addition to the solar sources of ultraviolet and visible energy. The reduction in the amount of CO_2 to organics by H_2 is an exergonic reaction and methanogenesis is considered a likely early metabolism. In a primitive syntrophic relationship, energetic reactions of metabolism could have also worked in the other direction to oxidize methane back to CO_2 in analogy with methylatrophic catabolism with a suitable oxidizer.

If sulfates and nitrates were produced by photooxidation processes, there would be ample redox couples as electron acceptors with any H_2 or CH_4 in the atmosphere. Sulfur, having multiple oxidation states and with -2, 0, $+2$, $+4$ and $+6$ all being relatively stable, has numerous energy-releasing reaction pathways with end products ranging from H_2S to native sulfur to sulfate, and various microbes that can take advantage of these transitions to drive their metabolic functions.

The transition metal elements, with their partially filled d orbitals, can occur in two or more oxidation states, providing well-known couples with Fe and Mn. Nitrate-utilizing microbes can combine that electron acceptor with Fe^{2+} as the electron donor as one energetic pathway analogous to the metabolism of a variety of iron-oxidizing microbes on Earth [228]. Iron meteorite plus an oxygen source has been shown to provide an ample source of energy for metabolism and the growth of acidophilic chemolithoautotrophic microorganisms [229].

Given that hydrothermal processes are available, due to local volcanic activity or deep-seated thermal transients induced by the conversion of kinetic energy of the larger hypervelocity impactors, there could be an even larger range of potential redox couples [81].

3.3. Settings on Mars

The geologic processes of Mars are far from fully understood [230], although many of the igneous and sedimentary features have analogs that are well known on Earth. Unlike Earth, much of the earliest geology of Mars is preserved in the intercrater regions. Far less known and understood are the early atmospheric and hydrologic environments, and it has been challenging for climate modelers to find parameter sets that make plausible the temperatures that would have been needed for the availability of liquid water, rather than ice, to form the geomorphic modifications [4,114] and aqueous-mediated geochemical concentrations [5] that are widespread across the planet.

3.3.1. Early Mars as Compared to Early Earth

Early Earth was very wet, with a possibly globe-encircling ocean and little if any exposed land other than the summits of island arc volcanoes and micro-continents [231]. This scenario is based on the modeled slow emergence of plate tectonics, from which the continents were formed, although the timing of the rise of the continents remains uncertain and highly controversial, with recent evidence from zircon trace elements of the existence of felsic crust within the first 500 Myr of Earth's history [232]. For all the OoL hypotheses requiring land and subaerial exposures, the expectations for a successful origin would be constrained if Earth's ocean were global and tectonic activity subdued [233].

For wet–dry cycling to be possible, the planetary body would need exposed land as well as shallow ponds, lakes, and seashores [234]. Mars could be far more favorable in terms of the amount of exposed land because even the most optimistic estimate of the size of an ocean in the lowlands of the northern hemisphere is less than one-third of the total surface area of Mars and would require a global equivalent layer (GEL) of ~550 m of H_2O [111]. This is at the high end compared to estimates that, by the late Noachian, there was only approximately one-tenth of this amount of water available, and which would, therefore, prevent fully filled ocean-sized bodies of water [235].

If, for example, 2% of the surface of the early Earth were exposed volcanic land masses, and 30% of Mars was submerged, then Mars would have 10× more subaerial land to enable the advantages of factors such as wet–dry cycling, UV irradiation, atmospheric stimulation, concentration of ingredients, etc. If only 15% of Mars were submerged and 25% of Earth was land, the planets would have equal amounts of subaerial terrain. However, with what is perhaps a more likely situation at early times (5% Mars submerged and 95% of Earth submerged) Mars would still have a 3× greater exposure of land in spite of being a smaller planet.

Punctuated Climate. Detailed climate models currently indicate that even with a supply of H_2 gas to the primitive atmosphere, the pressure-broadening of absorption lines for a CO_2 greenhouse effect [112,130] is insufficient to maintain a perennially warm climate to prevent widespread freezing [4,114,120,236].

Although methane is also a potent greenhouse gas, and would be a welcome addition as a feedstock for PCE, its photochemical lifetime would be short. Furthermore, its production attendant with the serpentinization reaction would be low compared to hydrogen [110].

Although there is no evidence of the quantity of H_2 that would have been present on Earth in its Hadean eon, or on early Mars in its early Hesperian phase, or even definitive evidence of its presence at all, there are many possible sources of H_2 that lend credence to its likely contribution as the key component for more tightly closing the early strong greenhouses on both planets. In addition to the hydrothermal serpentinization reaction, dihydrogen can be produced by several processes, including magmatic devolatilization; radiolysis of H_2O by ionizing radiation from mineral K, U, and Th; and H_2O reaction with dangling bonds or radicals on fresh mineral surfaces formed by rock abrasion and fractures [116]. Magnetite is ubiquitous on Mars [237] and the direct reaction of magnetite with H_2O has been shown to provide yet another de novo source of H_2 to enable a warmer early Mars [238].

Models of the effects during the creation of even the ~5000 "medium-sized" craters (>30 km diameter) on Mars [239] predict craters of this size and larger can produce major transient warm periods due to the greenhouse effects of the release of CO_2 and H_2O from the target material, and which can last from months to decades and centuries [240], but not sufficiently for longer-term warming [241]. Some models question whether such events are sufficient to cause the formation of the valley networks because the global effects are too short-lived [236]. However, other models conclude that 100 km impactors could create enough H_2 and heat to raise a cold Noachian temperature to above melting for millions of years [242] and recent models suggest ample temporary climate change to produce the valley networks and other fluvial as well as lacustrine features that have been observed [146].

Irrespective of models, numerous examples of diagenetic episodes of aqueous alteration have been discovered in Gale crater, including potassic sandstone in the Kimberley area [243] and sediments in the Vera Rubin Ridge (formerly "hematite ridge") [244,245]. Compositions of various diverse mudstones in Gale crater, combined with the inspection of sedimentary relationships, have led to a model of alkaline fluids in the Yellowknife Bay area (Mg-rich concretions), acidic fluids in Pahrump Hills (jarosite, and mobility of Zn, Ni, Mn, Mg, Ni, and S), hydrothermal fluids at Ouudam (gray hematite, opal-CT), high-redox, S-rich fluids elsewhere, as well as $CaSO_4$ fracture-fills crosscutting earlier diagenetic features [246]. Each observed diagenetic episode can be the result of numerous wetting, dry-out, and re-wetting events. The local environments for lithified sediments most relevant to biotic evolution may be where there are subaerial exposures.

Estimates of the effectiveness of splash erosion on Mars is used to infer widespread rainfall for the formation of the valley networks, once infiltration losses could be minimized by the fine particulates from clay formation [247]. An examination of 13 open- and closed-basin lakes resulted in estimates for precipitation minimums (rain and snowfall) of 4 to 159 m for their catchment-averaged runoff [248]. However, differing models suggest that rainfall on Mars may have been much rarer than on Earth [249]. Snowfall and/or cold trapping of H_2O vapor as ice could build reservoirs which, during warmer intervals, could be melted to produce runoff for the formation of valley networks.

Geomorphological analyses of the Kasei Valles region indicate episodic flooding with at least five periods of channel flows during 3.7 to 2 Ga, evidencing an active hydrological cycle well into the Amazonian [250]. Features implying thermokarst lakes and ponds also dating from the late Amazonian have been observed in Utopia and Elysium Planitiae [251]. Fan units of Amazonian age in Gale crater provide evidence of surface flow [252]. An example of extreme formation of diverse ponds in warming permafrost areas on Earth is seen in Figure 4.

Cold temperatures. Although there is great emphasis and interest in hydrothermal regimes, a case has also been made for a cold OoL [106]. A cool or cold early Mars has often been predicted [114,146,254]. As seen in Figure 5, if the predominant temperature for exposed bodies of H_2O were +20 °C maximum (southern summer), and only ~1% of the estimated inventory of buried CO_2 (as carbonates) were released, it could provide an atmospheric pressure of 23 mbar to prevent H_2O from boiling (only a three-fold increase over the present 6 mbar). Thus, an initial Martian atmosphere at 2 bars total pressure could decay by a factor of ~100× due to carbonate formation and escape to space, while still all the time enabling Martian H_2O to be liquid in specific locations without excessive loss rates and dispersal due to boiling.

Figure 4. Diverse thermokarst ponds in Sheldrake River valley near Nunavik, Quebec in the Canadian subarctic (56° N) [253]. Photo-credit, J. Comte (Institut national de la recherche and Centre for Northern Studies).

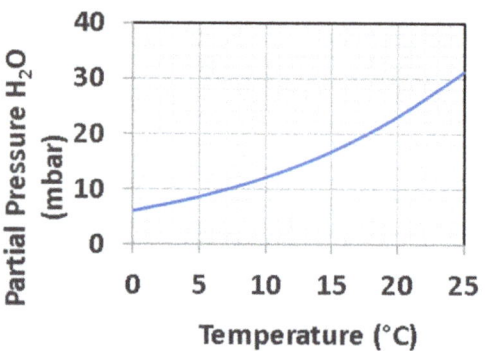

Figure 5. Partial pressure of pure H_2O, setting the limit on boiling (brines will be stable to higher temperatures).

Alternatively, the average temperature could be much lower but surface patches with pro-solar slopes, modest-to-low albedo, and low thermal inertia could be heated beyond the ice melting point and perhaps up to +20 °C peak temperature during daytime. Repeated transient cycles of wetting and drying can be especially advantageous to PCE formation of polymers. Due to obliquity cycling, and an analysis which finds that only 500,000 years ago Mars was at its lowest obliquity and with a predicted rise in atmospheric pressure to 31 mb [255], it might be possible that OoL processes could be ongoing in the most recent epoch.

If brines are formed, especially those containing halides, the liquid regime is extended to yet lower temperatures. This implies higher ionic strength in the milieu in which prebiotic syntheses and processes must occur. In Figure 6, some candidate salts on Mars which can depress the freezing point of their brines through formation of eutectics are shown. Sulfates are poor performers, but chlorides and the oxychlorines readily block entry into the solid state. These brine media also have greatly increased viscosity, which slows diffusion rates and hence promotes the spatial heterogeneities in a pond that can aid the semi-sequestered development of different key functions needed for the comprehensive set of proto-metabolic activities of life forms (nutrient acquisition, component synthesis, energy management, waste management).

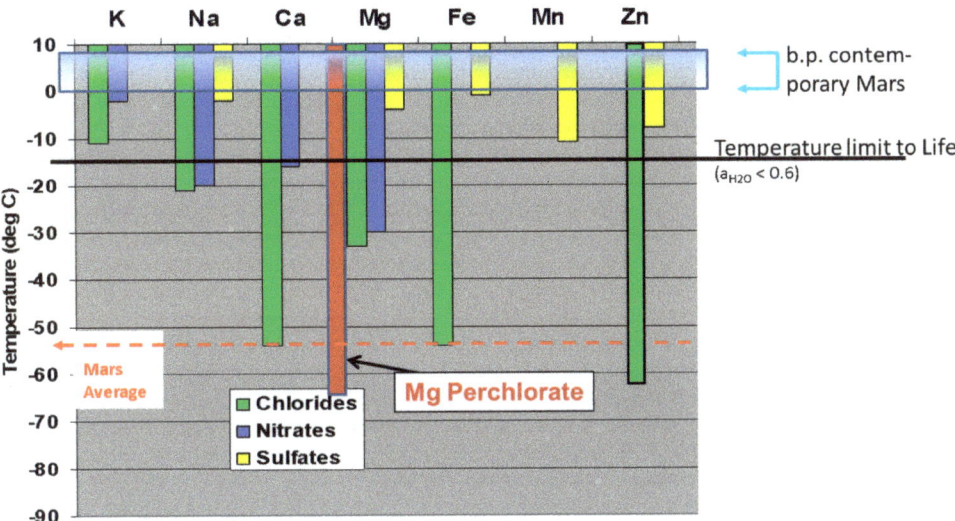

Figure 6. Freezing-point depression for salty brines, which extend the mobility of aqueous media in a sub-zero environment, but do not fully extend the temperature range for metabolic functionalities.

Occasional traces of chloride enrichments by rover instruments, especially on rock surfaces [256], have given way to more numerous detections, including hundreds from orbit [223,257]. In situ investigations in Gale crater [258] tend to indicate Na as the chloride salt. Even when the a_w at highly depressed freezing points may be too low for cellular growth and reproduction, the frigid environment might provide a unique and favorable intermediate environment for some processes of PCE.

Although there are psychrophilic organisms which can conduct metabolic activities and reproduce at temperatures somewhat below 0 °C (to about −10 or −15 °C) [259], some prebiotic reaction pathways may actually be strengthened or enabled by cycling above and below freezing, i.e., the freeze–thaw process analogous to wet–dry cycling [102–106].

3.3.2. Subaerial Terrain Proto-Macrobionts

Mars has abundant locales on its surface which could host the origin of life. These include both ambient temperature regimes and hydrothermal settings.

Ambient settings. In addition to the natural undulations of surface topography due to endogenic processes, there is the exogenous influence that creates abundant basins in the earliest history of planets, i.e., the terminal accretion phase which leaves the scars of impact craters as evidence of its progression.

From a study of world occurrences [260], there are ~250 million ponds and lakes with equivalent diameter of 30 m or larger on the present surface of the Earth.

The number of craters on Mars >1 km diameter, is a minimum of 380,000 [239,261] (by actual counts, but this does not include craters modified beyond recognition after their original formation). From the fitted slope exponent of −1.46 for the cumulative size distribution at the lowest sizes, it is projected that there could have been a minimum of 50 million primary craters greater than 30 m in diameter, and perhaps ten times this many craters due to secondaries created by larger primary impactors (fitted exponent of −2.48).

In a separate analysis, focused just on the Meridiani Planum area [262], the measured number of craters >30 m in diameter is 1.1 per km^2, which extrapolates to over 150 million craters of this size or larger on Mars, with 90% of those being between 30 and 300 m in diameter. These estimates are of the same order of magnitude as the number of ponds this size or larger on Earth today.

Unlike impact craters on Earth, these observed Martian craters have survived >3 Gyr's of geologic history since their formation. Craters smaller than 1 km will not have as significant heat energy density (J/kg) imparted to their vicinity, since they are evidence of a lesser deposition of kinetic energy, but they do form natural depressions for ponds of diameter up to the size of their rims. Other geologic forces will create additional natural basins in the remaining intercrater terrains.

The number of potential subaerial ponds is, therefore, extremely large on both planets, thereby facilitating the possibility of an OoL. However, they must be supplied with water to be effective as proto-macrobiont settings. On a very wet planet such as Earth, the large majority will be wet, or submerged beneath the ocean, whereas on Mars that essential condition is a function of not just location but also of climates and geologic time.

The Martian climate is now too cold to avoid the freezing of even large, exposed bodies of water. In spite of the faint early sun, however, the greenhouse was sufficiently effective that valley networks could be carved and perhaps even major bodies of water could form [111,263]. Topographic analyses of geomorphologic features indicate flooding of the large plains in the northern hemisphere to create a small ocean, followed (in time) by a smaller sea and accompanied by episodic occurrences of distributed lakes [263], shallow sediments [173], and thereby, by extension, of ponds. These major inundations span the Noachian to the late Hesperian, the end of which, ~3.7 Ga, coincides with the range of evidence that the establishment of a biosphere had already begun on Earth [264]. Lakes fed and discharged by the valley systems were comparable in volumes of H_2O to the small seas on Earth [265], although many might have been short-lived, judging by the general lack of detectable chemical alteration products by orbital spectroscopy [266].

From MSL's in situ exploration, abundant evidence of past activity of liquid water includes alteration chemistries (clays, salts) [6,267]. The Mg-Fe carbonate (Comanche outcrop) in Columbia Hills of Gusev, a former crater lake, is evidence of the ephemeral, mostly low-temperature alteration of mafic rocks [268].

The history of Vera Rubin Ridge at Gale crater indicates multiple episodes of groundwater interactions [269], evidenced in part by elevated concentrations of Mn. Accompanying long-term episodes of aqueous activity would have been undoubtedly shorter durations of superposed fluctuations which enhance the opportunities for concentration/dilution events and wet–dry cycles. Alternating episodes of wet–dry environments have been implicated by the chemostratigraphy of Mt. Sharp [270].

Although Gale itself is ancient, and its sedimentary load formed during the Hesperian, the measured low ^{36}Ar abundance also suggests that water–rock interactions continued to occur well into the Amazonian [271]. Stratigraphic sections observed from orbit indicated enrichments in hematite, phyllosilicate, and sulfates, which implied extensive aqueous alteration [272]. Occurrences of phyllosilicates, sulfates (Mg, Ca), as well as sulfate-independent Mg enrichments at differing concentrations among individual samples taken by the Curiosity rover, show that environments were dynamic on a small-scale [6]. The indications of a redox-stratified lake with iron precipitates is consistent with magnetite levels progressing to hematite at higher stratigraphic levels [273], and is indicative of variable redox conditions which could support a diverse community of chemolithoautotrophs, as in terrestrial redox-stratified lakes [274].

Foreshore mudcracks. A potential feature of special interest for the OoL is mudcrack patterns [13]. On Mars, there has so far been discovered a clear example of an area of mudcracks, which is indicative of repeated wet–dry cycling along the shoreline of an oscillating lake level [275], as seen in Figure 7 of the "Old Soaker" unit.

Figure 7. Mudcrack pattern of the "Old Soaker" slab (Sutton Island member of Murray Formation), a red mudstone overlying gray sandstone as imaged by the primary camera of the MSL Curiosity rover, sol 1555) [275].

Although this may seem rare, the sum total of Martian terrain that has been imaged so far by the four landers and five rovers at the minimum resolution needed to detect such patterns (3 mm), amounts to only 3×10^{-8} of the total surface area of the planet. Thus, there could be tens of millions of mudcrack units on Mars that have not yet been imaged but may have provided the heterogenous environmental conditions conducive to prebiotic chemical evolution [13]. Additionally, some mudcrack patterns will have become filled in by eolian-mobilized dust fallout and surface saltation or degraded beyond recognition by eolian abrasion. It is hypothesized that Gale crater was often wet, but on the basis of the presence of highly soluble perchlorate, there were dry periods that extended well into the Amazonian Period [276].

Hot springs hydrothermal activity. In spite of the presence of giant shield volcanoes on Mars, a tally of features nonetheless concludes there has been only a fraction (~2%) of the cumulative magmatic extrusive volumetric activity on Earth [277]. As another source of thermal energy, the buried heat from each hypervelocity impact on Mars, could also generate hydrothermal activity if H_2O were available (e.g., as permafrost ice) [96,278].

A plethora of geochemical evidence of localized aqueous activity has come from the in situ exploration in Gusev and Gale craters. At Home Plate in Columbia Hills, there is high SiO_2 with morphologic evidence for silica sinter similar to that at the hot springs at El Tatio, Chile [279]. High silica (~90% opal-A) enrichments have also been detected at Gale crater [280]. From orbit, it is difficult to detect the silica deposits expected from hot springs [281], but several light-toned deposits in Valles Marineris, the chaotic terrain, and some large craters has been interpreted as indicators for large-scale spring deposits [282].

A highly salt-enriched soil at Paso Robles on Husband Hill has been interpreted to have a hydrothermal origin [170]. In the Kimberly formation at Gale crater, there are occurrences of sanidine, the high temperature polymorph of K-feldspar that forms above 100 °C, which is evidence for hydrothermal activity similar to occurrences in the summit areas of Maunakea volcano in Hawaii [204]. Other minerals, such as tridymite, provide additional evidence of a hydrothermal history at Gale [194] and there is also evidence from the high concentrations of Ge and Zn detected in some sedimentary rocks [283].

It has been argued that subsurface hydrothermal activities were "abundant" on Mars as potential locales for the OoL, based on nearly three hundred exhumed sites detected by orbital remote sensing of relevant compositions (silica, carbonate, serpentine, and certain clays) which could be indicative of high-temperature geochemical alteration [284].

Evidence also comes from Martian meteorites, including the nakhlites, wherein phyllosilicates and Fe-rich carbonates were formed by high temperature processes [285]. Within the Tissint Martian meteorite are features that have been described as miniature vent-like morphologies and contain anhydrite, pyrrhotite, and magnetite nanophases with montmorillonite and associated organic nitrogen and oxygen compounds [185]. These features are in the 10 s of microns size range with obvious redox conditions existing on the scale of only tens of nanometers. If ubiquitous on Mars, their aggregate opportunities could provide significant opportunities for PCE.

3.3.3. Suboceanic Hydrothermal Proto-Macrobionts

A spectroscopic analog to the Lost City hydrothermal field is claimed for the Nili Fossae region on Mars [286], based on the occurrence of Mg-rich serpentine, Ca carbonates, talc, and amphiboles.

Many observations of mineralogy from Mars orbit have been interpreted as indicating individual areas of former hydrothermal activity [287–290], including the Eridania region which may even have once been an undersea setting [291].

Given the smaller size and possibly shorter lifetime of one or more oceans on Mars [111,250,263], coupled with the lower level of volcanic activity [277] and apparent lack of tectonic plate activity on Mars [292], the inferred likelihood of an origin of life by the pathway of oceanic hydrothermal vents must be much lower than for Earth.

4. Discussion

Because we have a minimum date for the OoL on Earth [264,293], we can compare the array of suitable settings on Mars and Earth in those earliest times to gauge whether it is reasonable to expect that life could or should have also arisen on Mars.

4.1. When Would Be an OoL on Mars: Past, Present, Future

Based on the early appearance of life on planet Earth, the likelihood that life also arose on Mars could be high since both planets had liquid water coexisting with similar basaltic surfaces and reduced greenhouse gases. After the cessation of extremely adverse conditions at the time of the formation of the Earth and its Moon, appropriate environments conducive to the formation of suitable settings, i.e., proto-macrobionts [13], there would elapse periods until the rise of the first life forms, which could accomplish wide-spread colonization. From age dating relative to the formation of refractory material (calcium aluminum inclusions) and from accretion models, it can be inferred that the moon-forming Theia impactor occurred at 20 to 100 Myr after the formation of the Earth itself [294], giving Mars that first interval of time as a "head start" for an origin of life since the moon formation event would have destroyed any early PCE or OoL. Although the Theia impact created a molten silicate surface, it would have cooled very rapidly (~ kyr) down to a 100 °C temperature.

During the decay of the heavy bombardment, the earliest time an OoL could have begun on Earth is variously estimated somewhere between 4.5 and 3.9 Ga [264]. However, evidence for a biosphere-scale abundance of life is by ~3.7 Ga [264,293], or perhaps even earlier [295]. This implies about 500 ± 300 Myr for the progression from a lifeless planet to one that is widely inhabited. A case has also been made for life to have arisen during an even shorter interval of ~100 Myr [295], following the last globally sterilizing large impact [97]. From Mars' location and size, its final global-scale sterilizing event by some giant impactor could have been even earlier than for Earth.

However, many investigators have pointed out that the duration for forming a living organism in any given setting could be much shorter, perhaps on the scale of a thousand years, or even less. This is because an assemblage of organic molecules and activities incipient to metabolism and RNA propagation will tend toward chaos, degradation, and decay unless a Darwinian evolutionary advantage is established. Once genetic properties become established, progression toward competent cellular life can be much more rapid. This is important because environmental settings can change on short time scales, and suitable conditions must persist for a sufficiently time for the complex sequence of PCE events to occur and to transition to entities that are alive before conditions become locally unsuitable.

Even if favorable climates were shorter on Mars, they could have been more than adequate because the OoL may be a rapid process for any given proto-macrobiont.

Nature changes on all time scales. A Martian lake that reaches the overflow or a breakout point may take years or decades to build up its inventory of brine, whether by runoff of precipitation or melting of cold-trapped ice. Once the discharge begins, however, it may complete its course in only hours or a few days [263].

A light photon emitted from the solar corona will take 8.3 min to reach the Earth (or 12.6 min to Mars), yet interact with its target atom, molecule or crystal lattice to release its energy to cause fundamental electrochemical changes in times measured in just nanoseconds. Geologic eons and eras are measured in millions of years, yet geysers erupt and subside in minutes.

How fast could be the rise of reproductive entities in a favorable proto-macrobiont setting? Some bacterial cells can fully reproduce themselves in less than 20 min [130] (primitive RNAzymes can replicate even faster). This biological feat requires the new synthesis of thousands of different molecules, made possible only because thousands of genes, proteins, and ribosomes are working in parallel and catalyzing reactions at high speeds [130]. In a prebiotic environment, the chemistries may also be complex, but are not coordinated or regulated, except to the degree that some molecules are autocatalytic for their own synthesis from available precursor molecules. Once a replicating system with heredity does form, such as an RNAzyme, the replication of programmed molecules can transition to exponential increases in rates of manufacture of useful products and improvements.

Furthermore, the natural lifetimes of most suggested settings for the OoL are not necessarily long compared to geological time scales of millions of years. How long can a pond avoid losing its water in any extended period of dry weather? How long can a given chimney formation of a hydrothermal vent remain active before it loses its connection with source water or becomes clogged due to excessive precipitation? How long can a hot spring continue to be hot? For planet Earth, virtually all the settings posited for macrobiont status are susceptible to having short lifetimes compared to geological time itself. Local activities are generally more realistically scaled as decades or centuries, but seldom multi-millennia. For planet Mars, these may be much longer, given its lower levels of almost all types of geologic activity [277].

Given that life on Earth did clearly start very early, and that early Mars was just as clement, if not more so, there is no justification in concluding that an OoL is a low-probability event. At this time, there is still no clear path susceptible to the de novo calculation of the likelihood for an origin of life on either planet, due to our still-nascent understanding of all the viable pathways by which prebiotic chemical evolution could occur, in addition to also not understanding all the suitable local environments that may have been abundant in those early millennia. These gaps in knowledge, especially for the former, have previously led to the general conclusion that the OoL may be a low-probability event.

An origin of life on small planetary bodies, such as the moons and minor planets of our solar system, is extremely unlikely for LAWKI because of a lack of a sustainable liquid form for H_2O, except for the case of tidally pumped but ice-capped H_2O-rich moons close to their host planet. For subaerial macrobionts, the planetary body must be large enough that its gravity sustains an atmospheric pressure greater than 6 mb. It is important to recognize that at these planetary scales, thousands of OHV's or millions of lakes can be hosted, and, therefore, the likelihood of one or more of these settings becoming transformed to a macrobiont can overcome the low probability of any single setting becoming so, especially considering the magnitude of geologic time over which various PCE's would have the opportunity to reach fruition.

It is clear that, in the mid-Noachian, there were abundant opportunities for the rise of life on Mars. Given the evidence of a slow progression of climate conditions toward those of the Hesperian Period [296], there was additional abundant time for the OoL, although the episodic freeze–thaw climate cycling envisioned by various groups [114,120,146] significantly reduces the aggregate available time, perhaps by one or more orders of magnitude. In spite of climate models, a variety of evidence indicates that local aqueous activity has occurred in Gale Crater well into the Amazonian, up until at least 2 Ga [271,297,298] and perhaps as recent as a few hundred Myr [299].

Studies of Garu crater and its vicinity suggest multiple crater lakes interconnected by hydrologic systems, including Gale crater, in late Hesperian times [106]. Modeling of the effects of obliquity cycling indicate recurring transient liquid wet conditions are possible on the Myr timescale [300]. Even if a new OoL did occur in the foreseeable future, however, the continued change in obliquity would revert to the conditions of today, which are considered inclement except at km scale depths [301,302].

4.2. Where an OoL Would Have Been on Mars

Crater-forming impacts continue to this day [303], at the rate of at least 700 per decade, some revealing shallow, extensive ice [304]. For contemporaneous Mars, however, there appears to be a major lack of opportunities for an OoL, and even daunting challenges for the survival of the most highly evolved extremophiles, because of the sub-freezing environment and the oxidative reactants in the atmosphere and soil [304]. Hydrothermal zones would be promising candidates, but would need to have a mechanism of recharge of liquid H_2O, which seems difficult under the broad-scale current thermal conditions, even with a substantial subsurface hydrosphere [141].

Volcanism persisted into the Amazonian and may still occur, making the detection of geothermal anomalies, such as those mapped from orbit at Earth's Yellowstone caldera [305], of particular interest. However, systematic observation campaigns by the Thermal Emission Imaging System (THEMIS) on the Mars Odyssey mission with its 100 m scale footprint, 2 K thermal sensitivity, and complete global coverage, have not revealed any locations on Mars where elevated surface temperatures might indicate the local availability of geothermal heat (V. E. Hamilton, personal communication, 2021).

After the discovery by the Viking missions of the unexpected paucity of organic compounds in the surface soil on Mars, it was realized that photochemically generated oxidants, including H_2O_2, OH·radical, atomic O_1, peroxy radicals (HO_2), superoxide ions (HO^{2-}), and less reactive O_2 itself, provide a significantly oxidative environment that can destroy [109,111] or degrade organics to unreactive carboxylate derivatives [306].

Contemporary Mars is presumably inhabitable at km-scale depths where temperatures can be high enough due to the planetary geothermal gradient to support a liquid hydrosphere [141]. However, habitability is not expected in the near surface where it is too cold, too dry, and too susceptible to damaging GCR radiation from space [301]. Future Mars actually holds some promise of an OoL because of upcoming favorable obliquity cycles, with the possibility of cold traps becoming sufficiently warmed to melt ice during daytime [300,307]. Could a macrobiont or nascent biosphere survive long-term when obliquity returns to unfavorably low values and the near-surface again becomes frozen and generally uninhabitable?

A most restrictive factor is the dearth of reducing power on Mars today. Hydrogen gas in the atmosphere is now only 15 ppm [308], which is too low for exergonic reactions with the abundant sulfates or atmospheric CO_2, and the organics in soils are also measured in ppm, as noted above.

4.3. Likelihood for Origin of Life

With the present state of knowledge, it is difficult to assign which planet, Mars or Earth, originally provided the greater a priori likelihood for an OoL. Mars was perhaps simply too dry and too cold, for too much of the time. Or, it was too small and therefore too inactive (volcanically and geomagnetically). In contrast, perhaps the Earth's surface was too wet and had too little sulfur, for life to arise in the first geological instant after the sterilizing bombardments waned. Perhaps Earth had too little boron available in its early history [309], whereas Mars did, which reinforces previous speculations for lithopanspermia, manifested as a Martian origin for life transplanted life to Earth [10,16,310–314].

At the most fundamental level, there is the "H_2O Problem." All scenarios for the origin of the form of life we know about have a requirement for the significant availability of H_2O in the liquid phase. An excess of water, however, can result in too extreme a dilution of ingredients to support a successful PCE because of the much lower reaction rates needed to achieve transitions to avoid deleterious degradation rates of labile ingredients (e.g., hydrolysis).

A Chicxulub-class cometary impactor of diameter 30 km, with specific density of 1.0 and 10% soluble organics would provide only an average 4 µM concentration of organic molecules (at 100 g/mol) into a 3.5 km global equivalent layer (GEL) ocean on Earth. In comparison, the same bolide onto Mars where the total surface water inventory is, say, 0.5 km GEL [111], could produce a 100 µM concentration if those same organics were taken up by the water.

Or, perhaps neither planet qualifies as fully optimized for the rise of life, but something more intermediate between the two planets would have been even more favorable. If so, the very fact that it did arise gives hope, in the Bayesian sense, that the origin of life is not a formidable task, considering the panoply of settings that would be possible on any planet whose expanse is vast, endowed with essential elements and organics, and with the suitable environments for the formation of macrobionts of one class or another.

4.3.1. Expected Value for the OoL

Expected value, and not just a probability, is the gauge for the likelihood of an origin of life on any given body. We currently cannot rule out any of the major hypotheses for the types of settings most suitable for life to have originated. Rather, it is possible that all are somewhat likely to have provided a pathway to life. Perhaps life began on one planet via one route, and on another planet by one of the other routes, depending on the relative prevalence of the various settings, and on happenstance.

Our agnostic approach, then, is to consider multiple plausible possibilities. If P_J is the probability of life beginning in a Jth type of setting, and N_J is the total number of settings of type J, then the expected value for an OoL in such a setting, $E_J[O]$, is simply.

$$E_J[O] = N_J \times P_J \qquad (1)$$

Even if the possibility of the chain of events leading to life is small yet non-negligible, but the number of settings in which it could occur is extremely large, then the expected value for an origin could be of order 1.0 or even higher.

Let us also assume, for sake of analysis, that there are four mutually independent *types* of proto-macrobionts where life could begin. Since they are disjoint, their probabilities and expected values are simply additive:

$$E[O] = E_{PAT}[O] + E_{GHS}[O] + E_{OHV}[O] + E_X[O] \qquad (2)$$

where PAT denotes a pond at ambient temperatures, GHS is for geothermal hot springs, OHV is oceanic hydrothermal vent, and X is "other" settings we have not yet considered or have been even conceived. Note that both impact crater lakes with hydrothermal consequences and magmatically heated groundwater could be subsets of GHS. Likewise, two or more types of smokers, or a vent field in general, could make up the OHV category. Thus, the number of suitable loci for an OoL could be even greater than the four indicated above.

The above considerations neglect the parameter of time. Proto-macrobionts come and go, e.g., because of the limited active lifetimes of ponds, springs, and hydrothermal vents. The planet-wide occurrences and lifetimes are a function of climate and magmatic changes, and hence vary. If we take $p_J(n,t)$ as the probability per unit time that the nth setting of type J will become a macrobiont that seeds a biosphere, then more explicitly,

$$P_J = p_J(n,t)\, dt \tag{3}$$

and

$$E_J[O] = \sum_{N=1}^{N_J} \int_0^{L_J} p_J(n,t)\, dt \tag{4}$$

where N_J is the number of sites of type J, each with its own probability function, and L_J is the average lifetime for that type of setting before becoming a macrobiont that achieves colonization. One other aspect is that although L_J is typically small compared to geologic time, a given setting may be "re-used."

Although this is the simplest possible model, it does emphasize that, for J-type settings, which are quite numerous (N_J) and have sufficient typical lifetimes, L_J, the probability can be quite small yet yield a reasonable expectation that one of more biosphere-seeding macrobionts can succeed, albeit rarely so on an individual basis. Including finite lifetimes in the formulation also emphasizes that, even when a given setting proceeds on a non-productive course, it will eventually be changed, sometimes for the better. For example, ponds can be covered over, springs can dry up, and vent chimneys clogged. However, they can also be rejuvenated, just as a test tube in the lab can be rinsed and re-used, which multiplies the opportunities for repeated "experiments" along the pathway to life and a biosphere.

4.3.2. Lithopanspermia

If the expected value is so extremely low that the OoL on Earth is an outlier event for planets of our general type and located favorably in their planetary system, then any expectation for an additional, independent origin of life on Mars must be negligible (multiplication of two very small probabilities). However, given the possibility that life arose on Mars early, could it have also seeded Earth (or vice versa)?

When Mars-to-Earth lithopanspermia was first proposed, as a result of the confirmation that SNC meteorites had been successfully transported from Mars without melting from the shock acceleration to escape the gravity of Mars, it was also realized that there could be several impediments that would render the transfer of life a very low probability event [310,311]. These impediments include the energetics of launch at Mars, the statistics of capture by Earth, and the insults of space radiation from solar particle events and GCR [312,313]. The rocks launched during the spallation process must generally be highly competent, as evidenced by the population of known Martian meteorites, which implies difficulty if not impossibility in launching the weaker sediments which would typically contain the much higher and more diverse bioloads. Once the spall phenomenon explanation was established, for the transfer mainly of kinetic rather than thermal and disruptive energy by an impacting bolide into rock in a spall zone [315], it became clear that seeding the Solar System was possible. However, detailed calculations of the statistics of interplanetary transfer [316] showed that such events required significant transfer times, such that the disruption of biological organization and processes by ionizing space radiation (penetrating GCR) could be too severe in all but the less likely cases of multi-meter scale meteorites [313]. Transfers from Earth to Mars are significantly less likely [316] because of Earth's much higher gravity field and thicker atmosphere, and the smaller target provided by Mars (smaller diameter, larger orbit).

From a probabilistic standpoint, Mars-to-Earth panspermia seems extraordinarily unlikely because it presumably is the multiplicative product of potentially two very small numbers, i.e., the probability of an OoL on Mars and the probability of successful transfer and colonization of Earth.

However, even if it is an extremely rare outcome, once life started on Mars, the large number of transfers that intrinsically occur, especially in the earlier high bombardment rate history of the solar system, could balance against the low probability of successful transfer, to combine for a likelihood for seeding Earth that is not necessarily out of the question [313]. Given that the opportunities for an OoL on Mars were highest in the time interval just as life appeared on Earth could be a coincidence, or it could be because Mars had the greater a priori advantage for the OoL. Detecting past life on Mars would be extraordinarily important for many reasons, but including the capability for comparative genomics to assess whether there were two OoL's which were truly independent, and if not, to determine the locus in the evolutionary tree of life where the branching took place and to constrain when the migration occurred based on the genomic clock.

4.4. Future Research

Mars orbiter missions will continue to make major contributions, but rovers on the surface are particularly well-suited for discoveries that can be confirmed only with the analyses that are possible in situ, as well as mineralogical confluences and aqueous settings that are subscale or otherwise not detectable from orbit.

Much remains to be learned of Mars from the bountiful number of operational and new missions (NASA's Perseverance and CNSA's Zhurong rovers) to the planet. These missions can take opportunities to explore with high relevance to the field of OoL, depending on how operations are implemented, especially with respect to the selection of samples to be returned for extremely detailed analyses in laboratories on Earth.

Mars sample return missions are often justified because of the lack of geologic context for the hundreds of Martian meteorites that have now been found on Earth. These meteorites are generally composed of igneous minerals, with only occasional minor or trace quantities of alteration products. Because the ejection process is energetic, converting plagioclase to maskelynite, with indications of peak shock pressures of 20 to 80 GPa [317], it is unlikely that sedimentary rocks such as mudstones and sandstones can be successfully ejected without becoming disaggregated. Sediments can not only establish the record of aqueous processes, and hence the past habitability of Mars [8,318,319], but are also the medium in which the secondary minerals and organics for PCE and the OoL itself reside. Hence, it is advocated that promising sediments be given high priority in the upcoming selection of samples by the M2020 Perseverance mission for future return to Earth because they are much more likely not only to preserve life or biosignatures, but also the history of aqueous processes and PCE. This would be especially important if a sediment could be associated with one of the scenarios proposed for the OoL, such as hydrothermal or cyclic wet–dry settings. Laboratory investigations of such samples should include study of the composition of aqueous extracts for various pH/Eh conditions.

5. Summary

We have endeavored to demonstrate the numerous suitable circumstances at Mars for an origin of life. A synthesis combining the discoveries from the exploration of Mars with terrestrial analogs, relevant laboratory experiments, and theoretical models point towards an OoL on Mars as being likely to the same degree, and even more so in many respects, than the origin of life on Earth itself. Several of the discoveries of the MSL Curiosity rover mission in Gale crater, such as enrichments of key elements proposed for the OoL (Cu, B, etc.), direct evidence of cyclic wet–dry cycling (e.g., mudcrack patterns, episodic wetting), and the preservation of organics at the surface, are directly favorable to the likelihood of an origin of life.

Mars not only has all the building-block elements (CHNOPS) for biochemical molecules but also other key elements critical for metabolic functions, in many cases enriched over their abundances in rocks and soils on Earth. Several extremely important elements of biology, including sulfur, iron, and magnesium, are especially highly abundant and mobile on Mars, more so than on terrestrial continents. Furthermore, several transition trace elements which serve as co-factors in important metalloenzymes, such as Mn, Ni, and Zn, are also unusually abundant. Ubiquitous amorphous components in Martian sediments are additional evidence of element mobility. Whether each of these elements are available in their most suitable form (solubility and redox state) depends on the pH and Eh of the contemporaneous environment, but these can be modulated by the intensity and duration of local mineral alteration or magmatic activity and attendant release of volatiles.

Settings suitable for the various OoL hypotheses are abundant. The extreme population of crater basins fulfills pond scenarios for macrobiont formation. The proximity to the Kuiper and outer asteroid belts assures equal or greater contribution than for Earth of the carbonaceous matter and pre-formed biotic precursors these contain, as well as accessible sources of phosphorus and nickel, while the shallower regolith, less water, and less active surface processes allow for greater concentrations of these components.

Although suboceanic hydrothermal vents were undoubtedly much less common on Mars because of less water, little or no plate tectonics, less magmatic activity, and uncertainties about the extent of an early ocean, if present at all, there is nonetheless widespread evidence of hydrothermal environments in the past. Furthermore, the global distribution of large craters provides the basis for meaningful durations of buried hydrothermal regimes created from impacts by large bolides, even if the initial regolith inventory of H_2O was as ice.

Late Noachian and early Hesperian Mars were sufficiently endowed with periods of liquid water that life could have begun. This time period overlaps Earth's Hadean and early Archean eon's, during which life appeared. If wet–dry and/or freeze–thaw cycling are indeed critical environments to enable the prebiotic chemical evolution needed to achieve the reproduction of protocells and beyond, then Mars would have a significant advantage in its area of land compared to the surface area of rare volcanic islands in a global ocean envisioned at the time of early Earth. For freeze–thaw cycling, the arid environments which may have been much more prevalent on Mars than Earth generally experience significantly larger diurnal temperature swings, promoting the freezing of shallow streams and foreshore areas at night, but with melt-out in daytime.

The period for an OoL on Mars from late Hesperian to current epochs would be less favorable on the basis of the rarity of liquid H_2O. However, it was in earlier times that life was already apparent on Earth. Given that life could have arisen on either planet, and with the interchange of ejected material from hypervelocity impact, it is possible that one planet seeded life on the other. For a variety of reasons, the expected probability of Mars-to-Earth lithopanspermia is greater than for the opposite direction.

Although the outlook for a future OoL is bleak, Mars is serving as a window into plausible conditions on early Earth, a time period in our geologic history which has been erased by subsequent processes. It also is providing support for hypotheses which view suitable exoplanets as candidates for their own origin of life. Further exploration by the 2020 and subsequent rovers will undoubtedly expand the list of relevant conditions and constituents that have occurred on Mars.

Given that sediments are generally too weak to be ejected from Mars by natural impact processes, sample-return missions could greatly enhance the value of laboratory analyses if sediments significantly populate the samples taken for potential return. Based on the range of settings hypothesized for the OoL, sediments collected from areas where spatially heterogeneous or time-variable conditions are in evidence may be especially beneficial for gaining insights into prebiotic chemical evolution and the steps leading to life.

Author Contributions: Conceptualization: B.C.C., V.M.K.; formal analysis: B.C.C.; validation: all; writing—original, review: all authors. B.C.C. authored the overall manuscript and conducted the analyses; V.M.K. contributed to the overall manuscript and Sections 2, 2.1, 3.1, 3.1.1, 4.3.2 and 5; A.S. contributed to Section 1, Section 2.3, Section 3.1, Section 3.1.1, and Section 3.3.2; C.H.H. to Section 2.1, Section 3.1.1, and Section 4.1; N.L.L. to Section 3.1.2, Section 3.2, and Section 3.3.1; P.J.G. to Sections 2.1 and 3.1.3; S.J.V. to Sections 3.1.1 and 3.1.2; H.E.N. to Sections 2.3 and 3.3.1; J.M.-F. to Section 3.2. All authors have read and agreed to the published version of the manuscript.

Funding: We gratefully acknowledge NASA and JPL/CalTech for their participation in the Mars Science Laboratory (Curiosity rover) mission, the previous missions of Mars Exploration Rover missions (Spirit, Opportunity), and the Phoenix Mars Lander.

Institutional Review Board Statement: Not applicable.

Informed Consent Statement: Not applicable.

Acknowledgments: Helpful comments by reviewers are greatly appreciated.

Conflicts of Interest: The authors declare no conflict of interest. The funders had no role in the design of the study, in the analyses or interpretation of data, in the writing of the manuscript, or in the decision to publish the results.

Abbreviations

AmC	Amorphous Component (ChemCam XRD)
APXS	Alpha Particle X-ray Spectrometer (XRF on MER, MSL rovers)
a_w	Chemical activity of H_2O
CCAM	ChemCam (MSL/LIBS on MSL rover)
CheMin	Chemical and Mineralogy instrument (XRD on MSL rover)
CRISM	Compact Recon Imaging Spectrometer for Mars (MRO orbiter)
OMEGA	Vis/IR Mineralogical Mapping (ESA/Mars Express orbiter)
EGA	Evolved Gas Analyzer (part of SAM on MSL rover)
GHS	Geothermal Hot Spring
GCMS	Gas Chromatograph and Mass Spectrometer (Viking, MSL/SAM)
GCR	Galactic Cosmic Radiation
LAWKI	Life As We Know It (biochemically)
LIBS	Laser-Induced Breakdown Spectroscopy
MER	Mars Exploration Rover (Spirit and Opportunity rovers)
MORB	Mid-Ocean Ridge Basalt
MS	Mass Spectrometer
MSL	Mars Science Laboratory (Curiosity rover)
M2020	Mars 2020 Mission (Perseverance rover)
NWA	Northwest Africa (meteorite identifier)
OoL	Origin of Life
OHV	Oceanic Hydrothermal Vent
PAT	Pond in Ambient Temperature
PIXL	Planetary Instrument for X-ray Lithochemistry (M2020 rover)
PCE	Prebiotic chemical evolution
RNAzyme	RNA enzyme (ribozyme)
SAM	Sample Analysis at Mars (GCMS, EGA on MSL rover)
THEMIS	Thermal Emission Imaging System (Odyssey orbiter)
WCL	Wet Chemistry Laboratory (instrument on Phoenix lander)
XRD	X-ray Diffraction
XRF	X-ray Fluorescence

References

1. Pimentel, G.C.; Forney, P.B.; Herr, K.C. Evidence about hydrate and solid water in the Martian surface from the 1969 Mariner infrared spectrometer. *J. Geophys. Res.* **1974**, *79*, 1623–1634. [CrossRef]
2. Milton, D.J. Water and processes of degradation in the Martian landscape. *J. Geophys. Res.* **1973**, *78*, 4037–4047. [CrossRef]
3. Klein, H.P. The search for life on Mars: What we learned from Viking. *J. Geophys. Res. Planets* **1998**, *103*, 28463–284636. [CrossRef]
4. Wordsworth, R.D. The climate of early Mars. *Annu. Rev. Earth Planet. Sci.* **2016**, *44*, 381–408. [CrossRef]

5. Ehlmann, B.L.; Edwards, C.S. Mineralogy of the Martian surface. *Annu. Rev. Earth Planet. Sci.* **2014**, *42*, 291–315. [CrossRef]
6. Rampe, E.B.; Blake, D.F.; Bristow, T.F.; Ming, D.W.; Vaniman, D.T.; Morris, R.V.; Achilles, C.N.; Chipera, S.J.; Morrison, S.M.; Tu, V.M.; et al. Mineralogy and geochemistry of sedimentary rocks and eolian sediments in Gale crater, Mars: A review after six Earth years of exploration with Curiosity. *Geochemistry* **2020**, *80*, 125605. [CrossRef]
7. McLennan, S.M.; Grotzinger, J.P.; Hurowitz, J.A.; Tosca, N.J. The sedimentary cycle on early Mars. *Annu. Rev. Earth Planet. Sci.* **2019**, *47*, 91–118. [CrossRef]
8. Grotzinger, J.P.; Sumner, D.Y.; Kah, L.C.; Stack, K.; Gupta, S.; Edgar, L.; Rubin, D.; Lewis, K.; Schieber, J.; Mangold, N.; et al. A habitable fluvio-lacustrine environment at Yellowknife Bay, Gale Crater, Mars. *Science* **2014**, *24*, 343. [CrossRef] [PubMed]
9. Knoll, A.H.; Carr, M.; Clark, B.; Marais, D.J.D.; Farmer, J.D.; Fischer, W.W.; Grotzinger, J.P.; McLennan, S.M.; Malin, M.; Schröder, C.; et al. Astrobiological implications of Meridiani sediments. *Earth Planet. Sci. Lett.* **2005**, *240*, 179–189. [CrossRef]
10. Benner, S.A.; Kim, H.J. The Case for a Martian Origin for Earth Life. In Proceedings of the Instruments, Methods, and Missions for Astrobiology XVII, San Diego, CA, USA, 28 September 2015; International Society for Optics and Photonics: Bellingham, WA, USA; Volume 9606, p. 96060C. [CrossRef]
11. Fornaro, T.; Steele, A.; Brucato, J.R. Catalytic/protective properties of martian minerals and implications for possible origin of life on Mars. *Life* **2018**, *8*, 56. [CrossRef]
12. Sasselov, D.D.; Grotzinger, J.P.; Sutherland, J.D. The origin of life as a planetary phenomenon. *Sci. Adv.* **2020**, *6*, eaax3419. [CrossRef] [PubMed]
13. Clark, B.C.; Kolb, V.M. Macrobiont: Cradle for the origin of life and creation of a biosphere. *Life* **2020**, *10*, 278. [CrossRef] [PubMed]
14. Mojarro, A.; Jin, L.; Szostak, J.W.; Head, J.W.; Zuber, M.T. In search of the RNA world on Mars. *Geobiology* **2021**, *19*, 307–321. [CrossRef] [PubMed]
15. Damer, B.; Deamer, D. The hot spring hypothesis for an origin of life. *Astrobiology* **2020**, *20*, 429–452. [CrossRef] [PubMed]
16. McKay, C.P. An origin of life on Mars. *Cold Spring Harb. Perspect. Biol.* **2010**, *2*, a003509. [CrossRef]
17. Frenkel-Pinter, M.; Rajaei, V.; Glass, J.B.; Hud, N.V.; Williams, L.D. Water and life: The medium is the message. *J. Mol. Evol.* **2021**, *11*, 1–10. [CrossRef]
18. Miller, S.L. A production of amino acids under possible primitive earth conditions. *Science* **1953**, *117*, 528–529. [CrossRef]
19. de Duve, C. Clues from Present-Day Biology: The Thioester World. In *The Molecular Origins of Life: Assembling Pieces of the Puzzle*; Brack, A., Ed.; Cambridge University Press: Cambridge, UK, 2000; pp. 219–236.
20. Muchowska, K.B.; Varma, S.J.; Moran, J. Nonenzymatic metabolic reactions and life's origins. *Chem. Rev.* **2020**, *120*, 7708–7744. [CrossRef]
21. Chevallot-Beroux, E.; Gorges, J.; Moran, J. Energy conservation via thioesters in a non-enzymatic metabolism-like reaction network. *ChemRxiv* **2019**, *9*, 8832425.
22. Goldford, J.E.; Hartman, H.; Marsland, R.; Segrè, D. Environmental boundary conditions for the origin of life converge to an organo-sulfur metabolism. *Nat. Ecol. Evol.* **2019**, *3*, 1715–1724. [CrossRef]
23. Muchowska, K.B.; Moran, J. Peptide synthesis at the origin of life. *Science* **2020**, *370*, 767–768. [CrossRef] [PubMed]
24. Wächtershäuser, G. The origin of life and its methodological challenge. *J. Theor. Biol.* **1997**, *187*, 483–494. [CrossRef] [PubMed]
25. Wächtershäuser, G. From volcanic origins of chemoautotrophic life to Bacteria, Archaea and Eukarya. *Philos. Trans. R. B Biol. Sci.* **2006**, *361*, 1787–1808. [CrossRef] [PubMed]
26. Wächtershäuser, G. On the chemistry and evolution of the pioneer organism. *Chem. Biodivers.* **2007**, *4*, 584–602. [CrossRef] [PubMed]
27. Ragsdale, S.W.; Pierce, E. Acetogenesis and the Wood-Ljungdahl pathway of CO_2 fixation. *Biochim. Biophys. Acta (BBA) Proteins Proteom.* **2008**, *1784*, 1873–1898. [CrossRef]
28. Todd, Z.R.; House, C.H. Vesicles protect activated acetic acid. *Astrobiology* **2014**, *14*, 859–865. [CrossRef]
29. Russell, M.J.; Martin, W. The rocky roots of the acetyl-CoA pathway. *Trends Biochem. Sci.* **2004**, *29*, 358–363. [CrossRef]
30. Ferry, J.G.; House, C.H. The stepwise evolution of early life driven by energy conservation. *Mol. Biol. Evol.* **2005**, *23*, 1286–1292. [CrossRef]
31. Patel, B.H.; Percivalle, C.; Ritson, D.J.; Duffy, C.D.; Sutherland, J.D. Common origins of RNA, protein and lipid precursors in a cyanosulfidic protometabolism. *Nat. Chem.* **2015**, *7*, 301–307. [CrossRef]
32. Liu, Z.; Wu, L.F.; Xu, J.; Bonfio, C.; Russell, D.A.; Sutherland, J.D. Harnessing chemical energy for the activation and joining of prebiotic building blocks. *Nat. Chem.* **2020**, *12*, 1–6. [CrossRef]
33. Ritson, D.J.; Battilocchio, C.; Ley, S.V.; Sutherland, J.D. Mimicking the surface and prebiotic chemistry of early Earth using flow chemistry. *Nat. Commun.* **2018**, *9*, 1–10.
34. Ferus, M.; Kubelík, P.; Knížek, A.; Pastorek, A.; Sutherland, J.; Civiš, S. High energy radical chemistry formation of HCN-rich atmospheres on early Earth. *Sci. Rep.* **2017**, *7*, 1–9. [CrossRef]
35. Xu, J.; Ritson, D.J.; Ranjan, S.; Todd, Z.R.; Sasselov, D.D.; Sutherland, J.D. Photochemical reductive homologation of hydrogen cyanide using sulfite and ferrocyanide. *Chem. Commun.* **2018**, *54*, 5566–5569. [CrossRef]
36. Toner, J.D.; Catling, D.C. Alkaline lake settings for concentrated prebiotic cyanide and the origin of life. *Geochim. Cosmochim. Acta* **2019**, *260*, 124–132. [CrossRef]

37. Foden, C.S.; Islam, S.; Fernández-García, C.; Maugeri, L.; Sheppard, T.D.; Powner, M.W. Prebiotic synthesis of cysteine peptides that catalyze peptide ligation in neutral water. *Science* **2020**, *370*, 865–869. [CrossRef]
38. Becker, S.; Feldmann, J.; Wiedemann, S.; Okamura, H.; Schneider, C.; Iwan, K.; Crisp, A.; Rossa, M.; Amatov, T.; Carell, T. Unified prebiotically plausible synthesis of pyrimidine and purine RNA ribonucleotides. *Science* **2019**, *366*, 76–82. [CrossRef]
39. Becker, S.; Schneider, C.; Okamura, H.; Crisp, A.; Amatov, T.; Dejmek, M.; Carell, T. Wet-dry cycles enable the parallel origin of canonical and non-canonical nucleosides by continuous synthesis. *Nat. Commun.* **2018**, *9*, 1–9. [CrossRef]
40. Becker, S.; Thoma, I.; Deutsch, A.; Gehrke, T.; Mayer, P.; Zipse, H.; Carell, T. A high-yielding, strictly regioselective prebiotic purine nucleoside formation pathway. *Science* **2016**, *352*, 833–836. [CrossRef]
41. Becker, S.; Schneider, C.; Crisp, A.; Carell, T. Non-canonical nucleosides and chemistry of the emergence of life. *Nat. Commun.* **2018**, *9*, 1–4. [CrossRef]
42. Silva, J.D.; Franco, A. Boron in prebiological evolution. *Angew. Chem.* **2020**, *133*, 10550–10560. [CrossRef]
43. Ricardo, A.; Carrigan, M.A.; Olcott, A.N.; Benner, S.A. Borate minerals stabilize ribose. *Science* **2004**, *303*, 196. [CrossRef]
44. Kim, H.J.; Furukawa, Y.; Kakegawa, T.; Bita, A.; Scorei, R.; Benner, S.A. Evaporite borate-containing mineral ensembles make phosphate available and regiospecifically phosphorylate ribonucleosides: Borate as a multifaceted problem solver in prebiotic chemistry. *Angew. Chem.* **2016**, *128*, 16048–16052. [CrossRef]
45. Kim, H.J.; Ricardo, A.; Illangkoon, H.I.; Kim, M.J.; Carrigan, M.A.; Frye, F.; Benner, S.A. Synthesis of carbohydrates in mineral-guided prebiotic cycles. *J. Am. Chem. Soc.* **2011**, *133*, 9457–9468. [CrossRef]
46. Kitadai, N.; Maruyama, S. Origins of building blocks of life: A review. *Geosci. Front.* **2018**, *9*, 1117–1153. [CrossRef]
47. Kim, H.J.; Benner, S.A. Abiotic synthesis of nucleoside 5′-triphosphates with nickel borate and cyclic trimetaphosphate (CTMP). *Astrobiology* **2021**, *21*, 298–306. [CrossRef] [PubMed]
48. Kawai, J.; McLendon, D.C.; Kim, H.J.; Benner, S.A. Hydroxymethanesulfonate from volcanic sulfur dioxide: A "mineral" reservoir for formaldehyde and other simple carbohydrates in prebiotic chemistry. *Astrobiology* **2019**, *19*, 506–516. [CrossRef] [PubMed]
49. Ferris, J.P.; Hill, A.R.; Liu, R.; Orgel, L.E. Synthesis of long prebiotic oligomers on mineral surfaces. *Nature* **1996**, *381*, 59–61. [CrossRef] [PubMed]
50. Huang, W.; Ferris, J.P. One-step, regioselective synthesis of up to 50-mers of RNA oligomers by montmorillonite catalysis. *J. Am. Chem. Soc.* **2006**, *128*, 8914–8919. [CrossRef]
51. Varma, S.J.; Muchowska, K.B.; Chatelain, P.; Moran, J. Native iron reduces CO_2 to intermediates and end-products of the acetyl-CoA pathway. *Nat. Ecol. Evol.* **2018**, *2*, 1019–1024. [CrossRef]
52. Martin, W.F. Older than genes: The acetyl CoA pathway and origins. *Front. Microbiol.* **2020**, *11*, 817. [CrossRef]
53. Deamer, D.W. Membrane Compartments in Prebiotic Evolution. In *The Molecular Origins of Life: Assembling Pieces of the Puzzle*; Brack, A., Ed.; Cambridge University Press: Cambridge, UK, 2000; pp. 189–205.
54. Deamer, D.; Damer, B.; Kompanichenko, V. Hydrothermal chemistry and the origin of cellular life. *Astrobiology* **2019**, *19*, 1523–1537. [CrossRef] [PubMed]
55. Cornell, C.E.; Black, R.A.; Xue, M.; Litz, H.E.; Ramsay, A.; Gordon, M.; Mileant, A.; Cohen, Z.R.; Williams, J.A.; Lee, K.K.; et al. Prebiotic amino acids bind to and stabilize prebiotic fatty acid membranes. *Proc. Natl. Acad. Sci. USA* **2019**, *116*, 17239–172344. [CrossRef]
56. Jin, L.; Engelhart, A.E.; Zhang, W.; Adamala, K.; Szostak, J.W. Catalysis of template-directed nonenzymatic RNA copying by Iron (II). *J. Am. Chem. Soc.* **2018**, *140*, 15016–15021. [CrossRef]
57. O'Flaherty, D.K.; Kamat, N.P.; Mirza, F.N.; Li, L.; Prywes, N.; Szostak, J.W. Copying of mixed-sequence RNA templates inside model protocells. *J. Am. Chem. Soc.* **2018**, *140*, 5171–5178. [CrossRef]
58. Li, M.; Huang, X.; Tang, T.D.; Mann, S. Synthetic cellularity based on non-lipid micro-compartments and protocell models. *Curr. Opin. Chem. Biol.* **2014**, *22*, 1–11. [CrossRef]
59. Deamer, D.; Singaram, S.; Rajamani, S.; Kompanichenko, V.; Guggenheim, S. Self-assembly processes in the prebiotic environment. *Philos. Trans. R. Soc. B Biol. Sci.* **2006**, *361*, 1809–1818. [CrossRef]
60. Ross, D.S.; Deamer, D. Dry/wet cycling and the thermodynamics and kinetics of prebiotic polymer synthesis. *Life* **2016**, *6*, 28. [CrossRef]
61. Milshteyn, D.; Damer, B.; Havig, J.; Deamer, D. Amphiphilic compounds assemble into membranous vesicles in hydrothermal hot spring water but not in seawater. *Life* **2018**, *8*, 11. [CrossRef]
62. Tjhung, K.F.; Sczepanski, J.T.; Murtfeldt, E.R.; Joyce, G.F. RNA-catalyzed cross-chiral polymerization of RNA. *J. Am. Chem. Soc.* **2020**, *142*, 15331–15339. [CrossRef]
63. Joyce, G.F.; Szostak, J.W. Protocells and RNA self-replication. *Cold Spring Harb. Perspect. Biol.* **2018**, *10*, a034801. [CrossRef]
64. Horning, D.P.; Joyce, G.F. Amplification of RNA by an RNA polymerase ribozyme. *Proc. Natl. Acad. Sci. USA* **2016**, *113*, 9786–9791. [CrossRef]
65. Joyce, G.F.; Orgel, L.E. Prospects for Understanding of RNA World. In *The RNA World: The Nature of Modern RNA Suggests a Prebiotic RNA World*; Gesteland, R.F., Atkins, J.F., Eds.; Cold Spring Harbor University Press: Cold Spring Harbor, NY, USA, 1993; pp. 1–25.
66. Robertson, M.P.; Joyce, G.F. The origins of the RNA World. *Cold Spring Harb. Perspect. Biol.* **2012**, *4*, a003608. [CrossRef]

67. Tjhung, K.F.; Shokhirev, M.N.; Horning, D.P.; Joyce, G.F. An RNA polymerase ribozyme that synthesizes its own ancestor. *Proc. Natl. Acad. Sci. USA* **2020**, *117*, 2906–2913. [CrossRef]
68. Kaim, W.; Schwederski, B.; Klein, A. *Bioinorganic Chemistry—Inorganic Elements in the Chemistry of Life: An Introduction and Guide*; John Wiley & Sons: Hoboken, NJ, USA, 2013.
69. Silva, J.F.D.; Williams, R.J. *The Biological Chemistry of the Elements: The Inorganic Chemistry of Life*; Oxford University Press: Oxford, UK, 2001.
70. Morowitz, H.J.; Srinivasan, V.; Smith, E. Ligand field theory and the origin of life as an emergent feature of the periodic table of elements. *Biol. Bull.* **2010**, *219*, 1–6. [CrossRef] [PubMed]
71. Muchowska, K.B.; Varma, S.J.; Chevallot-Beroux, E.; Lethuillier-Karl, L.; Li, G.; Moran, J. Metals promote sequences of the reverse Krebs cycle. *Nat. Ecol. Evol.* **2017**, *1*, 1716–1721. [CrossRef]
72. Degtyarenko, K. Bioinorganic motifs: Towards functional classification of metalloproteins. *Bioinformatics* **2000**, *16*, 851–864. [CrossRef]
73. Hao, J.; Mokhtari, M.; Pedreira-Segade, U.; Michot, L.J.; Daniel, I. Transition metals enhance the adsorption of nucleotides onto clays: Implications for the origin of life. *ACS Earth Space Chem.* **2018**, *3*, 109–119. [CrossRef]
74. Moore, E.K.; Jelen, B.I.; Giovannelli, D.; Raanan, H.; Falkowski, P.G. Metal availability and the expanding network of microbial metabolisms in the Archaean eon. *Nat. Geosci.* **2017**, *10*, 629–636. [CrossRef]
75. Waldron, K.J.; Rutherford, J.C.; Ford, D.; Robinson, N.J. Metalloproteins and metal sensing. *Nature* **2009**, *460*, 823–830. [CrossRef]
76. Frenkel-Pinter, M.; Samanta, M.; Ashkenasy, G.; Leman, L.J. Prebiotic peptides: Molecular hubs in the origin of life. *Chem. Rev.* **2020**, *11*, 4707–4765. [CrossRef] [PubMed]
77. Van Der Gulik, P.; Massar, S.; Gilis, D.; Buhrman, H.; Rooman, M. The first peptides: The evolutionary transition between prebiotic amino acids and early proteins. *J. Theor. Biol.* **2009**, *261*, 531–539. [CrossRef]
78. Presta, L.; Fondi, M.; Emiliani, G.; Fani, R. Molybdenum Availability in the Ecosystems (Geochemistry Aspects, When and How Did It Appear?). In *Molybdenum Cofactors and Their role in the Evolution of Metabolic Pathways*; Springer: Dordrecht, The Netherlands, 2015. [CrossRef]
79. Hickman-Lewis, K.; Cavalazzi, B.; Sorieul, S.; Gautret, P.; Foucher, F.; Whitehouse, M.J.; Jeon, H.; Georgelin, T.; Cockell, C.S.; Westall, F. Metallomics in deep time and the influence of ocean chemistry on the metabolic landscapes on Earth's earliest ecosystems. *Sci. Rep.* **2020**, *10*, 1–16. [CrossRef] [PubMed]
80. Mulkidjanian, A.Y.; Bychkov, A.Y.; Dibrova, D.V.; Galperin, M.Y.; Koonin, E.V. Origin of first cells at terrestrial, anoxic geothermal fields. *Proc. Natl. Acad. Sci. USA* **2012**, *109*, E821–E830. [CrossRef]
81. Lu, G.S.; LaRowe, D.E.; Fike, D.A.; Druschel, G.K.; Gilhooly, W.P., III; Price, R.E.; Amend, J.P. Bioenergetic characterization of a shallow-sea hydrothermal vent system: Milos Island, Greece. *PLoS ONE* **2020**, *15*, e0234175. [CrossRef] [PubMed]
82. Clark, B.C.; Kolb, V.M. Comet pond II: Synergistic intersection of concentrated extraterrestrial materials and planetary environments to form procreative Darwinian ponds. *Life* **2018**, *8*, 12. [CrossRef]
83. Pearce, B.K.; Pudritz, R.E.; Ralph, E.; Semenov, D.A.; Henning, T.K. Origin of the RNA world: The fate of nucleobases in warm little ponds. *Proc. Natl. Acad. Sci. USA* **2017**, *114*, 11327–11332. [CrossRef] [PubMed]
84. Baross, J.A. The rocky road to biomolecules. *Nature* **2018**, *564*, 42–43. [CrossRef] [PubMed]
85. Baross, J.A.; Anderson, R.E.; Stüeken, E.E. The Environmental Roots of the Origin of Life. In *Planetary Astrobiology*; Meadows, V.S., Arney, G.N., Schmidt, B.E., Marais, D.J.D., Eds.; University of Arizona: Tucson, AZ, USA, 2020; pp. 71–92.
86. Westall, F.; Hickman-Lewis, K.; Hinman, N.; Gautret, P.; Campbell, K.A.; Bréhéret, J.G.; Foucher, F.; Hubert, A.; Sorieul, S.; Dass, A.V.; et al. A hydrothermal-sedimentary context for the origin of life. *Astrobiology* **2018**, *18*, 259–293. [CrossRef]
87. Sojo, V.; Herschy, B.; Whicher, A.; Camprubi, E.; Lane, N. The origin of life in alkaline hydrothermal vents. *Astrobiology* **2016**, *16*, 181–197. [CrossRef] [PubMed]
88. Egel, R. Primal eukaryogenesis: On the communal nature of precellular states, ancestral to modern life. *Life* **2012**, *2*, 170–212. [CrossRef]
89. Russell, M.J. The alkaline solution to the emergence of life: Energy, entropy and early evolution. *Acta Biotheor.* **2007**, *55*, 133–179. [CrossRef] [PubMed]
90. Russell, M.J.; Barge, L.M.; Bhartia, R.; Bocanegra, D.; Bracher, P.J.; Branscomb, E.; Kidd, R.; McGlynn, S.; Meier, D.H.; Nitschke, W.; et al. The drive to life on wet and icy worlds. *Astrobiology* **2014**, *14*, 308–343. [CrossRef]
91. Russell, M.J.; Hall, A.J. The emergence of life from iron monosulphide bubbles at a submarine hydrothermal redox and pH front. *J. Geol. Soc.* **1997**, *154*, 377–402. [CrossRef] [PubMed]
92. Maslennikov, V.V.; Maslennikova, S.P.; Large, R.R.; Danyushevsky, L.V.; Herrington, R.J.; Ayupova, N.R.; Zaykov, V.V.; Lein, A.Y.; Tseluyko, A.S.; Melekestseva, I.Y.; et al. Chimneys in Paleozoic massive sulfide mounds of the Urals VMS deposits: Mineral and trace element comparison with modern black, grey, white and clear smokers. *Ore Geol. Rev.* **2017**, *85*, 64–106. [CrossRef]
93. Deamer, D.W. *Assembling Life: How Can Life Begin on Earth and Other Habitable Planets?* Oxford University Press: Oxford, UK, 2018.
94. Chatterjee, S. A symbiotic view of the origin of life at hydrothermal impact crater-lakes. *Phys. Chem. Chem. Phys.* **2016**, *18*, 20033–20046. [CrossRef] [PubMed]
95. Kring, D.A.; Whitehouse, M.J.; Schmieder, M. Microbial sulfur isotope fractionation in the Chicxulub hydrothermal system. *Astrobiology* **2020**, *21*, 103–114. [CrossRef]

96. Osinski, G.R.; Cockell, C.S.; Pontefract, A.; Sapers, H.M. The role of meteorite impacts in the origin of life. *Astrobiology* **2020**, *20*, 1121–1149. [CrossRef] [PubMed]
97. Benner, S.A.; Bell, E.A.; Biondi, E.; Brasser, R.; Carell, T.; Kim, H.J.; Trail, D. When did life likely emerge on Earth in an RNA-first process? *Chem. Syst. Chem.* **2019**, *2*, e1900035. [CrossRef]
98. Lahav, N.; White, D.; Chang, S. Peptide formation in the prebiotic era: Thermal condensation of glycine in fluctuating clay environments. *Science* **1978**, *201*, 67–69. [CrossRef]
99. Higgs, P.S. The effect of limited diffusion and wet-dry cycling on reversible polymerization reactions: Implications for prebiotic synthesis of nucleic acids. *Life* **2016**, *6*, 24. [CrossRef]
100. Marais, D.J.D.; Walter, M.R. Terrestrial hot spring systems: Introduction. *Astrobiology* **2019**, *19*, 1419–1432. [CrossRef] [PubMed]
101. Joshi, M.P.; Sawant, A.A.; Rajamani, S. Spontaneous emergence of membrane-forming protoamphiphiles from a lipid-amino acid mixture under wet-dry cycles. *Chem. Sci.* **2021**, *12*, 2970–2978. [CrossRef]
102. Qiao, H.; Hu, N.; Bai, J.; Ren, L.; Liu, Q.; Fang, L.; Wang, Z. Encapsulation of nucleic acids into giant unilamellar vesicles by freeze-thaw: A way protocells may form. *Orig. Life Evol. Biosph.* **2017**, *47*, 499–510. [CrossRef]
103. Vlassov, A.V.; Johnston, B.H.; Landweber, L.F.; Kazakov, S.A. Ligation activity of fragmented ribozymes in frozen solution: Implication for the RNAWorld. *Nucleic Acids Res.* **2004**, *32*, 2966–2974. [CrossRef]
104. Mutschler, H.; Wochner, A.; Holliger, P. Freeze-thaw cycles as drivers of complex ribozyme assembly. *Nat. Chem.* **2015**, *7*, 502–508. [CrossRef] [PubMed]
105. Feller, G. Cryosphere and psychrophiles: Insights into a cold origin of life? *Life* **2017**, *7*, 25. [CrossRef]
106. Putnam, A.R.; Palucis, M.C. The hydrogeomorphic history of Garu crater: Implications and constraints on the timing of large late-stage lakes in the Gale crater region. *J. Geophys. Res. Planets* **2020**, e2020JE006688. [CrossRef]
107. Salese, F.; Kleinhans, M.G.; Mangold, N.; Ansan, V.; McMahon, W.; Haas, T.D.; Dromart, G. Estimated minimum life span of the Jezero fluvial delta (Mars). *Astrobiology* **2020**, *20*, 977–993. [CrossRef]
108. Schwartz, A.W. Evaluating the plausibility of prebiotic multistage syntheses. *Astrobiology* **2013**, *13*, 784–789. [CrossRef]
109. Hunten, D.M. Possible oxidant sources in the atmosphere and surface of Mars. *J. Mol. Evol.* **1979**, *14*, 71–78. [CrossRef]
110. Ranjan, S.; Todd, Z.R.; Sutherland, J.D.; Sasselov, D.D. Sulfidic anion concentrations on early Earth for surficial origins-of-life chemistry. *Astrobiology* **2018**, *18*, 1023–1040. [CrossRef]
111. Achille, G.D.; Hynek, B.M. Ancient ocean on Mars supported by global distribution of deltas and valleys. *Nat. Geosci.* **2010**, *3*, 459–463. [CrossRef]
112. Lasne, J.; Noblet, A.; Szopa, C.; Navarro-González, R.; Cabane, M.; Poch, O.; Stalport, F.; François, P.; Atreya, S.K.; Coll, P. Oxidants at the surface of Mars: A review in light of recent exploration results. *Astrobiology* **2016**, *16*, 977–996. [CrossRef] [PubMed]
113. Ramirez, R.M.; Kopparapu, R.; Zugger, M.E.; Robinson, T.D.; Freedman, R.; Kasting, J.F. Warming early Mars with CO_2 and H_2. *Nat. Geosci.* **2014**, *7*, 59–63. [CrossRef]
114. Hayworth, B.P.; Kopparapu, R.K.; Haqq-Misra, J.; Batalha, N.E.; Payne, R.C.; Foley, B.J.; Ikwut-Ukwa, M.; Kasting, J.F. Warming early Mars with climate cycling: The effect of CO_2-H_2 collision-induced absorption. *Icarus* **2020**, *27*, 113770. [CrossRef]
115. Wordsworth, R.; Pierrehumbert, R. Hydrogen-nitrogen greenhouse warming in Earth's early atmosphere. *Science* **2013**, *339*, 64–67. [CrossRef]
116. Klein, F.; Tarnas, J.D.; Bach, W. Abiotic sources of molecular hydrogen on Earth. *Elem. Int. Mag. Mineral. Geoch. Petrol.* **2020**, *16*, 19–24. [CrossRef]
117. Jakosky, B.M. The CO_2 inventory on Mars. *Planet. Space Sci.* **2019**, *175*, 52–59. [CrossRef]
118. Kite, E.S.; Williams, J.P.; Lucas, A.; Aharonson, O. Low palaeopressure of the Martian atmosphere estimated from the size distribution of ancient craters. *Nat. Geosci.* **2014**, *7*, 335–339. [CrossRef]
119. Warren, A.O.; Kite, E.S.; Williams, J.P.; Horgan, B. Through the thick and thin: New constraints on Mars paleopressure history 3.8–4 Ga from small exhumed craters. *J. Geophys. Res. Planets.* **2019**, *124*, 2793–2818. [CrossRef]
120. Batalha, N.E.; Kopparapu, R.K.; Haqq-Misra, J.; Kasting, J.F. Climate cycling on early Mars caused by the carbonate–silicate cycle. *Earth Planet. Sci. Lett.* **2016**, *455*, 7–13. [CrossRef]
121. Falkowski, P.; Scholes, R.J.; Boyle, E.E.; Canadell, J.; Canfield, D.; Elser, J.; Gruber, N.; Hibbard, K.; Högberg, P.; Linder, S.; et al. The global carbon cycle: A test of our knowledge of earth as a system. *Science* **2000**, *290*, 291–296. [CrossRef] [PubMed]
122. Schmitt-Kopplin, P.; Harir, M.; Kanawati, B.; Gougeon, R.; Moritz, F.; Hertkorn, N.; Clary, S.; Gebefügi, I.; Gabelica, Z. Analysis of Extraterrestrial Organic Matter in Murchison Meteorite: A Progress Report. In *Astrobiology: An Evolutionary Approach*; Kolb, V.M., Ed.; CRC Press/Taylor &Francis: Boca Raton, FL, USA, 2014; pp. 63–82.
123. Yen, A.S.; Gellert, R.; Schröder, C.; Morris, R.V.; Bell, J.F.; Knudson, A.T.; Clark, B.C.; Ming, D.W.; Crisp, J.A.; Arvidson, R.E.; et al. An integrated view of the chemistry and mineralogy of Martian soils. *Nature* **2005**, *436*, 49–54. [CrossRef]
124. Humayun, M.; Nemchin, A.; Zanda, B.; Hewins, R.H.; Grange, M.; Kennedy, A.; Lorand, J.P.; Göpel, C.; Fieni, C.; Pont, S.; et al. Origin and age of the earliest Martian crust from meteorite NWA 7533. *Nature* **2013**, *503*, 513–516. [CrossRef]
125. Lodders, K. A survey of shergottite, nakhlite and chassigny meteorites whole-rock compositions. *Meteorit. Planet. Sci.* **1998**, *33*, A183–A190. [CrossRef]
126. Airapetian, V.S.; Glocer, A.; Gronoff, G.; Hebrard, E.; Danchi, W. Prebiotic chemistry and atmospheric warming of early Earth by an active young Sun. *Nat. Geosci.* **2016**, *9*, 452–455. [CrossRef]

127. Parkos, D.; Pikus, A.; Alexeenko, A.; Melosh, H.J. HCN production via impact ejecta reentry during the late heavy bombardment. *J. Geophys. Res. Planets.* **2018**, *123*, 892–909. [CrossRef]
128. Matthews, C.N.; Minard, R.D. Hydrogen cyanide polymers, comets and the origin of life. *Faraday Discuss.* **2006**, *133*, 393–401. [CrossRef]
129. Todd, Z.R.; Öberg, K.I. Cometary delivery of hydrogen cyanide to the early Earth. *Astrobiology* **2020**, *20*, 1109–1120. [CrossRef]
130. Milo, R.; Phillips, R. *Cell Biology by the Numbers*; Garland Science: New York, NY, USA, 2015.
131. Ramirez, R.M. A warmer and wetter solution for early Mars and the challenges with transient warming. *Icarus* **2017**, *297*, 71–82. [CrossRef]
132. Stern, J.C.; Sutter, B.; Freissinet, C.; Navarro-González, R.; McKay, C.P.; Archer, P.D.; Buch, A.; Brunner, A.E.; Coll, P.; Eigenbrode, J.L.; et al. Evidence for indigenous nitrogen in sedimentary and aeolian deposits from the Curiosity rover investigations at Gale crater, Mars. *Proc. Natl. Acad. Sci. USA* **2015**, *112*, 4245–4250. [CrossRef]
133. Stern, J.C.; Sutter, B.; Jackson, W.A.; Navarro-González, R.; McKay, C.P.; Ming, D.W.; Archer, P.D.; Mahaffy, P.R. The nitrate/(per)chlorate relationship on Mars. *Geophys. Res. Lett.* **2017**, *44*, 2643–2651. [CrossRef]
134. Ranjan, S.; Todd, Z.R.; Rimmer, P.B.; Sasselov, D.D.; Babbin, A.R. Nitrogen oxide concentrations in natural waters on early Earth. *Geochem. Geophys. Geosyst.* **2019**, *20*, 2021–2039. [CrossRef]
135. Kounaves, S.P.; Hecht, M.H.; Kapit, J.; Quinn, R.C.; Catling, D.C.; Clark, B.C.; Ming, D.W.; Gospodinova, K.; Hredzak, P.; McElhoney, K.; et al. Soluble sulfate in the Martian soil at the Phoenix landing site. *Geophys. Res. Lett.* **2010**, *37*. [CrossRef]
136. Stroble, S.T.; McElhoney, K.M.; Kounaves, S.P. Comparison of the Phoenix Mars lander WCL soil analyses with Antarctic Dry Valley soils, Mars meteorite EETA79001 sawdust, and a Mars simulant. *Icarus* **2013**, *225*, 933–939. [CrossRef]
137. Steele, A.; Benning, L.G.; Wirth, R.; Siljeström, S.; Fries, M.D.; Hauri, E.; Conrad, P.G.; Rogers, K.; Eigenbrode, J.; Schreiber, A.; et al. Organic synthesis on Mars by electrochemical reduction of CO. *Sci. Adv.* **2018**, *4*, eaat5118. [CrossRef]
138. Hu, L.; Khaniya, A.; Wang, J.; Chen, G.; Kaden, W.E.; Feng, X. Ambient electrochemical ammonia synthesis with high selectivity on Fe/Fe oxide catalyst. *ACS Catal.* **2018**, *8*, 9312–9319. [CrossRef]
139. Du, H.; Yang, C.; Pu, W.; Zeng, L.; Gong, J. Enhanced electrochemical reduction of N_2 to ammonia over pyrite FeS_2 with excellent selectivity. *ACS Sustain. Chem. Eng.* **2020**, *8*, 10572–10580. [CrossRef]
140. Fray, N.; Bardyn, A.; Cottin, H.; Baklouti, D.; Briois, C.; Engrand, C.; Fischer, H.; Hornung, K.; Isnard, R.; Langevin, Y.; et al. Nitrogen-to-carbon atomic ratio measured by COSIMA in the particles of comet 67P/Churyumov–Gerasimenko. *Mon. Not. R. Astron. Soc.* **2017**, *469*, S506–S516. [CrossRef]
141. Clifford, S.M.; Parker, T.J. The evolution of the Martian hydrosphere: Implications for the fate of a primordial ocean and the current state of the northern plains. *Icarus* **2001**, *154*, 40–79. [CrossRef]
142. Manning, C.V.; Zahnle, K.J.; McKay, C.P. Impact processing of nitrogen on early Mars. *Icarus* **2009**, *199*, 273–285. [CrossRef]
143. Fox, J.L. The production and escape of nitrogen atoms on Mars. *J. Geophys. Res. Planets.* **1993**, *8*, 3297–3310. [CrossRef]
144. McElroy, M.B.; Yung, Y.L.; Nier, A.O. Isotopic composition of nitrogen: Implications for the past history of Mars' atmosphere. *Science* **1976**, *194*, 70–72. [CrossRef]
145. Robinson, T.D.; Reinhard, C.T. Earth as an Exoplanet. In *Planetary Astrobiology*; University of Arizona: Tucson, AZ, USA, 2020; pp. 379–416.
146. Wordsworth, R.; Knoll, A.H.; Hurowitz, J.; Baum, M.; Ehlmann, B.L.; Head, J.W.; Steakley, K. A coupled model of episodic warming, oxidation and geochemical transitions on early Mars. *Nat. Geosci.* **2021**, *14*, 127–132. [CrossRef]
147. Dreibus, G.; Jagoutz, E.; Spettel, B.; Wanke, H. Phosphate-mobilization on Mars? Implication from leach experiments on SNC's. *LPI* **1996**, *27*, 323.
148. Adcock, C.T.; Hausrath, E.M.; Forster, P.M. Readily available phosphate from minerals in early aqueous environments on Mars. *Nat. Geosci.* **2013**, *6*, 824–827. [CrossRef]
149. Clark, B.C., III; Arvidson, R.E.; Gellert, R.; Morris, R.V.; Ming, D.W.; Richter, L.; Ruff, S.W.; Michalski, J.R.; Farrand, W.H.; Herkenhoff, K.E.; et al. Evidence for montmorillonite or its compositional equivalent in Columbia Hills, Mars. *J. Geophys. Res. Planets* **2007**, *112*. [CrossRef]
150. Ming, D.W.; Mittlefehldt, D.W.; Morris, R.V.; Golden, D.C.; Gellert, R.; Yen, A.; Clark, B.C.; Squyres, S.W.; Farrand, W.H.; Ruff, S.W.; et al. Geochemical and mineralogical indicators for aqueous processes in the Columbia Hills of Gusev crater, Mars. *J. Geophys. Res. Planets* **2006**, *111*, E2. [CrossRef]
151. McSween, H.Y.; Ruff, S.W.; Morris, R.V.; Bell, J.F., III; Herkenhoff, K.; Gellert, R.; Stockstill, K.R.; Tornabene, L.L.; Squyres, S.W.; Crisp, J.A.; et al. Alkaline volcanic rocks from the Columbia Hills, Gusev crater, Mars. *J. Geophys. Res. Planets* **2006**, *111*, E9. [CrossRef]
152. McSween, H.Y.; Arvidson, R.E.; Bell, J.F.; Blaney, D.; Cabrol, N.A.; Christensen, P.R.; Clark, B.C.; Crisp, J.A.; Crumpler, L.S.; Marais, D.J.D.; et al. Basaltic rocks analyzed by the Spirit rover in Gusev Crater. *Science* **2004**, *305*, 842–845. [CrossRef] [PubMed]
153. Berger, J.A.; Van Bommel, S.J.V.; Clark, B.C.; Gellert, R.; House, C.H.; King, P.L.; McCraig, M.A.; Ming, D.W.; O'Connell-Cooper, C.D.; Schmidt, M.E.; et al. Manganese and Phosphorus-Rich Nodules in Gale Crater, Mars: Results from the Groken Drill Site. In Proceedings of the 52nd Lunar and Planetary Science Conference, The Woodlands, TX, USA, 15–19 March 2021.
154. Santos, A.R.; Agee, C.B.; McCubbin, F.M.; Shearer, C.K.; Burger, P.V.; Tartese, R.; Anand, M. Petrology of igneous clasts in Northwest Africa 7034: Implications for the petrologic diversity of the Martian crust. *Geochim. Cosmochim. Acta* **2015**, *157*, 56–85. [CrossRef]

155. Liu, Y.; Fischer, W.W.; Ma, C.; Beckett, J.R.; Tschauner, O.; Guan, Y.; Lingappa, U.F.; Webb, S.M.; Prakapenka, V.B.; Lanza, N.L.; et al. Manganese oxides in Martian meteorites Northwest Africa (NWA) 7034 and 7533. *Icarus* **2021**, *364*, 114471. [CrossRef]
156. Wellington, D.F.; Meslin, P.Y.; Van Beek, J.; Johnson, J.R.; Wiens, R.C.; Calef, F.J.; Bell, J.F. Iron Meteorite Finds Across Lower Mt. Sharp, Gale Crater, Mars: Clustering and Implications. In Proceedings of the 50th Lunar and Planetary Society Conference, The Woodlands, TX, USA, 18–22 March 2019.
157. Fairen, A.G.; Dohm, J.M.; Baker, V.R.; Thompson, S.D.; Mahaney, W.C.; Herkenhoff, K.E.; Rodriguez, J.A.; Davila, A.F.; Schulze-makuch, D.; Maarry, M.R.E.; et al. Meteorites at Meridiani Planum provide evidence for significant amounts of surface and near-surface water on early Mars. *Meteorit. Planet. Sci.* **2011**, *46*, 1832–1841. [CrossRef]
158. Meslin, P.Y.; Wellington, D.; Wiens, R.C.; Johnson, J.R.; Van Beek, J.; Gasnault, O.; Sautter, V.; Maroger, I.; Lasue, J.; Beck, P.; et al. Diversity and Areal Density of Iron-Nickel Meteorites Analyzed by Chemcam in Gale Crater. In Proceedings of the Lunar and Planetary Science Conference, The Woodlands, TX, USA, 18–22 March 2019; No. 2132; p. 3179.
159. Lang, C.; Lago, J.; Pasek, M. Phosphorylation on the Early Earth: The Role of Phosphorus in Biochemistry and its Bioavailability. In *Handbook of Astrobiology*; Kolb, V.M., Ed.; CRC Press: Boca Raton, FL, USA, 2019; pp. 361–370.
160. Gibard, C.; Gorrell, I.B.; Jiménez, E.I.; Kee, T.P.; Pasek, M.A.; Krishnamurthy, R. Geochemical sources and availability of amidophosphates on the early Earth. *Angew. Chem.* **2019**, *131*, 8235–8239. [CrossRef]
161. Burcar, B.; Castañeda, A.; Lago, J.; Daniel, M.; Pasek, M.A.; Hud, N.V.; Orlando, T.M.; Menor-Salván, C. A stark contrast to modern earth: Phosphate mineral transformation and nucleoside phosphorylation in an iron-and cyanide-rich early earth scenario. *Angew. Chem. Int. Ed.* **2019**, *58*, 16981–16987. [CrossRef]
162. Clark, B.C.; Baird, A.K. Is the Martian lithosphere sulfur rich? *J. Geophys. Res. Solid Earth* **1979**, *84*, 8395–8403. [CrossRef]
163. Franz, H.B.; King, P.L.; Gaillard, F. Sulfur on Mars from the Atmosphere to the Core. In *Volatiles in the Martian Crust*; Elsevier: Amsterdam, The Netherlands, 2019; pp. 119–1483. [CrossRef]
164. Sutherland, J.D. The origin of life—Out of the blue. *Angew. Chem. Int. Ed.* **2016**, *55*, 104–121. [CrossRef]
165. Bonfio, C.; Valer, L.; Scintilla, S.; Shah, S.; Evans, D.J.; Jin, L.; Szostak, J.W.; Sasselov, D.D.; Sutherland, J.D.; Mansy, S.S. UV-light-driven prebiotic synthesis of iron–sulfur clusters. *Nat. Chem.* **2017**, *9*, 1229. [CrossRef] [PubMed]
166. Xu, J.; Tsanakopoulou, M.; Magnani, C.J.; Szabla, R.; Šponer, J.E.; Šponer, J.; Góra, R.W.; Sutherland, J.D. A prebiotically plausible synthesis of pyrimidine β-ribonucleosides and their phosphate derivatives involving photoanomerization. *Nat. Chem.* **2017**, *9*, 303–309. [CrossRef]
167. Styrt, M.M.; Brackmann, A.J.; Holland, H.D.; Clark, B.C.; Pisutha-Arnond, V.; Eldridge, C.S.; Ohmoto, H. The mineralogy and the isotopic composition of sulfur in hydrothermal sulfide/sulfate deposits on the East Pacific Rise, 21 N latitude. *Earth Planet. Sci. Lett.* **1981**, *53*, 382–390. [CrossRef]
168. Gale, A.; Dalton, C.A.; Langmuir, C.H.; Su, Y.; Schilling, J.G. The mean composition of ocean ridge basalts. *Geochem. Geophys. Geosyst.* **2013**, *14*, 489–518. [CrossRef]
169. Clark, B.C. Geochemical components in Martian soil. *Geochim. Cosmochim. Acta* **1993**, *57*, 4575–4581. [CrossRef]
170. Yen, A.S.; Morris, R.V.; Clark, B.C.; Gellert, R.; Knudson, A.T.; Squyres, S.; Mittlefehldt, D.W.; Ming, D.W.; Arvidson, R.; McCoy, T.; et al. Hydrothermal processes at Gusev Crater: An evaluation of Paso Robles class soils. *J. Geophys. Res. Planets* **2008**, *113*, E6. [CrossRef]
171. Clark, B.C.; Morris, R.V.; McLennan, S.M.; Gellert, R.; Jolliff, B.; Knoll, A.H.; Squyres, S.W.; Lowenstein, T.K.; Ming, D.W.; Tosca, N.J.; et al. Chemistry and mineralogy of outcrops at Meridiani Planum. *Earth Planet. Sci. Lett.* **2005**, *240*, 73–94. [CrossRef]
172. Wang, A.; Haskin, L.A.; Squyres, S.W.; Jolliff, B.L.; Crumpler, L.; Gellert, R.; Schröder, C.; Herkenhoff, K.; Hurowitz, J.; Tosca, N.J.; et al. Sulfate deposition in subsurface regolith in Gusev crater, Mars. *J. Geophys. Res. Planets* **2006**, *111*, E2. [CrossRef]
173. Rapin, W.; Ehlmann, B.L.; Dromart, G.; Schieber, J.; Thomas, N.H.; Fischer, W.W.; Fox, V.K.; Stein, N.T.; Nachon, M.; Clark, B.C.; et al. An interval of high salinity in ancient Gale crater lake on Mars. *Nat. Geosci.* **2019**, *12*, 889–895. [CrossRef]
174. Yen, A.S.; Gellert, R.; Achilles, C.N.; Berger, J.A.; Blake, D.F.; Clark, B.C.; McAdam, C.; Wing, D.M.; Morris, R.V.; Morrison, S.M.; et al. Origin and Speciation of Sulfur Compounds in the Murray Formation, Gale Crater, Mars. In Proceedings of the Lunar and Planetary Science Conference, The Woodlands, TX, USA, 18–22 March 2019; Volume 50, p. 2083.
175. Crisler, J.D.; Newville, T.M.; Chen, F.; Clark, B.C.; Schneegurt, M.A. Bacterial growth at the high concentrations of magnesium sulfate found in Martian soils. *Astrobiology* **2012**, *12*, 98–106. [CrossRef]
176. Morris, R.V.; Schröder, C.; Klingelhöfer, G.; Agresti, D.G. Mössbauer Spectroscopy at Gusev Crater and Meridiani Planum: Iron Mineralogy, Oxidation State, and Alteration on Mars. In *Remote Compositional Analysis: Techniques for Understanding Spectroscopy, Mineralogy, and Geochemistry of Planetary Surfaces*; Bishop, J.L., Bell, J.F., Moersch, J.E., Eds.; Cambridge Planetary Science, University Press: Cambridge, UK, 2019; pp. 538–554. [CrossRef]
177. Squyres, S.W.; Arvidson, R.E.; Bell, J.F.; Calef, F.; Clark, B.C.; Cohen, B.A.; Crumpler, L.A.; Souza, P.A.D.; Farrand, W.H.; Gellert, R.; et al. Ancient impact and aqueous processes at Endeavour Crater, Mars. *Science* **2012**, *336*, 570–576. [CrossRef]
178. Clark, B.C.; Van Hart, D.C. The salts of Mars. *Icarus* **1981**, *45*, 370–378. [CrossRef]
179. Banin, A.; Han, F.X.; Kan, I.; Cicelsky, A. Acidic volatiles and the Mars soil. *J. Geophys. Res. Planets* **1997**, *102*, 13341–13356. [CrossRef]
180. Settle, M. Formation and deposition of volcanic sulfate aerosols on Mars. *J. Geophys. Res. Solid Earth* **1979**, *84*, 8343–8354. [CrossRef]
181. Quinn, R.C.; Chittenden, J.D.; Kounaves, S.P.; Hecht, M.H. The oxidation-reduction potential of aqueous soil solutions at the Mars Phoenix landing site. *Geophys. Res. Lett.* **2011**, *38*. [CrossRef]

182. Bandfield, J.L.; Glotch, T.D.; Christensen, P.R. Spectroscopic identification of carbonate minerals in the Martian dust. *Science* **2003**, *301*, 1084–1087. [CrossRef] [PubMed]
183. Wong, G.M.; Lewis, J.M.; Knudson, C.A.; Millan, M.; McAdam, A.C.; Eigenbrode, J.L.; Andrejkovičová, S.; Gómez, F.; Navarro-González, R.; House, C.H. Detection of reduced sulfur on Vera Rubin ridge by quadratic discriminant analysis of volatiles observed during evolved gas analysis. *J. Geophys. Res. Planets* **2020**, *125*, e2019JE006304. [CrossRef]
184. Biemann, K.; Oro, J.I.; Toulmin, P., III; Orgel, L.E.; Nier, A.O.; Anderson, D.M.; Simmonds, P.G.; Flory, D.; Diaz, A.V.; Rushneck, D.R.; et al. The search for organic substances and inorganic volatile compounds in the surface of Mars. *J. Geophys. Res.* **1977**, *82*, 4641–4658. [CrossRef]
185. Hubbard, J.S.; Hardy, J.P.; Horowitz, N.H. Photocatalytic production of organic compounds from CO and H_2O in a simulated martian atmosphere. *Proc. Natl. Acad. Sci. USA* **1971**, *68*, 574–578. [CrossRef]
186. Szopa, C.; Freissinet, C.; Glavin, D.P.; Millan, M.; Buch, A.; Franz, H.B.; Summons, R.E.; Sumner, D.Y.; Sutter, B.; Eigenbrode, J.L.; et al. First detections of dichlorobenzene isomers and trichloromethylpropane from organic matter indigenous to mars mudstone in gale crater, Mars: Results from the sample analysis at Mars instrument onboard the curiosity rover. *Astrobiology* **2020**, *20*, 292–306. [CrossRef]
187. Eigenbrode, J.L.; Summons, R.E.; Steele, A.; Freissinet, C.; Millan, M.; Navarro-González, R.; Sutter, B.; McAdam, A.C.; Franz, H.B.; Glavin, D.P.; et al. Organic matter preserved in 3-billion-year-old mudstones at Gale crater, Mars. *Science* **2018**, *360*, 1096–1101. [CrossRef]
188. Steele, A.; McCubbin, F.M.; Fries, M.; Kater, L.; Boctor, N.Z.; Fogel, M.L.; Conrad, P.G.; Glamoclija, M.; Spencer, M.; Morrow, A.L.; et al. A reduced organic carbon component in martian basalts. *Science* **2012**, *337*, 212–215. [CrossRef]
189. Brown, P.G.; Hildebrand, A.R.; Zolensky, M.E.; Grady, M.; Clayton, R.N.; Mayeda, T.K.; Tagliaferri, E.; Spalding, R.; MacRae, N.D.; Hoffman, E.L.; et al. The fall, recovery, orbit, and composition of the Tagish Lake meteorite: A new type of carbonaceous chondrite. *Science* **2000**, *290*, 320–325. [CrossRef]
190. Mustard, J.F.; Murchie, S.L.; Pelkey, S.M.; Ehlmann, B.L.; Milliken, R.E.; Grant, J.A.; Bibring, J.P.; Poulet, F.; Bishop, J.; Dobrea, E.N.; et al. Hydrated silicate minerals on Mars observed by the Mars Reconnaissance Orbiter CRISM instrument. *Nature* **2008**, *454*, 305–309. [CrossRef]
191. Clark, B.C.; Morris, R.V.; Herkenhoff, K.E.; Farrand, W.H.; Gellert, R.; Jolliff, B.L.; Arvidson., R.E.; Squyres, S.W.; Mittlefehldt, D.W.; Ming, D.W.; et al. Esperance: Multiple episodes of aqueous alteration involving fracture fills and coatings at Matijevic Hill, Mars. *Am. Mineral.* **2016**, *101*, 1515–1526. [CrossRef]
192. Ferris, J.P. Catalyzed RNA synthesis for the RNA world. In *The Molecular Origins of Life: Assembling Pieces of the Puzzle*; Brack, A., Ed.; Cambridge University Press: Cambridge, UK, 2000; pp. 255–268.
193. Ming, D.W.; Morris, R.V.; Clark, B.C. Aqueous Alteration on Mars. In *The Martian Surface*; Bell, J., Ed.; Cambridge University Press: Cambridge, UK, 2008; Chapter 23; pp. 519–540.
194. Yen, A.S.; Morris, R.V.; Ming, D.W.; Schwenzer, S.P.; Sutter, B.; Vaniman, D.T.; Treiman, A.H.; Gellert, R.; Achilles, C.N.; Berger, J.A.; et al. Formation of tridymite and evidence for a hydrothermal history at gale crater, Mars. *J. Geophys. Res. Planets* **2021**. [CrossRef]
195. Goetz, W.; Wiens, R.C.; Dehouck, E.; Gasnault, O.; Lasue, J.; Payre, V.; Frydenvang, J.; Clark, B.; Clegg, S.M.; Forni, O.; et al. Tracking of Copper by the ChemCam Instrument in Gale Crater, Mars: Elevated Abundances in Glen Torridon. In Proceedings of the Lunar and Planetary Science Conference, The Woodlands, TX, USA, 18–22 March 2020; Volume 2326, p. 2974.
196. Payré, V.; Fabre, C.; Sautter, V.; Cousin, A.; Mangold, N.; Deit, L.L.; Forni, O.; Goetz, W.; Wiens, R.C.; Gasnault, O.; et al. Copper enrichments in the Kimberley formation in Gale crater, Mars: Evidence for a Cu deposit at the source. *Icarus* **2019**, *321*, 736–751. [CrossRef]
197. VanBommel, S.J.; Gellert, R.; Berger, J.A.; Yen, A.S.; Boyd, N.I. Mars science laboratory alpha particle X-ray spectrometer trace elements: Situational sensitivity to Co, Ni, Cu, Zn, Ga, Ge, and Br. *Acta Astronaut.* **2019**, *165*, 32–42. [CrossRef]
198. Gasda, P.J.; Haldeman, E.B.; Wiens, R.C.; Rapin, W.; Bristow, T.F.; Bridges, J.C.; Schwenzer, S.P.; Clark, B.; Herkenhoff, K.; Frydenvang, J.; et al. In situ detection of boron by ChemCam on Mars. *Geophys. Res. Lett.* **2017**, *44*, 8739–8748. [CrossRef]
199. Lanza, N.L.; Fischer, W.W.; Wiens, R.C.; Grotzinger, J.; Ollila, A.M.; Cousin, A.; Anderson, R.B.; Clark, B.C.; Gellert, R.; Mangold, N.; et al. High manganese concentrations in rocks at Gale crater, Mars. *Geophys. Res. Lett.* **2014**, *41*, 5755–5763. [CrossRef]
200. Lanza, N.L.; Wiens, R.C.; Arvidson, R.E.; Clark, B.C.; Fischer, W.W.; Gellert, R.; Grotzinger, J.P.; Hurowitz, J.A.; McLennan, S.M.; Morris, R.V.; et al. Oxidation of manganese in an ancient aquifer, Kimberley formation, Gale crater, Mars. *Geophys. Res. Lett.* **2016**, *43*, 7398–73407. [CrossRef]
201. Arvidson, R.E.; Squyres, S.W.; Morris, R.V.; Knoll, A.H.; Gellert, R.; Clark, B.C.; Catalano, J.G.; Jolliff, B.L.; McLennan, S.M.; Herkenhoff, K.E.; et al. High concentrations of manganese and sulfur in deposits on Murray Ridge, Endeavour Crater, Mars. *Am. Mineral.* **2016**, *101*, 1389–1405. [CrossRef]
202. Noda, N.; Imamura, S.; Sekine, Y.; Kurisu, M.; Fukushi, K.; Terada, N.; Uesugi, S.; Numako, C.; Takahashi, Y.; Hartmann, J. Highly oxidizing aqueous environments on early Mars inferred from scavenging pattern of trace metals on manganese oxides. *J. Geophys. Res. Planets* **2019**, *124*, 1282–1295. [CrossRef]
203. Berger, J.A.; King, P.L.; Gellert, R.; Clark, B.C.; O'Connell-Cooper, C.D.; Thompson, L.M.; Van Bommel, S.J.; Yen, A.S. Manganese enrichment pathways relevant to Gale crater, Mars: Evaporative concentration and chlorine-induced precipitation. In Proceedings of the Lunar and Planetary Science Conference, The Woodlands, TX, USA, 18–22 March 2019; Volume 50, p. 2487.

204. Salminen, R.; Batista, M.J.; Bidovec, M.; Demetriades, A.; Vivo, B.; Vos, W.D.; Duris, M.; Gilucis, A.; Gregorauskiene, V.; Halamić, J.; et al. *Geochemical Atlas of Europe*; Part 1, Background Information, Methodology and Maps; Geological Survey of Finland: Espoo, Finland, 2005.
205. Marshall, C.P.; Marshall, A.O.; Aitken, J.B.; Lai, B.; Vogt, S.; Breuer, P.; Steemans, P.; Lay, P.A. Imaging of vanadium in microfossils: A new potential biosignature. *Astrobiology* **2017**, *17*, 1069–1076. [CrossRef] [PubMed]
206. Carpentier, W.; Sandra, K.; Smet, I.; Brige, A.; Smet, L.D.; Van Beeumen, J. Microbial reduction and precipitation of vanadium by *Shewanella oneidensis*. *Appl. Environ. Microbiol.* **2003**, *69*, 3636–3639. [CrossRef] [PubMed]
207. Jochum, K.P.; Weis, U.; Schwager, B.; Stoll, B.; Wilson, S.A.; Haug, G.H.; Andreae, M.O.; Enzweiler, J. Reference values following ISO guidelines for frequently requested rock reference materials. *Geostand. Geoanalytical Res.* **2015**, *40*, 333–350. [CrossRef]
208. Campbell, T.D.; Febrian, R.; McCarthy, J.T.; Kleinschmidt, H.E.; Forsythe, J.G.; Bracher, P.J. Prebiotic condensation through wet–dry cycling regulated by deliquescence. *Nat. Commun.* **2019**, *10*, 1–7. [CrossRef]
209. Morris, R.V.; Rampe, E.B.; Vaniman, D.T.; Christoffersen, R.; Yen, A.S.; Morrison, S.M.; Ming, D.W.; Achilles, C.N.; Fraeman, A.A.; Le, L.; et al. Hydrothermal precipitation of sanidine (adularia) having full Al, Si structural disorder and specular hematite at Maunakea volcano (Hawai'i) and at Gale Crater (Mars). *J. Geophys. Res. Planets* **2020**, *125*, e2019JE006324. [CrossRef]
210. Cousin, M.; Desjardins, E.; Dehouck, O.; Forni, G.; David, G.; Berger, G.; Caravaca, P.; Meslin, J.; Lasue, A.; Ollila, W.; et al. Wiens5, K-Rich Rubbly Bedrock at Glen Torridon, Gale Crater, Mars: Investigating the Possible Presence of Illite. In Proceedings of the 52nd Lunar and Planetary Science Conference 2021, The Woodlands, TX, USA, 15–19 March 2021.
211. Bowman, J.C.; Petrov, A.S.; Frenkel-Pinter, M.; Penev, P.I.; Williams, L.D. Root of the tree: The significance, evolution, and origins of the ribosome. *Chem. Rev.* **2020**, *120*, 4848–4878. [CrossRef]
212. Bray, M.S.; Lenz, T.K.; Haynes, J.W.; Bowman, J.C.; Petrov, A.S.; Reddi, A.R.; Hud, N.V.; Williams, L.D.; Glass, J.B. Multiple prebiotic metals mediate translation. *Proc. Natl. Acad. Sci. USA* **2018**, *115*, 12164–12169. [CrossRef]
213. Kounaves, S.P.; Oberlin, E.A. Volatiles Measured by the Phoenix Lander at the Northern Plains of Mars. In *Volatiles in the Martian Crust*; Elsevier: Amsterdam, The Netherlands, 2019; pp. 265–283. [CrossRef]
214. Bibring, J.P.; Langevin, Y.; Mustard, J.F.; Poulet, F.; Arvidson, R.; Gendrin, A.; Gondet, B.; Mangold, N.; Pinet, P.; Forget, F.; et al. Global mineralogical and aqueous Mars history derived from OMEGA/Mars Express data. *Science* **2006**, *312*, 400–404. [CrossRef]
215. Das, D.; Gasda, P.J.; Wiens, R.C.; Berlo, K.; Leveille, R.J.; Frydenvang, J.; Mangold, N.; Kronyak, R.E.; Schwenzer, S.P.; Forni, O.; et al. Boron and lithium in calcium sulfate veins: Tracking precipitation of diagenetic materials in Vera Rubin Ridge, Gale crater. *J. Geophys. Res. Planets* **2020**, *125*, e2019JE006301. [CrossRef]
216. Stephenson, J.D.; Hallis, L.J.; Nagashima, K.; Freeland, S.J. Boron enrichment in martian clay. *PLoS ONE* **2013**, *8*, e64624. [CrossRef]
217. Palmer, M.R.; Sturchio, N.C. The boron isotope systematics of the Yellowstone National Park (Wyoming) hydrothermal system: A reconnaissance. *Geochim. Cosmochim. Acta* **1990**, *54*, 2811–2815. [CrossRef]
218. Anbar, A.D. Elements and evolution. *Science* **2008**, *5*, 1481–1483. [CrossRef]
219. Chemtob, S.M.; Nickerson, R.D.; Morris, R.V.; Agresti, D.G.; Catalano, J.G. Oxidative alteration of ferrous smectites and implications for the redox evolution of early Mars. *J. Geophys. Res. Planets* **2017**, *122*, 2469–2488. [CrossRef] [PubMed]
220. Bullock, M.A.; Moore, J.M. Atmospheric conditions on early Mars and the missing layered carbonates. *Geophys. Res. Lett.* **2007**, *34*. [CrossRef]
221. Baird, A.K.; Clark, B.C. On the original igneous source of Martian fines. *Icarus* **1981**, *45*, 113–123. [CrossRef]
222. Zolotov, M.Y.; Mironenko, M.V. Timing of acid weathering on Mars: A kinetic-thermodynamic assessment. *J. Geophys. Res. Planets* **2007**, *112*, E7. [CrossRef]
223. Thomas, N.H.; Ehlmann, B.L.; Meslin, P.Y.; Rapin, W.; Anderson, D.E.; Rivera-Hernández, F.; Forni, O.; Schröder, S.; Cousin, A.; Mangold, N.; et al. Mars science laboratory observations of chloride salts in Gale crater, Mars. *Geophys. Res. Lett.* **2019**, *46*, 10754–10763. [CrossRef]
224. Smith, R.J.; McLennan, S.M.; Achilles, C.N.; Dehouck, E.; Horgan, B.H.; Mangold, N.; Rampe, E.B.; Salvatore, M.; Siebach, K.L.; Sun, V. X-ray amorphous components in sedimentary rocks of Gale crater, Mars: Evidence for ancient formation and long-lived aqueous activity. *J. Geophys. Res. Planets* **2021**, *126*, e2020JE006782. [CrossRef]
225. Jackson, R.S.; Wiens, R.C.; Vaniman, D.T.; Beegle, L.; Gasnault, O.; Newsom, H.E.; Maurice, S.; Meslin, P.Y.; Clegg, S.; Cousin, A.; et al. ChemCam investigation of the John Klein and Cumberland drill holes and tailings, Gale crater, Mars. *Icarus* **2016**, *277*, 330–341. [CrossRef]
226. Yoshimura, Y. The Search for Life on Mars. In *Astrobiology*; Yamagishi, A., Kakegawa, T., Usui, T., Eds.; Springer: Singapore, 2019; pp. 367–381. [CrossRef]
227. Sholes, S.F.; Krissansen-Totton, J.; Catling, D.C. A maximum subsurface biomass on Mars from untapped free energy: CO and H_2 as potential antibiosignatures. *Astrobiology* **2019**, *19*, 655–668. [CrossRef]
228. Price, A.; Pearson, V.K.; Schwenzer, S.P.; Miot, J.; Olsson-Francis, K. Nitrate-dependent iron oxidation: A potential Mars metabolism. *Front. Microbiol.* **2018**, *9*, 513. [CrossRef] [PubMed]
229. Gonzalez-Toril, E.; Martínez-Frías, J.; Gomez, J.M.; Rull, F.; Amils, R. Iron meteorites can support the growth of acidophilic chemolithoautotrophic microorganisms. *Astrobiology* **2005**, *5*, 406–414. [CrossRef] [PubMed]
230. Soare, R.; Conway, S.; Williams, J.-P.; Oehler, D. (Eds.) *Mars Geological Enigmas: From the Late Noachian Epoch to the Present Day*, 1st ed.; Elsevier: Amsterdam, The Netherlands, 2021; ISBN 9780128202456.

231. Ernst, W.G. Earth's thermal evolution, mantle convection, and Hadean onset of plate tectonics. *J. Asian Earth Sci.* **2017**, *145*, 334–348. [CrossRef]
232. Faltys, J.P.; Wielicki, M.M. Inclusions in impact-formed zircon as a tracer of target rock lithology: Implications for Hadean continental crust composition and abundance. *Lithos* **2020**, *376*, 105761. [CrossRef]
233. Bada, J.L.; Korenaga, J. Exposed areas above sea level on Earth >3.5 Gyr ago: Implications for prebiotic and primitive biotic chemistry. *Life* **2018**, *8*, 55. [CrossRef]
234. Marshall, M. The water paradox and the origins of life. *Nature* **2020**, *588*, 210–213. [CrossRef]
235. Carr, M.H.; Head, J.W. Martian surface/near-surface water inventory: Sources, sinks, and changes with time. *Geophys. Res. Lett.* **2015**, *42*, 726–732. [CrossRef]
236. Palumbo, A.M.; Head, J.W.; Wordsworth, R.D. Late Noachian Icy Highlands climate model: Exploring the possibility of transient melting and fluvial/lacustrine activity through peak annual and seasonal temperatures. *Icarus* **2018**, *300*, 261–286. [CrossRef]
237. Morris, R.V.; Klingelhoefer, G.; Schröder, C.; Fleischer, I.; Ming, D.W.; Yen, A.S.; Gellert, R.; Arvidson, R.E.; Rodionov, D.S.; Crumpler, L.S.; et al. Iron mineralogy and aqueous alteration from Husband Hill through Home Plate at Gusev crater, Mars: Results from the Mössbauer instrument on the Spirit Mars Exploration Rover. *J. Geophys. Res. Planets* **2008**, *113*, E12. [CrossRef]
238. Tosca, N.J.; Ahmed, I.A.; Tutolo, B.M.; Ashpitel, A.; Hurowitz, J.A. Magnetite authigenesis and the warming of early Mars. *Nat. Geosci.* **2018**, *11*, 635–639. [CrossRef]
239. Robbins, S.J.; Hynek, B.M. The secondary crater population of Mars. *Earth Planet. Sci. Lett.* **2014**, *400*, 66–76. [CrossRef]
240. Segura, T.L.; Toon, O.B.; Colaprete, A. Modeling the environmental effects of moderate-sized impacts on Mars. *J. Geophys. Res. Planets* **2008**, *113*, E11. [CrossRef]
241. Turbet, M.; Forget, F.; Head, J.W.; Wordsworth, R. 3D modelling of the climatic impact of outflow channel formation events on early Mars. *Icarus* **2017**, *288*, 10–36. [CrossRef]
242. Haberle, R.M.; Zahnle, K.; Barlow, N.G.; Steakley, K.E. Impact degassing of H_2 on early Mars and its effect on the climate system. *Geophys. Res. Lett.* **2019**, *46*, 13355–13362. [CrossRef]
243. Treiman, A.H.; Bish, D.L.; Vaniman, D.T.; Chipera, S.J.; Blake, D.F.; Ming, D.W.; Morris, R.V.; Bristow, T.F.; Morrison, S.M.; Baker, M.B.; et al. Mineralogy, provenance, and diagenesis of a potassic basaltic sandstone on Mars: CheMin X-ray diffraction of the Windjana sample (Kimberley area, Gale Crater). *J. Geophys. Res. Planets* **2016**, *121*, 75–106. [CrossRef]
244. Fraeman, A.A.; Edgar, L.A.; Rampe, E.B.; Thompson, L.M.; Frydenvang, J.; Fedo, C.M.; Catalano, J.G.; Dietrich, W.E.; Gabriel, T.S.; Vasavada, A.R.; et al. Evidence for a diagenetic origin of Vera Rubin ridge, Gale crater, Mars: Summary and synthesis of Curiosity's exploration campaign. *J. Geophys. Res. Planets* **2020**, *125*, e2020JE006527. [CrossRef] [PubMed]
245. David, G.; Cousin, A.; Forni, O.; Meslin, P.Y.; Dehouck, E.; Mangold, N.; L'Haridon, J.; Rapin, W.; Gasnault, O.; Johnson, J.R.; et al. Analyses of high-iron sedimentary bedrock and diagenetic features observed with ChemCam at Vera Rubin ridge, Gale crater, Mars: Calibration and characterization. *J. Geophys. Res. Planets* **2020**, *125*, e2019JE006314. [CrossRef]
246. Achilles, C.N.; Rampe, E.B.; Downs, R.T.; Bristow, T.F.; Ming, D.W.; Morris, R.V.; Vaniman, D.T.; Blake, D.F.; Yen, A.S.; McAdam, A.C.; et al. Evidence for multiple diagenetic episodes in ancient fluvial-lacustrine sedimentary rocks in Gale crater, Mars. *J. Geophys. Res. Planets* **2020**, *125*, e2019JE006295. [CrossRef] [PubMed]
247. Craddock, R.A.; Lorenz, R.D. The changing nature of rainfall during the early history of Mars. *Icarus* **2017**, *293*, 172–179. [CrossRef]
248. Quay, G.S.D.; Goudge, T.A.; Fassett, C.I. Precipitation and aridity constraints from paleolakes on early Mars. *Geology* **2020**, *48*, 1189–1193. [CrossRef]
249. Palumbo, A.M.; Head, J.W.; Wilson, L. Rainfall on Noachian Mars: Nature, timing, and influence on geologic processes and climate history. *Icarus* **2020**, *11*, 113782. [CrossRef]
250. Duran, S.; Coulthard, T.J. The Kasei Valles, Mars: A unified record of episodic channel flows and ancient ocean levels. *Sci. Rep.* **2020**, *10*, 1–7. [CrossRef] [PubMed]
251. Soare, R.J.; Osinski, G.R.; Roehm, C.L. Thermokarst lakes and ponds on Mars in the very recent (late Amazonian) past. *Earth Planet. Sci. Lett.* **2008**, *272*, 382–393. [CrossRef]
252. Grant, J.A.; Wilson, S.A.; Mangold, N.; Calef III, F. and Grotzinger, J.P. The timing of alluvial activity in Gale crater, Mars. *Geophysical Res. Lett.* **2014**, *41*, 1142–1149. [CrossRef]
253. Comte, J.; Monier, A.; Crevecoeur, S.; Lovejoy, C.; Vincent, W.F. Microbial biogeography of permafrost thaw ponds across the changing northern landscape. *Ecography* **2016**, *39*, 609–618. [CrossRef]
254. Kadoya, S.; Krissansen-Totton, J.; Catling, D.C. Probable cold and alkaline surface environment of the Hadean Earth caused by impact ejecta weathering. *Geochem. Geophys. Syst.* **2019**, *21*, e2019GC008734. [CrossRef]
255. Buhler, P.B.; Piqueux, S. Mars Obliquity-Driven CO_2 Exchange between the Atmosphere, Regolith, and Polar Cap. *J. Geophy. Res. Planets* **2021**, e2020JE006759. [CrossRef]
256. Knoll, A.H.; Jolliff, B.L.; Farrand, W.H.; Bell, I.I.I.J.F.; Clark, B.C.; Gellert, R.; Golombek, M.P.; Grotzinger, J.P.; Herkenhoff, K.E.; Johnson, J.R.; et al. Veneers, rinds, and fracture fills: Relatively late alteration of sedimentary rocks at Meridiani Planum, Mars. *J. Geophys. Res. Planets* **2008**, *113*, E6. [CrossRef]
257. Osterloo, M.M.; Hamilton, V.E.; Bandfield, J.L.; Glotch, T.D.; Baldridge, A.M.; Christensen, P.R.; Tornabene, L.L.; Anderson, F.S. Chloride-bearing materials in the southern highlands of Mars. *Science* **2008**, *319*, 1651–1654. [CrossRef]
258. Hynek, B.M.; Osterloo, M.K.; Kierein-Young, K.S. Late-stage formation of Martian chloride salts through ponding and evaporation. *Geology* **2015**, *43*, 787–790. [CrossRef]

259. Rummel, J.D.; Beaty, D.W.; Jones, M.A.; Bakermans, C.; Barlow, N.G.; Boston, P.J.; Chevrier, V.F.; Clark, B.C.; Vera, J.P.D.; Gough, R.V.; et al. A new analysis of Mars "special regions": Findings of the second MEPAG Special Regions Science Analysis Group (SR-SAG2). *Astrobiology* **2014**, *14*, 887–968. [CrossRef]
260. Downing, J.A.; Prairie, Y.T.; Cole, J.J.; Duarte, C.M.; Tranvik, L.J.; Striegl, R.G.; McDowell, W.H.; Kortelainen, P.; Caraco, N.F.; Melack, J.M.; et al. The global abundance and size distribution of lakes, ponds, and impoundments. *Limnol. Oceanogr.* **2006**, *51*, 2388–2397. [CrossRef]
261. Robbins, S.J.; Hynek, B.M. A new global database of Mars impact craters ≥ 1 km: Database creation, properties, and parameters. *J. Geophys. Res. Planets* **2012**, *117*, E5. [CrossRef]
262. Golombek, M.P.; Warner, N.H.; Ganti, V.; Lamb, M.P.; Parker, T.J.; Fergason, R.L.; Sullivan, R. Small crater modification on Meridiani Planum and implications for erosion rates and climate change on Mars. *J. Geophys. Res. Planets* **2014**, *119*, 2522–2547. [CrossRef]
263. Fairén, A.G.; Dohm, J.M.; Baker, V.R.; Pablo, M.A.D.; Ruiz, J.; Ferris, J.C.; Anderson, R.C. Episodic flood inundations of the northern plains of Mars. *Icarus* **2003**, *165*, 53–67. [CrossRef]
264. Pearce, B.K.; Tupper, A.S.; Pudritz, R.E.; Higgs, P.G. Constraining the time interval for the origin of life on Earth. *Astrobiology* **2018**, *18*, 343–364. [CrossRef] [PubMed]
265. Fassett, C.I.; Head, I.I.I.J.W. Valley network-fed, open-basin lakes on Mars: Distribution and implications for Noachian surface and subsurface hydrology. *Icarus* **2008**, *198*, 37–56. [CrossRef]
266. Goudge, T.A.; Head, J.W.; Mustard, J.F.; Fassett, C.I. An analysis of open-basin lake deposits on Mars: Evidence for the nature of associated lacustrine deposits and post-lacustrine modification processes. *Icarus* **2012**, *219*, 211–229. [CrossRef]
267. Bristow, T.F.; Rampe, E.B.; Achilles, C.N.; Blake, D.F.; Chipera, S.J.; Craig, P.; Crisp, J.A.; Marais, D.J.D.; Downs, R.T.; Gellert, R.; et al. Clay mineral diversity and abundance in sedimentary rocks of Gale crater, Mars. *Sci. Adv.* **2018**, *4*, eaar3330. [CrossRef] [PubMed]
268. Ruff, S.W.; Niles, P.B.; Alfano, F.; Clarke, A.B. Evidence for a Noachian-aged ephemeral lake in Gusev crater, Mars. *Geology* **2014**, *42*, 359–362. [CrossRef]
269. Frydenvang, J.; Mangold, N.; Wiens, R.C.; Fraeman, A.A.; Edgar, L.A.; Fedo, C.M.; L'Haridon, J.; Bedford, C.C.; Gupta, S.; Grotzinger, J.P.; et al. The chemostratigraphy of the Murray formation and role of diagenesis at Vera Rubin ridge in Gale crater, Mars, as observed by the ChemCam instrument. *J. Geophys. Res. Planets* **2020**, *125*, e2019JE006320. [CrossRef]
270. Rapin, W.; Dromart, G.; Rubin, D.; Deit, L.L.; Mangold, N.; Edgar, L.A.; Gasnault, O.; Herkenhoff, K.; Mouélic, S.L.; Anderson, R.B.; et al. Alternating wet and dry depositional environments recorded in the stratigraphy of Mount Sharp at Gale crater, Mars. *Geology* **2021**. [CrossRef]
271. Martin, P.E.; Farley, K.A.; Malespin, C.A.; Mahaffy, P.R.; Edgett, K.S.; Gupta, S.; Dietrich, W.E.; Malin, M.C.; Stack, K.M.; Vasconcelos, P.M. Billion-year exposure ages in Gale crater (Mars) indicate Mount Sharp formed before the Amazonian period. *Earth Planet. Sci. Lett.* **2021**, *554*, 116667. [CrossRef]
272. Milliken, R.E.; Grotzinger, J.P.; Wiens, R.; Gellert, R.; Thompson, L.M.; Sheppard, R.; Vasavada, A.; Bristow, T.; Mangold, N. The chemistry and mineralogy of an ancient lacustrine sequence on Mars: Lessons learned from integrating rover and orbiter datasets. *LPI Contrib.* **2019**, *2089*, 6191.
273. Hurowitz, J.A.; Grotzinger, J.P.; Fischer, W.W.; McLennan, S.M.; Milliken, R.E.; Stein, N.; Vasavada, A.R.; Blake, D.F.; Dehouck, E.; Eigenbrode, J.L.; et al. Redox stratification of an ancient lake in Gale crater, Mars. *Science* **2017**, *356*, 6341. [CrossRef]
274. Alfreider, A.; Baumer, A.; Bogensperger, T.; Posch, T.; Salcher, M.M.; Summerer, M. CO_2 assimilation strategies in stratified lakes: Diversity and distribution patterns of chemolithoautotrophs. *Environ. Microbiol.* **2017**, *19*, 2754–2768. [CrossRef]
275. Stein, N.; Grotzinger, J.P.; Schieber, J.; Mangold, N.; Hallet, B.; Newsom, H.; Stack, K.M.; Berger, J.A.; Thompson, L.; Siebach, K.L.; et al. Desiccation cracks provide evidence of lake drying on Mars, Sutton Island member, Murray formation, Gale Crater. *Geology* **2018**, *46*, 515–518. [CrossRef]
276. Martin, P.E.; Farley, K.A.; Archer, D.P., Jr.; Hogancamp, J.V.; Siebach, K.L.; Grotzinger, J.P.; McLennan, S.M. Reevaluation of perchlorate in Gale crater rocks suggests geologically recent perchlorate addition. *J. Geophys. Res. Planets* **2020**, *125*, e2019JE006156. [CrossRef]
277. Greeley, R.; Schneid, B.D. Magma generation on Mars: Amounts, rates, and comparisons with Earth, Moon, and Venus. *Science* **1991**, *254*, 996–998. [CrossRef]
278. Newsom, H.E. Hydrothermal alteration of impact melt sheets with implications for Mars. *Icarus* **1980**, *44*, 207–216. [CrossRef]
279. Ruff, S.W.; Campbell, K.A.; Van Kranendonk, M.J.; Rice, M.S.; Farmer, J.D. The case for ancient hot springs in Gusev crater, Mars. *Astrobiology* **2020**, *20*, 475–499. [CrossRef] [PubMed]
280. Rapin, W.; Chauviré, B.; Gabriel, T.S.; McAdam, A.C.; Ehlmann, B.L.; Hardgrove, C.; Meslin, P.Y.; Rondeau, B.; Dehouck, E.; Franz, H.B.; et al. In situ analysis of opal in Gale Crater, Mars. *J. Geophys. Res. Planets* **2018**, *123*, 1955–1972. [CrossRef]
281. Sun, V.Z.; Milliken, R.E. Characterizing the mineral assemblages of hot spring environments and applications to Mars orbital data. *Astrobiology* **2020**, *20*, 453–474. [CrossRef]
282. Rossi, A.P.; Neukum, G.; Pondrelli, M.; Van Gasselt, S.; Zegers, T.; Hauber, E.; Chicarro, A.; Foing, B. Large-scale spring deposits on Mars? *J. Geophys. Res. Planets* **2008**, *113*, E8. [CrossRef]

283. Berger, J.A.; Schmidt, M.E.; Gellert, R.; Boyd, N.I.; Desouza, E.D.; Flemming, R.L.; Izawa, M.R.; Ming, D.W.; Perrett, G.M.; Rampe, E.B.; et al. Zinc and germanium in the sedimentary rocks of Gale Crater on Mars indicate hydrothermal enrichment followed by diagenetic fractionation. *J. Geophys. Res. Planets* **2017**, *122*, 1747–1772. [CrossRef]
284. Michalski, J.R.; Onstott, T.C.; Mojzsis, S.J.; Mustard, J.; Chan, Q.H.; Niles, P.B.; Johnson, S.S. The Martian subsurface as a potential window into the origin of life. *Nat. Geosci.* **2018**, *11*, 21–26. [CrossRef]
285. Bridges, J.C.; Schwenzer, S.P. The nakhlite hydrothermal brine on Mars. *Earth Planet. Sci. Lett.* **2012**, *359*, 117–123. [CrossRef]
286. Amador, E.S.; Bandfield, J.L.; Brazelton, W.J.; Kelley, D. The lost city hydrothermal field: A spectroscopic and astrobiological analogue for Nili Fossae, Mars. *Astrobiology* **2017**, *17*, 1138–1160. [CrossRef]
287. Brown, A.J.; Hook, S.J.; Baldridge, A.M.; Crowley, J.K.; Bridges, N.T.; Thomson, B.J.; Marion, G.M.; Filho, C.R.D.S.; Bishop, J.L. Hydrothermal formation of clay-carbonate alteration assemblages in the Nili Fossae region of Mars. *Earth Planet. Sci. Lett.* **2010**, *297*, 174–182. [CrossRef]
288. Turner, S.M.; Bridges, J.C.; Grebby, S.; Ehlmann, B.L. Hydrothermal activity recorded in post Noachian-aged impact craters on Mars. *J. Geophys. Res. Planets* **2016**, *121*, 608–625. [CrossRef]
289. Carrozzo, F.G.; Achille, G.D.; Salese, F.; Altieri, F.; Bellucci, G. Geology and mineralogy of the Auki Crater, Tyrrhena Terra, Mars: A possible post impact-induced hydrothermal system. *Icarus* **2017**, *281*, 228–239. [CrossRef]
290. Hamilton, C.W.; Mouginis-Mark, P.J.; Sori, M.M.; Scheidt, S.P.; Bramson, A.M. Episodes of aqueous flooding and effusive volcanism associated with Hrad Vallis, Mars. *J. Geophys. Res. Planets* **2018**, *123*, 1484–1510. [CrossRef]
291. Michalski, J.R.; Dobrea, E.Z.; Niles, P.B.; Cuadros, J. Ancient hydrothermal seafloor deposits in Eridania basin on Mars. *Nat. Commun.* **2017**, *8*, 1–10. [CrossRef] [PubMed]
292. Azuma, S.; Katayama, I. Evolution of the rheological structure of Mars. *Earth Planets Space* **2017**, *69*, 1–3. [CrossRef]
293. Quarles, B.L.; Lissauer, J.J. Dynamical evolution of the Earth–Moon progenitors–Whence Theia? *Icarus* **2015**, *248*, 318–339. [CrossRef]
294. Morrison, P.R.; Mojzsis, S.J. Tracing the early emergence of microbial sulfur metabolisms. *Geomicrobiol. J.* **2020**, *10*, 1–21. [CrossRef]
295. Bouvier, L.C.; Costa, M.M.; Connelly, J.N.; Jensen, N.K.; Wielandt, D.; Storey, M.; Nemchin, A.A.; Whitehouse, M.J.; Snape, J.F.; Bellucci, J.J.; et al. Evidence for extremely rapid magma ocean crystallization and crust formation on Mars. *Nature* **2018**, *558*, 586–589. [CrossRef] [PubMed]
296. Quantin-Nataf, C.; Craddock, R.A.; Dubuffet, F.; Lozac'h, L.; Martinot, M. Decline of crater obliteration rates during early martian history. *Icarus* **2019**, *317*, 427–433. [CrossRef]
297. Grant, J.A.; Wilson, S.A. Evidence for late alluvial activity in Gale crater, Mars. *Geophys. Res. Lett.* **2019**, *46*, 7287–7294. [CrossRef]
298. Martin, P.E.; Farley, K.A.; Baker, M.B.; Malespin, C.A.; Schwenzer, S.P.; Cohen, B.A.; Mahaffy, P.R.; McAdam, A.C.; Ming, D.W.; Vasconcelos, P.M.; et al. A two-step K-Ar experiment on Mars: Dating the diagenetic formation of jarosite from Amazonian groundwaters. *J. Geophys. Res. Planets* **2017**, *122*, 2803–2818. [CrossRef]
299. Guitreau, M.; Flahaut, J. Record of low-temperature aqueous alteration of Martian zircon during the late Amazonian. *Nat. Commun.* **2019**, *10*, 1–9. [CrossRef]
300. Richardson, M.I.; Mischna, M.A. Long-term evolution of transient liquid water on Mars. *J. Geophys. Res. Planets* **2005**, *110*, E3. [CrossRef]
301. Davila, A.; Kahre, M.A.; Quinn, R.; Marais, D.J.D. The biological potential of present-day Mars. *Planetary Astrobiology* **2020**, *16*, 169.
302. Meadows, V.; Arney, G.; Schmidt, B.; Marais, D.J.D. (Eds.) *Planetary Astrobiology*; University of Arizona Press: San Diego, CA, USA, 2020.
303. Malin, M.C.; Edgett, K.S.; Posiolova, L.V.; McColley, S.M.; Dobrea, E.Z. Present-day impact cratering rate and contemporary gully activity on Mars. *Science* **2006**, *314*, 1573–1577. [CrossRef] [PubMed]
304. Dundas, C.M.; Byrne, S.; McEwen, A.S.; Mellon, M.T.; Kennedy, M.R.; Daubar, I.J.; Saper, L. HiRISE observations of new impact craters exposing Martian ground ice. *J. Geophys. Res. Planets* **2014**, *119*, 109–127. [CrossRef]
305. Vaughan, R.G.; Heasler, H.; Jaworowski, C.; Lowenstern, J.B.; Keszthelyi, L.P. *Provisional Map of Thermal Areas in Yellowstone National Park Based on Satellite Thermal Infrared Imaging and Field Observations*; US Geological Survey Scientific Investigations Report, No. 5137; US Geological Survey: Washington, DC, USA, 2014; p. 2. [CrossRef]
306. Benner, S.A.; Devine, K.G.; Matveeva, L.N.; Powell, D.H. The missing organic molecules on Mars. *Proc. Natl. Acad. Sci. USA* **2000**, *97*, 2425–2430. [CrossRef] [PubMed]
307. Zent, A. A historical search for habitable ice at the Phoenix landing site. *Icarus* **2008**, *196*, 385–408. [CrossRef]
308. Krasnopolsky, V.A.; Feldman, P.D. Detection of molecular hydrogen in the atmosphere of Mars. *Science* **2001**, *294*, 1914–1917. [CrossRef] [PubMed]
309. Grew, E.S.; Bada, J.L.; Hazen, R.M. Borate minerals and origin of the RNA world. *Origins Life Evol. Biosph.* **2011**, *41*, 307–316. [CrossRef] [PubMed]
310. Horneck, G.; Bäcker, H. Can microorganisms withstand the multistep trial of interplanetary transfer? Considerations and experimental approaches. *Origins Life Evol. Biosph.* **1986**, *16*, 414–415. [CrossRef]
311. Clark, B.C. Barriers to natural interchange of biologically active material between Earth and Mars. *Origins Life* **1986**, *16*, 410–411. [CrossRef]

312. Mileikowsky, C.; Cucinotta, F.A.; Wilson, J.W.; Gladman, B.; Horneck, G.; Lindegren, L.; Melosh, J.; Rickman, H.; Valtonen, M.; Zheng, J.Q. Natural transfer of viable microbes in space: From Mars to Earth and Earth to Mars. *Icarus* **2000**, *145*, 391–427. [CrossRef]
313. Clark, B.C. Planetary interchange of bioactive material: Probability factors and implications. *Origins Life Evol. Biosph.* **2001**, *31*, 185–197. [CrossRef] [PubMed]
314. Kirschvink, J.L.; Weiss, B.P. Mars, panspermia, and the origin of life: Where did it all begin. *Palaeontol. Electron.* **2002**, *4*, 8–15.
315. Melosh, H.J. The rocky road to panspermia. *Nature* **1988**, *332*, 687–688. [CrossRef] [PubMed]
316. Gladman, B.J.; Burns, J.A.; Duncan, M.; Lee, P.; Levison, H.F. The exchange of impact ejecta between terrestrial planets. *Science* **1996**, *271*, 1387–1392. [CrossRef]
317. El Goresy, A.; Gillet, P.; Miyahara, M.; Ohtani, E.; Ozawa, S.; Beck, P.; Montagnac, G. Shock-induced deformation of Shergottites: Shock-pressures and perturbations of magmatic ages on Mars. *Geochim. Cosmochim. Acta* **2013**, *101*, 233–262. [CrossRef]
318. Cabrol, N.A.; Grin, E.A. (Eds.) *From Habitability to Life on Mars*; Elsevier: Amsterdam, The Netherlands, 2018.
319. Westall, F.; Foucher, F.; Bost, N.; Bertrand, M.; Loizeau, D.; Vago, J.L.; Kminek, G.; Gaboyer, F.; Campbell, K.A.; Bréhéret, J.G.; et al. Biosignatures on Mars: What, where, and how? Implications for the search for martian life. *Astrobiology* **2015**, *15*, 998–1029. [CrossRef]

Review

Astrochemical Pathways to Complex Organic and Prebiotic Molecules: Experimental Perspectives for In Situ Solid-State Studies

Daniele Fulvio [1,2,*], **Alexey Potapov** [3], **Jiao He** [2] **and Thomas Henning** [2]

1. Istituto Nazionale di Astrofisica, Osservatorio Astronomico di Capodimonte, Salita Moiariello 16, 80131 Naples, Italy
2. Max Planck Institute for Astronomy, Königstuhl 17, D-69117 Heidelberg, Germany; he@mpia.de (J.H.); henning@mpia.de (T.H.)
3. Laboratory Astrophysics Group of the Max Planck Institute for Astronomy at the Friedrich Schiller University Jena, Institute of Solid State Physics, Helmholtzweg 3, 07743 Jena, Germany; alexey.potapov@uni-jena.de
* Correspondence: daniele.fulvio@inaf.it

Abstract: A deep understanding of the origin of life requires the physical, chemical, and biological study of prebiotic systems and the comprehension of the mechanisms underlying their evolutionary steps. In this context, great attention is paid to the class of interstellar molecules known as "Complex Organic Molecules" (COMs), considered as possible precursors of prebiotic species. Although COMs have already been detected in different astrophysical environments (such as interstellar clouds, protostars, and protoplanetary disks) and in comets, the physical–chemical mechanisms underlying their formation are not yet fully understood. In this framework, a unique contribution comes from laboratory experiments specifically designed to mimic the conditions found in space. We present a review of experimental studies on the formation and evolution of COMs in the solid state, i.e., within ices of astrophysical interest, devoting special attention to the in situ detection and analysis techniques commonly used in laboratory astrochemistry. We discuss their main strengths and weaknesses and provide a perspective view on novel techniques, which may help in overcoming the current experimental challenges.

Keywords: complex organic molecules; astrobiology; astrochemistry; interstellar medium; molecular ices; solid state; protoplanetary disks; star forming regions; comets

Citation: Fulvio, D.; Potapov, A.; He, J.; Henning, T. Astrochemical Pathways to Complex Organic and Prebiotic Molecules: Experimental Perspectives for In Situ Solid-State Studies. *Life* **2021**, *11*, 568. https://doi.org/10.3390/life11060568

Academic Editors: Michele Fiore and Emiliano Altamura

Received: 28 May 2021
Accepted: 14 June 2021
Published: 17 June 2021

Publisher's Note: MDPI stays neutral with regard to jurisdictional claims in published maps and institutional affiliations.

Copyright: © 2021 by the authors. Licensee MDPI, Basel, Switzerland. This article is an open access article distributed under the terms and conditions of the Creative Commons Attribution (CC BY) license (https://creativecommons.org/licenses/by/4.0/).

1. Introduction

Over the next decades, it is expected that one of the main challenges in science will be to gain a clear understanding on the processes and phenomena linked to the concept of life how we know it, its origin and early evolution. Key questions, which need to be answered, include: (i) How did life begin and evolve on Earth? (ii) What are the conditions for the origins of life? Does life exist elsewhere in the Universe? (iii) What is life's future on Earth and beyond? Answering these and similar questions requires the physical, chemical, and biological study of prebiotic systems and a deep understanding of the principles governing their evolution into more complex systems and, finally, into living matter. Astrochemistry combines astronomy and chemistry. It directly studies the aforementioned issues, with a special focus on the so-called "Complex Organic Molecules" (COMs), a term referring to those astrophysically relevant organic molecules consisting of six or more atoms, many of which are considered possible precursors of prebiotic molecules, such as amino acids (in turn, precursors of proteins) and nucleobases (in turn, precursors of DNA and RNA). Although COMs are considered of paramount importance from an astrochemical and astrobiological point of view, the reader should keep in mind that they are only relatively "complex" when studied from a chemical and biological perspective [1,2].

From the observational point of view, the main detection tool for molecules in space, in various astrophysical environments, is gas-phase radio astronomy supported by microwave, millimeter-wave, and terahertz spectroscopy. Complementary, solid-state materials are characterized by infrared (IR) spectroscopy. To date, more than 200 species have already been detected in the gas phase while only about 10 molecules have been spectroscopically identified in the solid state (e.g., Cologne Database for Molecular Spectroscopy and [3]). The variety, abundance, and distribution of gas-phase and solid-state COMs already detected or tentatively detected in space have been increasing in the last decades, touching various astronomical environments, from interstellar clouds, protostars, and protoplanetary disks (e.g., [1–19]) to the outer solar system, where COMs and prebiotic species have already been detected in several comets (e.g., [20–31]). To give the reader an idea, the list of these COMs includes but is not limited to: acetaldehyde (CH_3CHO), ethanol (CH_3CH_2OH), formamide ($HCONH_2$), glycine (NH_2CH_2COOH), and urea (H_2NCONH_2). Their structural formulas are shown in Figure 1. The detection of COMs in the solid state, i.e., in the icy component present in cold cosmic regions, such as dense molecular clouds of the interstellar medium (ISM), protostellar envelopes and protoplanetary disks (beyond the snowline), and on the surface of minor bodies of the solar system (comets, asteroids, satellites of planets, trans-Neptunian objects, . . .), is made difficult by the intrinsic problems related to IR spectroscopy. Simply said, the difficulties in identifying COMs in ices by IR spectroscopy are due to the unspecific nature of the signal coming from overlapping features belonging to different species having the same functional groups. Nevertheless, for solar system studies, space missions typically combine IR spectroscopy and mass spectrometry (MS). This allows the identification of COMs in situ (e.g., [27,28]), although MS carries its own intrinsic critical issues as well, such as the difficulty to distinguish molecules of the same mass and chiral isomers. A detailed discussion of these techniques can be found in the following sections of this review.

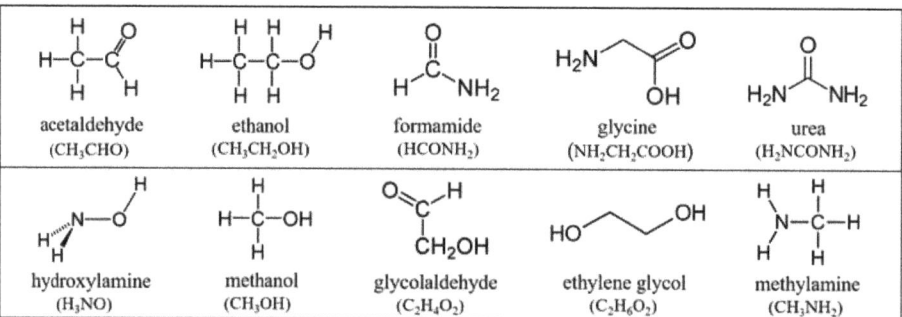

Figure 1. Structural formulas of some among the most astrophysically relevant COMs discussed in this review. The present figure was created by adapting original graphics taken from Wikipedia, the free encyclopedia (in particular, the graphic of acetaldehyde was created by UAwiki).

While most of the early astrochemistry models presumed that COM formation in the aforementioned astrophysical environments occurs through gas-phase reactions (e.g., [2,32–34]), the majority of current models assume that COM formation, especially the more complex ones, occurs mainly in the solid state, onto or within molecular ices found on cosmic dust grains and planetary surfaces. For the reader interested in understanding the nature and evolution of cosmic dust grains and their role in astrochemistry, we recommend a few review papers (e.g., [35–42]). Moving from earlier models to current ones has been favored by the comparison of theoretical simulations with new laboratory experiments and astronomical observations over the last few decades. Simply put, one of the main ideas beyond current astrochemistry models is that icy species may dissociate under energetic processing, such as UV irradiation and cosmic ray bombardment. Depending on the temperature of the ice, the so-formed fragments (radicals, ions, ...) will have a certain mobility and, as a consequence, a certain efficiency/probability to re-combine. This way, new species may appear directly within the ice and, among them, COMs may be formed. As the ice eventually warms up, the mobility of the fragments increases and the efficiency/probability to re-combine and form new molecules increases as well. Finally, once the temperature is high enough for the ice to sublimate, it will enrich the gas-phase with the species it contains, including the more complex molecules (e.g., [1,15,43–49]). Alternative and complementary models have also been proposed, such as those based on reactions between stable molecules requiring external energy input by warming up the sample and on atom addition reactions. Additional details on the various mechanisms, triggering the formation and evolution of new molecules in ices found on cosmic dust grains and planetary surfaces, are briefly discussed in the following Section 2. However, a detailed discussion of these processes and mechanisms is beyond the scope of the current review and we refer the interested reader to dedicated papers/reviews ([49–56] and references therein).

The discussion so far indicates the importance of a comprehensive understanding of the physical–chemical mechanisms and processes underlying COM formation and evolution, especially in molecular ices. Most processes and mechanisms are not yet fully understood, and in this framework, laboratory studies specifically designed to simulate astrophysically relevant conditions play a unique role in improving our comprehension. Thanks to dedicated experiments, realistic scenarios towards molecular complexity can be established.

In this contribution, after a brief review of the main formation and evolution processes of COMs in space (Section 2), we focus on a key aspect of COM studies, often not comprehensively covered in reviews: the main strengths and weaknesses of in situ detection and analysis techniques commonly used in experimental astrochemistry (Section 3). The current experimental challenges and novel techniques to overcome present limitations will be the focus of Section 4.

2. Formation of COMs in the Solid State

Chemical processes leading to the formation of molecules in interstellar and circumstellar environments can be divided into two groups, gas-phase and solid-state surface reactions. Gas-phase formation routes to COMs are out of the scope of the present review and we refer the interested reader to a recent review paper on this topic [57]. In the following, we will focus on solid-state reactions.

In colder astrophysical environments, such as diffuse and dense clouds of the ISM, protostars, protoplanetary disks, and the outer solar system, gas-phase atoms and molecules may condense onto the cold surface of cosmic dust grains and planetary surfaces, thereby forming molecular ice. As an example, the ice films covering the micron-sized cosmic dust grains in dense molecular clouds are predominantly made of H_2O; the other main components being CO_2 (up to about 35%), CO (up to about 30%), NH_3 (below 5%), CH_4 (below 5%), CH_3OH (below 5%), H_2CO (below 5%), OCN (below 2%), and OCS (below 2%). Values reported in parenthesis indicate abundances relative to H_2O ice. The abundance values for each species may vary substantially for specific interstellar, circumstellar, or planetary environments (see, e.g., [13] and references therein). Starting from these species, larger molecules including COMs are formed in the ice by solid-state reactions. These are reactions triggered by energetic photon (UV, EUV, X-rays) irradiation, ion and electron bombardment, thermal processing, and atom addition. In dedicated review papers (e.g., [15,47,48]), the interested reader can find details on the experimental techniques and setups commonly employed in laboratory astrochemistry studies to recreate the different conditions present in astronomical sources. Within the laboratory astrochemistry community, irradiation by energetic photons, ions, and electrons is often referred to as "energetic processing". In this case, energetic photons or particles impacting a solid-state target transfer energy into the system, causing the dissociation of stable molecules and leading to the formation of highly reactive radicals and ions. From these reactions, new species originally not present in the unprocessed ice can be formed.

By "thermal" reactions we typically mean reactions between stable molecules requiring additional external energy input by warming up the sample to overcome the reaction barrier. This energy input would depend on the specific solid-state system under study, ranging from a few dozen to hundreds of K. Atom addition reactions, typically studied at low temperatures (3–20 K), are often referred to in the literature as "non-energetic"; however, they have to also be considered as thermal reactions as atoms have a thermal distribution of speeds and are not excited by non-thermal processes. For these reactions, quantum tunneling may play an important role. There are a number of review papers devoted to the formation of COMs by the triggers mentioned above. We do not aim to duplicate them and, in the following, will refer the interested reader to these papers and will briefly discuss a few experimental examples of recent COM studies relevant to prebiotic chemistry. When possible, the importance of these COMs is also briefly put in the context of the direct comparison between laboratory results and astronomical observations.

2.1. Thermal Reactions

Atom addition reactions relevant to low-temperature astrophysical environments were reviewed by Linnartz et al. [53]. They discussed the pathways to simple and complex organic molecules of prebiotic interest, such as hydroxylamine (H_3NO), methanol (CH_3OH), and glycolaldehyde ($C_2H_4O_2$). Their structural formulas are shown in Figure 1. Further laboratory experiments led to the conclusion that there may be no need for energetic processing to synthesize COMs in the solid state. For instance, Chuang et al. [58] showed that the hydrogenation (H-addition) of simple ices may result in the formation of glycolaldehyde (the smallest sugar), ethylene glycol ($C_2H_6O_2$) (the simplest sugar alcohol), and methyl formate (an isomer of glycolaldehyde)—all these COMs are detected in astrophysical environments and in comets. Later, Ioppolo et al. [56] demonstrated that glycine ($C_2H_5NO_2$) (the simplest amino acid) and its direct precursor methylamine (CH_3NH_2), both detected on comet 67P/Churyumov–Gerasimenko by the Rosetta mission [59], can be synthesized through atom and radical–radical addition surface reactions. Krim and Mencos [60] showed that N-addition to acetonitrile (CH_3CN) (a potential amino acid precursor) leads to its chemical transformation into isomers CH_3NC and CH_2CNH with the abundance ratios explaining observations much better than the previous studies of such transformation induced by UV irradiation and particle bombardment. An alternative (to the hydrogenation of CO ice) route to methanol through H/O addition on the surface of laboratory analogues of cosmic carbonaceous grains was demonstrated by Potapov et al. [55]. Nguyen et al. [61] discussed the formation of amines (potential precursors of amino acids) through the hydrogenation of nitrile and isonitrile.

Thermal atom–molecule and molecule–molecule reactions as a step towards molecular complexity in astrophysical media were reviewed by Theulé et al. [49]. The authors classified molecules formed through thermal reactions according to their degree of complexity in an increasing order of generation. The largest molecules formed through molecule–molecule reactions initiated by warming up their samples (generation 2) included methylammonium methylcarbamate ($CH_3NH_3^+CH_3NHCOO^-$) and alpha-aminoethanol ($NH_2CH(CH_3)OH$), both of which may act as amino acid precursors in astrophysical environments. The latter is a chiral molecule—a molecule with an important property when considering that important biological molecules have chirality as a feature. Amino acids in proteins are "left-handed" (L) and sugars in nucleic acids are "right-handed" (D) (see, e.g., [62]). One of the important open questions for laboratory astrochemistry is whether we can detect chirality in COM chemistry. Such detection would allow for a better understanding of the conditions responsible for the formation of prebiotic molecules and the astronomical search for chiral molecules. This will provide a link between molecules in space and life on Earth. Until now, only one chiral molecule, propylene oxide, has been detected in the ISM [63] in addition to amino acids detected in meteorites [64].

Following earlier investigations, recently, new studies devoted to the formation of ammonium carbamate ($NH_4^+NH_2COO^-$) through the $CO_2 + 2NH_3$ reaction were presented. Ammonium carbamate can be converted into urea (a possible precursor of pyrimidine required for the synthesis of nucleobases) and water, giving a start to a network of prebiotic reactions. It was shown that the reaction can be driven by structural changes evolving in amorphous water ice, such as pore collapse and crystallization, at high temperatures relevant to protostars and protoplanetary disks (i.e., above 100 K) [65]. Potapov et al. [66,67] demonstrated that the formation of ammonium carbamate on dust grains in astrophysical environments can be catalyzed by the surface of grains. Alpha-aminoethanol, ammonium carbamate and other COMs of prebiotic interest have not been detected in astrophysical environments because their gas-phase spectra are not available due to the instability of these molecules in the gas phase at room temperature. This highlights another important question—how do we measure gas-phase spectra of such COMs for their detection in astrophysical environments? Section 4.2 discusses one of the potential experimental possibilities.

2.2. Energetic Processing

In the well-known experiments of Miller and Urey, mimicking the possible conditions of the primitive Earth atmosphere, amino acids were formed starting from simple molecules, such as CH_4, NH_3, H_2O, and H_2, after exposing the gas mixture to an electrical discharge producing reactive radicals [68,69]. The idea of triggering reactions by energetically formed radicals has been later followed by a number of groups, leading to the production of amino acids and other prebiotic species on various gas-phase mixtures of simple molecules (e.g., [70–72] and references therein). This idea has also been applied to the extraterrestrial context, where astrochemically relevant ice mixtures are energetically processed, as discussed in the following.

Reactions triggered by energetic photons (UV, EUV, X-rays) and cosmic rays/solar wind (i.e., ions and electrons) bombardment fall into the group of energetic processing. Details on the possible mechanisms of COM formation induced by irradiation with energetic photons and cosmic ray/solar wind particles, relevant to low-temperature astrophysical environments, can be found in recent review papers (e.g., [15,38,47,48]).

We point out that experiments simulating the effects of the energetic processing of simple ices (e.g., H_2O, CO, CO_2) of astrophysical interest have been carried out in different astrochemistry laboratories around the world, already starting several decades ago. Discussing these studies in detail is beyond the scope of this review and we refer the interested reader to a few relevant papers: [73–83]. From the experimental expertise and data accumulated over the years, it is obvious that the increasing complexity of the starting ice mixture, UV irradiation and ion and electron bombardment experiments shows great potential in providing formation routes for COMs. Nevertheless, one should take into account that increasing the complexity of the starting ices leads to challenges in molecule identification by commonly used techniques (see Section 3 for a detailed discussion on this topic).

Focusing on some examples of ion and electron bombardment experiments, interesting results on possible pathways of amino acid formation were shown after the ion processing of ices of astrochemical relevance containing CH_3CN or $CH_3CN:H_2O$ [84]. Similarly, the formation of glycine was observed in processing experiments of ices, after ion irradiation of $H_2O:NH_3:CO$ [85] as well as after electron irradiation of $NH_3:CH_3COOD$, $CH_3NH_2:CO_2$ and $CO_2:CH_4:NH_3$ mixtures ([86–88]). Another COM considered to have great potential in prebiotic chemistry—since it is the simplest molecule containing a peptide bond, known to be the "bridge" connecting amino acids in proteins and polypeptides—is formamide (NH_2HCO). This species has been formed in several processing experiments of solid-state mixtures containing key astrophysical ices such as H_2O, CO, CH_4, NH_3, HCN, or CH_3OH, irradiated by energetic ions (e.g., [89,90]) and electrons (e.g., [91,92]). Its astrophysical importance is given by the fact that formamide has already been detected in several astrophysical environments, such as molecular clouds, protostars (e.g., [10,16]) and cometary comae (e.g., [20,25]). As an additional example, methyl formate (already discussed in Section 2.1) and other COMs were also observed after ion and electron bombardment of methanol-based ices (e.g., [47,93,94]).

Focusing now on photoprocessing experiments, it has been shown that the irradiation of planetary and interstellar ice mixtures by energetic photons (UV, EUV, X-rays) may also induce COM formation. For instance, different amino acids (some of which found in proteins), sugars and nucleobases were identified in the organic residues left over at room temperature after the irradiation of solid-state mixtures containing a combination of key astrophysical ices such as H_2O, CH_3OH, NH_3, HCN, CO, and CO_2 (e.g., [95–101]). The formation of simpler COMs, such as formamide and methyl formate, was also reported in photoprocessing experiments starting from different ice mixtures (e.g., [47,102–106]).

We point out that the unambiguous identification of COMs formed in situ during processing experiments of astrophysically relevant ices is often hampered by the used analyzing techniques, mainly infrared spectroscopy and mass spectrometry. The main problems related to these techniques, together with interesting novel techniques which may overcome their limitations, are the focus of Sections 3 and 4.

3. Common Experimental Techniques Applied for In Situ Detection and Analysis of Solid-State COMs of Astrochemical Relevance

3.1. Infrared Spectroscopy

In the field of laboratory astrochemistry, Fourier transform infrared (FTIR) spectroscopy is one of the two main in situ experimental techniques, usually applied to detect molecules in the solid state. It uses a broadband thermal radiation source (e.g., globar, Hg lamp) and an interference pattern of a two-arm Michelson interferometer. In experiments with ices of astrophysical interest, typical resolutions, ranging from a few to 0.1 cm^{-1}, are achieved by varying the interferometer path length. FTIR spectroscopy is used in one of the two main configurations, transmission or reflection mode. In both configurations, the radiation beam of the IR source points to the surface of the ice sample. When going through the ice, part of this radiation is absorbed at specific wavelengths which depend on the molecules constituting the sample (while the intensity of the absorption depends on the amount of absorbing molecules). The remaining part of the radiation goes through the ice and, when working in transmission mode, also through the substrate on top of which the sample was grown, while when in reflection mode, it is reflected by the substrate. For reflection spectroscopy, the term Fourier transform reflection absorption infrared spectroscopy (FT-RAIRS or, simply, RAIRS) is usually used. The IR beam can be at near normal incidence to the substrate, but FT-RAIRS mainly works at large angles of incidence with parallel-polarized light and on metal substrates because it takes advantage of the high electric field strength on the metal surface, which is only true for the field component normal to the surface, increasing the sensitivity of the technique. The major advantage of FT-RAIRS over transmission spectroscopy is that it can be used to detect intra-adsorbate and adsorbate–substrate vibrations and adsorbate–adsorbate interactions at monolayer and submonolayer coverages on metal surfaces [107]. On the other hand, a problem of using reflection spectroscopy is the presence of optical interference effects which often lead to erroneous measurements of absorption band strengths and give an apparent dependence of this quantity on film thickness, the index of refraction and wavelength [108].

FTIR spectroscopy allows for in situ and non-destructive characterization of the evolution of the composition and structure of ices before, during and after their thermal or non-thermal processing at defined experimental conditions, such as a specific temperature, starting ice composition, and thickness. Besides this, FTIR spectroscopy provides many additional possibilities, such as multiplex recording, broadband coverage and easy wavelength calibration. It provides reference spectra that can be used for interpreting astronomical IR spectra (however, mainly taken by space-borne observatories due to the absorption of IR radiation by the Earth's atmosphere). FTIR spectroscopy is a very popular experimental technique worldwide. The reader can find many examples in the literature of experimental studies of COMs carried out by using FTIR spectroscopy (e.g., the review papers mentioned in the previous section and references therein).

However, the problem of IR spectroscopy as applied to the study of the formation and evolution of COMs in the solid state is that often several COMs show similar and overlapping broad spectral signatures belonging to different species, but having the same functional groups. An example is given in Figure 2, adapted from [104]. Moreover, an additional difficulty is typically related to the low efficiency of formation of new large and complex species within the ice, which leads to low amounts of COMs. In this case, the spectral features are very weak and the sensitivity of FTIR spectroscopy is not enough to detect them. Just to give the reader a reference value, the detection limit of FTIR spectrometers typically used in most laboratories is in the order of 0.1–1 monolayer of material, i.e., about 10^{14}–10^{15} molecules cm^{-2} of the species under study. All this complicates the analysis and makes the identification of COMs difficult and limited. At smaller wavenumbers, in the terahertz and millimeter-wave spectral ranges, which are the ranges of many ground-based observatories (e.g., Effelsberg, GBT, NOEMA and ALMA), the signal to noise of FTIR spectrometers is limited due to the low power density of the thermal radiation source. Additionally, the discrete FT procedure becomes problematic at low frequencies because of the poor spectral resolution.

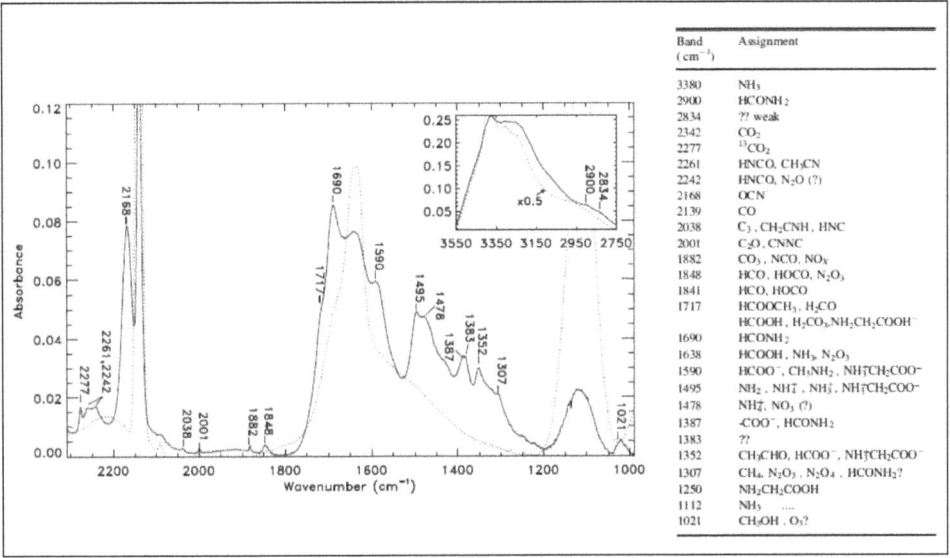

Figure 2. IR spectra of a H$_2$O:CO:NH$_3$ ice mixture (13 K) before (dotted line) and after (solid line) irradiation with soft X-ray for 2 h. In the figure, the upright inset panel shows the effects of photoprocessing in the range 2750–3550 cm^{-1}. Numerical labels refer to the wavenumber of the main products of irradiation, identified in the table to the right. Adapted from [104], © AAS. Reproduced with permission.

3.2. Mass Spectrometry

To acquire complementary information on the mass and composition of the "new" molecules formed in ices, solid-state FTIR spectroscopy is usually used in combination with gas-phase mass spectrometry (MS), the second key in situ experimental technique. In a typical MS experiment, species released from the solid state into the gas phase (process called desorption) reach the ionization region of a quadrupole mass spectrometer (QMS) by direct flight, are ionized and then ion signals for different m/z (mass to charge) ratios are detected by a mass spectrometer with a typical mass resolution m/Δm of several hundreds. MS allows one to follow molecular processes taking place within the ice before, during and after its processing by detecting and recording the species desorbed from the ice. As for FTIR spectroscopy, one can observe the evolution of ices at defined experimental conditions. MS has a very high sensitivity, capable of measuring partial pressures down to ~10^{-13} mbar [107]. MS in the laboratory allows for the direct comparison with in situ MS provided by solar system space missions such as Cassini–Huygens [109] and Rosetta [110]. The Rosetta mission has provided evidence for the presence of COMs and prebiotic species, e.g., glycine together with the precursor molecules methylamine and ethylamine, in cometary ices [27,59].

Mass spectrometry is usually applied to follow the kinetics of the desorption of molecules from ices in so-called temperature programmed desorption (TPD) experiments. In these experiments, after growing an ice sample with the molecules of interest (and its processing with one of the mechanisms discussed in Section 2), the ice is warmed up at a constant rate, inducing the desorption of molecules from its surface. Released molecules are detected in the gas phase by a mass spectrometer. Thus, one obtains information on the mass of a desorbed molecule and its desorption temperature and rate, which can be converted into a desorption energy. These quantities can then be used in astrochemical models. However, in the thermal desorption process, reactions may occur during the warm-up phase of the ice mixture. This makes it challenging to distinguish between COMs originally formed in the ice mixture at low temperature or during the heating process.

As for the case of FTIR, MS is a broadly used experimental technique. The reader can find many examples in studies on COM formation and evolution (see, e.g., the review papers mentioned in Section 2 and references therein).

As an example of a mass spectrum for COMs, we present in Figure 3 (left panel) the spectrum of carbamic acid (NH_2COOH), recorded at 265 K [111]. Masses 17, 18, 44, and 61 correspond to $^{14}NH_3$, $^{15}NH_3$, CO_2 and NH_2COOH, respectively. This example highlights one of the main problems of a typical MS experiment. In the given example, the peak at m/z = 18 may be attributed to H_2O instead of $^{15}NH_3$ and therefore some of the molecules at m/z = 17 are OH molecules. The same problem of uncertain identification of the peak signals is clearly visible in the right panel of Figure 3 [112], where another drawback of MS is also shown: electron impact ionization of desorbed molecules leads to the fragmentation of the molecule and ambiguous attribution of the fragments. This is especially true for larger COMs, which are fragile and, under electron impact, can be easily destroyed. This implies that, by using MS, we do not detect the intact desorbed COM, but only its fragments. Moreover, MS suffers from several additional drawbacks which limit its use. Thermal desorption in a conventional TPD experiment is only able to desorb relatively volatile molecules (i.e., simple ones); for refractory molecules that only desorb above room temperature, alternative desorption methods are needed. In addition, conventional QMS has a rather limited sensitivity and mass resolution. Larger COMs produced in standard experiments usually have a relatively low production yield, often below the detection limit of the technique, implying that they cannot be detected by QMS. Finally, as the size of COM increases, it becomes increasingly difficult to identify the chemical formula of the molecule because of the low mass resolution. Consequently, it is not always possible to distinguish high mass COMs, molecules with similar masses, and structural and chiral isomers of the same molecule. The problem of high- and similar-mass COMs (but not the isomer problem) can be solved by using high-resolution mass spectrometers, such as time-of-flight (TOF) MS and Orbitrap, the latter reaching a mass resolution of 10^6. However, specific structural isomers without their degradation (fragment-free) can be probed by using the tunable single photon vacuum ultraviolet photoionization in combination with reflectron time-of-flight mass spectrometry [113]. For details of these techniques, we refer the reader to Section 4.

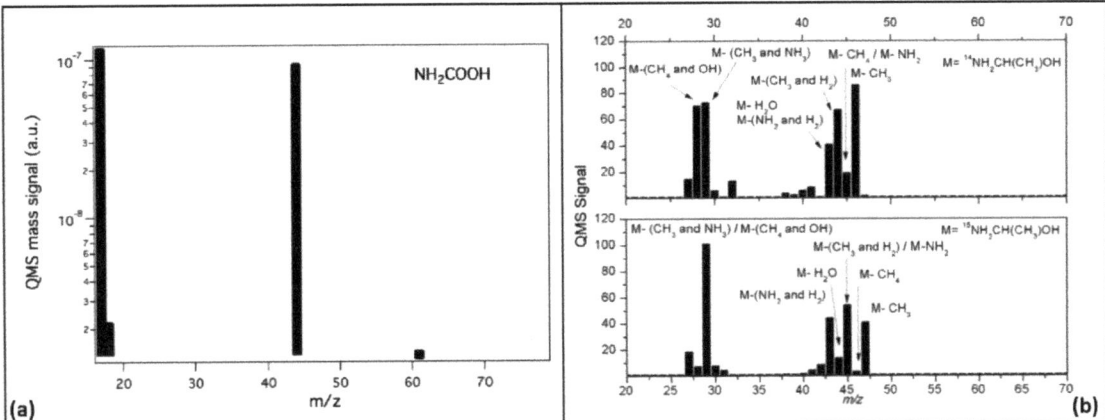

Figure 3. (a) Mass spectrum of NH_2COOH at T = 265 K obtained using a quadrupole mass spectrometer. Republished with permission of Royal Society of Chemistry, from [111]; permission conveyed through Copyright Clearance Center, Inc. (b) The $^{14}NH_2CH(CH_3)OH$ and $^{15}NH_2CH(CH_3)OH$ mass spectra at 200 K. M is the mass of the molecular ion of $NH_2CH(CH_3)OH$ m/z 61. Credit: adapted from [112], reproduced with permission © ESO.

3.3. Complementary Ex Situ Techniques

The main goal of in situ studies is to mimic physical–chemical processes taking place in space under particular conditions, such as low temperature and pressure, relevant to specific astrophysical environments. However, complimentary information can be obtained ex situ. Although outside the focus of this review, we want to mention some additional ex situ techniques used in complement to the techniques and methods discussed in the previous sections. As an example, two powerful ex situ methods to study COM produced by the processing of ices of astrophysical interest are gas chromatography–mass spectrometry (GC-MS) and high-performance liquid chromatography (HPLC). These methods were used, for instance, to identify prebiotic molecules produced by the energetic processing of ices such as ribose and other monosaccharides [99], amino acids [95,96] and nucleobases [114]. Other examples of relevant techniques are X-ray absorption near-edge structure (XANES) spectroscopy, energy-dispersive X-ray (EDX) spectroscopy, and transmission and scanning electron microscopy (TEM and SEM, respectively) used for the elemental analysis and chemical and structural characterization of samples. These and other ex situ methods are typically used when working with strongly bonded molecules, which pose a challenge to the thermal desorption technique. In such cases, after warming up the processed ices to room temperature, a refractory organic residue remains on the substrate. This residue can be constituted by COMs and prebiotic molecules, which cannot be fully characterized by conventional IR or MS techniques. After removing the residue from the vacuum chamber, it is ready to be analyzed by the aforementioned methods. However, the fact of working ex situ introduces additional problems, e.g., the opening for "undesirable" chemical reactions within samples during warming up and, after removing the residue from the vacuum chamber, with ambient air. This may alter the composition of the residue and affect, to some extent, the analysis results. A direct link to the astrophysical processes of interest taking place under low-temperature and low-pressure conditions can therefore be lost.

4. Experimental Challenges and Novel Techniques

In the previous section, we already discussed that infrared spectroscopy and mass spectrometry are the most widely used techniques in solid-state astrochemistry to characterize molecular ices and the desorption of molecules. Although important astrochemical pathways have been revealed by these techniques, their application is mostly limited to relatively simple molecules. As the size and number of COMs within the ice increase, it becomes more and more difficult to identify the exact molecules responsible for the observed IR features and MS signals. To overcome these difficulties and allow the identification of larger COMs, new techniques are being developed. Some of these techniques are discussed in this section.

4.1. Promising Mass Spectrometry Techniques

Molecules embedded in solid-state ices need to be desorbed into the gas phase for characterization by mass spectrometry. In addition to thermal desorption, laser desorption is a technique that gained popularity in chemical and biological studies (see, e.g., [115–117]) but started to be used in astrochemical studies only recently (e.g., [118,119]). Typically, a laser pulse, either in the UV or IR, is absorbed by a small column of the ice and generates a high intensity of heat in a relatively small spot size, increasing the temperature drastically. This leads to an instantaneous desorption of ice at the spot, releasing the molecules from the ice to the gas phase. As the desorption laser is usually pulsed, the technique works particularly well with mass spectrometers that rely on synchronization, such as time-of-flight mass spectrometers (TOF-MS). Although both UV and IR lasers are used for desorption, their performance varies with the specific system under study. UV lasers provide higher energy density than IR lasers, and may generate electronic vibrations of molecules. IR lasers are characterized by lower energy density and only excite the vibrational modes of molecules. For low absorption efficiencies of the molecules, the desorption yield would be relatively low. As a remedy to this problem, one could use an ice matrix that absorbs strongly at the laser wavelength. For this purpose, succinic acid, glycerol, urea, benzoic or cinnamic acid derivatives were used as matrices in prior studies [116,120,121]. However, in astrochemistry experiments, it makes sense to use an astrophysically relevant matrix. Water, being the main component of interstellar and planetary ices, is an ideal choice for the ice matrix. The possibility of using water ice as the matrix has already been demonstrated [119,122–124]. In these studies, an IR laser emitting at the frequency of the OH stretching mode is applied for efficient desorption. When designing laser desorption experiments, it is therefore important to choose the wavelength so that it overlaps with the infrared absorption peak of the ice matrix material. For this reason, one needs to know the wavelength dependence of the laser desorption yield for different matrix molecules. Such a study was carried out by Focsa et al. [125], who measured the laser desorption yield of water and ammonia. Similar measurements are desirable for other matrix candidates.

After desorption from the solid state, COMs need to be ionized before being detected by a mass spectrometer, which measures the mass to charge ratio. In QMS, molecules are typically ionized by electron impact. An electron energy of about 70–100 eV is usually chosen to have a balance between a high degree of ionization and a low degree of fragmentation. However, as the size of the molecule increases, the fragmentation probability increases. It is even possible that almost all the parent molecules are destroyed by electron impacts and one can only measure fragments. This makes the identification of the parent molecules a challenging task.

To solve the problems of destructive "hard" ionization by electron impacts, lasers have been employed as a source of less-destructive "soft" ionization. The energy of UV photons is much lower than the energy of 70–100 eV electrons and therefore UV light can ionize molecules without significant fragmentation [126]. When a UV laser is used, usually one desorbed molecule absorbs a photon, and is ionized to have a single charge with negligible fragmentation. The mass to charge ratio directly corresponds to the parent molecule. If a tunable UV laser is used, it is even possible to distinguish between molecules with the same mass or even the same chemical formula (i.e., isomers). This is because different molecules usually have different ionization energy thresholds. By selecting the photon energy to be above the threshold value, corresponding to one molecule while being below the value corresponding to the other molecule, one only ionizes the molecule with the lower ionization energy [113]. This provides extra information for the identification of COMs that is unavailable to conventional mass spectrometry. Resonance-enhanced multi-photon ionization (REMPI) is an alternative in situ method to ionize molecules (e.g., [127,128]). In REMPI experiments, molecules are first excited to an excited intermediate state by multiple photons, followed by ionization from the intermediate state by another photon or multiple photons. This technique has been applied by a number of groups to near ultraviolet spectroscopy of small molecules (H_2, H_2O) desorbed from ices of astrophysical interest in TPD experiments (e.g., [128,129]). However, so far, laser ionization has not gained wide application in the study of COM formation under conditions relevant to astrophysics. As the study of COM formation proceeds to larger and larger ones, laser ionization will have much to offer for the identification of larger COMs.

As discussed above, a laser is a powerful tool with tremendous potential for both the desorption and ionization of COMs. It is therefore natural to use a single laser for both aspects, thus exploiting their strength. Such a method, typically used in combination with a matrix, is called matrix-assisted laser desorption/ionization (MALDI). It has firstly gained popularity in chemistry and biomedicine after being introduced by Karas and co-workers [130]. Biological molecules as large as hundreds of thousands amu [122] are measured using MALDI without fragmentation. For a review of the application of MALDI to the identification of large biomolecules, we refer the reader to Clark et al. [131]. Figure 4 shows an example of the mass spectrum obtained from a MALDI experiment using a Linear Trap Quadrupole (LTQ) Orbitrap Mass Spectrometer as a detector. Here, a matrix which absorbs strongly at the wavelength of the laser is chosen. The intense heat generated by the laser pulse rapidly evaporates the matrix, forming a plume which contains the large biological molecules buried inside [121]. During the formation of the plume, the molecules are ionized. The exact mechanism of ionization is still debated, and several models are proposed as possible explanations of the ionization. For a detailed discussion, we refer the reader to dedicated reviews such as [132,133]. Both UV and IR lasers have been employed for MALDI. In the earlier years of MALDI, UV lasers were used almost exclusively [122]. Later on, since water is present in many biological samples and the IR beam absorbs strongly in the O–H stretching vibrational mode of water, there has been an increasing number of MALDI experiments using IR lasers. Niu et al. [134] compared the results of MALDI experiments performed by using IR and UV lasers, and showed that similar results were obtained from these two lasers. This suggests that the choice of the laser is largely dictated by the choice of the matrix material. In astrochemical studies of COM formation in astrophysically relevant ices, with water being the main component, the use of IR lasers would be a good choice.

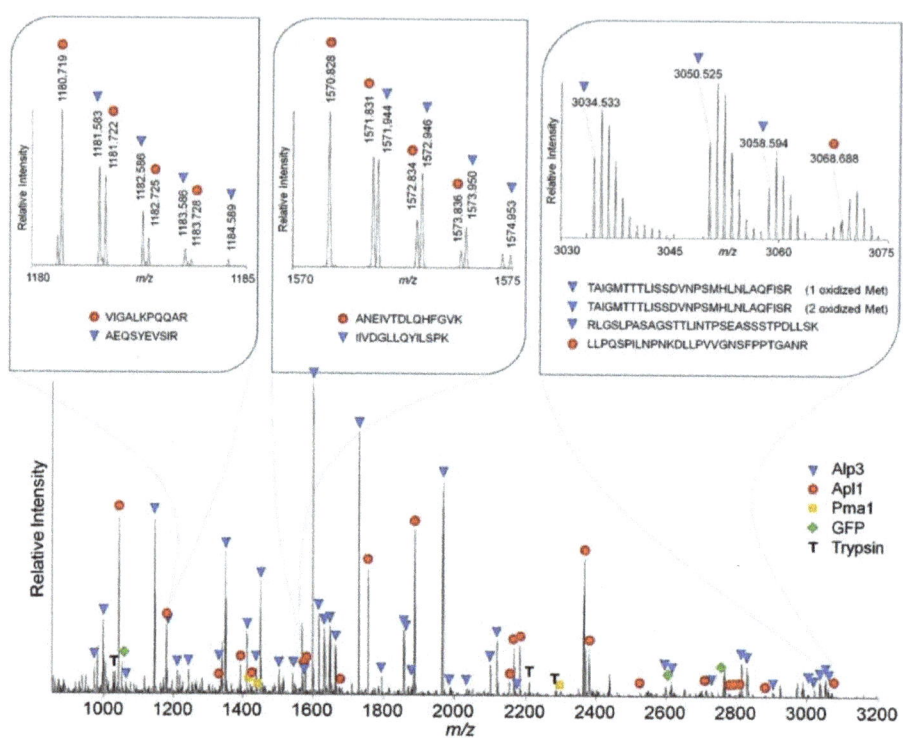

Figure 4. Representative MALDI LTQ Orbitrap MS of selected proteins. The panel insets show expanded MS regions containing Apl1 and Apl3 peptides. Adapted from [135], Copyright Elsevier (2010).

Although MALDI is widely used in biological studies, it gained little attention in the astrochemical community. As far as we know, so far, there is only one astrochemistry laboratory that used the MALDI technique to identify COMs that are formed in astrophysically relevant ices [103,119,136]. The setup is shown in Figure 5. Gudipati and co-workers grew an ice mixture on a substrate kept at a temperature as low as 5 K, followed by irradiation with 2 keV electrons or UV photons, producing new species, including COMs such as formamide, acetamide, and methyl formate [103]. Subsequently, the ice mixture is exposed to a pulsed IR laser beam which desorbs and ionizes the molecules in the ice. To have a better ionization efficiency, an additional pulsed UV laser is also employed; therefore, they named the technique "two color" (2C) MALDI. It is synchronized with the IR laser but with a time delay. The delay is tuned so that the signal of the detected ions is optimized. The detector is a TOF-MS, by means of which COMs are identified. Gudipati and co-workers used one or more of the following three molecules as the matrix: water, ammonia and methanol. All these three molecules represent key species for astrophysical ices and have a relatively strong absorption in the IR, around 3 microns.

Ions produced by the ionization process, either with or without a matrix, are analyzed by a mass spectrometer to identify the parent molecules. Conventional QMS has a relatively low mass resolution and therefore is not ideal for the identification of larger COMs. High-resolution mass spectrometry (HRMS, resolution better than 10^4) fills this gap and has been proven to be a powerful tool. A comprehensive review of HRMS can be found in [137]. There are three main types of HRMS: time-of-flight mass spectrometry (TOF-MS), Fourier transform ion cyclotron resonance mass spectrometry (FTICR-MS), and Orbitrap-MS. In TOF-MS, ions are accelerated by a high voltage, usually in the kV range, and travel in a flight tube. The time that the ions spent in the flight tube (i.e., time of flight) before reaching the detector depends on the mass to charge ratio of the ions. TOF-MS has a fast scan rate, but a lower sensitivity and lower mass resolution than the other two types of MS. Typically, high-resolution TOF-MS has a mass resolving power around 10^4. Modern TOF-MS uses additional techniques, such as reflectron or multi-pass, to yield an even higher resolving power of a few 10^5 ([137] and references therein). FTICR-MS and Orbitrap-MS are both Fourier transform mass analyzers. Both of them have a mass resolving power in the 10^5–10^6 range. FTICR-MS has a better performance in the lower mass range (less than a few hundred amu) [138]. In an FTICR-MS, a superconducting magnetic field confines the motion of ions to approximately cyclotron motion. The frequency of the cyclotron motion is related to the mass to charge ratio. The ions rotating at their resonance frequency induce an image current, which is measured and then Fourier transformed to the mass domain. Orbitrap-MS, instead of a superconducting magnet, uses an electrostatic field between a spindle-like central electrode and a barrel-like outer electrode for ion trapping. Due to the electrostatic force, ions cycle around the central electrode and simultaneously oscillate along its axis. Axial oscillations of ions are m/z-dependent and are detected by their image current produced on the outer electrode. Similar to FTICR-MS, this signal is measured and converted to the mass domain. Figure 4 shows an example of the mass spectrum obtained by an Orbitrap-MS. Since its introduction in 2004, Orbitrap-MS has seen a wide range of applications in the identification of large biomolecules. Due to its compact design of ion trapping, it is even being modified for future spaceflight missions [139]. For a more detailed discussion of Orbitrap-MS, we refer the interested reader to the reviews by [138,140].

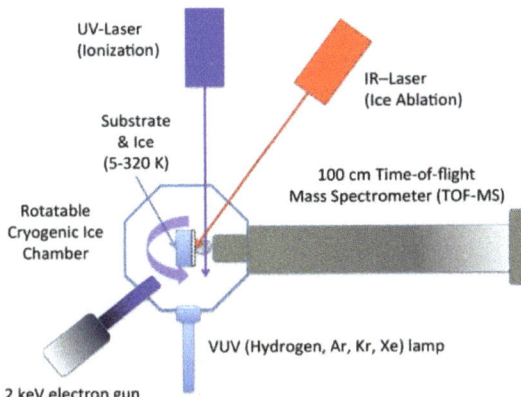

Figure 5. Schematic representation of the system developed at the Ice Spectroscopy Laboratory (ISL) at NASA Jet Propulsion Laboratory. The substrate is rotated for Vacuum Ultraviolet (VUV) irradiation and subsequent 2C-MALDI TOF-MS measurements, in situ, within the vacuum chamber. Adapted from [136], © AAS. Reproduced with permission.

To summarize, the combination of the aforementioned techniques, MALDI plus high-resolution mass spectrometry (TOF-MS, Orbitrap-MS, or FTICR-MS), shows great potential for the in situ detection and characterization of astrophysically relevant COMs. The final choice between the proper mass spectrometry technique to be coupled to MALDI would be defined by the resolution requirements of the specific system to be studied.

4.2. Promising Spectroscopy Techniques

As discussed in previous sections, even with a highly sensitive MS, it is not possible to identify structural and chiral isomers of COMs. However, there is an additional promising experimental approach to detect, in situ, COMs formed in solid-state or surface reactions and released into the gas phase. This possibility is provided by high-resolution gas-phase spectroscopy. High-resolution spectroscopy in the microwave (MW), millimeter-wave (MMW), terahertz (THz), and infrared (IR) spectral regions is a well-known and widely used tool to study rotational and ro-vibrational spectra of molecules and weakly bound molecular clusters. We refer the interested reader to the Handbook of high-resolution spectroscopy [141] and a couple of review papers [142,143]. High-resolution laboratory spectra are routinely used for the assignment of astronomical spectra obtained by ground-based radio telescopes.

High-resolution laboratory spectra provide unambiguous information on the molecular composition and structure of the analyzed sample because each molecule (including isomers) has its own, unique, spectroscopic signature due to the specific rotational and ro-vibrational energy levels defined by the composition and structure of the molecule. Importantly, it was shown that MW spectroscopy allows for the detection of molecular chirality [144,145]. Thus, high-resolution spectroscopy may also help in exploring the origin of the chirality of COMs.

Combing desorption techniques discussed above with high-resolution gas-phase spectroscopy may potentially allow for the unambiguous identification and characterization of solid-state COMs. Indeed, this approach led to the independent development of two new experimental setups. Figure 6 shows the schematics of these setups.

One setup (shown in the left panel of Figure 6) is based on a combination of a broad-band high-resolution THz spectrometer and a simple surface desorption experiment [146]. The setup allows for the direct absorption measurements of gas-phase spectra of desorbed species directly above the ice surface and is benchmarked on the detection of thermally desorbed H_2O, D_2O, and CH_3OH. It was demonstrated that the detection limit of the technique is about 10^9–10^{10} molecule cm^{-3}, which is several orders of magnitude worse than the MS detection limit [107]. However, the sensitivity of such a setup can be improved by implementing multi-pass optics, providing many passes of the radiation through the sample and, thus, increasing the absorption path length. Examples of multi-pass optics in high-resolution spectroscopic studies can be found in the literature (e.g., ([147,148]). The setup also includes a 670 nm laser diode for monitoring the ice thickness and a UV lamp for inducing energetic processing of the pure ices.

The other setup (shown in the right panel of Figure 6) is a combination of a chirped pulse Fourier transform microwave (CP-FTMW) spectrometer and a U-shaped waveguide as a molecular cell mounted in a high-vacuum chamber, where this waveguide could be cooled to cryogenic temperatures [149]. The development of the CP-FTMW technique was performed by Pate and co-workers [150] and has quickly been used by several groups to obtain high-resolution gas phase spectra (e.g., [151–154]) and to study reaction dynamics (e.g., [155–158]). The ability of a CP-FTMW spectrometer to create a phase-reproducible chirped excitation pulse of more than 10 GHz linear frequency sweep and microsecond duration with subsequent broadband signal detection makes it possible to simultaneously detect many transitions, belonging to a single or several species with meaningful relative intensities, thus allowing one to follow the time evolution of spectra and reaction dynamics. In the experiments discussed in [149], ice samples were deposited onto the waveguide inner walls, and the desorbed molecules were detected inside the waveguide by using the CP-FTMW spectrometer with a 1 GHz frequency sweep. The setup was tested by measuring the high-resolution inversion spectrum of NH_3 desorbed during the TPD experiments. The detection limit of the setup was estimated to be 6×10^{-6} mbar or 5 pmol of material. This is again several orders of magnitude worse than the MS detection limit; however, the setup has the advantage of providing broadband molecular spectra of desorbing volatiles. Potentially, the sensitivity of such a setup can be increased by increasing the power of the microwave excitation and decreasing the noise of the molecular response. However, the authors estimated that their setup should already be sensitive enough to study rotational spectra of COMs formed in the solid state and released into the gas phase. The authors also discussed possible modifications of the setup to perform UV-triggered surface chemistry.

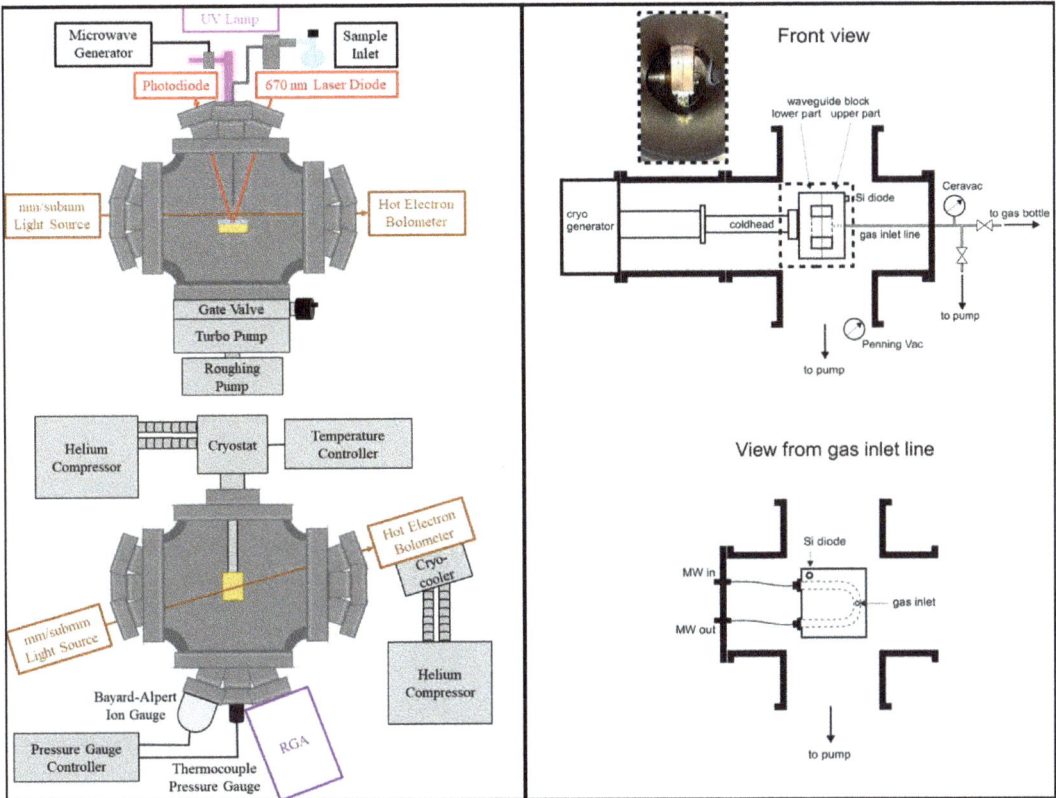

Figure 6. Left panel: Schematics of the side (upper left) and top view (lower left) of the experimental setup based on THz spectroscopy, adapted with permission from [146]. Copyright 2019, American Chemical Society. Right panel: Schematics of the front (upper right) and side view (lower right) of the experimental setup based on CP-FTMW spectroscopy, adapted with permission from [149]. Copyright 2020, American Chemical Society.

To summarize, a new experimental approach—the combination of gas-phase high-resolution spectroscopy with a surface desorption technique—has been presented by two groups. Both experimental setups show potential for in situ COM detection and characterization. However, neither has been directly tested, so far, on COMs synthesized in surface reactions. For this, proof of the applicability and suitability of this new experimental approach is necessary.

5. Conclusions

Complex organic molecules (COMs) are considered to be potential precursors of prebiotic species, such as amino acids, sugars, and nucleobases. In this review, we discussed the nature of COMs and their detection and occurrence in several astrophysical environments such as interstellar clouds, protostars, protoplanetary disks, and the outer solar system. We provided the processes and mechanisms underlying the formation and evolution of COMs in the solid state, i.e., within molecular ices of astrophysical interest. We reviewed several fundamental experimental studies, which have been performed by different groups and with various setups and facilities, resulting in a great experimental effort and advancement toward the understanding of the COM formation mechanisms. Nevertheless, many questions are still open, such as those involving the identification and analysis of more complex COMs formed in situ during processing experiments of astrophysically relevant ices.

In this context, we discussed the main problems and limitations of infrared spectroscopy and mass spectrometry, the most commonly used experimental techniques for the study of solid-state COMs. Although these techniques have greatly contributed to the advancement of knowledge in the field of astrochemistry, their further contribution to this area of research has reached its intrinsic limit. The main problems of infrared spectroscopy and mass spectrometry are related to the identification and characterization of large molecules as the main scientific goals in the field of astrochemistry are moving towards species with increased molecular complexity. However, knowledge brought from other research fields (such as gas-phase spectroscopy, chemistry, and biology) provides the possibility to initiate novel experimental techniques in the field of solid-state astrochemistry, which show great potential to overcome the main experimental limitations of present experiments. Examples have been provided in this review, focusing on important technical details and demonstrating the advantages and benefits of these novel techniques.

We believe that the experimental route described in the present review will be the next big step for laboratory astrochemistry towards a better understanding of the astrochemical pathways from simple molecules to complex organic and prebiotic molecules.

Author Contributions: All authors contributed to the conceptualization, resources, writing, review, and editing of the manuscript, according to their expertise. All authors have read and agreed to the published version of the manuscript.

Funding: T.H. acknowledges support from the European Research Council under the Horizon 2020 Framework Program via the ERC Advanced Grant Origins 83 24 28. A.P. acknowledges support from the Research Unit FOR 2285 "Debris Disks in Planetary Systems" of the Deutsche Forschungsgemeinschaft (grant JA 2107/3-2).

Conflicts of Interest: The authors declare no conflict of interest.

References

1. Herbst, E.; van Dishoeck, E.F. Complex Organic Interstellar Molecules. *Annu. Rev. Astron. Astrophys.* **2009**, *47*, 427–480. [CrossRef]
2. Jørgensen, J.K.; Belloche, A.; Garrod, R.T. Astrochemistry during the Formation of Stars. *Annu. Rev. Astron. Astrophys.* **2020**, *58*, 727–778. [CrossRef]
3. McGuire, B.A. 2018 Census of Interstellar, Circumstellar, Extragalactic, Protoplanetary Disk, and Exoplanetary Molecules. *Astrophys. J. Suppl. Ser.* **2018**, *239*. [CrossRef]
4. Gibb, E.L.; Whittet, D.C.B.; Boogert, A.C.A.; Tielens, A.G.G.M. Interstellar ice: The Infrared Space Observatory legacy. *Astrophys. J. Suppl. Ser.* **2004**, *151*, 35–73. [CrossRef]
5. Raunier, S.; Chiavassa, T.; Duvernay, F.; Borget, F.; Aycard, J.P.; Dartois, E.; d'Hendecourt, L. Tentative identification of urea and formamide in ISO-SWS infrared spectra of interstellar ices. *Astron. Astrophys.* **2004**, *416*, 165–169. [CrossRef]
6. Arce, H.G.; Santiago-García, J.; Jørgensen, J.K.; Tafalla, M.; Bachiller, R. Complex Molecules in the L1157 Molecular Outflow. *Astrophys. J.* **2008**, *681*, L21–L24. [CrossRef]
7. Oberg, K.I.; Boogert, A.C.A.; Pontoppidan, K.M.; van den Broek, S.; van Dishoeck, E.F.; Bottinelli, S.; Blake, G.A.; Evans, N.J. The Spitzer Ice Legacy: Ice Evolution from Cores to Protostars. *Astrophys. J.* **2011**, *740*. [CrossRef]
8. Oberg, K.I.; Bottinelli, S.; Jorgensen, J.K.; van Dishoeck, E.F. A Cold Complex Chemistry toward the Low-Mass Protostar B1-b: Evidence for Complex Molecule Production in Ices. *Astrophys. J.* **2010**, *716*, 825–834. [CrossRef]
9. Bacmann, A.; Taquet, V.; Faure, A.; Kahane, C.; Ceccarelli, C. Detection of complex organic molecules in a prestellar core: A new challenge for astrochemical models. *Astron. Astrophys.* **2012**, *541*, L12. [CrossRef]
10. Kahane, C.; Ceccarelli, C.; Faure, A.; Caux, E. Detection of Formamide, the Simplest but Crucial Amide, in a Solar-Type Protostar. *Astrophys. J. Lett.* **2013**, *763*, L38. [CrossRef]
11. Remijan, A.J.; Snyder, L.E.; McGuire, B.A.; Kuo, H.L.; Looney, L.W.; Friedel, D.N.; Golubiatnikov, G.Y.; Lovas, F.J.; Ilyushin, V.V.; Alekseev, E.A.; et al. Observational Results of a Multi-Telescope Campaign in Search of Interstellar Urea [$(Nh_2)_2Co$]. *Astrophys. J.* **2014**, *783*, 77. [CrossRef]
12. Fayolle, E.C.; Oberg, K.I.; Garrod, R.T.; van Dishoeck, E.F.; Bisschop, S.E. Complex organic molecules in organic-poor massive young stellar objects. *Astron. Astrophys.* **2015**, *576*, A45. [CrossRef]
13. Boogert, A.C.A.; Gerakines, P.A.; Whittet, D.C.B. Observations of the Icy Universe. *Annu. Rev. Astron. Astrophys.* **2015**, *53*, 541–581. [CrossRef]
14. Lopez-Sepulcre, A.; Jaber, A.A.; Mendoza, E.; Lefloch, B.; Ceccarelli, C.; Vastel, C.; Bachiller, R.; Cernicharo, J.; Codella, C.; Kahane, C.; et al. Shedding light on the formation of the pre-biotic molecule formamide with ASAI. *Mon. Not. R. Astron. Soc.* **2015**, *449*, 2438–2458. [CrossRef]

15. Oberg, K.I. Photochemistry and Astrochemistry: Photochemical Pathways to Interstellar Complex Organic Molecules. *Chem. Rev.* **2016**, *116*, 9631–9663. [CrossRef] [PubMed]
16. López-Sepulcre, A.; Balucani, N.; Ceccarelli, C.; Codella, C.; Dulieu, F.; Theulé, P. Interstellar Formamide (NH_2CHO), a Key Prebiotic Precursor. *ACS Earth Space Chem.* **2019**, *3*, 2122–2137. [CrossRef]
17. Thiel, V.; Belloche, A.; Menten, K.M.; Giannetti, A.; Wiesemeyer, H.; Winkel, B.; Gratier, P.; Müller, H.S.P.; Colombo, D.; Garrod, R.T. Small-scale physical and chemical structure of diffuse and translucent molecular clouds along the line of sight to Sgr B2. *Astron. Astrophys.* **2019**, *623*, A68. [CrossRef]
18. Bergner, J.B.; Martin-Domenech, R.; Oberg, K.I.; Jorgensen, J.K.; Artur de la Villarmois, E.; Brinch, C. Organic Complexity in Protostellar Disk Candidates. *ACS Earth Space Chem.* **2019**, *3*, 1564–1575. [CrossRef]
19. Kleiner, I. Spectroscopy of Interstellar Internal Rotors: An Important Tool for Investigating Interstellar Chemistry. *ACS Earth Space Chem.* **2019**, *3*, 1812–1842. [CrossRef]
20. Bockelee-Morvan, D.; Lis, D.C.; Wink, J.E.; Despois, D.; Crovisier, J.; Bachiller, R.; Benford, D.J.; Biver, N.; Colom, P.; Davies, J.K.; et al. New molecules found in comet C/1995 O1 (Hale-Bopp)—Investigating the link between cometary and interstellar material. *Astron. Astrophys.* **2000**, *353*, 1101–1114.
21. Elsila, J.E.; Glavin, D.P.; Dworkin, J.P. Cometary glycine detected in samples returned by Stardust. *Meteorit. Planet. Sci.* **2009**, *44*, 1323–1330. [CrossRef]
22. Mumma, M.J.; Charnley, S.B. The Chemical Composition of Comets-Emerging Taxonomies and Natal Heritage. *Annu. Rev. Astron. Astrophys.* **2011**, *49*, 471–524. [CrossRef]
23. Biver, N.; Bockelée-Morvan, D.; Moreno, R. Ethyl alcohol and sugar in comet C/2014 Q2 (Lovejoy). *Sci. Adv.* **2015**, *1*, e1500863. [CrossRef]
24. Biver, N.; Bockelee-Morvan, D.; Debout, V.; Crovisier, J.; Boissier, J.; Lis, D.C.; Dello Russo, N.; Moreno, R.; Colom, P.; Paubert, G.; et al. Complex organic molecules in comets C/2012 F6 (Lemmon) and C/2013 R1 (Lovejoy): Detection of ethylene glycol and formamide. *Astron. Astrophys.* **2014**, *566*, L5. [CrossRef]
25. Goesmann, F.; Rosenbauer, H.; Bredehöft, J.H.; Cabane, M.; Ehrenfreund, P.; Gautier, T.; Giri, C.; Krüger, H.; Le Roy, L.; MacDermott, A.J.; et al. Organic compounds on comet 67P/Churyumov-Gerasimenko revealed by COSAC mass spectrometry. *Science* **2015**, *349*, aab0689. [CrossRef]
26. Le Roy, L.; Altwegg, K.; Balsiger, H.; Berthelier, J.-J.; Bieler, A.; Briois, C.; Calmonte, U.; Combi, M.R.; De Keyser, J.; Dhooghe, F.; et al. Inventory of the volatiles on comet 67P/Churyumov-Gerasimenko from Rosetta/ROSINA. *Astron. Astrophys.* **2015**, *583*, A1. [CrossRef]
27. Altwegg, K.; Balsiger, H.; Berthelier, J.J.; Bieler, A.; Calmonte, U.; Fuselier, S.A.; Goesmann, F.; Gasc, S.; Gombosi, T.I.; Le Roy, L.; et al. Organics in comet 67P-a first comparative analysis of mass spectra from ROSINA-DFMS, COSAC and Ptolemy. *Mon. Not. R. Astron. Soc.* **2017**, *469*, S130–S141. [CrossRef]
28. Altwegg, K.; Balsiger, H.; Fuselier, S.A. Cometary Chemistry and the Origin of Icy Solar System Bodies: The View after Rosetta. *Annu. Rev. Astron. Astrophys.* **2019**, *57*, 113–155. [CrossRef]
29. Hadraoui, K.; Cottin, H.; Ivanovski, S.L.; Zapf, P.; Altwegg, K.; Benilan, Y.; Biver, N.; Della Corte, V.; Fray, N.; Lasue, J.; et al. Distributed glycine in comet 67P/Churyumov-Gerasimenko. *Astron. Astrophys.* **2019**, *630*. [CrossRef]
30. Schuhmann, M.; Altwegg, K.; Balsiger, H.; Berthelier, J.J.; De Keyser, J.; Fiethe, B.; Fuselier, S.A.; Gasc, S.; Gombosi, T.I.; Hänni, N.; et al. Aliphatic and aromatic hydrocarbons in comet 67P/Churyumov-Gerasimenko seen by ROSINA. *Astron. Astrophys.* **2019**, *630*, A31. [CrossRef]
31. Rubin, M.; Bekaert, D.V.; Broadley, M.W.; Drozdovskaya, M.N.; Wampfler, S.F. Volatile Species in Comet 67P/Churyumov-Gerasimenko: Investigating the Link from the ISM to the Terrestrial Planets. *ACS Earth Space Chem.* **2019**, *3*, 1792–1811. [CrossRef]
32. Charnley, S.B. Hot core chemistry. *Astrophys. Space Sci.* **1995**, *224*, 251–254. [CrossRef]
33. Geppert, W.D.; Hamberg, M.; Thomas, R.D.; Osterdahl, F.; Hellberg, F.; Zhaunerchyk, V.; Ehlerding, A.; Millar, T.; Roberts, H.; Semaniak, J. Dissociative Recombination of Protonated Methanol. *Faraday Discuss.* **2006**, *133*, 177. [CrossRef] [PubMed]
34. Garrod, R.T.; Herbst, E. Formation of methyl formate and other organic species in the warm-up phase of hot molecular cores. *Astron. Astrophys.* **2006**, *457*, 927–936. [CrossRef]
35. Henning, T.; Salama, F. Carbon in the Universe. *Science* **1998**, *282*, 2204–2210. [CrossRef]
36. Draine, B.T. Interstellar dust grains. *Annu. Rev. Astron. Astrophys.* **2003**, *41*, 241–289. [CrossRef]
37. Apai, D.; Lauretta, D.S. Planet formation and protoplanetary dust. In *Protoplanetary Dust: Astrophysical and Cosmochemical Perspectives*; Lauretta, D.S., Apai, D., Eds.; Cambridge Planetary Science/Cambridge University Press: Cambridge, UK, 2010; pp. 1–26.
38. Muñoz Caro, G.M.; Dartois, E. Prebiotic chemistry in icy grain mantles in space. An experimental and observational approach. *Chem. Soc. Rev.* **2013**, *42*, 2173–2185. [CrossRef]
39. Jäger, C.; Gail, H.-P.; Rietmeijer, F.J.M.; Nuth, J.A.; Mutschke, H.; Mennella, V. Formation of Nanoparticles and Solids. *Lab. Astrochem.* **2014**, 419–500. [CrossRef]
40. Jones, A.P.; Ysard, N.; Köhler, M.; Fanciullo, L.; Bocchio, M.; Micelotta, E.; Verstraete, L.; Guillet, V. The cycling of carbon into and out of dust. *Faraday Discuss.* **2014**, *168*, 313–326. [CrossRef]
41. Potapov, A.; McCoustra, M.R.S. Physics and chemistry on the surface of cosmic dust grains: A laboratory view. *Int. Rev. Phys. Chem.* **2021**, *40*, 299–364. [CrossRef]
42. Henning, T. Cosmic Silicates. *Annu. Rev. Astron. Astrophys.* **2010**, *48*, 21–46. [CrossRef]

43. Garrod, R.T.; Widicus Weaver, S.L.; Herbst, E. Complex chemistry in star-forming regions: An expanded gas-grain warm-up chemical model. *Astrophys. J.* **2008**, *682*, 283–302. [CrossRef]
44. Garrod, R.T. A Three-Phase Chemical Model of Hot Cores: The Formation of Glycine. *Astrophys. J.* **2013**, 765. [CrossRef]
45. Cazaux, S.; Tielens, A.G.G.M.; Ceccarelli, C.; Castets, A.; Wakelam, V.; Caux, E.; Parise, B.; Teyssier, D. The hot core around the low-mass protostar IRAS 16293-2422: Scoundrels rule! *Astrophys. J.* **2003**, *593*, L51. [CrossRef]
46. Jørgensen, J.K.; Bourke, T.L.; Nguyen Luong, Q.; Takakuwa, S. Arcsecond resolution images of the chemical structure of the low-mass protostar IRAS 16293-2422. *Astron. Astrophys.* **2011**, *534*, A100. [CrossRef]
47. Arumainayagam, C.R.; Garrod, R.T.; Boyer, M.C.; Hay, A.K.; Bao, S.T.; Campbell, J.S.; Wang, J.Q.; Nowak, C.M.; Arumainayagam, M.R.; Hodge, P.J. Extraterrestrial prebiotic molecules: Photochemistry vs. radiation chemistry of interstellar ices. *Chem. Soc. Rev.* **2019**, *48*, 2293–2314. [CrossRef]
48. Sandford, S.A.; Nuevo, M.; Bera, P.P.; Lee, T.J. Prebiotic Astrochemistry and the Formation of Molecules of Astrobiological Interest in Interstellar Clouds and Protostellar Disks. *Chem. Rev.* **2020**, *120*, 4616–4659. [CrossRef]
49. Theule, P.; Duvernay, F.; Danger, G.; Borget, F.; Bossa, J.B.; Vinogradoff, V.; Mispelaer, F.; Chiavassa, T. Thermal reactions in interstellar ice: A step towards molecular complexity in the interstellar medium. *Adv. Space Res.* **2013**, *52*, 1567–1579. [CrossRef]
50. Vidali, G. H-2 Formation on Interstellar Grains. *Chem. Rev.* **2013**, *113*, 8762–8782. [CrossRef]
51. Balucani, N.; Ceccarelli, C.; Taquet, V. Formation of complex organic molecules in cold objects: The role of gas-phase reactions. *Mon. Not. R. Astron. Soc.* **2015**, *449*, L16–L20. [CrossRef]
52. Fedoseev, G.; Cuppen, H.M.; Ioppolo, S.; Lamberts, T.; Linnartz, H. Experimental evidence for glycolaldehyde and ethylene glycol formation by surface hydrogenation of CO molecules under dense molecular cloud conditions. *Mon. Not. R. Astron. Soc.* **2015**, *448*, 1288–1297. [CrossRef]
53. Linnartz, H.; Ioppolo, S.; Fedoseev, G. Atom addition reactions in interstellar ice analogues. *Int. Rev. Phys. Chem.* **2015**, *34*, 205–237. [CrossRef]
54. Herbst, E. The synthesis of large interstellar molecules. *Int. Rev. Phys. Chem.* **2017**, *36*, 287–331. [CrossRef]
55. Potapov, A.; Jäger, C.; Henning, T.; Jonusas, M.; Krim, L. The Formation of Formaldehyde on Interstellar Carbonaceous Grain Analogs by O/H Atom Addition. *Astrophys. J.* **2017**, *846*, 131. [CrossRef]
56. Ioppolo, S.; Fedoseev, G.; Chuang, K.J.; Cuppen, H.M.; Clements, A.R.; Jin, M.; Garrod, R.T.; Qasim, D.; Kofman, V.; van Dishoeck, E.F.; et al. A non-energetic mechanism for glycine formation in the interstellar medium. *Nat. Astron.* **2020**, *5*, 197. [CrossRef]
57. Cooke, I.R.; Sims, I.R. Experimental Studies of Gas-Phase Reactivity in Relation to Complex Organic Molecules in Star-Forming Regions. *ACS Earth Space Chem.* **2019**, *3*, 1109–1134. [CrossRef]
58. Chuang, K.-J.; Fedoseev, G.; Qasim, D.; Ioppolo, S.; van Dishoeck, E.F.; Linnartz, H. Production of complex organic molecules: H-atom addition versus UV irradiation. *Mon. Not. R. Astron. Soc.* **2017**, *467*, 2552–2565. [CrossRef]
59. Altwegg, K.; Balsiger, H.; Bar-Nun, A.; Berthelier, J.J.; Bieler, A.; Bochsler, P.; Briois, C.; Calmonte, U.; Combi, M.R.; Cottin, H.; et al. Prebiotic chemicals-amino acid and phosphorus-in the coma of comet 67P/Churyumov-Gerasimenko. *Sci. Adv.* **2016**, *2*, e1600285. [CrossRef] [PubMed]
60. Krim, L.; Mencos, A. Determination of [CH3NC]/[H2C=C=NH] Abundance Ratios from N + CH3CN Solid Phase Reaction in the Temperature Range from 10 to 40 K: Application to the Complex Chemistry in Star-Forming Regions. *ACS Earth Space Chem.* **2019**, *3*, 973–979. [CrossRef]
61. Nguyen, T.; Fourre, I.; Favre, C.; Barois, C.; Congiu, E.; Baouche, S.; Guillemin, J.C.; Ellinger, Y.; Dulieu, F. Formation of amines: Hydrogenation of nitrile and isonitrile as selective routes in the interstellar medium. *Astron. Astrophys.* **2019**, *628*. [CrossRef]
62. Bonner, W.A. The Origin and Amplification of Biomolecular Chirality. *Orig. Life Evol. Biosph.* **1991**, *21*, 59–111. [CrossRef]
63. McGuire, B.A.; Carroll, P.B.; Loomis, R.A.; Finneran, I.A.; Jewell, P.R.; Remijan, A.J.; Blake, G.A. Discovery of the interstellar chiral molecule propylene oxide (CH_3CHCH_2O). *Science* **2016**, *352*, 1449–1452. [CrossRef]
64. Evans, A.C.; Meinert, C.; Giri, C.; Goesmann, F.; Meierhenrich, U.J. Chirality, photochemistry and the detection of amino acids in interstellar ice analogues and comets. *Chem. Soc. Rev.* **2012**, *41*, 5447–5458. [CrossRef] [PubMed]
65. Ghesquiere, P.; Ivlev, A.; Noble, J.A.; Theule, P. Reactivity in interstellar ice analogs: Role of the structural evolution. *Astron. Astrophys.* **2018**, *614*, A107. [CrossRef]
66. Potapov, A.; Jäger, C.; Henning, T. Thermal formation of ammonium carbamate on the surface of laboratory analogs of carbonaceous grains in protostellar envelopes and planet-forming disks. *Astrophys. J.* **2020**, *894*, 110. [CrossRef]
67. Potapov, A.; Theule, P.; Jäger, C.; Henning, T. Evidence of surface catalytic effect on cosmic dust grain analogues: The ammonia and carbon dioxide surface reaction. *Astrophys. J. Lett.* **2019**, *878*, L20. [CrossRef]
68. Miller, S.L. A Production of Amino Acids under Possible Primitive Earth Conditions. *Science* **1953**, *117*, 528–529. [CrossRef]
69. Miller, S.L.; Urey, H.C. Organic Compound Synthesis on the Primitive Earth. *Science* **1959**, *130*, 245–251. [CrossRef] [PubMed]
70. Parker, E.T.; Cleaves, H.J.; Callahan, M.P.; Dworkin, J.P.; Glavin, D.P.; Lazcano, A.; Bada, J.L. Enhanced Synthesis of Alkyl Amino Acids in Miller's 1958 H2S Experiment. *Orig. Life Evol. Biosph.* **2011**, *41*, 569–574. [CrossRef]
71. Wollrab, E.; Scherer, S.; Aubriet, F.; Carre, V.; Carlomagno, T.; Codutti, L.; Ott, A. Chemical Analysis of a "Miller-Type" Complex Prebiotic Broth Part I: Chemical Diversity, Oxygen and Nitrogen Based Polymers. *Orig. Life Evol. Biosph.* **2016**, *46*, 149–169. [CrossRef]
72. Scherer, S.; Wollrab, E.; Codutti, L.; Carlomagno, T.; da Costa, S.G.; Volkmer, A.; Bronja, A.; Schmitz, O.J.; Ott, A. Chemical Analysis of a "Miller-Type" Complex Prebiotic Broth. *Orig. Life Evol. Biosph.* **2017**, *47*, 381–403. [CrossRef] [PubMed]

73. Moore, M.H.; Donn, B.; Khanna, R.; A'Hearn, M.F. Studies of proton-irradiated cometary-type ice mixtures. *Icarus* **1983**, *54*, 388–405. [CrossRef]
74. Strazzulla, G.; Palumbo, M.E. Evolution of icy surfaces: An experimental approach. *Planet. Space Sci.* **1998**, *46*, 1339–1348. [CrossRef]
75. Gerakines, P.; Moore, M.; Hudson, R. Energetic Processing of Laboratory Ice Analogs: UV Photolysis versus Ion Bombardment. *J. Geophys. Res. Planets* **2001**, *106*, 33381. [CrossRef]
76. Hudson, R.L.; Moore, M.H. The N-3 radical as a discriminator between ion-irradiated and UV-photolyzed astronomical ices. *Astrophys. J.* **2002**, *568*, 1095–1099. [CrossRef]
77. Huels, M.A.; Parenteau, L.; Bass, A.D.; Sanche, L. Small steps on the slippery road to life: Molecular synthesis in astrophysical ices initiated by low energy electron impact. *Int. J. Mass Spectrom.* **2008**, *277*, 256–261. [CrossRef]
78. Muñoz Caro, G.M.; Dartois, E.; Boduch, P.; Rothard, H.; Domaracka, A.; Jimenez-Escobar, A. Comparison of UV and high-energy ion irradiation of methanol:ammonia ice. *Astron. Astrophys.* **2014**, *566*. [CrossRef]
79. Kim, Y.S.; Kaiser, R.I. Electron irradiation of Kuiper Belt surface ices: Ternary N_2–CH_4–CO mixtures as a case study. *Astrophys. J.* **2012**, *758*, 37. [CrossRef]
80. Materese, C.K.; Cruikshank, D.P.; Sandford, S.A.; Imanaka, H.; Nuevo, M. Ice chemistry on outer solar system bodies: Electron radiolysis of N_2-, CH_4-, and CO-containing ices. *Astrophys. J.* **2015**, *812*, 150. [CrossRef]
81. Boyer, M.C.; Rivas, N.; Tran, A.A.; Verish, C.A.; Arumainayagam, C.R. The role of low-energy ($\leq$20 eV) electrons in astrochemistry. *Surf. Sci.* **2016**, *652*, 7. [CrossRef]
82. Rothard, H.; Domaracka, A.; Boduch, P.; Palumbo, M.E.; Strazzulla, G.; da Silveira, E.F.; Dartois, E. Modification of ices by cosmic rays and solar wind. *J. Phys. B At. Mol. Opt. Phys.* **2017**, *50*, 062001. [CrossRef]
83. Ada Bibang, P.C.J.; Agnihotri, A.N.; Augé, B.; Boduch, P.; Desfrançois, C.; Domaracka, A.; Lecomte, F.; Manil, B.; Martinez, R.; Muniz, G.S.V.; et al. Ion radiation in icy space environments: Synthesis and radioresistance of complex organic molecules. *Low Temp. Phys.* **2019**, *45*, 590–597. [CrossRef]
84. Hudson, R.L.; Moore, M.H.; Dworkin, J.P.; Martin, M.P.; Pozun, Z.D. Amino Acids from Ion-Irradiated Nitrile-Containing Ices. *Astrobiology* **2008**, *8*, 771–779. [CrossRef] [PubMed]
85. Pilling, S.; Duarte, E.S.; da Silveira, E.F.; Balanzat, E.; Rothard, H.; Domaracka, A.; Boduch, P. Radiolysis of ammonia-containing ices by energetic, heavy, and highly charged ions inside dense astrophysical environments. *Astron. Astrophys.* **2010**, *509*. [CrossRef]
86. Holtom, P.D.; Bennett, C.J.; Osamura, Y.; Mason, N.J.; Kaiser, R.I. A combined experimental and theoretical study on the formation of the amino acid glycine (NH_2CH_2COOH) and its isomer ($CH_3NHCOOH$) in extraterrestrial ices. *Astrophys. J.* **2005**, *626*, 940–952. [CrossRef]
87. Lafosse, A.; Bertin, M.; Domaracka, A.; Pliszka, D.; Illenberger, E.; Azria, R. Reactivity induced at 25 K by low-energy electron irradiation of condensed NH_3-CH_3COOD (1:1) mixture. *Phys. Chem. Chem. Phys.* **2006**, *8*, 5564–5568. [CrossRef]
88. Esmaili, S.; Bass, A.D.; Cloutier, P.; Sanche, L.; Huels, M.A. Glycine formation in CO_2:CH_4:NH_3 ices induced by 0–70 eV electrons. *J. Chem. Phys.* **2018**, *148*. [CrossRef]
89. Gerakines, P.A.; Moore, M.H.; Hudson, R.L. Ultraviolet photolysis and proton irradiation of astrophysical ice analogs containing hydrogen cyanide. *Icarus* **2004**, *170*, 202–213. [CrossRef]
90. Kanuchova, Z.; Urso, R.G.; Baratta, G.A.; Brucato, J.R.; Palumbo, M.E.; Strazzulla, G. Synthesis of formamide and isocyanic acid after ion irradiation of frozen gas mixtures. *Astron. Astrophys.* **2016**, *585*. [CrossRef]
91. Jones, B.M.; Bennett, C.J.; Kaiser, R.I. Mechanistical Studies on the Production of Formamide (H_2NCHO) within Interstellar Ice Analogs. *Astrophys. J.* **2011**, *734*, 78. [CrossRef]
92. Frigge, R.; Zhu, C.; Turner, A.M.; Abplanalp, M.J.; Bergantini, A.; Sun, B.J.; Chen, Y.L.; Chang, A.H.H.; Kaiser, R.I. A Vacuum Ultraviolet Photoionization Study on the Formation of N-methyl Formamide ($HCONHCH_3$) in Deep Space: A Potential Interstellar Molecule with a Peptide Bond. *Astrophys. J.* **2018**, *862*. [CrossRef]
93. Bennett, C.J.; Kaiser, R.I. On the formation of glycolaldehyde ($HCOCH_2OH$) and methyl formate ($HCOOCH_3$) in interstellar ice analogs. *Astrophys. J.* **2007**, *661*, 899–909. [CrossRef]
94. Maity, S.; Kaiser, R.I.; Jones, B.M. Formation of complex organic molecules in methanol and methanol-carbon monoxide ices exposed to ionizing radiation—A combined FTIR and reflectron time-of-flight mass spectrometry study. *Phys. Chem. Chem. Phys.* **2015**, *17*, 3081–3114. [CrossRef]
95. Muñoz Caro, G.M.; Meierhenrich, U.J.; Schutte, W.A.; Barbier, B.; Segovia, A.A.; Rosenbauer, H.; Thiemann, W.H.P.; Brack, A.; Greenberg, J.M. Amino acids from ultraviolet irradiation of interstellar ice analogues. *Nature* **2002**, *416*, 403–406. [CrossRef] [PubMed]
96. Bernstein, M.P.; Dworkin, J.P.; Sandford, S.A.; Cooper, G.W.; Allamandola, L.J. Racemic amino acids from the ultraviolet photolysis of interstellar ice analogues. *Nature* **2002**, *416*, 401–403. [CrossRef] [PubMed]
97. Elsila, J.E.; Dworkin, J.P.; Bernstein, M.P.; Martin, M.P.; Sandford, S.A. Mechanisms of amino acid formation in interstellar ice analogs. *Astrophys. J.* **2007**, *660*, 911–918. [CrossRef]
98. Pilling, S.; Andrade, D.P.P.; Neto, A.C.; Rittner, R.; de Brito, A.N. DNA Nucleobase Synthesis at Titan Atmosphere Analog by Soft X-rays. *J. Phys. Chem. A* **2009**, *113*, 11161–11166. [CrossRef]
99. Meinert, C.; Myrgorodska, I.; de Marcellus, P.; Buhse, T.; Nahon, L.; Hoffmann, S.V.; d'Hendecourt, L.L.; Meierhenrich, U.J. Ribose and related sugars from ultraviolet irradiation of interstellar ice analogs. *Science* **2016**, *352*, 208–212. [CrossRef]

100. Nuevo, M.; Cooper, G.; Sandford, S.A. Deoxyribose and deoxysugar derivatives from photoprocessed astrophysical ice analogues and comparison to meteorites. *Nat. Commun.* **2018**, *9*. [CrossRef]
101. Oba, Y.; Takano, Y.; Naraoka, H.; Watanabe, N.; Kouchi, A. Nucleobase synthesis in interstellar ices. *Nat. Commun.* **2019**, *10*. [CrossRef]
102. Chen, Y.J.; Ciaravella, A.; Muñoz Caro, G.M.; Cecchi-Pestellini, C.; Jiménez-Escobar, A.; Juang, K.J.; Yih, T.S. Soft x-ray irradiation of methanol ice: Formation of products as a function of photon energy. *Astrophys. J.* **2013**, *778*, 162. [CrossRef]
103. Henderson, B.L.; Gudipati, M.S. Direct detection of complex organic products in ultraviolet (lyα) and electron-irradiated astrophysical and cometary ice analogs using two-step laser ablation and ionization mass spectrometry. *Astrophys. J.* **2015**, *800*, 66. [CrossRef]
104. Ciaravella, A.; Jimenez-Escobar, A.; Cecchi-Pestellini, C.; Huang, C.H.; Sie, N.E.; Caro, G.M.M.; Chen, Y.J. Synthesis of Complex Organic Molecules in Soft X-Ray Irradiated Ices. *Astrophys. J.* **2019**, *879*, 21. [CrossRef]
105. Martin-Domenech, R.; Oberg, K.; Rajappan, M. Formation of NH_2CHO and CH_3CHO upon UV Photoprocessing of Interstellar Ice Analogs. *Astrophys. J.* **2020**, *894*, 15. [CrossRef]
106. Ciaravella, A.; Muñoz Caro, G.M.; Jiménez-Escobar, A.; Cecchi-Pestellini, C.; Hsiao, L.-C.; Huang, C.-H.; Chen, Y.-J. X-ray processing of a realistic ice mantle can explain the gas abundances in protoplanetary disks. *Proc. Natl. Acad. Sci. USA* **2020**, *117*, 16149. [CrossRef]
107. Fraser, H.J.; Collings, M.P.; McCoustra, M.R.S. Laboratory surface astrophysics experiment. *Rev. Sci. Instrum.* **2002**, *73*, 2161–2170. [CrossRef]
108. Teolis, B.D.; Loeffler, M.J.; Raut, U.; Fama, A.; Baragiola, R.A. Infrared reflectance spectroscopy on thin films: Interference effects. *Icarus* **2007**, *190*, 274–279. [CrossRef]
109. Coustenis, A.; Hirtzig, M. Cassini-Huygens results on Titan's surface. *Res. Astron. Astrophys.* **2009**, *9*, 249–268. [CrossRef]
110. Krüger, H.; Stephan, T.; Engrand, C.; Briois, C.; Siljestrom, S.; Merouane, S.; Baklouti, D.; Fischer, H.; Fray, N.; Hornung, K.; et al. Cosima-Rosetta calibration for in situ characterization of 67P/Churyumov-Gerasimenko cometary inorganic compounds. *Planet. Space Sci.* **2015**, *117*, 35–44. [CrossRef]
111. Noble, J.A.; Theule, P.; Duvernay, F.; Danger, G.; Chiavassa, T.; Ghesquiere, P.; Mineva, T.; Talbi, D. Kinetics of the NH_3 and CO_2 solid-state reaction at low temperature. *Phys. Chem. Chem. Phys.* **2014**, *16*, 23604–23615. [CrossRef] [PubMed]
112. Duvernay, F.; Dufauret, V.; Danger, G.; Theule, P.; Borget, F.; Chiavassa, T. Chiral molecule formation in interstellar ice analogs: Alpha-aminoethanol $NH_2CH(CH_3)OH$. *Astron. Astrophys.* **2010**, *523*, A79. [CrossRef]
113. Abplanalp, M.J.; Forstel, M.; Kaiser, R.I. Exploiting single photon vacuum ultraviolet photoionization to unravel the synthesis of complex organic molecules in interstellar ices. *Chem. Phys. Lett.* **2016**, *644*, 79–98. [CrossRef]
114. Nuevo, M.; Milam, S.N.; Sandford, S.A. Nucleobases and Prebiotic Molecules in Organic Residues Produced from the Ultraviolet Photo-Irradiation of Pyrimidine in NH_3 and H_2O+NH_3 Ices. *Astrobiology* **2012**, *12*, 295–314. [CrossRef] [PubMed]
115. Karas, M.; Hillenkamp, F. Laser desorption ionization of proteins with molecular masses exceeding 10,000 daltons. *Anal. Chem.* **1988**, *60*, 2299–2301. [CrossRef] [PubMed]
116. Beavis, R.C.; Chait, B.T.; Fales, H.M. Cinnamic acid derivatives as matrices for ultraviolet laser desorption mass spectrometry of proteins. *Rapid Commun. Mass Spectrom.* **1989**, *3*, 432–435. [CrossRef]
117. Merchant, M.; Weinberger, S.R. Recent advancements in surface-enhanced laser desorption/ionization-time of flight-mass spectrometry. *Electrophoresis* **2000**, *21*, 1164–1177. [CrossRef]
118. Paardekooper, D.M.; Bossa, J.B.; Isokoski, K.; Linnartz, H. Laser desorption time-of-flight mass spectrometry of ultraviolet photo-processed ices. *Rev. Sci. Instrum.* **2014**, *85*, 104501. [CrossRef]
119. Yang, R.; Gudipati, M.S. Novel two-step laser ablation and ionization mass spectrometry (2S-LAIMS) of actor-spectator ice layers: Probing chemical composition of D_2O ice beneath a H_2O ice layer. *J. Chem. Phys.* **2014**, *140*, 104202. [CrossRef]
120. Strupat, K.; Karas, M.; Hillenkamp, F. 2,5-Dihydroxybenzoic acid: A new matrix for laser desorption—Ionization mass spectrometry. *Int. J. Mass Spectrom. Ion Process.* **1991**, *111*, 89–102. [CrossRef]
121. Dreisewerd, K. The Desorption Process in MALDI. *Chem. Rev.* **2003**, *103*, 395–426. [CrossRef]
122. Berkenkamp, S.; Menzel, C.; Karas, M.; Hillenkamp, F. Performance of Infrared Matrix-assisted Laser Desorption/Ionization Mass Spectrometry with Lasers Emitting in the 3 μm Wavelength Range. *Rapid Commun. Mass Spectrom.* **1997**, *11*, 8. [CrossRef]
123. Caldwell, K.L.; Murray, K.K. Mid-infrared matrix assisted laser desorption ionization with a water/glycerol matrix. *Appl. Surf. Sci.* **1998**, *127*, 242–247. [CrossRef]
124. Pirkl, A.; Soltwisch, J.; Draude, F.; Dreisewerd, K. Infrared Matrix-Assisted Laser Desorption/Ionization Orthogonal-Time-of-Flight Mass Spectrometry Employing a Cooling Stage and Water Ice as a Matrix. *Anal. Chem.* **2012**, *84*, 5669. [CrossRef]
125. Focsa, C.; Mihesan, C.; Ziskind, M.; Chazallon, B.; Therssen, E.; Desgroux, P.; Destombes, J.L. Wavelength-selective vibrationally excited photodesorption with tunable IR sources. *J. Phys. Condens. Matter* **2006**, *18*, S1357–S1387. [CrossRef]
126. Maity, S.; Kaiser, R.I.; Jones, B.M. Formation of ketene (H_2CCO) in interstellar analogous methane (CH_4)-carbon monoxide (CO) ices: A combined FTIR and reflectron time-of-flight mass spectroscopic study. *Astrophys. J.* **2014**, *789*, 36. [CrossRef]
127. Zhang, J.; Han, F.; Pei, L.; Kong, W.; Li, A. Far-Infrared Spectroscopy of Cationic Polycyclic Aromatic Hydrocarbons: Zero Kinetic Energy Photoelectron Spectroscopy of Pentacene Vaporized from Laser Desorption. *Astrophys. J.* **2010**, *715*, 485–492. [CrossRef]

128. Hama, T.; Watanabe, N.; Kouchi, A.; Yokoyama, M. Spin Temperature of Water Molecules Desorbed from the Surfaces of Amorphous Solid Water, Vapor-Deposited and Produced from Photolysis of a Ch_4/O_2 Solid Mixture. *Astrophys. J. Lett.* **2011**, *738*, L15. [CrossRef]
129. Gavilan, L.; Lemaire, J.L.; Vidali, G.; Sabri, T.; Jaeger, C. The Formation of Molecular Hydrogen on Silicate Dust Analogs: The Rotational Distribution. *Astrophys. J.* **2014**, *781*, 79. [CrossRef]
130. Karas, M.; Bachmann, D.; Hillenkamp, F. Influence of the wavelength in high-irradiance ultraviolet laser desorption mass spectrometry of organic molecules. *Anal. Chem.* **1985**, *57*, 2935. [CrossRef]
131. Clark, A.E.; Kaleta, E.J.; Arora, A.; Wolk, D.M. Matrix-Assisted Laser Desorption Ionization-Time of Flight Mass Spectrometry: A Fundamental Shift in the Routine Practice of Clinical Microbiology. *Clin. Microbiol. Rev.* **2013**, *26*, 547–603. [CrossRef] [PubMed]
132. Knochenmuss, R. Ion formation mechanisms in UV-MALDI. *Analyst* **2006**, *131*, 966–986. [CrossRef]
133. Trimpin, S.; Wang, B.; Inutan, E.D.; Li, J.; Lietz, C.B.; Harron, A.; Pagnotti, V.S.; Sardelis, D.; McEwen, C.N. A Mechanism for Ionization of Nonvolatile Compounds in Mass Spectrometry: Considerations from MALDI and Inlet Ionization. *J. Am. Soc. Mass Spectrom.* **2012**, *23*, 1644–1660. [CrossRef]
134. Niu, S.; Zhang, W.; Chait, B.T. Direct comparison of infrared and ultraviolet wavelength matrix-assisted laser desorption/ionization mass spectrometry of proteins. *J. Am. Soc. Mass Spectrom.* **1998**, *9*, 1–7. [CrossRef]
135. Luo, Y.; Li, T.; Yu, F.; Kramer, T.; Cristea, I.M. Resolving the composition of protein complexes using a MALDI LTQ Orbitrap. *J. Am. Soc. Mass Spectrom.* **2010**, *21*, 34–46. [CrossRef] [PubMed]
136. Gudipati, M.S.; Yang, R. In-situ probing of radiation-induced processing of organics in astrophysical ice analogs—Novel laser desorption laser ionization time-of-flight mass spectroscopic studies. *Astrophys. J. Lett.* **2012**, *756*, L24. [CrossRef]
137. Xian, F.; Hendrickson, C.L.; Marshall, A.G. High Resolution Mass Spectrometry. *Anal. Chem.* **2012**, *84*, 708–719. [CrossRef] [PubMed]
138. Zubarev, R.A.; Makarov, A. Orbitrap Mass Spectrometry. *Anal. Chem.* **2013**, *85*, 5288–5296. [CrossRef] [PubMed]
139. Arevalo Jr, R.; Selliez, L.; Briois, C.; Carrasco, N.; Thirkell, L.; Cherville, B.; Colin, F.; Gaubicher, B.; Farcy, B.; Li, X.; et al. An Orbitrap-based laser desorption/ablation mass spectrometer designed for spaceflight. *Rapid Commun. Mass Spectrom.* **2018**, *32*, 1875–1886. [CrossRef]
140. Eliuk, S.; Makarov, A. Evolution of Orbitrap Mass Spectrometry Instrumentation. *Annu. Rev. Anal. Chem.* **2015**, *8*, 61–80. [CrossRef]
141. Quack, M.; Merkt, F.E. *Handbook of High-Resolution Spectroscopy*; John Wiley & Sons Ltd.: Hoboken, NJ, USA, 2011.
142. Herman, M.; Georges, R.; Hepp, M.; Hurtmans, D. High resolution Fourier transform spectroscopy of jet-cooled molecules. *Int. Rev. Phys. Chem.* **2000**, *19*, 277–325. [CrossRef]
143. Potapov, A.; Asselin, P. High-resolution jet spectroscopy of weakly bound binary complexes involving water. *Int. Rev. Phys. Chem.* **2014**, *33*, 275–300. [CrossRef]
144. Grabow, J.U. Fourier Transform Microwave Spectroscopy: Handedness Caught by Rotational Coherence. *Angew. Chem. Int. Ed.* **2013**, *52*, 11698–11700. [CrossRef]
145. Patterson, D.; Schnell, M.; Doyle, J.M. Enantiomer-specific detection of chiral molecules via microwave spectroscopy. *Nature* **2013**, *497*, 475. [CrossRef]
146. Yocum, K.M.; Smith, H.H.; Todd, E.W.; Mora, L.; Gerakines, P.A.; Milam, S.N.; Weaver, S.L.W. Millimeter/Submillimeter Spectroscopic Detection of Desorbed Ices: A New Technique in Laboratory Astrochemistry. *J. Phys. Chem. A* **2019**, *123*, 8702–8708. [CrossRef] [PubMed]
147. Harada, K.; Tanaka, K.; Tanaka, T.; Nanbu, S.; Aoyagi, M. Millimeter-wave spectroscopy of the internal-rotation band of the He-HCN complex and the intermolecular potential energy surface. *J. Chem. Phys.* **2002**, *117*, 7041–7050. [CrossRef]
148. Potapov, A.V.; Surin, L.A.; Schlemmer, S.; Giesen, T.F. Submillimeter-wave spectroscopy of the K = 2−1 subband of the Ne-CO complex. *J. Mol. Spectrosc.* **2011**, *270*, 116–119. [CrossRef]
149. Theulé, P.; Endres, C.; Hermanns, M.; Zingsheim, O.; Bossa, J.B.; Potapov, A. High-Resolution Gas Phase Spectroscopy of Molecules Desorbed from an Ice Surface: A Proof-of-Principle Study. *ACS Earth Space Chem.* **2020**, *4*, 86–91. [CrossRef]
150. Brown, G.G.; Dian, B.C.; Douglass, K.O.; Geyer, S.M.; Shipman, S.T.; Pate, B.H. A broadband Fourier transform microwave spectrometer based on chirped pulse excitation. *Rev. Sci. Instrum.* **2008**, *79*, 053103. [CrossRef]
151. Jahn, M.K.; Dewald, D.A.; Wachsmuth, D.; Grabow, J.U.; Mehrotra, S.C. Rapid capture of large amplitude motions in 2,6-difluorophenol: High-resolution fast-passage FT-MW technique. *J. Mol. Spectrosc.* **2012**, *280*, 54–60. [CrossRef]
152. Perez, C.; Lobsiger, S.; Seifert, N.A.; Zaleski, D.P.; Temelso, B.; Shields, G.C.; Kisiel, Z.; Pate, B.H. Broadband Fourier transform rotational spectroscopy for structure determination: The water heptamer. *Chem. Phys. Lett.* **2013**, *571*, 1–15. [CrossRef]
153. Medcraft, C.; Wolf, R.; Schnell, M. High-Resolution Spectroscopy of the Chiral Metal Complex [CpRe(CH$_3$)(CO)(NO)]: A Potential Candidate for Probing Parity Violation. *Angew. Chem. Int. Ed.* **2014**, *53*, 11656–11659. [CrossRef] [PubMed]
154. Hermanns, M.; Wehres, N.; Lewen, F.; Muller, H.S.P.; Schlemmer, S. Rotational spectroscopy of the two higher energy conformers of 2-cyanobutane. *J. Mol. Spectrosc.* **2019**, *358*, 25–36. [CrossRef]
155. Dian, B.C.; Brown, G.G.; Douglass, K.O.; Pate, B.H. Measuring picosecond isomerization kinetics via broadband microwave spectroscopy. *Science* **2008**, *320*, 924–928. [CrossRef] [PubMed]

156. Abeysekera, C.; Zack, L.N.; Park, G.B.; Joalland, B.; Oldham, J.M.; Prozument, K.; Ariyasingha, N.M.; Sims, I.R.; Field, R.W.; Suits, A.G. A chirped-pulse Fourier-transform microwave/pulsed uniform flow spectrometer. II. Performance and applications for reaction dynamics. *J. Chem. Phys.* **2014**, *141*, 214203. [CrossRef] [PubMed]
157. Prozument, K.; Park, G.B.; Shaver, R.G.; Vasiliou, A.K.; Oldham, J.M.; David, D.E.; Muenter, J.S.; Stanton, J.F.; Suits, A.G.; Ellison, G.B.; et al. Chirped-pulse millimeter-wave spectroscopy for dynamics and kinetics studies of pyrolysis reactions. *Phys. Chem. Chem. Phys.* **2014**, *16*, 15739–15751. [CrossRef] [PubMed]
158. Abeysekera, C.; Joalland, B.; Ariyasingha, N.; Zack, L.N.; Sims, I.R.; Field, R.W.; Suits, A.G. Product Branching in the Low Temperature Reaction of CN with Propyne by Chirped-Pulse Microwave Spectroscopy in a Uniform Supersonic Flow. *J. Phys. Chem. Lett.* **2015**, *6*, 1599–1604. [CrossRef]

Perspective

Did Solid Surfaces Enable the Origin of Life?

İrep Gözen [1,2]

1 Centre for Molecular Medicine Norway, Faculty of Medicine, University of Oslo, 0318 Oslo, Norway; irep@uio.no
2 Department of Chemistry, Faculty of Mathematics and Natural Sciences, University of Oslo, 0315 Oslo, Norway

Abstract: In this perspective article, I discuss whether and how solid surfaces could have played a key role in the formation of membranous primitive cells on the early Earth. I argue why surface energy could have been used by prebiotic amphiphile assemblies for unique morphological transformations, and present recent experimental findings showing the surface-dependent formation and behavior of sophisticated lipid membrane structures. Finally, I discuss the possible unique contributions of such surface-adhered architectures to the transition from prebiotic matter to living systems.

Keywords: protocell; compartment; solid interface; lipid; origin of life

1. Introduction

Primitive cell formation and development is closely linked to the origin of life, which is considered to be one of the unsolved fundamental scientific problems. Researchers seek answers in the laboratory by simulating early Earth conditions as they existed approximately 3.8 billion years ago, with the aim to instantiate autonomously forming primitive compartments that can develop into self-sustaining and reproducing cells.

Different types of model structures are used to mimic protocells, the hypothetical precursors of living cells. One category is based on droplets, formed via liquid–liquid phase separation [1,2]. A prominent example is coacervates [3,4], droplets concentrated with macromolecules, dispersed in a more dilute liquid phase. The other category of protocell models comprises amphiphiles, such as fatty acids or phospholipids, in aqueous environments [3,5,6]. Amphiphiles can spontaneously self-assemble to form spherical membrane compartments freely suspended in an aqueous phase. Lipid membranes envelope modern cells, and are ubiquitous among almost all cellular structures today. It has been shown that lipids could have been synthesized under prebiotic conditions [7,8] or delivered to the early Earth by meteorites [9,10]. The potential availability of lipid species at that time, and their ability to spontaneously self-assemble to ordered membranous structures, are two arguments in favor of liposomal compartments as plausible building blocks of primitive cells. Naturally present were, besides water, rocks and minerals as fundamental constituents of the early Earth. The possible role of solid surfaces in the process of primitive cell formation and development might have been largely overlooked.

In this perspective article, I discuss the possible involvement of solid surfaces in the emergence of membranous primitive cells on the early Earth. I briefly explain why surfaces intrinsically possess energy, and how it can be used by biosurfactants for morphological transformations. I then progress to recent experimental observations of solid surface-dependent membrane transformation pathways that consistently lead to robust model protocells. Finally, I discuss the variety of resulting unique primitive structures, their potential advantages, and possible contributions to the transition from non-living to animated matter.

2. Intrinsic Energy of Surfaces

Surface energy emerges from the interactions of individual atoms at interfaces. When a block of a solid material is cut into two pieces (Figure 1A), two new solid interfaces are being created. In the bulk (unaltered) phase of the material, atoms maintain bonds with all surrounding atoms, and experience attractive interactions in all directions (Figure 1B). Interatomic bonds at the broken interface, however, are now disrupted, and the surface atoms are lacking favorable interactions (Figure 1C).

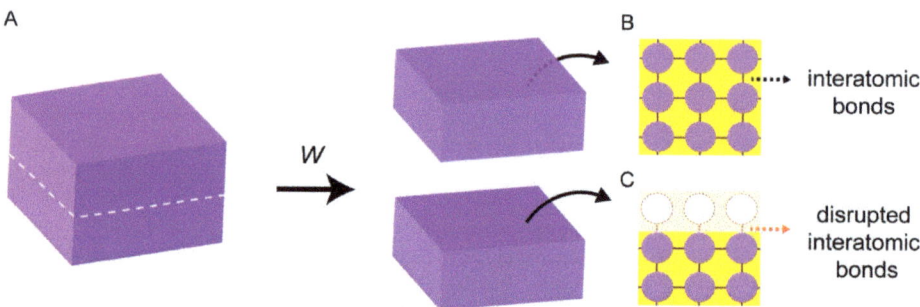

Figure 1. Surface energy. (**A**) Block of solid material. Work (*W*) is performed to split the solid block into two pieces. Inside the bulk phase of the block, the atoms experience attractive forces established with all their neighboring atoms (**B**). Atoms located at the newly created interface have disrupted, unbalanced cohesive interactions (**C**).

Work (*W*) needs to be performed to break the bonds between the atoms, and to create new surfaces. The work that is performed to create each unit area of a new surface is equivalent of surface energy or surface tension (σ) [11,12]. The magnitude of the work is associated with the nature of the bonds in the material. High energy surfaces ($\sigma > 200$ mN/m [12,13]) are composed of atoms with strong bonds such as covalent, ionic, or metallic bonds. Most natural surfaces including metals, diamond, silica glasses, and ceramics fall under this category. The atoms at low energy surfaces ($\sigma < 50$ mN/m [12,13].) e.g., plastics or resins, are attached to each other with rather weak bonds, e.g., van der Waals or hydrogen bonds.

All high energy surfaces will tend to reduce their surface energy, hence the overall Gibbs free energy. If surfaces cannot establish bonds with atoms of their own kind, they will seek physical contact with other matter, e.g., water or surface-active molecules (surfactants), and reduce their surface energy. The energy of a solid surface can therefore be harnessed by other soft and deformable materials with the ability to cover the surface. In this context, the interaction of biological and biomimetic membranes, consisting largely of lipids, with solid high energy surfaces is of special interest for the autonomous formation of primitive protocells [14].

Solid surfaces were abundant on the early Earth in the form of minerals and rocks, which are defined as aggregates of several minerals. From the beginning of planet formation until today, there have been 10 stages of mineral evolution, and each of them increased mineral diversity on the Earth [15,16]. Stages 1–6 occurred during Accretion, the Hadean eon and the following Eoarchean era, all together the 'early Earth' (4.6–3.6 Gya). During this period, about 1500 different mineral species appeared [15,16]. Several minerals have been shown to be able to catalyze peptide, lipid, and nucleic acid synthesis [17]. Two specific minerals which have been described as potentially important in the context of the origin of life, are clay [18,19] and quartz (SiO_2) [16,20]. Natural quartz specimens, along with synthetic SiO_2 surfaces, have been used as solid substrates in the key studies that constitute the experimental foundation of my perspective.

3. Biomembrane Transformations on Solid Surfaces

Giant unilamellar vesicles, encapsulating an aqueous volume within a spherical lipid bilayer, are a common experimental model system for protocell studies [21]. Although much more complex both in structure and function, contemporary biological cells are also enveloped in a lipid-based membrane, and are of similar size. The potential availability of lipid species in the prebiotic environment, and their ability to spontaneously self-assemble to cell-like compartments, make lipid vesicles plausible biomimetic architectures for studies of primitive cells. Among different lipid species, the structurally simpler fatty acids are considered to have been more prevalent on the early Earth. Experimental evidence for prebiotic pathways to simple amphiphiles is well-established [22]. Monocarboxylic acids were also found in extraterrestrial sources, e.g., carbonaceous meteorites [9,10]. However, experimental findings suggest that phospholipids could have also been present under prebiotic conditions [7,8,23].

When a giant unilamellar lipid compartment, a spherical continuous bilayer membrane, is brought in contact with a high energy solid substrate in an aqueous medium (Figure 2A), the vesicular membrane initially partially adheres onto the substrate, adopting a dome-like shape (Figure 2B). This process, in which a fluid material makes and maintains physical contact with a solid substrate, is termed 'wetting'. Wetting of surfaces by lipid membranes, which display two-dimensional fluid properties, occurs in various forms [24] depending on the surface material, the overall surface energy, and the composition of the lipid membrane.

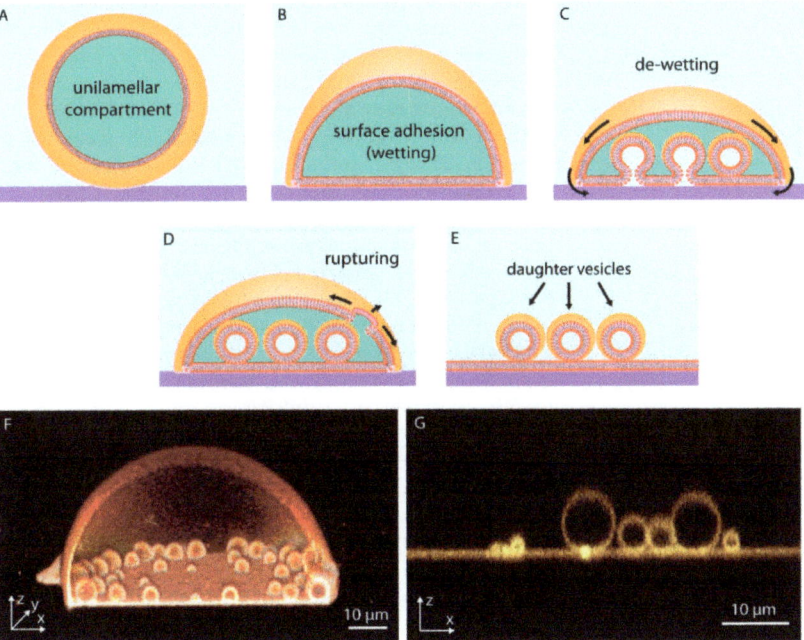

Figure 2. Surface-enhanced subcompartmentalization and pseudo-division of model protocells. An isolated giant unilamellar compartment (**A**), adheres on a solid substrate upon contact (**B**). Reversed adhesion leads to the de-wetting of the substrate, release of small membrane regions from the surface, and formation of small subcompartments (**C**). The disintegration of the upper (with respect to the surface) protocell membrane due to an increase in tension (**D**), leads to the transformation of daughter cells to independent surface-adhered compartments (**E**–**G**) shows confocal micrographs corresponding to (**D,E**).

The membrane-substrate adhesion can be enhanced by the presence of multivalent ions, e.g., Ca^{2+}, Mg^{2+}, in the aqueous environment surrounding the compartment. If the adhesion weakens, for example due to a decrease in the concentration of the ions in the surrounding solution, the membrane starts to partially de-wet and lift off from the surface (Figure 2C). The released membrane regions and newly formed invaginations possess nanoscale membrane curvature. Membrane tension rises because at these highly curved regions, the distance between the individual lipid molecules increases and their hydrophobic moieties become exposed to water [25]. The locally high tension initiates flow of lipid material from membrane areas of lower tension towards high tension regions (Marangoni flow) [26,27] (Figure 2C). This newly arriving material transforms the membrane invaginations over time into giant spherical compartments. These membranous subcompartments show resemblance to membrane-enclosed organelles inside modern cells; they were shown to take up, and concentrate ambient compounds, similar to cellular organelles [27].

The enveloping membrane can rupture and completely disintegrate, leaving behind the subcompartments (Figure 2D), each of which can thus be considered an individual daughter protocell (Figure 2E). This phenomenon may well be viewed as a pseudo-division mechanism of primitive cells, which were certainly lacking the complex biological machinery required to enable contraction and fission. Figure 2F shows a 3D cross-sectional confocal micrograph of a subcompartmentalized protocell and Figure 2G the daughter cells after disintegration of the enveloping membrane.

Unilamellar compartments are not the only known kind of self-assembled lipid structures. If there is sufficient lipid material available, they can form multilamellar vesicles (MLV), in which several lipid bilayers are packed on top of each other in an onion shell-like fashion (Figure 3A). The individual layers in such structure are interconnected through membrane defects, and can thus be viewed as a single large fluid reservoir. When a MLV is brought in contact with a high energy substrate in an aqueous surrounding (Figure 3A), it spreads isotropically and wets the entire available surface until the accessible lipid material in the reservoir is completely depleted. On glass, the membranes generally spread as a bilayer [28,29] (Figure 3B), while on hydrophobic surfaces as a monolayer [11,12,30] (Figure 3C). The most commonly experimentally investigated form of bio (mimetic) membranes is the bilayer, as the plasma membrane of contemporary cells as well as most of the intracellular membranes are single lipid bilayers.

On SiO_2 and some metal oxides, the lipid reservoirs spread as a double bilayer membrane [26,31,32] (Figure 3D), where the proximal bilayer (in closer proximity to the surface) is immobilized on the solid substrate as it spreads. The distal (of greater distance to the surface) membrane is simultaneously expanding along the spreading edge in a tank thread-like motion (Figure 3D). The spreading continues until the membrane tension exceeds lysis tension (5–10 mN/m), resulting in rupturing of the distal membrane (Figure 3D) [32,33]. Some of the distal membrane areas however, especially the regions pinned to the proximal bilayer, remain intact. The curved edge of these distal membrane regions grows as the ruptures propagate, leading to an increase in membrane edge (line) energy [34]. To avoid the increasing edge energy cost, these distal membrane regions rapidly wrap into lipid nanotubes (Figure 3F,G). Lipid nanotubes are highly curved cylindrical membrane structures. They represent a local energy minimum, yet remain costly to the overall system due to high bending energy. To reduce bending energy, fragments of the nanotubes swell into spherical vesicular compartments over time (Figure 3H). A protocell network emerges in this process, consisting of several lipid compartments interconnected by lipid nanotubes [11,12,33] (Figure 3I).

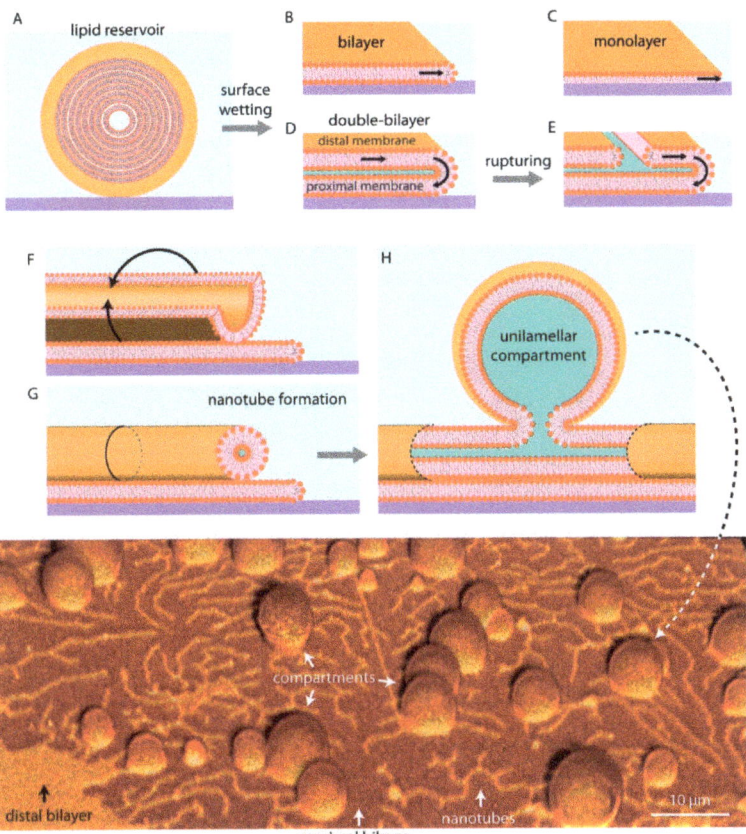

Figure 3. Lipid reservoir-surface interactions. When multilamellar vesicles, i.e., lipid reservoirs, (**A**) come in physical contact with solid surfaces, they can wet the surface in different ways and can spread as a bilayer (**B**), monolayer (**C**), or a double bilayer (**D**). The distal membrane (upper bilayer of a double bilayer membrane with respect to the surface) ruptures due to an increase in membrane tension (**E**). The distal membrane regions remaining on the proximal (lower) membrane wrap up (**F**) to form lipid nanotubes (**G**), alleviating tension at the membrane edges. Fragments of nanotubes swell over time (**H**), resulting in protocell-nanotube networks. (**I**) Confocal micrograph showing protocells connected via lipid nanotubes. Please note that the schematic drawings are not to scale: the lipid nanotubes are typically 100 nm in diameter, where a lipid bilayer is about 5 nm thick.

The protocell networks can be formed from membranes consisting of phospholipids only, or from mixtures of phospholipids and fatty acids [35]. The networks have been shown to encapsulate water soluble compounds, even RNA and DNA [35–37]. Especially RNA has been a molecule of interest in origin of life studies, as it can act uniquely both as a genetic information carrier and an enzyme analog, catalyzing its own replication [38]. In an environment where molecules such as proteins were not yet available, the capability of RNA to replicate without the need for an enzyme would be an advantageous feature. The ability of model protocells to encapsulate and maintain RNA has therefore been widely investigated. The surface-adhered compartments can grow rapidly and fuse with merging of contents [36], passively replicate by transporting material through the nanotubes [39], divide, migrate, and re-attach to a surface in remote locations [33]. The protocell networks can also grow in groups in a colony-like manner, where the membranes of the individual compartments are in physical contact with each other [35,37].

4. Possible Implications of Self-Forming Surface-Based Protocells for the Origin of Life

Although the autonomous formation and development of unique protocell structures is not the direct equivalent of the origin of life, a strong relation has been established between self-forming and -developing protocells, and abiogenesis. For example, according to Ganti's chemoton model [40], one of the three essential components of an elementary unit to be considered alive is a bilayer-enclosed compartment. The others are metabolism and self-replication. Nutrients are adsorbed through the membrane and incorporated into the metabolic cycle, where the waste products are released through the membrane again. Key molecular constituents of the membrane are produced by metabolic reactions as well as the "replicator unit".

Most of the current studies on membranous compartmentalization focus on bulk lipid vesicle suspensions as the initial step. The examples discussed in this article comprise alternative structural starting points, all of which are based on surface pathways (Figure 4).

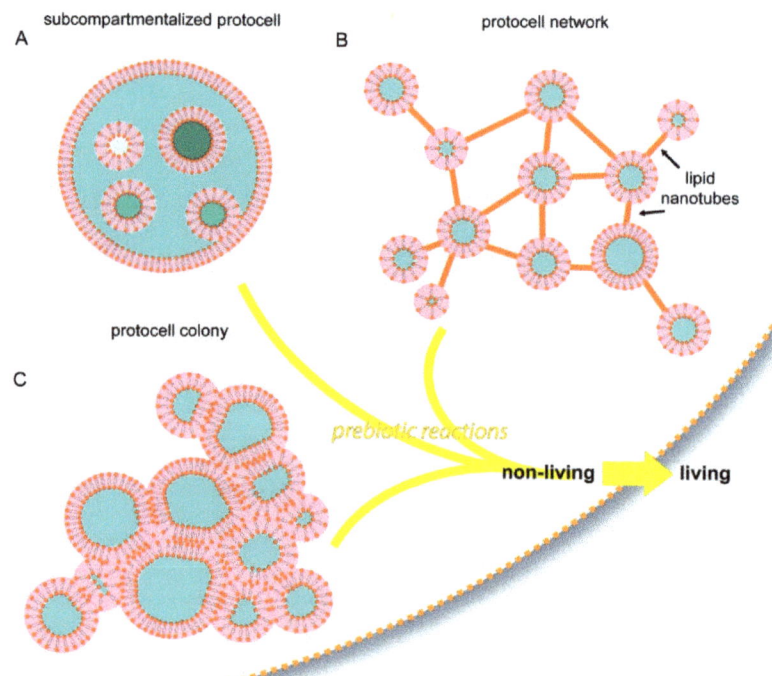

Figure 4. Schematic drawings summarizing the experimentally observed protocell model structures which autonomously and consistently form due to surface interactions. (**A**) Subcompartmentalized protocells with organelle-like membranous subunits, (**B**) Protocell-nanotube networks, (**C**) Protocell colonies. The possible contribution of such unique structures to the transition from non-living to living, compared to the 'bulk hypothesis' in which a spherical compartment freely suspended in water is assumed as first step towards life, is discussed in this article.

With the involvement of surface-membrane interactions, membranous protocellular subcompartments can rapidly, autonomously, and consistently form inside a protocell [27] (Figures 2 and 4A). The subcompartments are able to encapsulate, segregate and maintain ambient compounds at different concentrations in isolated environments [27], similar to their biological counterparts in modern cells: the organelles. Until the last decade, membrane-enclosed organelles were associated exclusively with eukaryotic cells. Recent evidence indicates that membranous subunits with specific functions also exist in Prokaryota [41], e.g., bacteria [42] and archaea [43]. These new reports, and the recent findings showing the ability of surface-adhered model protocells to easily form subcompartments, support the possibility that primitive protocells, leading to the last common universal ancestor, already had separate structural subunits to support different prebiotic reactions. Furthermore, as described above (Figure 2D–G), after a protocell disintegrates, subunits may become independent daughter cells [27]. It could be that one or some of the subcompartments inside an original protocell preserved a particular chemical reaction, which provided an evolutionary advantage. An enveloping protocellular membrane would impede the impact of detrimental environmental factors on the subcompartments, which later become the daughter cells. Such system could persist, despite an often unpredictable and changing environment.

Another type of biomembrane morphology occurring as a result of surface interactions is the interconnected micro-container network [35,36] (Figures 3 and 4B). The containers in such physical network may have accommodated different prebiotic chemical reaction networks [44]. Chemical reaction networks comprise multiple chemical reactions which are connected to each other through chemical compounds that participate both as reactants and products. The nanotube protocell networks have the necessary features to serve as physical reactors for prebiotic chemical reaction networks. In this setting, it is possible that the product of one reaction is transferred to a nearby reactor via the tunneling lipid nanotubes, where it becomes a reactant in the new reaction node, in a new compartment. The protocell-nanotube networks allow the continuous or discontinuous versions of chemical reactions in networks [44], as well as both the 'genetics first' or 'metabolism first' models of the origin of life problem. Discontinuous chemical reactions follow a specific order of added reagents, or require specific steps in which chemical compounds are segregated, concentrated, purified or eliminated leading to conditions permitting the subsequent reactions [44].

The third interesting form of assembly of primitive membranes on solid substrates is the protocell colony (Figure 4C). A key feature of biological colonies is their ability to adopt properties and capabilities greater than the individual units, for example to cope with environmental conditions unsuitable for survival as individuals. It is plausible that prebiotic protocell colonies, if they existed, exhibited increased mechanical stability, and had the ability to perform complex chemical interactions and exchange information. Vesicle colonies were hypothesized earlier as possible prebiotic compartments [5], and have been shown experimentally [45,46]. However, the mechanisms of formation involve fusogenic materials that were most likely not present on the early Earth, e.g., poly(arginine) [45], or colony formation requires directed assembly [46]. Vesicle agglomerates were previously reported to assemble from lipid monomers in the presence of mineral micro- and nanoparticles [18,47,48]. New research then showed the ability of protocell colonies to spontaneously form on synthetic surfaces [35], on Hadean Earth minerals, and on a rare Martian meteorite specimen [37]. In these new reports, protocell colonies were derived from the same lipid reservoir and consisted of a membrane of identical composition.

5. Conclusions

The necessary assumptions for the surface-mediated processes described above are minimal: the presence of lipid assemblies in an aqueous environment, and the presence of solid surfaces, all of which are thought to have existed on the early Earth.

Solid surfaces naturally possess energy which can be used by soft materials to perform astonishing shape transformations. Both synthetic and natural forms of solid surfaces have been recently experimentally explored in this context under laboratory conditions, which led to the discovery of unique lipid assemblies: subcompartmentalized protocells, protocell-nanotube networks, and protocell colonies. Considering the necessary minimal assumptions, it is well plausible that the completely autonomous transformations described above could have occurred on rock and mineral surfaces on the early Earth, or even in similar environments on other planets. Solid natural surfaces can possibly be the key enabling factor for the emergence of primitive cells. The morphologies generated by the recently identified surface-mediated pathways are fit to be combined with prebiotic chemical reactions. As a first step towards that goal, exploration of chemical reactions involving lipids and other amphiphiles might be a suitable route, extending the 'Lipid World hypothesis' [49,50]. Dry-wet cycles, or fusion of new multilamellar reservoirs to existing structures, can likely provide further advancement of our understanding of the relevance of these special prebiotic morphologies for the origin of life.

I have shared here my perspective on the role of solid surfaces as potential key components for the formation of non-trivial protocell structures, networks, and colonies. Since protocells are considered a stepping stone towards life's origin, I conclude that protocell morphologies enabled by solid surfaces might have been relevant and advantageous. I note that the emergence of life can hardly be understood from a single, isolated point of view, and I strongly advocate a transdisciplinary approach to eventually generate a coherent theory.

Funding: I gratefully acknowledge financial support from the Research Council of Norway (Forskningsrådet) project grant 274433, the UiO: Life Sciences Convergence Environment and the startup funding provided by the Centre for Molecular Medicine Norway (RCN 187615) and the Faculty of Mathematics and Natural Sciences at the University of Oslo.

Institutional Review Board Statement: Not applicable.

Informed Consent Statement: Not applicable.

Conflicts of Interest: The author declares no conflict of interest.

References

1. Poudyal, R.R.; Pir Cakmak, F.; Keating, C.D.; Bevilacqua, P.C. Physical Principles and Extant Biology Reveal Roles for RNA-Containing Membraneless Compartments in Origins of Life Chemistry. *Biochemistry* **2018**, *57*, 2509–2519. [CrossRef] [PubMed]
2. Jia, T.Z.; Chandru, K.; Hongo, Y.; Afrin, R.; Usui, T.; Myojo, K.; Cleaves, H.J. Membraneless polyester microdroplets as primordial compartments at the origins of life. *Proc. Natl. Acad. Sci. USA* **2019**, *116*, 15830. [CrossRef] [PubMed]
3. Abbas, M.; Lipiński, W.P.; Wang, J.; Spruijt, E. Peptide-based coacervates as biomimetic protocells. *Chem. Soc. Rev.* **2021**, *50*, 3690–3705. [CrossRef] [PubMed]
4. Donau, C.; Späth, F.; Sosson, M.; Kriebisch, B.A.K.; Schnitter, F.; Tena-Solsona, M.; Kang, H.-S.; Salibi, E.; Sattler, M.; Mutschler, H.; et al. Active coacervate droplets as a model for membraneless organelles and protocells. *Nat. Commun.* **2020**, *11*, 5167. [CrossRef]
5. Monnard, P.-A.; Walde, P. Current Ideas about Prebiological Compartmentalization. *Life* **2015**, *5*, 1239–1263. [CrossRef]
6. Wang, A.; Szostak, J.W. Lipid constituents of model protocell membranes. *Emerg. Top. Life Sci.* **2019**, *3*, 537–542. [PubMed]
7. Rao, M.; Eichberg, J.; Oró, J. Synthesis of phosphatidylcholine under possible primitive Earth conditions. *J. Mol. Evol.* **1982**, *18*, 196–202. [CrossRef]
8. Hargreaves, W.R.; Mulvihill, S.J.; Deamer, D.W. Synthesis of phospholipids and membranes in prebiotic conditions. *Nature* **1977**, *266*, 78–80. [CrossRef]
9. Dworkin, J.P.; Deamer, D.W.; Sandford, S.A.; Allamandola, L.J. Self-assembling amphiphilic molecules: Synthesis in simulated interstellar/precometary ices. *Proc. Natl. Acad. Sci. USA* **2001**, *98*, 815. [CrossRef]
10. Lawless, J.G.; Yuen, G.U. Quantification of monocarboxylic acids in the Murchison carbonaceous meteorite. *Nature* **1979**, *282*, 396–398. [CrossRef]
11. Israelachvili, J.N. *Intermolecular and Surface Forces*, 3rd ed.; Academic Press: Cambridge, MA, USA, 2011.
12. Czolkos, I.; Jesorka, A.; Orwar, O. Molecular phospholipid films on solid supports. *Soft Matter* **2011**, *7*, 4562–4576. [CrossRef]
13. Lewin, M.; Mey-Marom, A.; Frank, R. Surface free energies of polymeric materials, additives and minerals. *Polym. Adv. Technol.* **2005**, *16*, 429–441. [CrossRef]
14. Damer, B.; Deamer, D. Coupled phases and combinatorial selection in fluctuating hydrothermal pools: A scenario to guide experimental approaches to the origin of cellular life. *Life* **2015**, *5*, 872–887. [CrossRef] [PubMed]
15. Hazen, R.M.; Ferry, J.M. Mineral Evolution: Mineralogy in the Fourth Dimension. *Elements* **2010**, *6*, 9–12. [CrossRef]
16. Hazen, R.M.; Papineau, D.; Bleeker, W.; Downs, R.T.; Ferry, J.M.; McCoy, T.J.; Sverjensky, D.A.; Yang, H. Mineral evolution. *Am. Mineral.* **2008**, *93*, 1693–1720. [CrossRef]
17. James Cleaves, H., II; Michalkova Scott, A.; Hill, F.C.; Leszczynski, J.; Sahai, N.; Hazen, R. Mineral–organic interfacial processes: Potential roles in the origins of life. *Chem. Soc. Rev.* **2012**, *41*, 5502–5525. [CrossRef]
18. Hanczyc, M.M.; Fujikawa, S.M.; Szostak, J.W. Experimental Models of Primitive Cellular Compartments: Encapsulation, Growth, and Division. *Science* **2003**, *302*, 618–622. [CrossRef] [PubMed]

19. Brack, A. Chapter 10.4—Clay Minerals and the Origin of Life. In *Developments in Clay Science*; Bergaya, F., Lagaly, G., Eds.; Elsevier: Amsterdam, The Netherlands, 2013; Volume 5, pp. 507–521.
20. Hazen, R.M.; Sverjensky, D.A. Mineral surfaces, geochemical complexities, and the origins of life. *Cold Spring Harb. Perspect. Biol.* **2010**, *2*, a002162. [CrossRef]
21. Kindt, J.T.; Szostak, J.W.; Wang, A. Bulk self-assembly of giant, unilamellar vesicles. *ACS Nano* **2020**, *14*, 14627–14634. [CrossRef] [PubMed]
22. Fiore, M.; Strazewski, P. Prebiotic Lipidic Amphiphiles and Condensing Agents on the Early Earth. *Life* **2016**, *6*, 17. [CrossRef]
23. Liu, L.; Zou, Y.; Bhattacharya, A.; Zhang, D.; Lang, S.Q.; Houk, K.N.; Devaraj, N.K. Enzyme-free synthesis of natural phospholipids in water. *Nat. Chem.* **2020**, *12*, 1029–1034. [CrossRef]
24. Jõemetsa, S.; Spustova, K.; Kustanovich, K.; Ainla, A.; Schindler, S.; Eigler, S.; Lobovkina, T.; Lara-Avila, S.; Jesorka, A.; Gözen, I. Molecular Lipid Films on Microengineering Materials. *Langmuir* **2019**, *35*, 10286–10298. [CrossRef] [PubMed]
25. Melcrová, A.; Pokorna, S.; Pullanchery, S.; Kohagen, M.; Jurkiewicz, P.; Hof, M.; Jungwirth, P.; Cremer, P.S.; Cwiklik, L. The complex nature of calcium cation interactions with phospholipid bilayers. *Sci. Rep.* **2016**, *6*, 38035. [CrossRef] [PubMed]
26. Lobovkina, T.; Gözen, I.; Erkan, Y.; Olofsson, J.; Weber, S.G.; Orwar, O. Protrusive growth and periodic contractile motion in surface-adhered vesicles induced by Ca^{2+}-gradients. *Soft Matter* **2010**, *6*, 268–272. [CrossRef]
27. Spustova, K.; Köksal, E.S.; Ainla, A.; Gözen, I. Subcompartmentalization and Pseudo-Division of Model Protocells. *Small* **2021**, *17*, 2005320. [CrossRef] [PubMed]
28. Cremer, P.S.; Boxer, S.G. Formation and Spreading of Lipid Bilayers on Planar Glass Supports. *J. Phys. Chem. B* **1999**, *103*, 2554–2559. [CrossRef]
29. Nissen, J.; Jacobs, K.; Rädler, J.O. Interface Dynamics of Lipid Membrane Spreading on Solid Surfaces. *Phys. Rev. Lett.* **2001**, *86*, 1904–1907. [CrossRef]
30. Czolkos, I.; Guan, J.; Orwar, O.; Jesorka, A. Flow control of thermotropic lipid monolayers. *Soft Matter* **2011**, *7*, 6926–6933. [CrossRef]
31. Raedler, J.; Strey, H.; Sackmann, E. Phenomenology and Kinetics of Lipid Bilayer Spreading on Hydrophilic Surfaces. *Langmuir* **1995**, *11*, 4539–4548. [CrossRef]
32. Gözen, I.; Dommersnes, P.; Czolkos, I.; Jesorka, A.; Lobovkina, T.; Orwar, O. Fractal avalanche ruptures in biological membranes. *Nat. Mater.* **2010**, *9*, 908–912. [CrossRef]
33. Köksal, E.S.; Liese, S.; Kantarci, I.; Olsson, R.; Carlson, A.; Gözen, I. Nanotube-Mediated Path to Protocell Formation. *ACS Nano* **2019**, *13*, 6867–6878. [CrossRef] [PubMed]
34. Karatekin, E.; Sandre, O.; Guitouni, H.; Borghi, N.; Puech, P.-H.; Brochard-Wyart, F. Cascades of Transient Pores in Giant Vesicles: Line Tension and Transport. *Biophys. J.* **2003**, *84*, 1734–1749. [CrossRef]
35. Põldsalu, I.; Köksal, E.S.; Gözen, I. Mixed fatty acid-phospholipid protocell networks. *BioRxiv* **2021**. [CrossRef]
36. Köksal, E.S.; Liese, S.; Xue, L.; Ryskulov, R.; Viitala, L.; Carlson, A.; Gözen, I. Rapid Growth and Fusion of Protocells in Surface-Adhered Membrane Networks. *Small* **2020**, *16*, 2002569. [CrossRef] [PubMed]
37. Köksal, E.S.; Põldsalu, I.; Friis, H.; Mojzsis, S.; Bizzarro, M.; Gözen, I. Spontaneous formation of prebiotic compartment colonies on Hadean Earth and pre-Noachian Mars. *BioRxiv* **2021**. [CrossRef]
38. Alberts, B.; Johnson, A.; Lewis, J.; Raff, M.; Roberts, K.; Walter, P. *The RNA World and the Origins of Life*, 4th ed.; Garland Science: New York, NY, USA, 2002.
39. Gözen, İ. A Hypothesis for Protocell Division on the Early Earth. *ACS Nano* **2019**, *13*, 10869–10871. [CrossRef]
40. Ganti, T. *The Principles of Life*; OUP Oxford: Oxford, UK, 2003; p. 201.
41. Diekmann, Y.; Pereira-Leal, J.B. Evolution of intracellular compartmentalization. *Biochem. J.* **2012**, *449*, 319–331. [CrossRef] [PubMed]
42. Cornejo, E.; Abreu, N.; Komeili, A. Compartmentalization and organelle formation in bacteria. *Curr. Opin. Cell Biol.* **2014**, *26*, 132–138. [CrossRef]
43. Seufferheld, M.J.; Kim, K.M.; Whitfield, J.; Valerio, A.; Caetano-Anollés, G. Evolution of vacuolar proton pyrophosphatase domains and volutin granules: Clues into the early evolutionary origin of the acidocalcisome. *Biol. Direct* **2011**, *6*, 50. [CrossRef]
44. Tran, Q.P.; Adam, Z.R.; Fahrenbach, A.C. Prebiotic Reaction Networks in Water. *Life* **2020**, *10*, 352. [CrossRef]
45. Carrara, P.; Stano, P.; Luisi, P.L. Giant Vesicles "Colonies": A Model for Primitive Cell Communities. *ChemBioChem* **2012**, *13*, 1497–1502. [CrossRef] [PubMed]
46. Wang, X.; Tian, L.; Du, H.; Li, M.; Mu, W.; Drinkwater, B.W.; Han, X.; Mann, S. Chemical communication in spatially organized protocell colonies and protocell/living cell micro-arrays. *Chem. Sci.* **2019**, *10*, 9446–9453. [CrossRef]
47. Hanczyc, M.M.; Mansy, S.S.; Szostak, J.W. Mineral Surface Directed Membrane Assembly. *Orig. Life Evol. Biosph.* **2007**, *37*, 67–82. [CrossRef] [PubMed]
48. Sahai, N.; Kaddour, H.; Dalai, P.; Wang, Z.; Bass, G.; Gao, M. Mineral Surface Chemistry and Nanoparticle-aggregation Control Membrane Self-Assembly. *Sci. Rep.* **2017**, *7*, 43418. [CrossRef] [PubMed]
49. Segré, D.; Ben-Eli, D.; Deamer, D.W.; Lancet, D. The Lipid World. *Orig. Life Evol. Biosph.* **2001**, *31*, 119–145. [CrossRef] [PubMed]
50. Lancet, D.; Zidovetzki, R.; Markovitch, O. Systems protobiology: Origin of life in lipid catalytic networks. *J. R. Soc. Interface* **2018**, *15*, 20180159. [CrossRef]

Article

Effect of the Membrane Composition of Giant Unilamellar Vesicles on Their Budding Probability: A Trade-Off between Elasticity and Preferred Area Difference

Ylenia Miele [1,†], Gábor Holló [2,†], István Lagzi [2,3,*] and Federico Rossi [4,*]

1. Department of Chemistry and Biology "A. Zambelli", University of Salerno, Via Giovanni Paolo II 132, 84084 Fisciano, SA, Italy; ymiele@unisa.it
2. MTA-BME Condensed Matter Physics Research Group, Budapest University of Technology and Economics, Budafoki ut 8, 1111 Budapest, Hungary; hollo88@gmail.com
3. Department of Physics, Budapest University of Technology and Economics, Budafoki ut 8, 1111 Budapest, Hungary
4. Department of Earth, Environmental and Physical Sciences—DEEP Sciences, University of Siena, Pian dei Mantellini 44, 53100 Siena, Italy
* Correspondence: lagzi.istvan.laszlo@ttk.bme.hu (I.L.); federico.rossi2@unisi.it (F.R.); Tel.: +36-14631341 (I.L.); +39-0577232215 (F.R.)
† These authors contributed equally to this work.

Abstract: The budding and division of artificial cells engineered from vesicles and droplets have gained much attention in the past few decades due to an increased interest in designing stimuli-responsive synthetic systems. Proper control of the division process is one of the main challenges in the field of synthetic biology and, especially in the context of the origin of life studies, it would be helpful to look for the simplest chemical and physical processes likely at play in prebiotic conditions. Here we show that pH-sensitive giant unilamellar vesicles composed of mixed phospholipid/fatty acid membranes undergo a budding process, internally fuelled by the urea–urease enzymatic reaction, only for a given range of the membrane composition. A gentle interplay between the effects of the membrane composition on the elasticity and the preferred area difference of the bilayer is responsible for the existence of a narrow range of membrane composition yielding a high probability for budding of the vesicles.

Keywords: vesicles; division; urea–urease enzymatic reaction; bending modulus; budding; ADE theory

1. Introduction

One of the most ambitious goals of the recently born discipline systems chemistry is the creation of synthetic life following a bottom-up approach [1–4]. Such research and investigations are important for the design of stimuli-responsive materials and their usage in several applications (e.g., targeted drug delivery) [5,6] and to understand fundamental open questions in the origin of life studies [2,7]. A general approach followed in this context is to recreate life-like functions within a minimal artificial cell-like container (water-in-oil droplets, emulsions, protocells, etc.) rather than replicate the complex biological environment typical of the living cells [8,9]. Several attempts have been made to imitate distinctive processes that characterize modern cells, such as intercellular communication [10–14], energy harvesting [15,16], metabolism [17,18] and self-replication or division [19–29].

The most common model for the investigation of the division and budding processes are giant unilamellar vesicles (GUVs) stabilized by an amphiphilic bilayer made of phospholipids and/or fatty acids. In general, division occurs when the volume of the vesicles decreases (mostly due to an osmotic shock) and concurrently the inner surface area of the inner bilayer leaflet decreases in respect to the outer one. Recently, we reported that a pH change occurring inside the lumen of mixed 1-palmitoyl-2-oleoyl-*sn*-glycero-3-phosphocholine (POPC)/oleic acid (HOA) GUVs can be utilized to drive and govern the

division process [30,31]. In particular, the urea–urease enzymatic reaction [32–34] inside the lumen of the vesicles, coupled to the cross-membrane transport of urea, can trigger the shape deformation of the pH-sensitive mixed membrane by promoting the solubilization of oleate molecules in the inner water phase. In the framework of the Area Difference Elasticity (ADE) theory [23,35–43] we demonstrated that the shape deformation is driven by the synergic action of osmosis, acting on the volume of the vesicle, and pH change, the latter responsible for the increase of the preferred area difference of the bilayer leaflets. However, the ADE model predicts that the equilibrium shape of an elastic membrane undergoing a transformation, also depends on a third factor: the mechanical properties of the bilayer accounted by the ratio of the local (intrinsic fluidity of the membrane) and the non-local (tension generated by the different curvature of the two leaflets) bending moduli. In this study, we investigate the effect of the membrane composition on the division probability of the mixed GUVs. We present some experiments where we varied the total amphiphiles concentration (s = [POPC] + [HOA]) and the relative ratio between the two components (α = [HOA]/[POPC]) to show that α has an important role in determining the final shape of the vesicles. A general finding of this work is to assess the role of the mechanical properties of the membrane, such as the elasticity and the bending modulus, in driving the protocells division.

2. Materials and Methods

2.1. Experimental

For GUVs preparation and encapsulation of chemical species, the phase transfer method [33,44–46] was used, which allowed full encapsulation of chemical species. Briefly, the method consists in the preparation of two isotonic solutions: outer solution (O-solution) and inner solution (I-solution) separated by an interface made of amphiphilic molecules. The outer solution includes water and glucose (Sigma Aldrich, St. Louis, Missouri, USA), above this phase a mixture of 1-palmitoyl-2-oleoyl-*sn*-glycero-3-phosphocholine, POPC (Lipoid, Ludwigshafen, Germany) and oleic acid, HOA (Sigma Aldrich) in mineral oil (M5904 Sigma Aldrich) is allowed to incubate to make a monolayer. The inner solution is prepared pipetting by hand a water in oil microemulsion. The aqueous phase of the inner solution is composed of the enzyme urease (Sigma Aldrich, typical activity 34310 U/g), sucrose (Sigma Aldrich), the fluorescence probe pyranine (Sigma Aldrich) and water, while the apolar phase contains POPC and HOA in mineral oil (we used the same concentration of interface). After the period of incubation, the water in oil microemulsion is poured over the interface and centrifuged. The formation of the vesicles takes place at the interface of the two phases, where emulsion droplets sink by gravity in the outer phase and are coated by the amphiphilic molecules to form bilayers (sucrose and glucose are used to enhance the difference of density between the droplets and the outer solution). Details are reported in SI.

The lipid content in vesicles samples was determined by an HPLC apparatus (Agilent 1200 Infinity) by using a C18 column (Agilent Zorbax Eclipse Plus, 4.6 × 100 mm, 3.5 µm) and a mobile phase composed of acetonitrile, methanol and formic acid at 0.1% (all solvents from Sigma-Aldrich) in the proportions 50:45:5 (isocratic mode) [47]. HPLC experiments were done in three replicates by collecting together 4 pellets at a time. The yield was calculated as the average mass of POPC in a pellet over the total mass of POPC used in a typical preparation prior the vesicles formation.

2.2. Numerical Calculation

2.2.1. Surface Evolver

In the area difference elasticity model [40,43] a vesicle is considered as an elastic membrane and its equilibrium shape can be found with the minimization of the Helfrich energy [48,49] (W). In this model not just the local principal curvatures (C_1 and C_2) can differ from their equilibrium value, i.e., from the local spontaneous curvature (C_0), but an elastic membrane is taken into account, so the area difference between the inner and outer leaflets (ΔA) can be different from its equilibrium value (from the preferred area difference ΔA_0) as well. The relative significance of these terms can be fine tuned with the local and the non-local bending moduli (κ_c and κ_r). The energy can be expressed as

$$W = \frac{1}{2}\kappa_c \oint_S (C_1 + C_2 - C_0)^2 dA + \frac{\kappa_r}{2A_0 h^2}(\Delta A - \Delta A_0)^2 \quad (1)$$

where A_0 is the area of the vesicle, h is the distance between the neutral surfaces of the leaflets and the integral is calculated on the surface of the vesicle (S) [35,36]. The area difference can be calculated from the integral of the mean curvatures

$$\Delta A = h \oint_S (C_1 + C_2) dA \quad (2)$$

and ΔA values can be determined experimentally from images as well [30,31].

For simplicity, the energy of the vesicle is often expressed in a reduced form which can be observed after the division with the bending energy of the sphere ($8\pi\kappa_c$)

$$w = \frac{1}{4}\oint_S (c_1 + c_2)^2 da + K(\Delta a - \Delta a_0)^2 \quad (3)$$

where $\Delta a = \Delta A / 8\pi h R_s$ is the reduced area difference, $\Delta a_0 = \Delta A_0 / 8\pi h R_s$ and R_s is the radius of the sphere. $K = \frac{\kappa_r}{\kappa_c}$ represents the relative weight of the energy terms and it represents an intrinsic property of the membrane, the other transformed variables are $da = dA/4\pi R_s$, $c_1 = C_1 R_s$, $c_2 = C_2 R_s$.

The energy minimization was performed with Surface Evolver (SE) 2.70 [50]. During the minimization the reduced volume

$$v = \frac{V}{4/3\pi R^3} \quad (4)$$

was constrained where $R = \sqrt{A/4\pi}$ is defined as the radius of an equivalent sphere and A is the surface of the vesicle.

The shape of the vesicle is determined by three variables ($\Delta a_0, v, K$) in the energy minimization. During the simulation—besides the shape of the vesicle—the value of the reduced area could be changed, since an elastic membrane is considered so it can be different from Δa_0. The value of the reduced volume (0.7) was determined experimentally and it was kept constant in every simulation and K was adjusted according to a given bilayer composition.

2.2.2. ODE Model

To describe the pH change inside the GUVs and obtain Δa_0 in time induced by the urea–urease enzymatic reaction, we developed a seven chemical species kinetic model, which can be represented mathematically by a set of ordinary differential equations (ODEs)

$$\frac{d[X]}{dt} = r([X]) + k_X ([X]_o - [X]) \quad (5)$$

where $[X]$ denotes the concentration of the chemical species X, $r([X])$ represents the set of reaction rates involving X, and $[X]_o$ is the concentration of the chemical species in the outer phase. The transfer rate k_X (s^{-1}) is proportional to the surface-to-volume ratio of the vesicle and to the specific membrane permeability of the species.

The change of the number of molecules of oleate in the inner and outer leaflets is linked to the pH dynamics described by (5) through the following differential equations

$$\frac{dN_{outer}}{dt} = \frac{dN_{outerHOA}}{dt} + \frac{dN_{outerOA^-}}{dt} \tag{6}$$

$$\frac{dN_{outer}}{dt} = -k_f\left(N_{outer} - N_{inner}\frac{R_s^2}{(R_s-h)^2}\right) \tag{7}$$

$$\frac{dN_{inner}}{dt} = \frac{dN_{innerHOA}}{dt} + \frac{dN_{innerOA^-}}{dt} \tag{8}$$

$$\frac{dN_{inner}}{dt} = +k_f\left(N_{outer} - N_{inner}\frac{R_s^2}{(R_s-h)^2}\right)$$
$$- N_{innerOA^-}\frac{k_{off}}{1+e^{(-k_t(pH-pH_{thres}))}} \tag{9}$$

where N_{inner} and N_{outer} represent the total number of molecules (POPC + HOA + OA$^-$) in the inner and outer leaflet, repsectively; $N_{innerHOA}$ and $N_{innerOA^-}$ indicate the number of oleic acid molecules (unionized and deionized form) present in the inner leaflet, $N_{outerHOA}$ and $N_{outerOA^-}$ are the oleic acid molecules in the outer leaflet; k_f is the flip-flop constant for HOA and k_{off} is the kinetic constant for the solubilization of OA$^-$; R_s is the radius of the initial spherical vesicle, h is the distance between the two neutral surfaces. The number of molecules of POPC is considered constant. The logistic function in the last term of Equation (9) accounts for the pH-dependent formation of oleate aggregates in the vesicles lumen when a critical pH is reached (pH$_{thres}$).

The preferred area difference (ΔA_0) and the normalized preferred area difference (Δa_0) were finally calculated in time as

$$\Delta A_0(t) = N_{outer}(t)<\tilde{a}> - N_{inner}(t)<\tilde{a}> \tag{10}$$

$$\Delta a_0(t) = \frac{\Delta A_0(t)}{8\pi R_0(t)h} \tag{11}$$

where $<\tilde{a}>$ is the average cross section of the amphiphiles and $R_{0(t)} = \sqrt{A_{0(t)}/(4\pi)}$ [23].

The model was described in details in ref. [31] and a short description can be found in the Supplementary Information (SI).

3. Results and Discussion

In a typical experiment, a fast enzymatic reaction inducing a pH change inside the GUVs with the consequent division of the vesicles, was obtained in the following conditions: urease (0.5 U/mL), pyranine (5.0 ×10^{-5} M) and acetic acid (1 ×10^{-6} M) were encapsulated in hybrid giant POPC/HOA vesicles with various amphiphiles total concentration [POPC]$_0$ + [HOA]$_0$ (s) and proportions [HOA]$_0$/[POPC]$_0$ (α). A urea solution ([CO(NH$_2$)$_2$]$_0$ = 6.0 ×10^{-2} M) was added to the vesicles suspension to initiate the reaction. Urea penetrated through the membrane of GUVs and initiated the urea–urease enzymatic reaction, which caused a pH increase (from ∼6 to ∼6.5, measured through the fluorescence of the pH-sensitive probe pyranine encapsulated in the vesicles lumen [30]) due to the produced ammonia. A shift towards basic conditions inside the vesicles lumen causes the deprotonation of HOA to form oleate molecules readily dissolved in the aqueous medium. In a previous work [31], we demonstrated that the addition of urea in the outer solution generates an hypertonic condition, which, in turn, drives a deflation of the vesicles due to the osmotic shock. Thus, the addition of urea triggers a synergistic dynamics between the reduction of the vesicles volume (ν changes from 1 to ∼0.7 at s = 5 mM and α ∼ 1) and the increase of the area difference between the lipid leaflets in the bilayer, caused by the dissolution of oleate molecules in the aqueous lumen (Δa_0 changes from 1 to ∼1.7 at s = 5 mM and α ∼ 1). According to the ADE theory the final values of ν and Δa_0 correspond to a budded shape of the vesicles in our experimental conditions.

After spanning different compositions of the membranes, we identified s = 5 mM and α = 0.92 ([HOA] = 2.4 mM, [POPC] = 2.6 mM) as the optimal condition to obtain the highest probability of successful budding. Actually, the budding rate was found substantially independent from the parameter s (~16% in the interval 0.3 mM < s < 10 mM, (see Figure S1a in SI), while α was found to be determinant. This behaviour can be expected taking into account that the yield of the droplet transfer method, i.e., the mass of amphiphiles in a vesicles pellet over the total mass of amphiphiles initially present in the Eppendorf tube (see SI for preparation details), is substantially independent from s as revealed by HPLC analysis (see Figure S1b in SI); as previously reported [51], the total quantity of amphiphiles has a major impact on the number of generated vesicles rather then on their properties (size, relative composition, etc.).

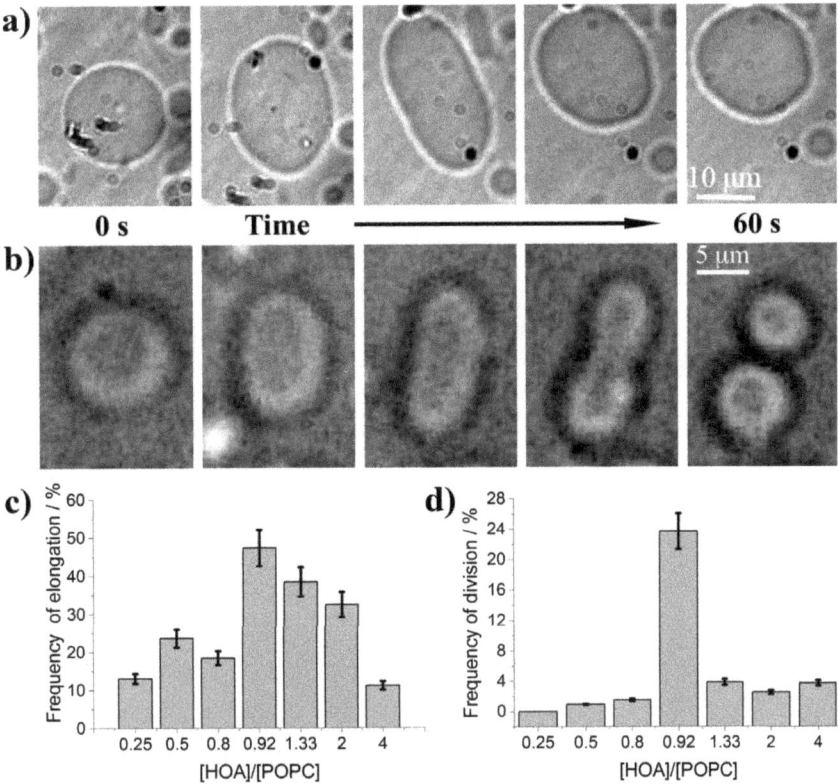

Figure 1. (**a**) Reversible shape transformation of a GUV from sphere to prolate and back to spherical shape triggered by the urea–urease enzymatic reaction, s = 5 mM, α = 0.5; (**b**) shape transformation of GUV leading to a final budded conformation, s = 5 mM, α = 0.92; (**c**) frequency of elongation as a function of the ratio $[HOA]_0/[POPC]_0$, s = 5 mM; (**d**) frequency of budding as a function of the ratio $[HOA]_0/[POPC]_0$, s = 5 mM.

In contrast, the relative composition of the membrane, α, impacted both the frequency of simple elongations (vesicles that evolved in a prolate or oblate shape but did not bud at the end of the process, Figure 1a,c) and the frequency of successful budding (Figure 1b,d). α was varied between 0.25 and 4 for a total amount of amphiphiles at fixed s = 5 mM and the ratio α = 2.4 mM/2.6 mM = 0.92 was found to correspond to the best budding rate, lower ratios or higher ratios are less effective. At a first glance, the optimal composition α = 0.92 might be considered a good compromise between the pH sensitivity given by the oleic acid and the stability guaranteed by POPC. In fact, high content of POPC made the vesicles less sensitive to pH changes and less prone to bud, in contrast high amounts of oleic acid might act as a buffer and interfere with the enzymatic reaction, though the reduced preferred area difference must in principle increase being more deprotonable molecules in the inner leaflet.

Therefore, to fully understand the effect of α, we should also consider as important parameter the elasticity of the membrane, expressed by the ratio of the local and non-local bending moduli K and determined by the composition of the membrane. In general, in mixed HOA/phospholipids membrane, the local bending rigidity (κ_c) increases whilst the non-local elastic modulus (κ_r) decreases with the oleic acid content, as found for 1,2-dioleoyl-sn-glycero-3-phosphocholine (DOPC)/HOA [52] and 1,2-dipalmitoyl-sn-glycero-3-phosphocholine (DPPC)/HOA [53] mixtures.

To gain more insight into the dependence of the equilibrium limiting shapes of a vesicle, and hence the budding probability, on the membrane composition, we studied the influence of α on two parameters that drive the minimization of the energy functional W: Δa_0, which depends on the pH and K, which accounts for the elasticity of the membrane. The reduced volume v was fixed to 0.7 since it was not influenced by α but only by the osmolarity difference between the lumen of a vesicle and the outer solution. The pH course inside a GUV and the variation of Δa_0 were calculated using the ODE model reported in the SI, running a time-course simulation (Figure 2a,b) for each α value in the interval 0.3–1.8, i.e., the one comprising the highest budding probability. From each simulation was then extracted the maximum value reached by Δa_0 (Δa_0^{max}) which was finally plotted as a function of α (Figure 2c). As expected, despite a buffering effect of HOA, the net result of adding deprotonable molecules in the membrane is to increase Δa_0^{max}.

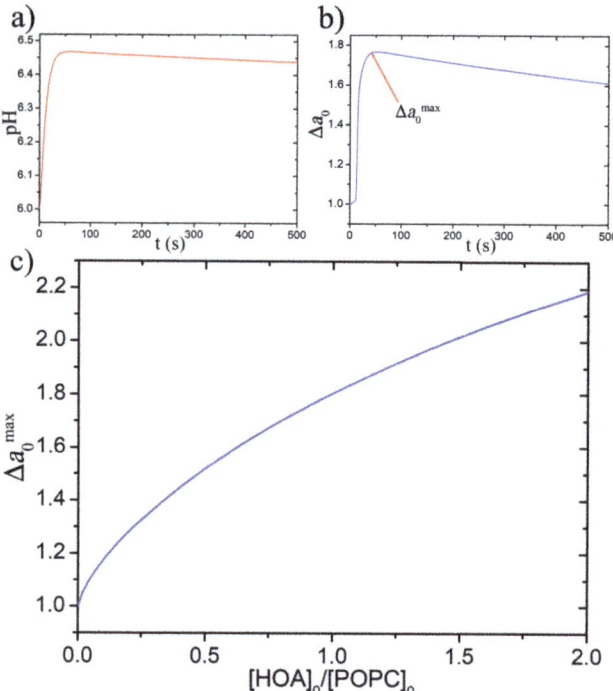

Figure 2. Examples of time course simulations of (**a**) pH and (**b**) Δa_0 inside a vesicle with [urease]$_0$ = 1.1 U/mL, [CH$_3$COOH]$_0$ = 1 × 10^{-6} mol/L, [urea]$_0$ = 60 mmol/L, s = 5 mmol/L and α = 1; (**c**) dependence of the maximum values of Δa_0 on the membrane composition, for each simulation at different α initial reactants concentration and s were kept constant.

To account for the dependence of K on the membrane composition we defined an arbitrary linear function, $K = 5 - 2.5\alpha$, which reflects the fact that by increasing the molecules of oleic acid the combined effect on the bending moduli causes K to decrease. Thus, K varies in the interval 4.25–0.85 that represents a reliable range for phospholipids-based bilayers [23,40,54].

According to the experimental observations, the budding limiting shape is a compromise between the increase of the reduced preferred area difference caused by the pH increase, with the consequent deprotonation of the inner leaflet, and an optimal elasticity of the membrane. We then used the values of K and Δa_0 as parameters in SE software to simulate the equilibrium limiting shape of a vesicle according to the methodology sketched in Figure 3. At a given α, the corresponding Δa_0^{max} and K, were calculated and used as input in SE together with the fixed ν.

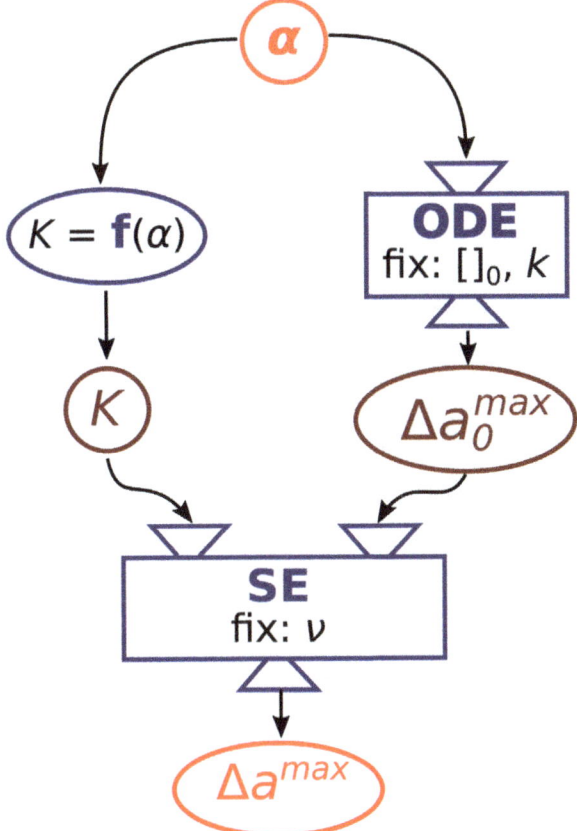

Figure 3. Sketch of the numerical simulations strategy. The membrane composition parameter, α, is used to calculate the corresponding elasticity ratio K through the linear empiric function $K = \mathbf{f}(\alpha)$ and the value of Δa_0^{max} through the ODE model with fixed initial concentrations of reactants ($[\,]_0$) and kinetic constants (k) for each run; K and Δa_0^{max} are then used in SE with fixed ν to yield the maximum reduced area difference Δa^{max}.

The minimization of the energy functional W yields the reduced area difference Δa^{max} as output parameter, which, together with the fixed ν, determines the shape of the vesicle. According to the ADE theory, for $\nu = 0.7$ the corresponding value of Δa for the equilibrium budded shape has to be $\sim$1.4 [36,38,40,42].

Figure 4 reports the phase diagram $\Delta a^{max} - \alpha$ where the output values of the energy minimization procedure are reported for each membrane composition. Simulations confirmed that a successful budding takes place when the membrane composition lies in a small interval around $\alpha = 1$ (HOA:POPC = 1:1); this represents the best trade off between the mechanical properties of the membrane and the capability of the enzymatic reaction to increase the preferred area difference. When $\alpha < 0.8$ the deprotonated molecules in the inner leaflet are not enough to bring the vesicle to a budded shape. In contrast, when $\alpha > 1.2$, the enzymatic reaction causes the dissolution of a high amount of oleate but the membrane oppose more resistance to be effectively deformed.

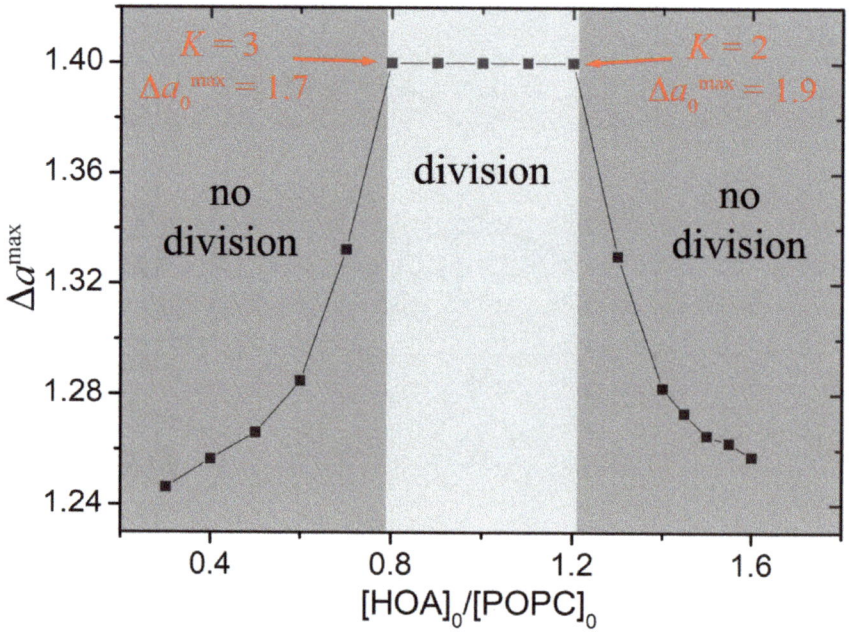

Figure 4. Phase diagram $\Delta a^{max} - \alpha$ showing the effect of the membrane composition on the successful budding for a vesicle, calculated through Surface Evolver software.

4. Conclusions

The division process is one of the fundamental function necessary for the self-replication and self-reproduction of biological entities. In modern cells replication is driven by a complex mechanism of growth and division, regulated by proteins and involving several chemico-physical dynamics (mass transfer, phospholipids synthesis, enzymatic reactions, etc.) [55,56]. However, in the context of abiogenesis, to understand the basic mechanisms at the basis of protocells division can be helpful to look for the simplest chemical and physical process likely at play in prebiotic conditions. For a few years our group has been studying a simple cell model having a diameter larger than 5 μm (GUV) and based on a mixed phospholipid/fatty acid membrane encapsulating a pH clock reaction. We demonstrated that a simple pH change inside the vesicles lumen, coupled with an osmotic stress, is enough to induce shape transformations of the vesicles eventually leading to the fission of a mother GUV into two daughter GUVs [30,31]. Interestingly, our operative conditions are compatible with the slightly acidic pH suggested for the early Archean age oceans [57].

In this paper we presented some experiments that showed how the mechanical properties of the membrane are another ingredient necessary to understand the physical basis of protocells division. In fact, by varying the ratio between phospholipids and fatty acids, we could control both the ratio of the membrane's bending moduli and the number of mobile molecules in the inner leaflet of the bilayer (i.e., the preferred area difference). We thus demonstrated that, for a mixed POPC/HOA vesicle, an effective budded shape could be attained only in a narrow range of the membrane composition, where the elasticity and the preferred area difference of the bilayer assume optimal values. According to experiments and simulations the budding region is comprised in the interval $0.8 < \alpha < 1.2$, which corresponds to $2 < K < 3$ and $1.7 < \Delta a_0^{max} < 1.9$.

Supplementary Materials: The following are available online at www.mdpi.com/xxx/s1, Figure S1: (a) Frequency of divisions as a function of the sum [POPC] + [HOA] at a fixed a = [HOA]/[POPC] = 1; (b) Percentage of POPC in the final vesicles vs initial concentration of POPC in mineral oil during preparation. Data were extrapolated from HPLC analysis (chromatogram in the inset); error bars represent the standard deviation of three replicates, Table S1: Initial concen-trations and parameters used for the kinetic simulations, Table S2: Kinetic constants used in the model, Table S3: Initial conditions and parameters used for the case [POPC]: [HOA] = 2.6 mM:2.4.

Author Contributions: Conceptualization, I.L. and F.R.; methodology, Y.M., G.H., I.L. and F.R.; software, G.H. and Y.M.; experiments, Y.M.; writing—original draft preparation, I.L. and F.R.; writing—review and editing, Y.M., G.H., I.L. and F.R.; supervision, I.L. and F.R. All authors have read and agreed to the published version of the manuscript.

Funding: This research was funded by the National Research, Development and Innovation Office of Hungary, grant number NN125752, K131425 and the NRDI Fund (TKP2020 IES, Grant No. BME-IE-NAT) based on the charter of bolster issued by the NRDI Office under the auspices of the Ministry for Innovation and Technology.

Acknowledgments: Thanks are due to David Lemaitre, Jairo Sanz Corbacho and Ada Fasulo for experimental help.

Conflicts of Interest: The authors declare no conflict of interest.

References

1. Ludlow, R.F.; Otto, S. Systems Chemistry. *Chem. Soc. Rev.* **2008**, *37*, 101–108. [CrossRef] [PubMed]
2. Ruiz-Mirazo, K.; Briones, C.; de la Escosura, A. Prebiotic Systems Chemistry: New Perspectives for the Origins of Life. *Chem. Rev.* **2014**, *114*, 285–366. [CrossRef] [PubMed]
3. Ashkenasy, G.; Hermans, T.M.; Otto, S.; Taylor, A.F. Systems Chemistry. *Chem. Soc. Rev.* **2017**, *46*, 2543–2554. [CrossRef]
4. Stano, P.; Rampioni, G.; D'Angelo, F.; Altamura, E.; Mavelli, F.; Marangoni, R.; Rossi, F.; Damiano, L. Current Directions in Synthetic Cell Research. In *Advances in Bionanomaterials*; Lecture Notes in Bioengineering; Springer: Cham, Switzerland, 2018; pp. 141–154. [CrossRef]
5. Meng, F.; Zhong, Z.; Feijen, J. Stimuli-Responsive Polymersomes for Programmed Drug Delivery. *Biomacromolecules* **2009**, *10*, 197–209. [CrossRef] [PubMed]
6. Giuseppone, N. Toward Self-Constructing Materials: A Systems Chemistry Approach. *Acc. Chem. Res.* **2012**, *45*, 2178–2188. [CrossRef] [PubMed]
7. Lopez, A.; Fiore, M. Investigating Prebiotic Protocells for a Comprehensive Understanding of the Origins of Life: A Prebiotic Systems Chemistry Perspective. *Life* **2019**, *9*, 49. [CrossRef]
8. Blain, J.C.; Szostak, J.W. Progress Toward Synthetic Cells. *Annu. Rev. Biochem.* **2014**, *83*, 615–640. [CrossRef] [PubMed]
9. Buddingh', B.C.; van Hest, J.C.M. Artificial Cells: Synthetic Compartments with Life-like Functionality and Adaptivity. *Acc. Chem. Res.* **2017**, *50*, 769–777. [CrossRef]
10. Rampioni, G.; Damiano, L.; Messina, M.; D'Angelo, F.; Leoni, L.; Stano, P. Chemical Communication between Synthetic and Natural Cells: A Possible Experimental Design. *Electron. Proc. Theor. Comput. Sci.* **2013**, *130*, 14–26. [CrossRef]
11. Tomasi, R.; Noel, J.M.; Zenati, A.; Ristori, S.; Rossi, F.; Cabuil, V.; Kanoufi, F.; Abou-Hassan, A. Chemical Communication between Liposomes Encapsulating a Chemical Oscillatory Reaction. *Chem. Sci.* **2014**, *5*, 1854–1859. [CrossRef]
12. Niederholtmeyer, H.; Chaggan, C.; Devaraj, N.K. Communication and Quorum Sensing in Non-Living Mimics of Eukaryotic Cells. *Nat. Commun.* **2018**, *9*, 5027. [CrossRef]
13. Aufinger, L.; Simmel, F.C. Establishing Communication Between Artificial Cells. *Chem. Eur. J.* **2019**, *25*, 12659–12670. [CrossRef]
14. Budroni, M.A.; Torbensen, K.; Ristori, S.; Abou-Hassan, A.; Rossi, F. Membrane Structure Drives Synchronization Patterns in Arrays of Diffusively Coupled Self-Oscillating Droplets. *J. Phys. Chem. Lett.* **2020**, *11*, 2014–2020. [CrossRef]
15. Altamura, E.; Milano, F.; Tangorra, R.R.; Trotta, M.; Omar, O.H.; Stano, P.; Mavelli, F. Highly Oriented Photosynthetic Reaction Centers Generate a Proton Gradient in Synthetic Protocells. *Proc. Natl. Acad. Sci. USA* **2017**, *114*, 3837–3842. [CrossRef] [PubMed]

16. Altamura, E.; Albanese, P.; Marotta, R.; Milano, F.; Fiore, M.; Trotta, M.; Stano, P.; Mavelli, F. Chromatophores Efficiently Promote Light-Driven ATP Synthesis and DNA Transcription inside Hybrid Multicompartment Artificial Cells. *Proc. Natl. Acad. Sci. USA* **2021**, *118*. [CrossRef] [PubMed]
17. de Souza, T.P.; Steiniger, F.; Stano, P.; Fahr, A.; Luisi, P.L. Spontaneous Crowding of Ribosomes and Proteins inside Vesicles: A Possible Mechanism for the Origin of Cell Metabolism. *ChemBioChem* **2011**, *12*, 2325–2330. [CrossRef]
18. van Roekel, H.W.H.; Rosier, B.J.H.M.; Meijer, L.H.H.; Hilbers, P.A.J.; Markvoort, A.J.; Huck, W.T.S.; de Greef, T.F.A. Programmable Chemical Reaction Networks: Emulating Regulatory Functions in Living Cells Using a Bottom-up Approach. *Chem. Soc. Rev.* **2015**, *44*, 7465–7483. [CrossRef] [PubMed]
19. Zhu, T.F.; Szostak, J.W. Coupled growth and division of model protocell membranes. *J. Am. Chem. Soc.* **2009**, *131*, 5705–5713. [CrossRef]
20. Peterlin, P.; Arrigler, V.; Kogej, K.; Svetina, S.; Walde, P. Growth and shape transformations of giant phospholipid vesicles upon interaction with an aqueous oleic acid suspension. *Chem. Phys. Lipids* **2009**, *159*, 67–76. [CrossRef]
21. Kurihara, K.; Tamura, M.; Shohda, K.I.; Toyota, T.; Suzuki, K.; Sugawara, T. Self-reproduction of supramolecular giant vesicles combined with the amplification of encapsulated DNA. *Nat. Chem.* **2011**, *3*, 775–781. [CrossRef] [PubMed]
22. Sakuma, Y.; Imai, M. From Vesicles to Protocells: The Roles of Amphiphilic Molecules. *Life* **2015**, *5*, 651–675. [CrossRef]
23. Jimbo, T.; Sakuma, Y.; Urakami, N.; Ziherl, P.; Imai, M. Role of Inverse-Cone-Shape Lipids in Temperature-Controlled Self-Reproduction of Binary Vesicles. *Biophys. J.* **2016**, *110*, 1551–1562. [CrossRef]
24. Dervaux, J.; Noireaux, V.; Libchaber, A.J. Growth and instability of a phospholipid vesicle in a bath of fatty acids. *Eur. Phys. J. Plus* **2017**, *132*, 284. [CrossRef]
25. Litschel, T.; Ramm, B.; Maas, R.; Heymann, M.; Schwille, P. Beating Vesicles: Encapsulated Protein Oscillations Cause Dynamic Membrane Deformations. *Angew. Chem. Int. Ed.* **2018**, *57*, 16286–16290. [CrossRef]
26. Kurisu, M.; Aoki, H.; Jimbo, T.; Sakuma, Y.; Imai, M.; Serrano-Luginbühl, S.; Walde, P. Reproduction of vesicles coupled with a vesicle surface-confined enzymatic polymerisation. *Commun. Chem.* **2019**, *2*, 117. [CrossRef]
27. Li, Y.; ten Wolde, P.R. Shape Transformations of Vesicles Induced by Swim Pressure. *Phys. Rev. Lett.* **2019**, *123*, 148003. [CrossRef] [PubMed]
28. Vutukuri, H.R.; Hoore, M.; Abaurrea-Velasco, C.; van Buren, L.; Dutto, A.; Auth, T.; Fedosov, D.A.; Gompper, G.; Vermant, J. Active particles induce large shape deformations in giant lipid vesicles. *Nature* **2020**, *586*, 52–56. [CrossRef] [PubMed]
29. Dreher, Y.; Jahnke, K.; Bobkova, E.; Spatz, J.P.; Göpfrich, K. Division and Regrowth of Phase-Separated Giant Unilamellar Vesicles. *Angew. Chem. Int. Ed.* **2021**, *133*, 10756–10764. doi:10.1002/anie.202014174. [CrossRef]
30. Miele, Y.; Medveczky, Z.; Hollo, G.; Tegze, B.; Derenyi, I.; Horvolgyi, Z.; Altamura, E.; Lagzi, I.; Rossi, F. Self-Division of Giant Vesicles Driven by an Internal Enzymatic Reaction. *Chem. Sci.* **2020**, *11*, 3228–3235. [CrossRef]
31. Holló, G.; Miele, Y.; Rossi, F.; Lagzi, I. Shape Changes and Budding of Giant Vesicles Induced by an Internal Chemical Trigger: An Interplay between Osmosis and pH Change. *Phys. Chem. Chem. Phys.* **2021**, *23*, 4262–4270. [CrossRef]
32. Hu, G.; Pojman, J.A.; Scott, S.K.; Wrobel, M.M.; Taylor, A.F. Base-Catalyzed Feedback in the Urea-Urease Reaction. *J. Phys. Chem. B* **2010**, *114*, 14059–14063. [CrossRef] [PubMed]
33. Miele, Y.; Bánsági, T.; Taylor, A.; Stano, P.; Rossi, F. Engineering Enzyme-Driven Dynamic Behaviour in Lipid Vesicles. In *Advances in Artificial Life, Evolutionary Computation and Systems Chemistry*; Rossi, F., Mavelli, F., Stano, P., Caivano, D., Eds.; Number 587 in Communications in Computer and Information Science; Springer International Publishing: Cham, Switzerland, 2016; pp. 197–208. [CrossRef]
34. Miele, Y.; Bánsági, T.; Taylor, A.; Rossi, F. Modelling Approach to Enzymatic pH Oscillators in Giant Lipid Vesicles. In *Advances in Bionanomaterials I*; Piotto, S., Rossi, F., Concilio, S., Reverchon, E., Cattaneo, G., Eds.; Lecture Notes in Bioengineering; Springer International Publishing: Cham, Switzerland, 2018; pp. 63–74. [CrossRef]
35. Svetina, S.; Žekš, B. Membrane bending energy and shape determination of phospholipid vesicles and red blood cells. *Eur. Biophys. J.* **1989**, *17*, 101–111. [CrossRef]
36. Käs, J.; Sackmann, E. Shape transitions and shape stability of giant phospholipid vesicles in pure water induced by area-to-volume changes. *Biophys. J.* **1991**, *60*, 825–844. [CrossRef]
37. Wiese, W.; Harbich, W.; Helfrich, W. Budding of lipid bilayer vesicles and flat membranes. *J. Phys. Condens. Matter* **1992**, *4*, 1647. [CrossRef]
38. Miao, L.; Seifert, U.; Wortis, M.; Döbereiner, H.G. Budding transitions of fluid-bilayer vesicles: The effect of area-difference elasticity. *Phys. Rev. E* **1994**, *49*, 5389. [CrossRef]
39. Seifert, U. Configurations of fluid membranes and vesicles. *Adv. Phys.* **1997**, *46*, 13–137. [CrossRef]
40. Svetina, S.; Žekš, B. Shape behavior of lipid vesicles as the basis of some cellular processes. *Anat. Rec. Off. Publ. Am. Assoc. Anat.* **2002**, *268*, 215–225. [CrossRef]
41. Heinrich, V.; Svetina, S.; Žekš, B. Nonaxisymmetric vesicle shapes in a generalized bilayer-couple model and the transition between oblate and prolate axisymmetric shapes. *Phys. Rev. E* **1993**, *48*, 3112. [CrossRef] [PubMed]
42. Ikari, K.; Sakuma, Y.; Jimbo, T.; Kodama, A.; Imai, M.; Monnard, P.A.; Rasmussen, S. Dynamics of fatty acid vesicles in response to pH stimuli. *Soft Matter* **2015**, *11*, 6327–6334. [CrossRef] [PubMed]
43. Bian, X.; Litvinov, S.; Koumoutsakos, P. Bending Models of Lipid Bilayer Membranes: Spontaneous Curvature and Area-Difference Elasticity. *Comput. Methods Appl. Mech. Eng.* **2020**, *359*, 112758. [CrossRef]

44. Pautot, S.; Frisken, B.J.; Weitz, D.A. Production of Unilamellar Vesicles Using an Inverted Emulsion. *Langmuir* **2003**, *19*, 2870–2879. [CrossRef]
45. Carrara, P.; Stano, P.; Luisi, P.L. Giant Vesicles Colonies: A Model for Primitive Cell Communities. *ChemBioChem* **2012**, *13*, 1497–1502. [CrossRef] [PubMed]
46. Stano, P.; Wodlei, F.; Carrara, P.; Ristori, S.; Marchettini, N.; Rossi, F. Approaches to Molecular Communication Between Synthetic Compartments Based on Encapsulated Chemical Oscillators. In *Advances in Artificial Life and Evolutionary Computation*; Pizzuti, C., Spezzano, G., Eds.; Number 445 in Communications in Computer and Information Science; Springer International Publishing: Cham, Switzerland, 2014; pp. 58–74. [CrossRef]
47. Fiore, M.; Maniti, O.; Girard-Egrot, A.; Monnard, P.A.; Strazewski, P. Glass Microsphere-Supported Giant Vesicles for the Observation of Self-Reproduction of Lipid Boundaries. *Angew. Chem. Int. Ed.* **2018**, *57*, 282–286. [CrossRef]
48. Sakashita, A.; Urakami, N.; Ziherl, P.; Imai, M. Three-dimensional analysis of lipid vesicle transformations. *Soft Matter* **2012**, *8*, 8569–8581. [CrossRef]
49. Helfrich, W. The size of bilayer vesicles generated by sonication. *Phys. Lett. A* **1974**, *50*, 115–116. [CrossRef]
50. Brakke, K.A. The Surface Evolver. *Exp. Math.* **1992**, *1*, 141–165. [CrossRef]
51. Carrara, P. Constructing a Minimal Cell. Ph.D. Thesis, University of Roma Tre, Rome, Italy, 2011.
52. Tyler, A.I.I.; Greenfield, J.L.; Seddon, J.M.; Brooks, N.J.; Purushothaman, S. Coupling Phase Behavior of Fatty Acid Containing Membranes to Membrane Bio-Mechanics. *Front. Cell Dev. Biol.* **2019**, *7*, 187. [CrossRef] [PubMed]
53. Kurniawan, J.; Suga, K.; Kuhl, T.L. Interaction Forces and Membrane Charge Tunability: Oleic Acid Containing Membranes in Different pH Conditions. *Biochim. Biophys. Acta (BBA) Biomembr.* **2017**, *1859*, 211–217. [CrossRef]
54. Majhenc, J.; Božič, B.; Svetina, S.; Žekš, B. Phospholipid Membrane Bending as Assessed by the Shape Sequence of Giant Oblate Phospholipid Vesicles. *Biochim. Biophys. Acta (BBA) Biomembr.* **2004**, *1664*, 257–266. [CrossRef]
55. Novák, B.; Tyson, J.J. Design principles of biochemical oscillators. *Nat. Rev. Mol. Cell Biol.* **2008**, *9*, 981–991. [CrossRef]
56. Murtas, G. Early self-reproduction, the emergence of division mechanisms in protocells. *Mol. Biosyst.* **2013**, *9*, 195–204. [CrossRef] [PubMed]
57. Halevy, I.; Bachan, A. The Geologic History of Seawater pH. *Science* **2017**, *355*, 1069–1071. [CrossRef] [PubMed]

Article

Racemate Resolution of Alanine and Leucine on Homochiral Quartz, and Its Alteration by Strong Radiation Damage

Adrien D. Garcia [1], Cornelia Meinert [1,*], Friedrich Finger [2], Uwe J. Meierhenrich [1] and Ewald Hejl [3,*]

1. Institut de Chimie de Nice, Université Côte d'Azur, CNRS, UMR 7272, 06108 Nice, France; adrien.garcia@univ-cotedazur.fr (A.D.G.); uwe.meierhenrich@univ-cotedazur.fr (U.J.M.)
2. Fachbereich Chemie und Physik der Materialien, Universität Salzburg, Hellbrunnerstraße 34, 5020 Salzburg, Austria; friedrich.finger@sbg.ac.at
3. Fachbereich für Geographie und Geologie, Universität Salzburg, Hellbrunnerstraße 34, 5020 Salzburg, Austria
* Correspondence: cornelia.meinert@univ-cotedazur.fr (C.M.); ewald.hejl@sbg.ac.at (E.H.)

Citation: Garcia, A.D.; Meinert, C.; Finger, F.; Meierhenrich, U.J.; Hejl, E. Racemate Resolution of Alanine and Leucine on Homochiral Quartz, and Its Alteration by Strong Radiation Damage. *Life* **2021**, *11*, 1222. https://doi.org/10.3390/life11111222

Academic Editors: Michele Fiore and Emiliano Altamura

Received: 23 September 2021
Accepted: 8 November 2021
Published: 11 November 2021

Publisher's Note: MDPI stays neutral with regard to jurisdictional claims in published maps and institutional affiliations.

Copyright: © 2021 by the authors. Licensee MDPI, Basel, Switzerland. This article is an open access article distributed under the terms and conditions of the Creative Commons Attribution (CC BY) license (https://creativecommons.org/licenses/by/4.0/).

Abstract: Homochiral proteins orchestrate biological functions throughout all domains of life, but the origin of the uniform L-stereochemistry of amino acids remains unknown. Here, we describe enantioselective adsorption experiments of racemic alanine and leucine onto homochiral *d*- and *l*-quartz as a possible mechanism for the abiotic emergence of biological homochirality. Substantial racemate resolution with enantiomeric excesses of up to 55% are demonstrated to potentially occur in interstitial pores, along grain boundaries or small fractures in local quartz-bearing environments. Our previous hypothesis on the enhanced enantioselectivity due to uranium-induced fission tracks could not be validated. Such capillary tubes in the near-surface structure of quartz have been proposed to increase the overall chromatographic separation of enantiomers, but no systematic positive correlation of accumulated radiation damage and enantioselective adsorption was observed in this study. In general, the natural *l*-quartz showed stronger enantioselective adsorption affinities than synthetic *d*-quartz without any significant trend in amino acid selectivity. Moreover, the L-enantiomer of both investigated amino acids alanine and leucine was preferably adsorbed regardless of the handedness of the enantiomorphic quartz sand. This lack of mirror symmetry breaking is probably due to the different crystal habitus of the synthetic z-bar of *d*-quartz and the natural mountain crystals of *l*-quartz used in our experiments.

Keywords: chirogenesis; quartz; amino acids; radiation damage; origin-of-life; GC×GC-TOFMS

1. Introduction

Homochiral polymers must have emerged rather early during abiogenesis, most probably in a pre-RNA/DNA/metabolism world [1–3]. Because of enantiomeric cross-inhibition, the synthesis of polypeptides and the self-replication of polynucleotides is repressed or even impossible in a racemic environment [4–6]. Consequently, homochirality or at least a substantial enantiomeric enrichment (*ee*) in the chemical building blocks of the first functional biopolymers should have developed early during chemical evolution. While the chronological details are still unclear, it is well accepted that chemical evolution did not occur suddenly; instead, a gradual transition from simple organic building blocks, including chiral molecules toward polymeric and supramolecular assemblies, has built the fundaments for self-assembling, self-sustaining interactive systems with emerging patterns that have ultimately evolved into what could be considered as living entities [7,8].

Abiotic synthesis of asymmetric molecules, such as amino acids or sugars, usually produces racemic compositions. Several mechanisms capable to entail molecular symmetry breaking leading to biomolecular homochirality on Earth, so-called chirogenesis, have been proposed and investigated. The most popular hypotheses comprise symmetry breaking by crystallization [9], selective adsorption of enantiomers on mineral surfaces [10–12], molecular parity violation [13], and enantioselective chemistries induced by circularly polarized light [14–18] or spin-polarized electrons [19].

Regarding terrestrial chirogenesis due to enantioselective adsorption on minerals, the putative role of naturally etched nuclear particle tracks, for example fission tracks under early Archean weathering conditions, has not yet been evaluated experimentally. Such capillary tubes in enantiomorphic minerals can eventually produce a local chiral enrichment of amino acids or other pre-existing biomolecules. Hejl has proposed a hypothesis for racemate resolution by chemical etching of nuclear particle tracks, as for example fission tracks originating from the spontaneous or induced fission of ^{238}U or ^{235}U, respectively [20]. The hypothesis relies on the principle of liquid chromatography, i.e., on the chemical separation of a mixture of dissolved chemical components (*mobile phase*) passing through a capillary structure (*stationary phase*). Molecular separation is supposed to occur in etched fission tracks, serving as a *stationary phase* with respect to a mobile solution of molecules. According to this hypothesis, fission tracks in enantiomorphic minerals could produce chiral separation of a racemic mixture, in addition to the chiral separation on their external crystal faces, cleavage planes, and fractures. In other words, accumulated radiation damage by nuclear particle tracks could increase the overall chiral resolution effect of enantiomorphic minerals—especially when they are chirally enriched on a local scale. Despite its low uranium content (<1 µg g^{-1}), natural occurring quartz may contain clusters of fission tracks around uranium-rich inclusions. Such track clusters in quartz were used in the past for fission-track dating [21–23]. Adequate etching conditions for fission tracks in quartz have already been published by Fleischer and Price [24]. Note, however, that quartz does not incorporate uranium in its crystal lattice.

Besides quartz, most enantiomorphic mineral species are either rare or have probably not occurred on the early Earth. With regard to eventual chiral enrichment by autocatalytic secondary nucleation, Hejl and Finger have investigated chiral proportions of nepheline in rocks undersaturated in SiO_2 which had crystallized from melts with low viscosity [25]. They found that nepheline mainly occurs in racemic proportions, even in small volumes of rock. Furthermore, nepheline-bearing rocks are not known from Archean terrains. They first appeared on Earth about 2.7 Ga ago [26], i.e., about 1 Ga later than the eldest traces of life. Enantiomorphic zeolites were considered for racemate resolution [27], but their ubiquitous presence on the early Earth is doubtful. Other enantiomorphic mineral species, as for example berlinite, pinnoite, wardite, or leucophanite are extremely rare and not known from the early Archean crust.

Therefore, quartz is a more promising mineral for terrestrial chirogenesis to have potentially generated racemate resolution. The presence of a granitic crust and, thus, the widespread occurrence of quartz on the early Earth are testified by detrital zircons in gneisses of the Yilgarn craton in Western Australia. The oldest ever zircon has a U-Pb age of 4.404 ± 0.008 Ga [28]. Chemical zoning and oxygen isotope composition of this zircon indicate that it has crystallized from a granitic magma that had been molten from supracrustal rocks in contact with an early hydrosphere [28].

Bonner et al. have first demonstrated asymmetric adsorption of alanine on crystal faces of quartz [11], but it was not considered as an effective process for regional or global racemate resolution because *d*- and *l*-quartz occur in equal proportions on a global scale [29–31]. However, sizeable volumes of homochiral quartz were reported from hydrothermal veins [32], and from graphic granite [33]. Thus, local enrichments of either *d*- or *l*-quartz on the early Earth may have occurred.

Even non-enantiomorphic minerals can exhibit surfaces with a chiral atomic pattern [34–36], but such chiral crystal faces occur pairwise with opposite handedness on the same crystal. Thus, crystal faces with opposite handedness are equally frequent in granulates of a non-enantiomorphic mineral species. Any chiral separation effect on crystal faces or cleavage planes will be compensated in a relatively small volume of rock, granulated sediment, or soil. On the other hand, chiral faces do not occur pairwise with opposite handedness on a single enantiomorphic crystal. Its whole surface has a biased chiral pattern, and a granulated enantiomorphic mineral with similar handedness (either *d* or *l*) may have an enantioselective separation effect on a percolating racemic solution of chiral molecules.

Here, we experimentally demonstrate significant racemate resolution in percolating solutions of racemic alanine and leucine on homochiral quartz sand with randomly orientated surfaces (i.e., without well-defined crystallographic orientation). Moreover, our results partly disprove the amplifying role of etched nuclear particle tracks according to the hypothesis of Hejl on the enantioselective adsorption potential of enantiomorphic quartz [20].

2. Materials and Methods

2.1. Characterization and Preparation of Homochiral Quartz Samples

Homochiral quartz sand was produced in two different ways. A first sample was obtained by crushing and sieving a synthetic z-bar quartz from a commercial provider (Hausen GmbH, Telfs, Austria). Such quartz crystals grow under hydrothermal conditions on so-called seed plates that are produced by cutting already existing z-bar crystals in thin plates of well-defined crystallographic orientation. A commercial laboratory will, thus, always duplicate z-bars of the same handedness. They are not twinned and strictly homochiral *l*- or *d*-quartz. Our z-bar quartz crystal had a mass of 1473 g (Figure 1a). Prior to crushing and sieving, we determined its sense of optical rotation for light propagation parallel to the crystallographic c-axis. For this purpose, we produced a quartz plate by cutting the crystal with a diamond saw perpendicular to the crystallographic c-axis. The plate was ground with silicon carbide, and polished with diamond paste down to a grain size of 1 µm. Its thickness of 3.16 mm was measured with a vernier calliper with a precision of ± 0.01 mm. The sense and angle of optical rotation of the transparent plate were determined with an optical polarimeter (Modular Compact Polarimeter MCP 150, Anton Paar GmbH, Graz, Austria). The temperature of the sample during the measurement was kept constant using Peltier elements. Clockwise optical rotation corresponds to *d*-quartz, and counter-clockwise optical rotation corresponds to *l*-quartz when rotation is observed in the sense opposite to the light propagation [37,38]. Our z-bar crystal was found to be *d*-quartz with an optical rotatory power of +21.68 ° mm^{-1} at 20 °C.

The pieces of z-bar quartz were treated by alternating crushing and sieving, in order to produce sand with a grain size of 0.25–0.50 mm. The quartz pieces were first broken with a hammer on a steel plate, and then shortly crumbled with a disk mill for 5 s only. After each sieving step, the coarser fraction (>0.50 mm) was crushed and sieved again. Finally, we obtained about 400 g of homochiral *d*-quartz sand.

This quartz sand was washed with acetone, isopropanol, and double-distilled water, and dried in a drying chamber at 105 °C. The dried sand was purified with a Frantz magnetic separator to remove small steel particles that may have been produced by abrasion of the steel hammer or the mill. Afterwards, this purified sand was heated for 3 h at 500 °C in a muffle furnace to minimize biological contaminations. An aliquot of the glowed sand was then powdered in an agate mill and chemically analyzed by X-ray fluorescence, revealing the composition to be of 99.88% SiO_2 with minor contaminations of Fe and Cr, probably from abrasion of the steel hammer or the mill.

Unfortunately, we could not purchase synthetic *l*-quartz due to the method of production of z-bar quartz. We found only z-bar providers which offer either *d*-quartz or give no specification on the handedness of the quartz. Therefore, we decided to produce *l*-quartz sand by crushing and sieving crystallographically well-defined mountain crystals, which had grown under natural conditions. First, we evaluated 20 natural quartz crystals from an open tension joint in Madagascar (Figure 1b). These transparent crystals of fairly similar size and habitus had a total mass of 507 g with individual masses between 13.7 and 40.1 g, and individual lengths ranging from 5 to 9 cm. The longitudinal axes were found to be the crystallographic c-axis. Ten of these crystals were rejected because they were either twinned or intergrown with smaller quartz crystals. The remaining ten crystals were found to be single crystals without twinning. They had a total mass of 227 g and were chosen for further treatment.

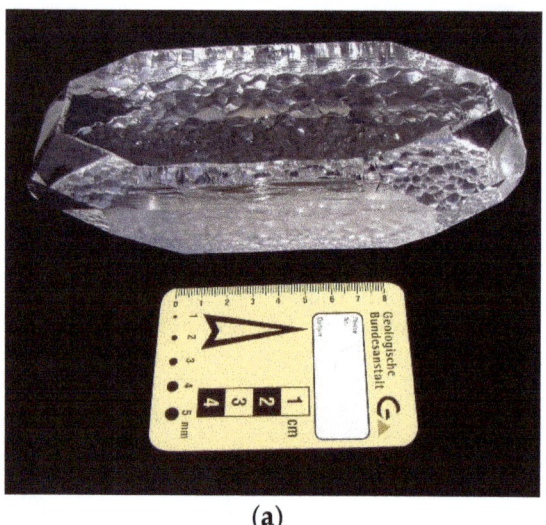 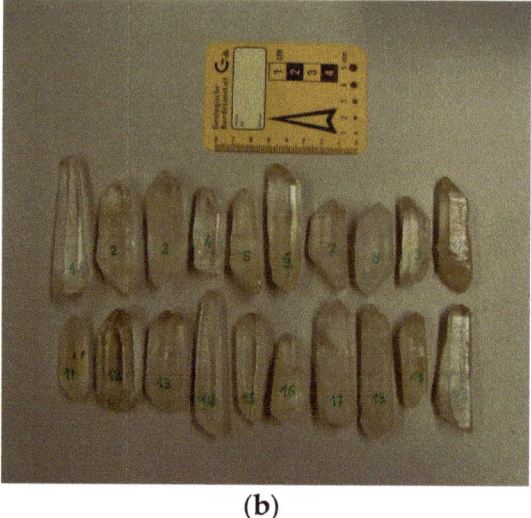

Figure 1. Enantiomorphic crystals of *d*- and *l*-quartz. (**a**) Synthetic z-bar of *d*-quartz produced by hydrothermal growth on a rectangular seed plate with a total mass of 1.473 kg. Its crystallographic c-axis is orientated perpendicular to the main elongation of the crystal and perpendicular to the table plane.; (**b**) Twenty natural quartz crystals from an open fissure in Madagascar. Ten of them were found to be single crystals without twinning nor intergrown with smaller crystals.

These single crystals were cut with a diamond saw perpendicular to their c-axis, to produce plates with a thickness of about 3 mm (Figure 2). These plates were ground with silicon carbide and polished with diamond paste, in a similar way as for the plate from the z-bar. Six crystals were found to be *l*-quartz with optical rotatory powers between 21.26 and $-22.74°$ mm^{-1} at 20 °C (Table 1). The small variations of the optical rotatory power are probably due to a slight imprecision of the cutting angle relative to the c-axis (ca. $90° \pm 2°$). The other four crystals were found to be *d*-quartz with a positive optical rotatory power and, therefore, rejected. The material from the six *l*-quartz crystals was treated by alternating crushing and sieving, washed, purified, and glowed in a similar way as the z-bar sample. Finally, we obtained 44 g of homochiral *l*-quartz sand with a grain size of 0.25–0.50 mm. The specific surface area (SSA) of both kinds of quartz sand (*l* and *d*) can be deduced from their quasi equigranular grain size (0.25–0.5 mm), using published data of granulometric investigations [39]. Recovered amino acid quantities normalized to the specific surface area were calculated with a SSA of 60 m^2 g^{-1} and a porosity of 40% that is typical for equigranular sand with angular grains.

2.2. Thermal Neutron Irradiation of Homochiral Quartz Sand Mixed with Uraniferous Zircon Powder

Synthetic z-bar quartz nor mountain quartz contain any uraniferous inclusions. In order to produce fission tracks on the fractured surfaces of quartz grains, aliquots of the homochiral sand samples were mixed with uraniferous zircon powder. For this purpose, zircon sand from a placer deposit in New South Wales (Australia) was ground to fine powder in an agate mill. The chemical composition of the zircon powder was analyzed by X-ray fluorescence using a Bruker Pioneer S4 crystal spectrometer at the Department for Chemistry and Physics of Materials (University of Salzburg). Obtained net count rates on single X-ray lines were recast into concentration values (wt% and ppm) based on an in-house calibration routine that involves measurements of ca. 30 international geo-standards (USGS and GSJ). The calibration was based on the Bruker AXS software SPECTRAplus FQUANT (v1.7) which corrects absorption, fluorescence, and line overlap effects. Concentrations of U and Th were 0.049 and 0.018 wt%, respectively (Table 2).

Figure 2. Some non-twinned mountain crystals with ca. 3-mm-thick sheets that were cut perpendicular to the crystallographic c-axis.

Table 1. Polarimetric measurements on quartz slices perpendicular to the crystallographic c-axes. Slices 4, 7, 11, 12, 14 and 20 were found to be *l*-quartz. The corresponding crystals were used to produce *l*-quartz sand.

Quartz Slices Normal to C Axis	Thickness [nm]	Optical Rotation [°]	$[\alpha]_D^{20}$ [° mm^{-1}]
Slice 4	2.99	−67.998	−22.74
Slice 6	3.35	+72.545	+21.66
Slice 7	3.02	−64.213	−21.26
Slice 9	3.16	+70.282	+22.24
Slice 11	2.91	−62.330	−21.42
Slice 12	3.07	−67.043	−21.84
Slice 14	3.16	−69.470	−21.98
Slice 15	3.22	+72.254	+22.44
Slice 16	3.17	not measurable [1]	-
Slice 20	2.92	−63.292	−21.68
z-bar quartz	3.16	+68.693	+21.74

[1] Slice 16 could not be measured because of internal fractures.

Table 2. Chemical composition of the zircon powder measured by X-ray fluorescence with major elements in oxide concentrations. Components are listed in descending order of concentrations.

Chemical Components	Concentration [wt %]	LLD [1] [ppm]
ZrO_2	64.12	91.8
SiO_2	33.44	403.1
Fe_2O_3	0.539	191.0
TiO_2	0.372	61.5
Ce	0.284	51.6
P_2O_5	0.249	45.1
Al_2O_3	0.222	48.7
SO_3	0.138	39.6
Cr	0.120	24.3
Na_2O	0.084	37.4
CaO	0.062	45.9
U	0.049	22.9
MgO	0.049	50.0
MnO	0.028	14.2
Th	0.028	17.8
K_2O	0.019	37.1
Sc	0.013	20.0
La	0.009	37.2
Pb	0.009	31.6
Ba	0.008	54.5
Cl	0.004	12.6
Ni	0.004	9.8
V	0.003	21.9
Sum	99.85	

[1] Lower limit of detection.

The zircon powder was glowed in a muffle furnace at 500 °C for 3 h to reduce the risk of biological contaminants, such as natural L-amino acids. Afterwards, it was mixed with aliquots of either homochiral *d*- or *l*-quartz sand. Thus, 125 g of *d*-quartz sand were mixed with 25 g of zircon powder, and 12.5 g of *l*-quartz sand were mixed with 2.5 g of zircon powder corresponding to an identical mass ratio of 88.33% quartz and 16.67% zircon. Both mixtures were filled in polyethylene tubes. When the surface of quartz grains is completely covered by zircon powder with a uranium concentration of 0.049 wt%, a thermal neutron fluence of 5×10^{15} neutrons cm^{-2} will induce a fission-track density of about 10^7 tracks per cm^2 onto the quartz surface [40]. The addition of zircon powder alone would not lead to enough fission tracks within an adequate time scale as the spontaneous fission of ^{238}U is a very rare event. It would take several million years to produce a significant quantity of fission tracks on the surface of quartz. We, thus, induced the fission of ^{235}U by subsequent neutron irradiation of the mixed quartz sand/zircon powder.

Thermal neutron irradiation of the filled tubes was performed in the TRIGA Mark-II Reactor of the TU Vienna (Atominstitut der TU Vienna, Austria). This research reactor of the swimming pool type has a nominal power of 250 kW. The reactor is controlled by three control rods which contain boron carbide as an absorber material. The thermal neutron flow rate is in the order of 10^{13} cm^{-2} s^{-1} in a central position. Desired neutron fluences of about 5×10^{15} neutrons cm^{-2} were obtained after an irradiation time of about 1 h, with irradiation position at the border of the reactor core. After a decay time of two months, the capsules possessed total activities of up to 200 kBc—mainly due to the activities of ^{46}Sc, ^{95}Zr, ^{95}Nb, ^{181}Hf, and ^{175}Hf. Besides uranium fission tracks, the samples must have accumulated a rather high radiation damage by beta decay and gamma radiation.

2.3. Sand Etching and Subsequent Infiltration with Aqueous Solutions of dl-alanine and dl-Leucine

Irradiated *d*- and *l*-quartz sand, as well as nonirradiated *d*- and *l*-quartz, were etched with 10% HF. This etching transforms superficial latent fission tracks into capillary tubes [24,40]. Identical etching conditions were applied to the nonirradiated *d*- and *l*-quartz samples to avoid any differences in the adsorption properties caused by radiation damage vs. etching, allowing for a better comparison of the enantioselective adsorption with the irradiated samples.

The sand fractions were filled in plastic Erlenmeyer flasks together with 10% HF and were etched at 20 °C for 2 h under permanent agitation on a shaking table. Afterwards, the sand was rinsed several times with double-distilled water. A second etching with 10% HCl was applied for 20 min, to dissolve potential fluorine components. While repeatedly rinsing the etched sand with double-distilled water, the zircon powder was removed by decantation. Finally, the sand was desiccated at 95 °C in a drying chamber.

Subsequent adsorption experiments of both *d*- (synthetic) and *l*-quartz (natural), respectively, with diluted aqueous solutions of individual racemic amino acids (DL-Ala and DL-Leu) were conducted as follows. The glassware (Erlenmeyer flasks, beakers, watch glasses, and volumetric flasks) as well as needles, spatulas, and aluminum foil were cleaned with acetone and isopropanol, and then glowed for 3 h in a muffle furnace at 500 °C. Solutions of racemic amino acids with concentrations of 10^{-3} M were prepared in volumetric flasks. This amino acid concentration was chosen to allow for accurate quantitation of recovered amino acids. Glass funnels were placed on the Erlenmeyer flasks, and conically folded aluminum foils were placed in the funnels. These foils were checked for impermeability prior to utilization. First, 4 mL of 10^{-3} M racemic amino acid solutions (either DL-Ala or DL-Leu) were dropwise added to the funnels, followed by homochiral quartz sand (either *d*-quartz or *l*-quartz) up to the liquid surface. This procedure avoided the formation of air bubbles, and the interstices between the sand grains became entirely filled with the racemic solutions. Eight funnels were prepared in this way, i.e., DL-Ala + nonirradiated *d*-quartz (1), DL-Ala + irradiated *d*-quartz (2), DL-Leu + nonirradiated *d*-quartz (3), DL-Leu + irradiated *d*-quartz (4), DL-Ala + nonirradiated *l*-quartz (5), DL-Ala + irradiated *l*-quartz (6), DL-Leu + nonirradiated *l*-quartz (7), and DL-Leu + irradiated *l*-quartz (8). In addition, four blanks with nonirradiated *d*- and *l*-quartz as well as irradiated *d*- and *l*-quartz were filled with 4 mL of double-distilled water.

The funnels filled with DL-amino acids and homochiral sand were covered with watch glasses and were allowed to stand for 72 h at 20 °C. Afterwards, a clean needle—with a thickness smaller than the grain size of the sand—was inserted vertically in the sand body and pushed downward to the bottom of the foil to perforate it. Then, 8 mL of double-distilled water was dropwise added to the top of the sand. In this way, the aqueous solution containing the diluted amino acid depleted in one adsorbed enantiomer was rinsed out from the pores between the sand grains and recovered in a small beaker and then filled in vials with silicon screw caps.

In a previous experiment, thin glass tubes were used instead of funnels. The lower end of these tubes was closed with aluminum foil and silicone. Mixed solutions containing 10^{-3} M DL-Ala and 10^{-3} M DL-Leu were used in the following manner: DL-Ala + DL-Leu + nonirradiated *d*-quartz (9) and DL-Ala + DL-Leu + irradiated *d*-quartz (10). In addition, two blanks with nonirradiated and irradiated *d*-quartz, both with 4 mL of double-distilled water, were prepared.

2.4. GC×GC-TOFMS Analysis—Sample Preparation and Derivatization

All sample-handling glassware was wrapped in aluminum foil and heated at 500 °C for 3 h prior to usage. Eppendorf tips® were sterile, and the water used for standard solutions, reagent solutions, and blanks was prepared with a Direct-Q® water purification system (<3 ppb total organic carbon). All solvents, reagents, and amino acid standards were purchased from Sigma-Aldrich and stored according to the respective instructions.

Specific volumes of the eluate of each infiltration experiment described above were transferred into a 1 mL of Reacti-Vial™ and dried under a gentle flow of nitrogen. These volumes were chosen to reach final concentrations that resulted in adequate peak intensities and areas for reliable enantiomeric excess quantification. The chosen volumes were as follows: 75 µL of (1) DL-Ala + nonirradiated d-quartz, 100 µL of (2) DL-Ala + irradiated d-quartz, 75 µL of (3) DL-Leu + nonirradiated d-quartz, 100 µL of (4) DL-Leu + irradiated d-quartz, 75 µL of (5) DL-Ala + nonirradiated l-quartz, 200 µL of (6) DL-Ala + irradiated l-quartz, 75 µL of (7) DL-Leu + nonirradiated l-quartz, 200 µL of (8) DL-Leu + irradiated l-quartz, 50 µL of (9) DL-Ala + DL-Leu + nonirradiated d-quartz, and 50 µL of (10) DL-Ala + DL-Leu + irradiated d-quartz.

All samples were re-dissolved in 50 µL of 0.1 M HCl, and individually derivatized to form N-ethylchloroformate heptafluorobutyryl esters (ECHFBE) by adding first 6 µL of pyridine and 19 µL of heptafluorobutanol followed by 5 µL of ethylchloroformate. The reaction mixtures were vortexed for 10 s, and the derivatives were extracted and transferred into GC vials by adding 50 µL of chloroform containing methyl myristate as an internal standard at a concentration of 10^{-6} M. Three aliquots of each incubated quartz sample (1) to (10) were derivatized, and each derivatized sample was analyzed (1 µL) as triplicates by enantioselective two-dimensional gas chromatography coupled to a time-of-flight mass spectrometer GC×GC-TOFMS [41,42], i.e., each quartz sample was analyzed nine times by preparing three derivatized aliquots with n = 3 to obtain sufficient replicates for the statistical analysis.

Racemic standard solutions of alanine and leucine—identical to the standards used for the adsorption experiments—with approximately the same concentration than the final quartz samples were used to determine the enantiomeric bias of the commercially available racemic standards. These reference standards were injected three times in between each triplicate injections of the three aliquots of the same quartz infiltration experiment to follow any chromatographic bias over time. In total, nine injections of the alanine and leucine reference standard, respectively, were performed for one adsorption sample, i.e., three individually derivatized reference standards with n = 3 injections. To ensure that no contamination occurred during the analytical protocol, nonirradiated and irradiated d- and l-quartz blanks were separately dried and derivatized following the same procedure, as described for the above samples and analyzed (1 µL) by GC×GC-TOFMS. The volume of the blank aliquots was the same as for the corresponding infiltration experiment. No contamination was observed.

2.5. GC×GC-TOFMS Analysis–Instrumental Conditions

The enantioselective multidimensional analysis was carried out by a GC×GC Pegasus instrument coupled to a TOFMS (LECO, St. Joseph, MI, USA). The TOFMS system operated at a storage rate of 150 Hz, with a 50–400 amu mass range, a detector voltage of 1650 V, and a solvent delay of 15 min. Ion source and injector temperatures were set to 230 °C and the transfer line to 240 °C. The column set consisted of a Varian–Chrompack Chirasil-L-Val column (25 m × 0.25 mm, 0.12 µm film thickness, Agilent-Varian, Santa Clara, California, US) in the first dimension and a DB-Wax in the second dimension (1.3 m × 0.1 mm, 0.1 µm film thickness) connected by a siltite µ-union. 2-propanol and chloroform (sample solvents) were used as washing solvents for the injection needle. Helium was used as carrier gas with a constant flow of 1 mL min^{-1}. All samples were injected in a 1:20 split mode.

For the analyses of samples containing alanine, the temperature of the primary oven was held at 40 °C for 1 min, increased to 80 °C at a rate of 10 °C min^{-1}, and held for 5 min, followed by an increase to 120 °C at 2 °C min^{-1} and finally to 190 °C at 10 °C min^{-1} and held for 6 min. The temperature of the secondary oven was held at 65 °C for 1 min, increased to 105 °C at a rate of 10 ° C min^{-1} with a 5-min hold time, followed by an increase to 145 °C at 2 °C min^{-1} and, finally, to 190 °C at 8 °C min^{-1}.

For the analyses of samples containing leucine, the temperature of the primary column was held at 40 °C for 1 min, then increased to 105 °C at a rate of 10 °C min^{-1} and held for

5 min, followed by an increase to 140 °C at 2 °C min^{-1} and finally to 190 °C at 10 °C min^{-1} and held for 5 min. The secondary oven used the same temperature program with a constant temperature offset of 20 °C.

For the mixed alanine–leucine samples, the temperature of the primary column was held at 40 °C for 1 min, increased to 80 °C at a rate of 10 °C min^{-1} and held for 5 min, then by an increase to 170 °C at 2 °C min^{-1}, and finally to 190 °C at 10 °C min^{-1} and held for 5 min. The secondary oven used the same temperature program with a constant temperature offset of 25 °C. The modulator used a temperature offset of 40 °C compared to the primary oven temperature and a modulation period of 5 s was applied to all samples.

2.6. Enantiomeric Excess Determination

The enantiomeric excess *ee* which expresses the excess of one enantiomer over the other is defined as $ee = (E_1 - E_2)/(E_1 + E_2)$ where E_1 and E_2 are the quantities of the enantiomers. In practice, *ee* is often quoted as a percentage, defined as $\%ee = 100 \times (E_1 - E_2)/(E_1 + E_2)$. Thus, a racemate is expressed as *ee* = 0%. While amino acid standards are commercially available as racemic mixtures, their expected equimolar ratios determined by chromatographic techniques are generally unequal to 0% ($ee \neq 0\%$) due to (*i*) intrinsic manufacture effects of the racemic standards itself; (*ii*) differential instrument responses for each enantiomer such as stereoselective signal suppression; or (*iii*) erroneous chromatographic quantitation caused by the co-elution of chemical impurities with only one enantiomer spuriously increasing peak areas, poor enantioresolution ($R_S < 1.5$), or peak tailing. To best account for any non-racemic composition measurements for analytes which are, in fact, racemic and vice versa, the non-zero *ee* of the racemic alanine and leucine standard, respectively, were determined as follows:

$$\%ee_{\text{ref}} = 100 \times [A(\text{L-}AA_{\text{ref}}) - A(\text{D-}AA_{\text{ref}})]/[A(\text{L-}AA_{\text{ref}}) + A(\text{D-}AA_{\text{ref}})] \quad (1)$$

where *A* is the peak area of the GC×GC signal and *AA* is the corresponding amino acid. The %*ee* values of the alanine and leucine samples after the adsorption experiments on homochiral quartz sand were calculated analogously by subtracting the %ee_{ref} of the respective amino acid in the standard solutions used as references:

$$\%ee_{AA} = 100 \times [A(\text{L-}AA) - A(\text{D-}AA)]/[A(\text{L-}AA) + A(\text{D-}AA)] - \%ee_{\text{ref}} \quad (2)$$

3. Results

A summary of the results of the GC×GC-TOFMS analyses of recovered interstitial liquids is given in Table 3 and Figure 3. Both amino acids, Ala and Leu, were enriched in the D-enantiomer ($ee_L < 0$), regardless of the handedness of the quartz sand. Moreover, the original hypothesis that capillary tubes on the quartz surface, produced by etching of uranium fission tracks, may enhance the overall enantioselective adsorption is partly disproved by our experiments.

Table 3. Enantiomeric excesses %ee_L of alanine and leucine after adsorption on nonirradiated and irradiated enantiomorphic d- and l-quartz. The recovered amino acid quantity is specified for each adsorption experiment as well as the recovered quantity normalized to the estimated total specific surface area of quartz (954 m^2).

Amino Acid	%ee_L-d-Quartz		%ee_L-l-Quartz	
	Nonirradiated	Irradiated	Nonirradiated	Irradiated
alanine	−2.92 ± 0.26 [2]	−1.09 ± 0.10 [3]	−6.75 ± 0.32 [2]	−12.82 ± 0.48 [5]
alanine [1]	−2.34 ± 0.45 [3]	−15.21 ± 0.74 [4]	-	-
leucine	−23.93 ± 0.35 [2]	−10.86 ± 1.24 [3]	−55.63 ± 0.46 [2]	−12.58 ± 0.50 [5]
leucine [1]	−6.75 ± 0.52 [3]	−3.57 ± 0.55 [4]	-	-

[1] Competitive adsorption experiment using a mixture of alanine and leucine. [2] $c \approx 4.3 \times 10^{-4}$ M $\approx 4.5 \times 10^{-7}$ M m^{-2}. [3] $c \approx 3.2 \times 10^{-4}$ M $\approx 3.4 \times 10^{-7}$ M m^{-2}. [4] $c \approx 4.0 \times 10^{-4}$ M $\approx 4.2 \times 10^{-7}$ M m^{-2}. [5] $c \approx 1.6 \times 10^{-4}$ M $\approx 1.7 \times 10^{-7}$ M m^{-2}.

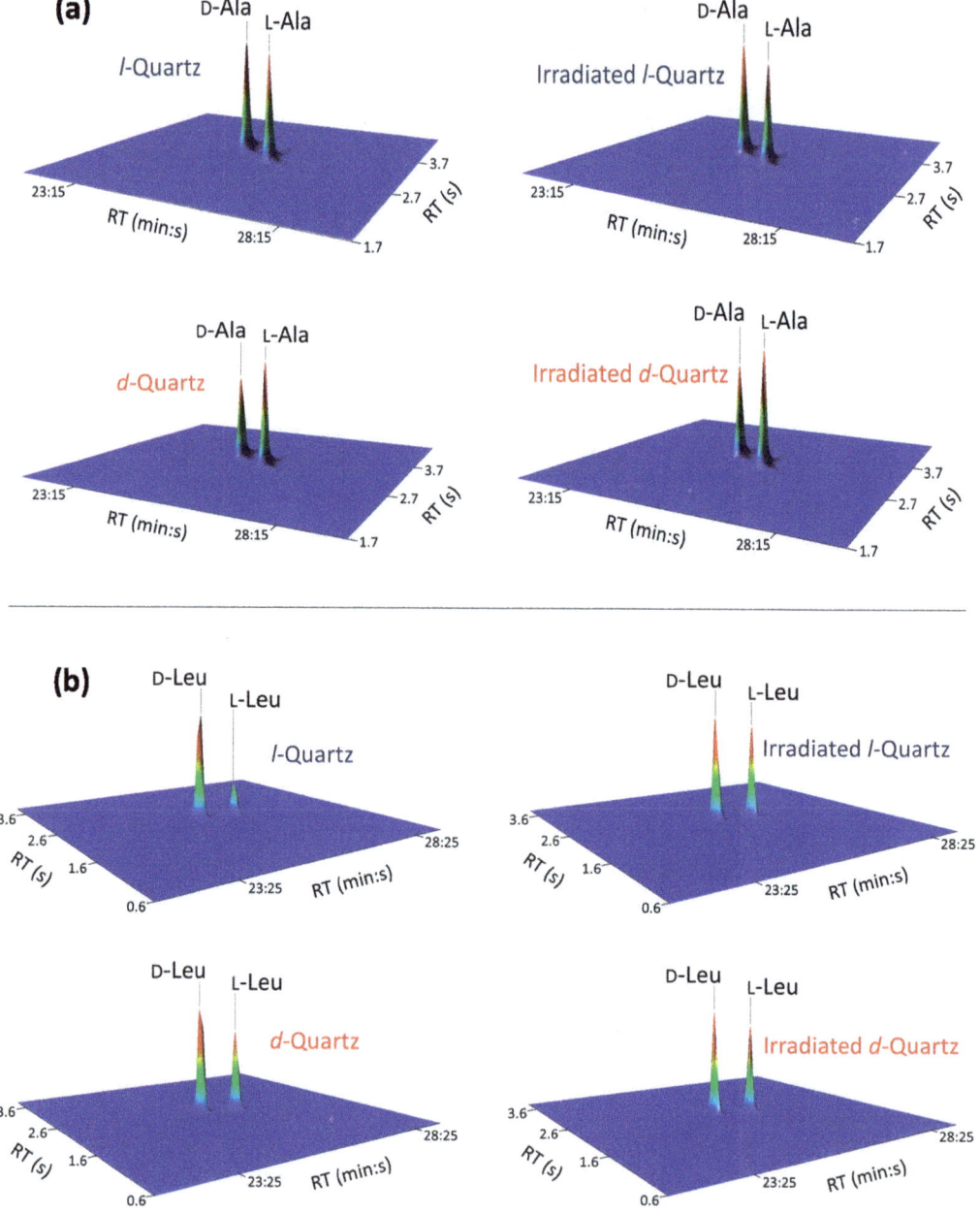

Figure 3. Enantiomorphic quartz induced enantiomeric excesses. Close-up view of the multidimensional enantioselective gas chromatographic analysis of (**a**) alanine and (**b**) leucine enantiomers after adsorption on nonirradiated and irradiated *l*-quartz (blue) and *d*-quartz (red).

First, we examined the absolute values of %ee_L, regardless of their algebraic sign. We measured six %ee_L values for Ala, and additional six %ee_L values for Leu. The |%ee| for Ala in six independent experiments ranges from 1.1 to 15.2%. These %ee values are significantly above the 3σ standard deviations (n = 3). They demonstrate that substantial racemate resolution can be expected in interstitial pores, along grain boundaries or small fractures in a local quartz-bearing environment. The |%ee| for Leu in six independent experiments ranges from 3.5 to 55.6%. These values are even higher than those for Ala. They substantiate the foregoing conclusion that random quartz fractures may produce significant racemate resolution on a local scale. This statement holds true for polygranular quartz occurrences with a chiral enrichment of either *d*- or *l*-quartz.

Second, we examined the influence of thermal neutron irradiation, induced fission tracks, and radiation damage, in general, on the observed ee values. In the case of Ala, the mean %ee_L value in the three experiments using nonirradiated quartz is −4.0% (spread between −2.3% and −6.8%), and the mean %ee_L value with irradiated quartz is −9.7% (spread between −1.1 and −15.2%). These results indicate an overall stronger enantioselective adsorption of L-Ala on irradiated quartz compared with nonirradiated quartz. One exception is the experiment of the pure racemic alanine solution absorbed on *d*-quartz were the %ee_L of −2.92% for nonirradiated quartz decreased to −1.09% for irradiated quartz. Thus, the results for Ala are ambiguous and will be discussed later.

In the case of Leu, the mean %ee_L value of three independent experiments with nonirradiated quartz is −28.8% (spread between −6.8 and −55.6%), and the mean %ee_L value with irradiated quartz is −9.0% (spread between −3.5 and −12.6%). The L-enantiomer of Leu adsorbs distinctly stronger on nonirradiated quartz than on irradiated quartz, without any exception. In all experiments with either *d*- or *l*-quartz, the adsorption of the L-enantiomers on nonirradiated quartz was about two or even four times higher than on irradiated quartz. Strong radiation damage clearly reduces enantioselective adsorption of L-Leu onto the quartz surface or in any superficial structures. Thus, the hypothesis of Hejl is disproved for leucine [20]. This observed reduction in enantioselective adsorption on irradiated quartz might be due to metamictization of the crystal lattice. Radiation dose-dependent destruction of the crystalline structure, up to amorphous consistency, is well known for zircon [43]. Atomic scale observations by transmission electron microscopy (TEM) have shown that the core zone of latent fission tracks in zircon is amorphous [44]. If this amorphous domain is not completely removed by etching, the walls of the fission track capillary tube will keep a certain metamictization after etching. Although the atomic structure of latent fission tracks in quartz has not yet been investigated, a superficial metamictization by heavy ion bombardment from adjacent uraniferous minerals, such as zircon, cannot be excluded. This kind of radiation damage would explain the alteration of enantioselective adsorption properties.

Comparing the ratios of the peak areas of the amino acid and the internal standard (A_{AA}/A_{IS}) before and after the adsorption experiments on enantiomorphic quartz allowed us to evaluate the adsorption efficiencies of each experiment (Table 3). Starting from an initial amino acid concentration of 10^{-3} M (sum of both enantiomers), the recovered quantities were found to range from 1.6×10^{-4} M for the experiments using irradiated *l*-quartz up to 4.3×10^{-4} M for the nonirradiated *d*- and *l*-quartz samples. Thus, less than 15% to 25% of the amino acids were recovered from irradiated *l*- and *d*-quartz, respectively, whereas almost 50% of amino acids were desorbed from the nonirradiated enantiomorphic quartz samples when using individual amino acid solutions. While these results indicate an increased adsorption affinity of irradiated over nonirradiated enantiomorphic quartz due to irradiation-induced fission tracks—that was not found to directly correlate with enhanced enantioselective adsorption, no significant adsorption selectivity between alanine and leucine was observed. The results of the competitive adsorption experiments are less obvious with very similar adsorption affinities of the irradiated and nonirradiated *d*-quartz but inverse trends in enantioselectivity for alanine and leucine. While the %ee_L of alanine significantly decreased to −15% on irradiated *d*-quartz, i.e., enhanced adsorption of D-alanine on irradiated *d*-quartz, D-leucine adsorbed more efficiently on nonirradiated *d*-quartz compared to irradiated *d*-quartz. It is currently uncertain whether these differences in enantioselectivity of alanine and leucine in the combined mixture are due to systematic or pure random effects. We note, however, that, in preparation of the incubation experiments of the alanine–leucine mixtures, the quartz sand was filled using thin tubes, whereas small funnels were used to prepare the adsorption experiments of the individual amino acids. This may have eventually entailed a difference in the packing density of the grains and thus differences in the cumulative pore volume. The grain packing in the funnels corresponds much better to the textural conditions in natural sediments and will be chosen as a preferred experimental design for future adsorption experiments. Further competitive enantioselective adsorption experiments are required to study the profound impact of intimate binding between enantiomorphic quartz and amino acids of different size, charge, and polarizability, as well as mutual effects between different molecular species during competitive adsorption.

4. Discussion

In our experimental investigation, the measured %*ee* values of both amino acids, Ala and Leu, were always enriched in the D-enantiomer (L-%*ee* < 0). Originally, we had assumed that sand grains have a nearly isometric form corresponding approximately to a sphere. With regard to the crystal lattice, all possible spatial orientations of crystalline planes occur with similar likelihood on the surface of the sphere. Even irregular planar orientations which cannot be denominated by Miller–Bravais indices would have the same frequency than any other plane. Indeed, sand grains of aeolian dunes, beach sands, or fluvial sediments are often well rounded [45]. Most quartz grains exhibit a spherical or ellipsoidal shape. If two monocrystalline quartz grains consist of quartz with opposite handedness (*d* and *l*, respectively), their overall chiral adsorption properties of their surfaces are expected to produce bulk-*ee* with opposite handedness on both grain surfaces. This assumption is still valid when a quasi-infinite number of homochiral grains with irregular forms is considered—provided that these irregular forms have no preferential orientation relative to the crystal lattice. Quartz has no preferential cleavage planes and mainly produces conchoidal fragments at rupture. This well-known fact seems to support the foregoing assumption.

However, the interstitial liquid of both enantiomorphic quartz crystals was always enriched in the D-enantiomer and no mirror effect was observed. Moreover, both L-Ala and L-Leu enantiomers tend to adsorb much stronger on *l*-quartz (natural) than on *d*-quartz (synthetic z-bar). We suppose that this phenomenon is due to different frequency distributions of the crystallographic orientation of fracture planes in the z-bar and the natural mountain quartz. The longitudinal axis of the z-bar is perpendicular to the crystallographic c-axis, while the longitudinal axis of the mountain crystals is parallel to the c-axis (Figure 4). These differences in the crystal habitus may have caused a dissimilar behavior at rupture. When a longitudinal homogenous material without preferential cleavage is broken with a hammer, one would expect that the predominant factures are mainly perpendicular to the longitudinal axis. These first fractures may have influenced the further ruptures in course of crumbling with the disk mill.

With regard to future adsorption experiments, we suggest and intend the use of *d*- and *l*-quartz with similar habitus for the production of homochiral sand—i.e., natural quartz crystals from the same location, with similar size and form. By this procedure, similar frequency distributions of crystallographic fractures orientations in both *d*- and *l*-quartz could be obtained. Random effects will be minimized, and the resulting bulk adsorption on the grain surfaces of *d*- or *l*-quartz will produce rather 'symmetrical' *ee* with opposite sign.

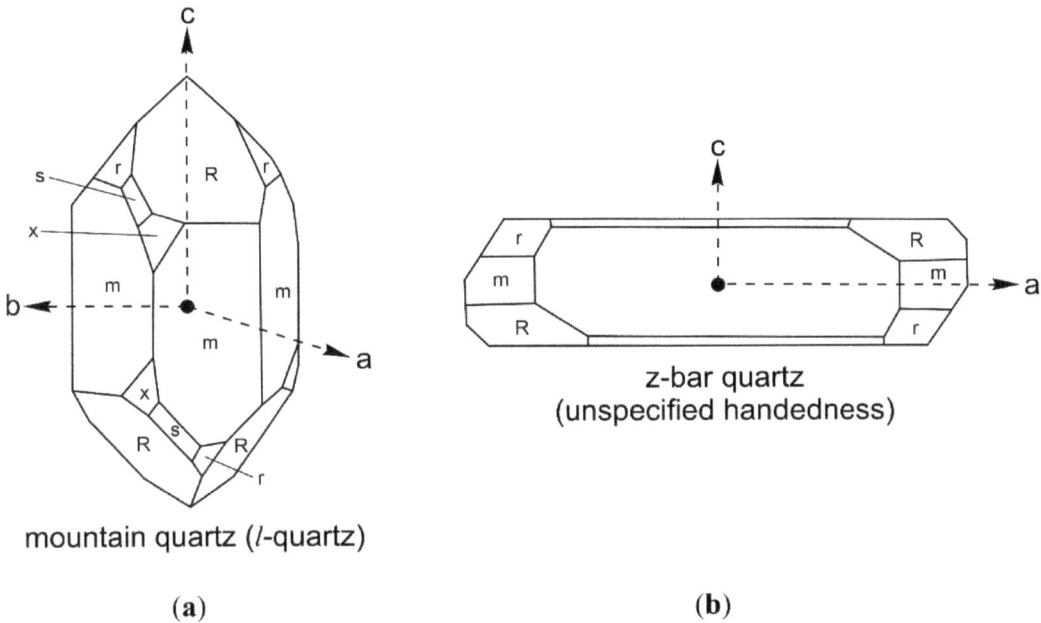

Figure 4. Orientations of crystallographic axes and principal crystal planes m, r, R, s and x after Cheng et al. [46]. (**a**) Schematic representation of l-quartz with space group P3121. The position of the faces of the trigonal trapezohedron x {6 $\bar{1}$ 5 1} and the trigonal bipyramid s {2 $\bar{1}$ $\bar{1}$ 1} reflects the handedness of the crystal; (**b**) The longitudinal axis a' of the z-bar quartz is orthogonal to a prism face m and forms an angle of 30° with the a-axis. The z-bar used for this investigation was *d*-quartz (Figure 1), but its handedness cannot be deduced from its morphology.

Overall, leucine showed higher racemate resolution on quartz with a maximum %*ee* of 55.6% compared with alanine in our experiments. Molecular size-related effects may be responsible for the observed trend and future investigations on various individual amino acid solutions covering different molecular sizes or branching, and different functional groups (aliphatic, aromatic, acidic, basic, etc.) are required to study size- and polarization-related effects on the enantioselective adsorption behavior. Moreover, additional competitive adsorption experiments are required for investigating the interplay of amino acid mixtures and the enantioselective adsorption forces of enantiomorphic minerals to illuminate prebiotically plausible systems.

5. Conclusions

The enantioselective adsorption of alanine and leucine with %*ee* values ranging from 1% to 55% was observed to occur on homochiral quartz sand. The L-enantiomer of both investigated amino acids was preferably adsorbed regardless of the handedness of the enantiomorphic quartz. We currently explain the lack of mirror symmetry breaking in our experiments with the different crystal habitus of the synthetic z-bar of *d*-quartz and the natural mountain crystals of *l*-quartz. Future experiments using *d*- and *l*-quartz with similar habitus to minimize random effects are required to conclude on the possible mirror symmetry breaking stimulus through amino acid adsorption on enantiomorphic minerals with opposite handedness.

Despite the absence of %*ee* values of the opposite sign, there are consistent trends in our collected data, such as natural *l*-quartz, showed stronger enantioselective adsorption affinities than synthetic *d*-quartz for both amino acids. Moreover, strong radiation damage clearly reduced the enantioselective adsorption of leucine, while no clear conclusion can be drawn for the results on alanine regarding nonirradiated vs. irradiated quartz. Finally, the desorption capacity of alanine and leucine was equally strong in each of the corresponding experiments using individual amino acid solutions as well as using the combined amino acid mixture, i.e., no amino acid selectivity in terms of pure adsorption of alanine vs. leucine onto identical quartz surfaces.

Author Contributions: Conceptualization and methodology, C.M., U.J.M. and E.H.; enantioselective adsorption experiments, E.H.; enantioselective GC×GC-TOFMS analyses, A.D.G. and C.M.; X-ray fluorescence, F.F.; data curation, C.M. and E.H.; writing—original draft preparation, C.M. and E.H.; funding acquisition, C.M. and E.H. All authors have read and agreed to the published version of the manuscript.

Funding: This research was funded by the Austrian Science Foundation (FWF—Der Wissenschaftsfonds, Fonds zur Förderung der Wissenschaftlichen Forschung), grant number P 30444-N28 and the European Research Council under the European Union's Horizon 2020 research and innovation program, grant agreement 804144. ADG received a PhD scholarship from the French Ministry of Science and Education.

Data Availability Statement: All data analyzed during this study are included in this published article. The datasets generated or analysed during the current study are available from the corresponding authors upon request.

Acknowledgments: We would like to thank the staff of the TRIGA Mark-II Reactor at the University Vienna, Austria for the operation of the thermal neutron irradiation experiments. Open Access Funding by the Austrian Science Fund (FWF) is acknowledged.

Conflicts of Interest: The authors declare no conflict of interest. The funders had no role in the design of the study; in the collection, analyses, or interpretation of data; in the writing of the manuscript, or in the decision to publish the results.

References

1. Wächtershäuser, G. Biomolecules: The origin of their optical activity. *Med. Hypotheses* **1991**, *36*, 307–311. [CrossRef]
2. Klabunovskii, E.I. Can Enantiomorphic Crystals Like Quartz Play a Role in the Origin of Homochirality on Earth? *Astrobiology* **2001**, *1*, 127–131. [CrossRef] [PubMed]
3. Meierhenrich, U.J. Amino Acids and the Asymmetry of Life. In *Advances in Astrobiology and Biogeophysics*; Springer: Berlin/Heidelberg, Germany, 2008; ISBN 978-3-540-76886-9.
4. Idelson, M.; Blout, E.R. Polypeptides. XVIII.1 A Kinetic Study of the Polymerization of Amino Acid N-Carboxyanhydrides Initiated by Strong Bases. *J. Am. Chem. Soc.* **1958**, *80*, 2387–2393. [CrossRef]
5. Joyce, G.F.; Visser, G.M.; van Boeckel, C.A.A.; van Boom, J.H.; Orgel, L.E.; van Westrenen, J. Chiral selection in poly(C)-directed synthesis of oligo(G). *Nat. Cell Biol.* **1984**, *310*, 602–604. [CrossRef]
6. Joyce, G.F.; Schwartz, A.W.; Miller, S.L.; Orgel, L.E. The case for an ancestral genetic system involving simple analogues of the nucleotides. *Proc. Natl. Acad. Sci. USA* **1987**, *84*, 4398–4402. [CrossRef]
7. Eschenmoser, A. The search for the chemistry of life's origin. *Tetrahedron* **2007**, *63*, 12821–12844. [CrossRef]
8. Krishnamurthy, R.; Hud, N.V. Introduction: Chemical Evolution and the Origins of Life. *Chem. Rev.* **2020**, *120*, 4613–4615. [CrossRef]
9. Kondepudi, D.K.; Kaufman, R.J.; Singh, N. Chiral Symmetry Breaking in Sodium Chlorate Crystallization. *Science* **1990**, *250*, 975–976. [CrossRef]
10. Karagunis, G.; Coumoulos, G. A New Method of Resolving a Racemic Compound. *Nat. Cell Biol.* **1938**, *142*, 162–163. [CrossRef]
11. Bonner, W.A.; Kavasmaneck, P.R.; Martin, F.S.; Flores, J.J. Asymmetric Adsorption of Alanine by Quartz. *Science* **1974**, *186*, 143–144. [CrossRef]
12. Hazen, R.M.; Filley, T.R.; Goodfriend, G.A. Selective adsorption of L- and D-amino acids on calcite: Implications for biochemical homochirality. *Proc. Natl. Acad. Sci. USA* **2001**, *98*, 5487–5490. [CrossRef] [PubMed]
13. Quack, M. How Important is Parity Violation for Molecular and Biomolecular Chirality? *Angew. Chem. Int. Ed.* **2002**, *41*, 4618–4630. [CrossRef] [PubMed]
14. Balavoine, G.; Moradpour, A.; Kagan, H.B. Preparation of chiral compounds with high optical purity by irradiation with circularly polarized light, a model reaction for the prebiotic generation of optical activity. *J. Am. Chem. Soc.* **1974**, *96*, 5152–5158. [CrossRef]
15. Meierhenrich, U.J.; Filippi, J.-J.; Meinert, C.; Hoffmann, S.V.; Bredehöft, J.H.; Nahon, L. Photolysis of rac-Leucine with Circularly Polarized Synchrotron Radiation. *Chem. Biodivers.* **2010**, *7*, 1651–1659. [CrossRef]
16. Meinert, C.; Hoffmann, S.V.; Cassam-Chenaï, P.; Evans, A.C.; Giri, C.; Nahon, L.; Meierhenrich, U.J. Photonenergy-Controlled Symmetry Breaking with Circularly Polarized Light. *Angew. Chem. Int. Ed.* **2014**, *53*, 210–214. [CrossRef] [PubMed]

17. Meinert, C.; Cassam-Chenaï, P.; Jones, N.C.; Nahon, L.; Hoffmann, S.V.; Meierhenrich, U.J. Anisotropy-Guided Enantiomeric Enhancement in AlanineUsing Far-UV Circularly Polarized Light. *Orig. Life Evol. Biosph.* **2015**, *45*, 149–161. [CrossRef]
18. Garcia, A.D.; Meinert, C.; Sugahara, H.; Jones, N.C.; Hoffmann, S.V.; Meierhenrich, U.J. The Astrophysical Formation of Asymmetric Molecules and the Emergence of a Chiral Bias. *Life* **2019**, *9*, 29. [CrossRef]
19. Rosenberg, R.A.; Abu Haija, M.; Ryan, P.J. Chiral-Selective Chemistry Induced by Spin-Polarized Secondary Electrons from a Magnetic Substrate. *Phys. Rev. Lett.* **2008**, *101*, 178301. [CrossRef]
20. Hejl, E. Are fission tracks in enantiomorphic minerals a key to the emergence of homochirality? *Neues Jahrb. Für Mineral. Abh.* **2017**, *194*, 97–106. [CrossRef]
21. Weiland, E.F.; Ludwig, K.R.; Naeser, C.W.; Simmons, E.C. Fission-Track Dating Applied to Uranium Mineralization. Master's Thesis, Colorado School of Mines, Golden, CO, USA, 1979.
22. Cunningham, C.G.; Ludwig, K.R.; Naeser, C.W.; Weiland, E.K.; Mehnert, H.H.; Steven, T.A.; Rasmussen, J.D. Geochronology of hydrothermal uranium deposits and associated igneous rocks in the eastern source area of the Mount Belknap Volcanics, Marysvale, Utah. *Econ. Geol.* **1982**, *77*, 453–463. [CrossRef]
23. Vincent, D.; Clocchiatti, R.; Langevin, Y. Fission-track dating of glass inclusions in volcanic quartz. *Earth Planet. Sci. Lett.* **1984**, *71*, 340–348. [CrossRef]
24. Fleischer, R.; Price, P. Techniques for geological dating of minerals by chemical etching of fission fragment tracks. *Geochim. Cosmochim. Acta* **1964**, *28*, 1705–1714. [CrossRef]
25. Hejl, E.; Finger, F. Chiral Proportions of Nepheline Originating from Low-Viscosity Alkaline Melts. A Pilot Study. *Symmetry* **2018**, *10*, 410. [CrossRef]
26. Blichert-Toft, J.; Arndt, N.; Ludden, J. Precambrian alkaline magmatism. *Lithos* **1996**, *37*, 97–111. [CrossRef]
27. Dryzun, C.; Mastai, Y.; Shvalb, A.; Avnir, D. Chiral silicate zeolites. *J. Mater. Chem.* **2009**, *19*, 2062–2069. [CrossRef]
28. Wilde, S.A.; Valley, J.; Peck, W.; Graham, C.M. Evidence from detrital zircons for the existence of continental crust and oceans on the Earth 4.4 Gyr ago. *Nat. Cell Biol.* **2001**, *409*, 175–178. [CrossRef]
29. Palache, C.; Berman, G.B.; Frondel, C. Relative frequencies of Left and Right Quartz. In *The System of Mineralogy*; J. Wiley New-York: New York, NY, USA, 1962; Volume 17.
30. Evgenii, K.; Wolfram, T. The role of quartz in the origin of optical activity on earth. *Orig. Life Evol. Biosph.* **2000**, *30*, 431–434. [CrossRef]
31. Pavlov, V.A.; Shushenachev, Y.V.; Zlotin, S.G. Chiral and Racemic Fields Concept for Understanding of the Homochirality Origin, Asymmetric Catalysis, Chiral Superstructure Formation from Achiral Molecules, and B-Z DNA Conformational Transition. *Symmetry* **2019**, *11*, 649. [CrossRef]
32. Frondel, C. Characters of Quartz Fibers. *Am. Miner.* **1978**, *63*, 17–27.
33. Heritsch, H. Die Verteilung von Rechts- und Linksquarzen in Schriftgraniten. *Miner. Pet.* **1953**, *3*, 115–125. [CrossRef]
34. Hazen, R.M. Chapter 11—Chiral Crystal Faces of Common Rock-Forming Minerals. In *Progress in Biological Chirality*; Pályi, G., Zucchi, C., Caglioti, L., Eds.; Elsevier Science Ltd.: Oxford, UK, 2004; pp. 137–151. ISBN 978-0-08-044396-6.
35. Churchill, H.; Teng, H.; Hazen, R.M. Correlation of pH-dependent surface interaction forces to amino acid adsorption: Implications for the origin of life. *Am. Miner.* **2004**, *89*, 1048–1055. [CrossRef]
36. Hazen, R.M. Presidential Address to the Mineralogical Society of America, Salt Lake City, October 18, 2005: Mineral surfaces and the prebiotic selection and organization of biomolecules. *Am. Miner.* **2006**, *91*, 1715–1729. [CrossRef]
37. Lowry, T.M. *Optical Rotatory Power*; Longmans, Green and Co.: London, UK, 1935.
38. Yogev-Einot, D.; Avnir, D. The temperature-dependent optical activity of quartz: From Le Châtelier to chirality measures. *Tetrahedron Asymmetry* **2006**, *17*, 2723–2725. [CrossRef]
39. Ersahin, S.; Gunal, H.; Kutlu, T.; Yetgin, B.; Coban, S. Estimating specific surface area and cation exchange capacity in soils using fractal dimension of particle-size distribution. *Geoderma* **2006**, *136*, 588–597. [CrossRef]
40. Wagner, G.; Van Den Haute, P. Fission-Track Dating. In *Solid Earth Sciences Library*; Springer: Berlin/Heidelberg, Germany, 1992; ISBN 978-94-010-5093-7.
41. Meinert, C.; Meierhenrich, U.J. A New Dimension in Separation Science: Comprehensive Two-Dimensional Gas Chromatography. *Angew. Chem. Int. Ed.* **2012**, *51*, 10460–10470. [CrossRef]
42. Myrgorodska, I.; Meinert, C.; Martins, Z.; D'Hendecourt, L.L.S.; Meierhenrich, U.J. Quantitative enantioseparation of amino acids by comprehensive two-dimensional gas chromatography applied to non-terrestrial samples. *J. Chromatogr. A* **2016**, *1433*, 131–136. [CrossRef]
43. Woodhead, J.A.; Rossman, G.R.; Silver, L.T. The Metamictization of Zircon: Radiation Dose-Dependent Structural Characteristics. *Am. Mineral.* **1991**, *76*, 74–82.
44. Li, W.; Wang, L.; Lang, M.; Trautmann, C.; Ewing, R.C. Thermal annealing mechanisms of latent fission tracks: Apatite vs. zircon. *Earth Planet. Sci. Lett.* **2011**, *302*, 227–235. [CrossRef]
45. Reineck, H.-E.; Singh, I.B. *Depositional Sedimentary Environments: With Reference to Terrigenous Clastics*, Springer Study Edition, 2nd ed.; Springer: Berlin/Heidelberg, Germany, 1980; ISBN 978-3-540-10189-5.
46. Cheng, D.; Sato, K.; Shikida, M.; Ono, A.; Sato, K.; Asaumi, K.; Iriye, Y. Characterization of Orientation-Dependent Etching Properties of Quartz: Application to 3-D Micromachining Simulation System. *Sens. Mater.* **2005**, *17*, 179–186.

 *life*

Article

Membrane Structure Obtained in an Experimental Evolution Process

María J. Dávila * and Christian Mayer

Institute of Physical Chemistry, CENIDE, University of Duisburg-Essen, 45141 Essen, Germany; christian.mayer@uni-due.de
* Correspondence: maria.davila@uni-due.de; Tel.: +49-(0)201-183-2439

Abstract: Recently, an evolution experiment was carried out in a cyclic process, which involved periodic vesicle formation in combination with peptide and vesicle selection. As an outcome, an octapeptide (KSPFPFAA) was identified which rapidly integrated into the vesicle membrane and, according to its significant accumulation, is obviously connected to a particular advantage of the corresponding functionalized vesicle. Here we report a molecular dynamics study of the structural details of the functionalized vesicle membrane, which was a product of this evolution process and is connected to several survival mechanisms. In order to elucidate the particular advantage of this structure, we performed all-atom molecular dynamics simulations to examine structural changes and interactions of the peptide (KSPFPFAA) with the given octadecanoic acid/octadecylamine (1:1) bilayer under acidic conditions. The calculations clearly demonstrate the specific interactions between the peptide and the membrane and reveal the mechanisms leading to the improved vesicle survival.

Keywords: origin of life; bilayer structure; molecular dynamics; aggregation process; selection; evolution

Citation: Dávila, M.J.; Mayer, C. Membrane Structure Obtained in an Experimental Evolution Process. *Life* **2022**, *12*, 145. https://doi.org/10.3390/life12020145

Academic Editors: Michele Fiore and Emiliano Altamura

Received: 30 December 2021
Accepted: 17 January 2022
Published: 20 January 2022

Publisher's Note: MDPI stays neutral with regard to jurisdictional claims in published maps and institutional affiliations.

Copyright: © 2022 by the authors. Licensee MDPI, Basel, Switzerland. This article is an open access article distributed under the terms and conditions of the Creative Commons Attribution (CC BY) license (https://creativecommons.org/licenses/by/4.0/).

1. Introduction

In early prebiotic evolution, membranous boundaries were probably formed by simple amphiphilic molecules, easier to synthesize than phospholipids and able to integrate and accumulate other amphiphilic molecules. They would give rise to aggregation and membrane disruption, as well as to the formation of primitive membrane pores, thus enhancing selective molecular exchange between the first cell-like compartments and their environment [1,2]. The most commonly accepted model of a prebiotic membrane is the one formed by fatty acids, mainly because of their high permeability and their ability to grow by the integration of new fatty acids [3–5]. However, fatty acid membranes may also contain other compounds present in prebiotic conditions, improving their stability and permeability [6–10]. It has been shown that vesicles composed of fatty acids together with other single-chain amphiphilic compounds, such as fatty alcohols, are formed over a wider pH range than vesicles composed of fatty acids only, which are stable at a pH close to their pKa [11]. The addition of other prebiotically plausible molecules to fatty acid/alcohol mixtures, such as isoprenoids, improves vesicle stability, even under extreme conditions of high temperatures, pH, and salinity, and even in the presence of divalent cations [12]. Maurer et al. also described an increase in the stability of fatty acid vesicles by adding their glycerol monoacyl amphiphile derivatives [13]. It was also observed that by incorporating oxidized polycyclic aromatic hydrocarbons into the membrane of fatty acid vesicles, an increase in their stability is produced. Moreover, this leads to a decrease in the membrane permeability, thus showing an effect similar to that of cholesterol or other sterols in current membranes [14]. Namani and Deamer reported a specific type of vesicles formed by fatty acids and their corresponding fatty amines, characterized by being stable in strongly basic and acidic pH ranges [15].

One of the most widespread hypotheses of the origin of life, in which the formation of membranous compartments plays a decisive role, is the hot spring hypothesis [16]. It is based on the assumption that lipid-encapsulated polymers with initially random sequences can be synthesized in wet-dry cycling processes. In these processes, protocells subject polymers to combinatorial selection, leading to the emergence of structural and catalytic functions [16,17].

A new hypothesis proposed by Schreiber et al. [18] proposes the early steps of the origin of life in deep-reaching tectonic faults. In this model, the formation of vesicular structures in these water- and CO_2-filled zones has been proposed. The mechanism of formation includes cyclic phase transition processes between supercritical CO_2 and subcritical gaseous CO_2 by periodic pressure variations at a suitable depth [19]. In water/carbon dioxide interface zones, a pool of oligopeptides of variable length is generated by condensation reactions from amino acids in a hydrothermal environment [20]. Those oligopeptides, which have an amphiphilicity similar to the membrane of vesicular structures, will integrate and accumulate, increasing in concentration over time [21]. Based on the above model, Mayer et al. performed experiments with cyclic pressure variations at the water/CO_2 interface, allowing vesicular structures to form from a mixture of long amine chains and long fatty acids previously proposed by Namani and Deamer [15]. Since the experiments were performed at 120 °C under acidic conditions, the octadecylamine/octadecanoic acid mixture was selected to improve the stability of the resulting vesicles. The continuous peptide accumulation combined with a selection of the vesicles for thermal stability led to an evolution-like process, finally resulting in functional peptides with the specific capability to make vesicles survive. One of the peptides that was particularly successful under these conditions turned out to be an octapeptide with the sequence H-Lys-Ser-Pro-Phe-Pro-Phe-Ala-Ala-OH (KSPFPFAA). In separate experiments, it could be shown that this peptide led to a reduced vesicle size, an increased membrane permeability, and to a significant increase in the thermal stability of the vesicles [22].

The main goal of this study was to understand the particular mechanism of stabilization which has evolved in the pressure-cycling process. More explicitly, we wanted to elucidate the structural changes and interactions that are induced by the presence of the given octapeptide in the vesicle membrane. To this end, we used all-atom molecular dynamics to study the aggregation ability of the peptide KSPFPFAA in the octadecanoic acid/octadecylamine (1:1) bilayer membrane under conditions that mimic an acidic environment.

2. Computational Methods

2.1. Simulation Systems

2.1.1. Peptide in Explicit Solvent

The conformation of the octapeptide KSPFPFAA was initially fully extended (backbone dihedral angles are 180° except for dihedral angle C–N–Cα–C of proline, which is determined by its ring structure) and generated by GaussView [23]. To mimic the experimental acidic conditions described above, the peptide carries a net charge of +1 and both termini are charged.

A biphasic octane/water solvent was used to reproduce the behaviour of the octadecylamine/octadecanoic acid bilayer. From the initial fully extended conformation, the peptide was inserted parallel to the z-axis in a hydrated octane slab, with the system's initial dimensions of 6.4 nm × 6.4 nm × 8.6 nm. The resulting system contained 423 octane molecules and 5120 water molecules (chloride and sodium counterions were added to neutralize the charge). For comparative purposes, the peptide molecule was also placed in the centre of an initial cubic box (4.0 nm × 4.0 nm × 4.0 nm) containing 2043 water molecules (again, chloride and sodium counterions were added to neutralize the charge).

2.1.2. Peptides in Octadecylamine/Octadecanoic Acid Bilayer

The peptide aggregation study was carried out with sixteen peptides located perpendicular to the membrane surface in both parallel and antiparallel arrangement at a distance of 2.8 nm. Thus, the peptides translocate in the xy plane forming a 4 × 4 network, ensuring that there is no prior interaction between adjacent peptides. The initial dimensions of the simulation box were 11.4 × 11.4 × 8.6 nm^3, containing 366 amphiphilic pairs, octadecylamine and octadecanoic acid, and 15,680 water molecules. Both membrane–peptide systems were neutralized with sodium and chloride counterions. The most representative peptide conformation, which was obtained in a clustering process after the peptide simulation in the previously described octane–water system, was used as the initial peptide conformation in each simulation system.

2.2. Simulation Setup

All-atom MD simulations were performed using GROMACS package [24], where CHARMM36 force field [25] and TIP3P [26,27] water models were used. Periodic boundary conditions were applied in all directions. The initial systems were minimized to ensure the elimination of any close contacts by using the steepest descent algorithm and equilibrated for 500 ps in the *NVT* ensemble. Afterwards, an equilibration process was carried out for 20 ns and 40 ns in the *NPT* ensemble, for the biphasic octane/water system and for octadecylamine–octadecanoic acid bilayer, respectively. The Berendsen thermostat method [28] was applied for temperature coupling to a temperature bath of 353 K. In the *NPT* ensemble, the Berendsen weak-coupling algorithm [28] was used to maintain the system to a pressure bath of 1 bar. Production dynamics were performed in the *NPT* ensemble at a temperature of 353 K, using the velocity rescaling thermostat [29] with a coupling time of 0.1 ps, and pressure of 1 bar, using the Parrinello-Rahman barostat [30] with a coupling time of 2 ps. All H-bonds were constrained using the LINCS algorithm [31]. The particle mesh Ewald method (PME) was used for computation of the electrostatic forces with a real space cut-off of 1.2 nm [32]. Van der Waals interactions were cut off at 1.2 nm with a switch function starting at 1.0 nm. GROMACS package was also used to analyse the trajectories. All system configurations were visualized using VMD software [33].

3. Results

3.1. Peptide Conformation into Biphasic Membrane Mimetic

In order to obtain the most representative conformation of the octapeptide KSPFPFAA and to start the simulations in each of the systems described in the previous section, we carried out a 2.5 µs simulation at 353 K, above the membrane transition temperature, with the peptide initially in the extended–coil conformation placed in a hydrated octane slab. For comparative purposes, a simulation was performed under the same conditions as described above, with the peptide in explicit water.

After performing a clustering analysis for the octapeptide into the hydrated octane slab with a cut-off of 0.13 nm, the most representative conformation was extracted, being the initial conformation for each of the simulations we performed in this work (Figure 1a). The most representative conformation of the peptide in water with a cut-off of 0.19 nm is shown in Figure 1b.

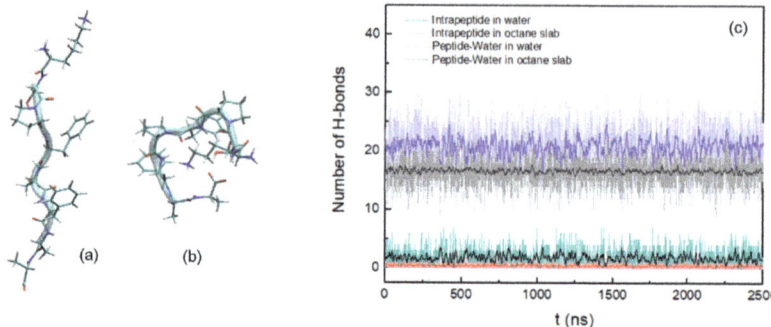

Figure 1. Most representative peptide conformations in the hydrated octane slab (**a**) and in water (**b**). (**c**) Number of hydrogen bonds as a function of time during 2.5 μs: intrapeptide in octane slab, light red; intrapeptide in water, cyan; peptide–water in octane slab, light grey; peptide–water in water, light blue. A running average (window of length 10 ns) was applied to smooth the time series (intrapeptide in octane slab, red; intrapeptide in water, black; peptide–water in octane slab, dark grey; peptide–water in water, dark blue).

In the following, the amino acids of the chain will be referred to by a combination of the letter code for the amino acid combined with its position in the peptide chain, so K1 means the first amino acid, A8 the last one. For both systems, the number of intramolecular and peptide–water hydrogen bonds was calculated during the entire simulation, considering their formation with cut-offs of 0.35 nm for the donor-acceptor distance and 30° for the hydrogen-donor-acceptor angle. Thus it was observed that for the peptide insert into the hydrated octane slab, the formation of H-bonds is less frequent than for the peptide in water (Figure 1c). Most of the hydrogen bonds are being formed between the hydroxyl hydrogen of S2 and the carbonyl oxygen of P3 in the first system, while in the second system is dominated by the H-bonds K1-A8, S2-A8, and S2-F4. The results of the formation of hydrogen bonds between peptide and water monitored over the course of the simulations primarily show H-bonds between the amine hydrogen of K1 and the oxygen of water and between the hydrogen of water and the carbonyl oxygen of A8, being predominant in the biphasic membrane mimetic (Table 1).

Table 1. Percentage of the most characteristic intra-peptide and amino acid residues–water hydrogen bond interactions in the MD trajectory.

			Peptide H-Bonds (hydrated octane slab)/%				
			S2-P3	F4-F6	F6-F8		
			87	4	5		
			Peptide H-Bonds (water)/%				
		K1-S2	K1-A8	S2-F4	S2-A8	F4-F6	F4-A7
		3	23	23	31	6	4
			Peptide–Water H-Bonds (hydrated octane slab)/%				
		K1	S2	F4	A7	A8	
		32	14	3	5	42	
			Peptide–Water H-Bonds (water)/%				
K1	S2	P3	F4	P5	F6	A7	A8
26	13	6	5	6	7	9	28

The initially fully extended structure was used as a reference to calculate the root mean square deviation (RMSD) of the backbone atoms over the course of the trajectory, so the lower RMSD values correspond to the most extended structures. As shown in Figure 2a, the RMSD levels of around 0.15 nm for the peptide in the hydrated octane slab suggest that within 2.5 µs of simulation, with the force field chosen to model the peptide, it did not reach a folded structure. The maximum conformational changes can be found in the peptide in water, increasing the RMSD values up to 0.5 nm. However, strong fluctuations are observed, suggesting a high flexibility and instability of the peptide in this system over the entire MD simulation.

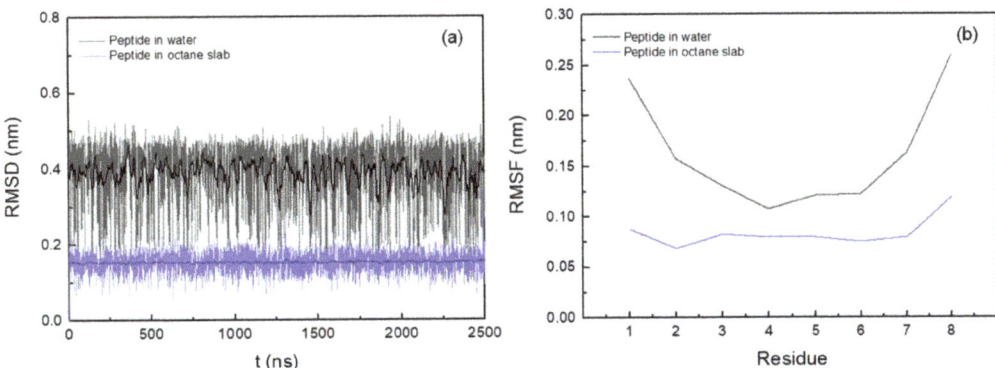

Figure 2. Backbone RMSD (**a**) and RMSF (**b**): peptide in octane slab, blue; peptide in water, grey. A running average (window of length 10 ns) was applied to smooth the time series (peptide in octane slab, dark blue; peptide in water, black).

In order to investigate the structural flexibility of the peptide in the octane slab and in water, the backbone root mean square fluctuation (RMSF) was calculated (Figure 2b). The values amount to 0.12 nm in the first system and to 0.26 nm in the second, with the highest RMSF values corresponding to the N- and C-terminal amino acid residues. In the biphasic membrane mimetic, the low RMSF values suggest conformational stability of the peptide in octane slab, showing low flexibility. However, RMSF values of the peptide in water are higher, varying between 0.11 nm and 0.16 nm from S2 to A7, and increasing its values for amino acids directly connected to the terminal residues. Thus, it may indicate that the high flexibility of the terminal residues may also affect the mobility of the adjacent amino acids.

3.2. Cluster Process and Pore Formation

In order to obtain information about aggregation and pore formation of the KSPFPFAA peptide, we studied the peptide–peptide and peptide–membrane interactions. This was done in a parallel arrangement of sixteen peptides located perpendicular to the membrane surface, and in an antiparallel arrangement with alternating orientation of the peptides (Figure 3). The method described by Thøgersen et al. [34] was applied where the peptides are positioned at a starting distance of 2.8 nm, thus ensuring no prior interaction between adjacent peptides.

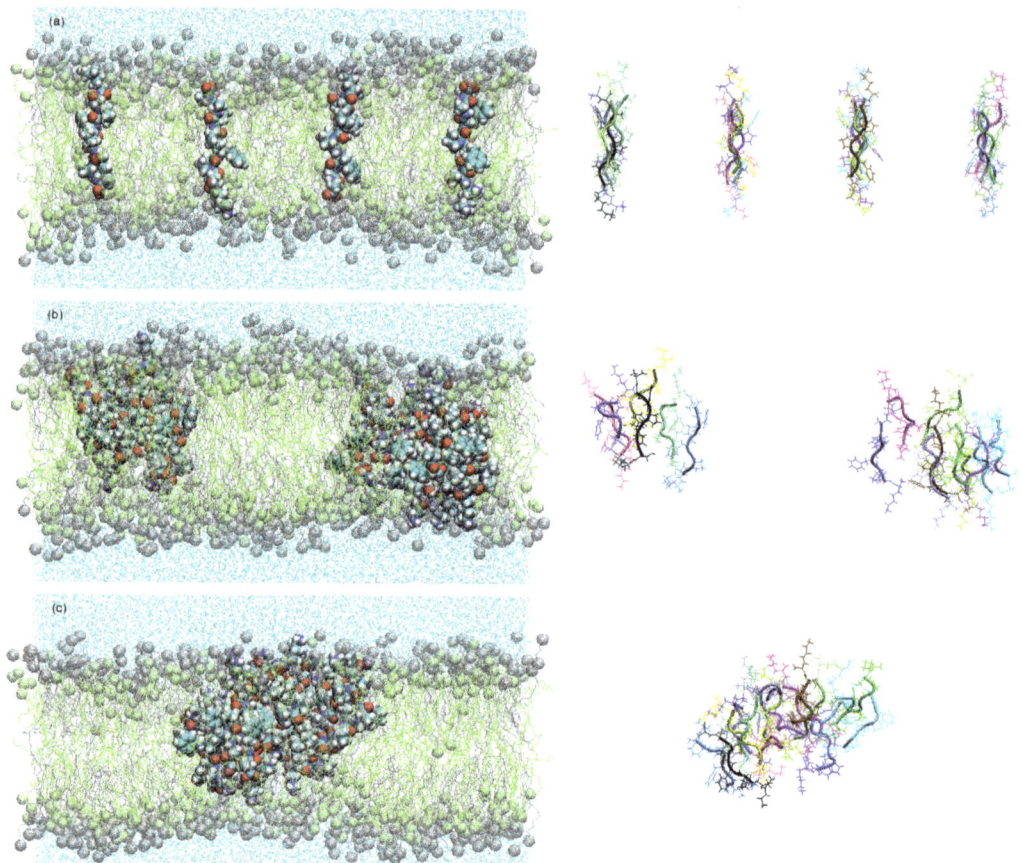

Figure 3. Snapshots of the simulation systems in the peptide aggregation process (peptides are shown as lines with cartoon representations on the right) in antiparallel arrangement at (**a**) 0 ns, (**b**) 100 ns, and (**c**) 450 ns.

It was observed that in the case of peptides located in antiparallel orientations, they remained embedded in the membrane during 450 ns simulation. Here, peptides are not exchanged between clusters over the entire MD simulation. In general, the membrane embedded peptides show a certain stability of its conformation, thus remaining extended structures (Figure 3). Peptides positioned in parallel arrangement, on the other hand, tend to emerge to the membrane surface, so that after only 100 ns there are no peptides in the transmembrane position. In this arrangement, the peptides rapidly emerge to the surface of the membrane while changing their conformation. Hence, in the transition from embedded peptides to surface-emerged peptides, probably due to their higher structural flexibility, a progressive increase of instability in their conformation is observed. (The results of the parallel configuration are not shown in this work).

Figure 4a shows the number of clusters formed in the antiparallel configuration during 450 ns MD simulation. The lateral diffusion of the octapeptide allows for rapid dimer formation in the first 8 ns and the stable cluster formation of up to five peptides at just 22 ns. Cluster size increases progressively and completes its total aggregation process at 428 ns.

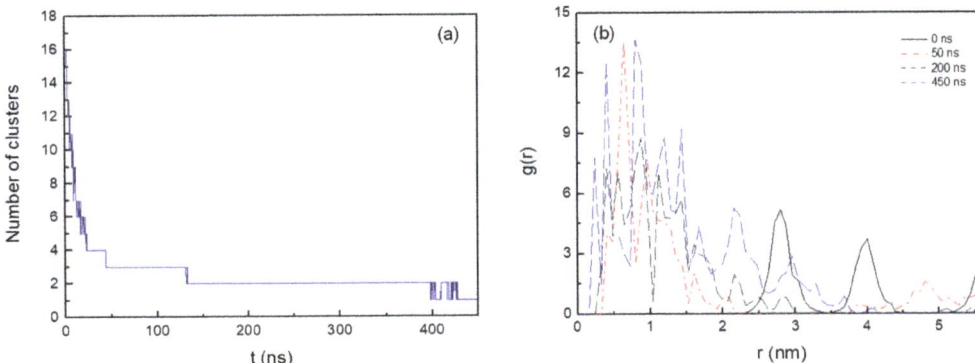

Figure 4. (a) Number of the clusters as a function of time formed during the 450 ns MD simulation with cut-off 0.6 nm. (b) Intermolecular radial distribution function, g(r), between the mass centres of the peptide pairings at 0 ns, 50 ns, 200 ns, and 450 ns (0.08 nm bin width).

The radial distribution function, g(r), was used to determine the peptide–peptide interactions during the cluster formation process. It was calculated in the xy plane at 0 ns, 50 ns, 200 ns, and 450 ns for the peptide pairs inserted in the membrane (Figure 4b). A high degree of clustering is observed as the aggregation process evolves, in relation to the initial configuration, with the maximum peaks positioned at 2.8 nm and 4 nm, corresponding to the distances between the mass centres of the peptides initially fixed. At 50 ns, a maximum peak in the radial distribution function appears at 0.65 nm, although, as indicated by its profile, the aggregation process is still incomplete. At 200 ns, the most likely distance between aggregated peptides is 0.93 nm, with several pronounced distribution peaks at 0.77 nm, 0.57 nm, and 0.41 nm. At the end of the aggregation process at 450 ns, several peaks are observed, with the most intense distribution peaks at 0.82 and 0.41 nm, as well as a narrow peak at 0.25 nm. Hence, the radial distribution function profiles obtained during the aggregation process suggest that there is a strong tendency for peptides to pack and cluster in the membrane octadecylamine/octadecanoic acid.

The H-bonds between the peptide pairings were also determined, considering a donor-acceptor distance within 0.35 nm and 30° for the hydrogen-donor-acceptor angle (Figure 5a). The number of hydrogen bonds continuously increases during the simulation until stabilizing at around 175 ns. Here it is worth noting the formation of hydrogen bond interactions between the residues K1 and A8 of the peptide pairings, which are present with an occurrence of 36% during the whole MD trajectory. More specifically, these hydrogen bonds are mainly established between the amine hydrogen of K1 and the carbonyl oxygen of A8, which contribute to a greater stabilization of the peptides inserted in the membrane.

The interactions between octadecylamine and octadecanoic acid hydrophilic head groups with the peptides were calculated to obtain information on how these interactions are affected by the peptide aggregation process. The number of hydrogen bonds between the fatty acid head groups and the peptides in the antiparallel arrangement system decreases rapidly up to 54 ns. After that, the interactions deteriorate more slowly until approximately 350 ns. At this point in time, the number of hydrogen bonds begins to stabilize, corresponding to the end of the aggregation process (Figure 5b). However, the decrease in the number of hydrogen bonds is less pronounced for the interactions between the octadecylamine head group and the peptide, showing a slight decrease in the number of H-bonds, decreasing up to 61 nm from the beginning of the aggregation process and stabilizing practically from there until the end of the simulation. Thus, the results seem to indicate that the amine head group forms hydrogen bonds with the peptides less frequently than the fatty acid head group.

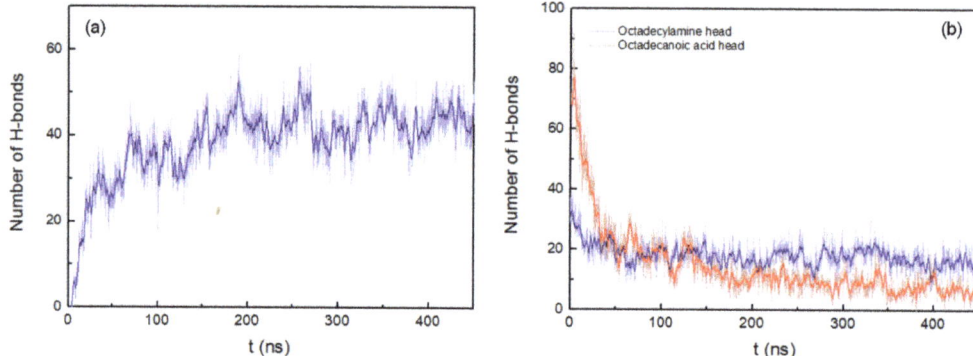

Figure 5. (**a**) Number of hydrogen bonds between peptides pairings. (**b**) Number of hydrogen bonds between octadecylamine and octadecanoic acid head groups and the peptides. Both representations show the results of the system with peptides of antiparallel arrangement up to 450 ns. A running average (window of length 2 ns) was applied to smooth the time series: (**a**), dark blue; (**b**), dark blue (octadecylamine head group–peptide interactions) and red (octadecanoic acid head group–peptide interactions).

The interactions between the amphiphilic membrane molecules and the peptides were also evaluated by calculating the number of atomic contacts for both systems (cutoff 0.6 nm). Figure 6 shows the number of atomic interactions of the hydrophilic head groups ($-NH_3^+$ and $-COOH$) and hydrophobic tails of octadecylamine and octadecanoic acid with the peptides.

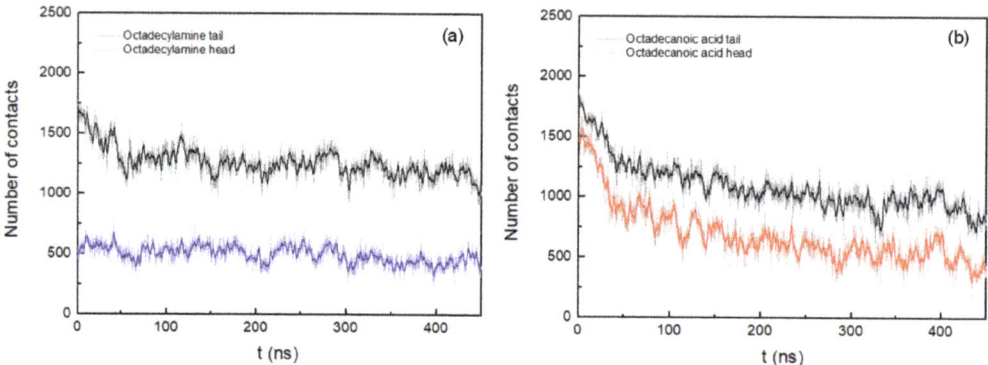

Figure 6. (**a**) Number of atomic contacts between octadecylamine molecules and peptides. (**b**) Number of atomic contacts between octadecanoic acid molecules and peptides. Both representations show the results of the system with peptides in antiparallel arrangement up to 450 ns. A running average (window of length 2 ns) was applied to smooth the time series: (**a**) dark blue (octadecylamine head group) and black (octadecylamine tail region); (**b**) red (octadecanoic acid head group) and black (octadecanoic acid tail region).

Figure 6a shows that the number of interacting atoms between octadecylamine tail regions and peptides decreases sharply up to 55 ns. After that, only a slight decrease is observed until the end of the simulation. However, the number of contacts between atoms of the octadecylamine head group and the peptides is practically constant during the whole simulation. Obviously, it is not being influenced by the aggregation process. In the interactions between the atoms of the octadecanoic acid and the peptides (Figure 6b), a progressive decrease is observed in most of the 450 ns of the MD simulation; F4 and F6 residues interact the most frequently with the atoms of the octadecanoic acid tail region (Table 2 shows the relative amounts of contacts of the various amino acid positions with the head and tail ends of the octadecylamine and the octadecanoic acid). The number of atomic interactions between the octapeptides and acid tail region also increase, being more pronounced at the beginning of the simulation, with strong fluctuations.

Table 2. Percentage of the number of contacts between octadecylamine and octadecanoic acid (head group or tail region) and amino acid residues over the course of the MD trajectory.

Number of Contacts–Octadecylamine (head)/%							
K1	S2	P3	F4	P5	F6	A7	A8
35	9	9	5	5	9	14	14
Number of Contacts–Octadecanoic Acid (head)/%							
K1	S2	P3	F4	P5	F6	A7	A8
16	10	14	16	6	17	7	13
Number of Contacts–Octadecylamine (tail)/%							
K1	S2	P3	F4	P5	F6	A7	A8
12	8	12	19	13	18	8	11
Number of Contacts–Octadecanoic Acid (tail)/%							
K1	S2	P3	F4	P5	F6	A7	A8
8	7	13	21	15	20	8	9

During the aggregation process, penetration of water molecules into the membrane was observed. Thus, in the pore formed by six peptides at 200 ns in the clustering process, MD simulations were carried out for another 100 ns to observe this effect more in detail (Figure 7).

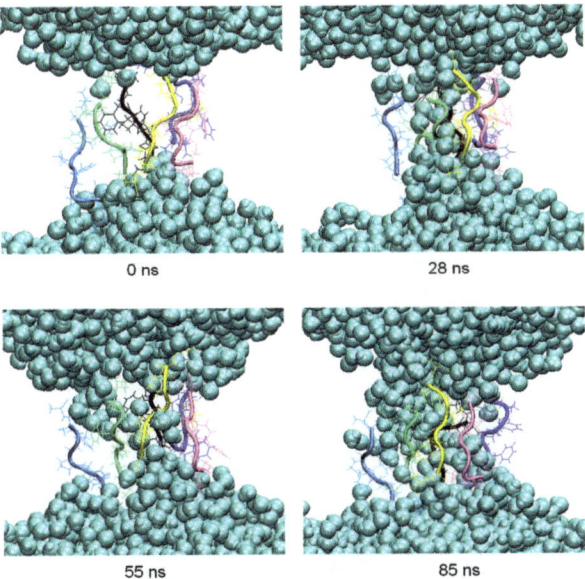

Figure 7. Snapshots of the penetration of the water molecules (blue spheres) into the membrane at relevant times of the 100 ns MD simulation.

The number of hydrogen bonds between the residues of the peptides forming the pore and the water molecules were analyzed in the 100 ns of simulation. The highest occurrences were observed in the terminal residues, K1 and A8, with 30% and 28%, respectively, and S2 with 18%. The difference in the occurrence of hydrogen bond formation between the residues of the octapeptide and the water molecules, and hence the ability of the pore to allow water molecules to flow inside the membrane, is influenced by the different orientations adopted by the side chains of the residues.

The water density inside the membrane was measured from the pore centre to 0.3 nm in both directions of the z-axis. (Figure 8). Over the period of simulation, three distinct events regarding the water content can be observed in the pore formed by the six octapeptides. The first one occurs up to approximately 34.2 ns, where the average water density is relatively low (0.2 molecules/nm^3). In the second event, there is a progressive increase in pore water density up to 85 ns, with an average density of 0.3 molecules/nm^3 (two maxima are observed at 67.6 nm and 84.9 nm). A third event shows a slight decrease in density down to 0.2 molecules/nm^3 near the end of the simulation. Thus, water molecules seem to penetrate the membrane continuously, but with significant fluctuations.

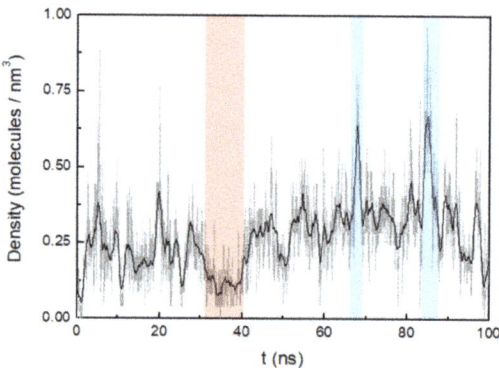

Figure 8. Density of water as a function of time. The most representative pore density maxima and minima have been indicated with blue and red zones, respectively. A running average (window of length 1 ns) was applied.

4. Discussion

In order to understand the obvious advantage of the octapeptide KSPFPFAA in an experimental selection process [22], a study of its conformational behaviour in the membrane was carried out. Initially, the peptide conformation in a biphasic system that mimics the octadecylamine–octadecanoic acid bilayer was determined. The octane/water solvent biphasic system has been used as a model as it has been shown to significantly accelerate the conformational relaxation of the peptide as compared to a lipid bilayer [35–38]. The results obtained in RMSF and RMSD suggest that, in contrast to the peptide–water system, the peptide in a hydrated octane slab exhibits high stability and low flexibility throughout the simulation. It shows a distinctly extended conformation along the bilayer normal and remains largely unfolded. In addition, it was determined that the hydrogen bonds formed by the two end group amino acids (K1 and A8) with the water molecules are principally responsible for keeping the peptide in the unfolded conformation.

In the aggregation study, the peptides in an antiparallel arrangement remained embedded in the membrane throughout the process. In contrast, the parallel arrangement is largely unstable; here, the peptides quickly began to emerge to the membrane surface as larger clusters formed. Obviously, the antiparallel peptide arrangement leads to a significant stabilization via hydrogen bond interactions between the terminal sequences. Particularly, the intermolecular interaction between the amine hydrogen of K1 and the carbonyl oxygen of A8 gives rise to strong and stable trans-membrane aggregates. In addition, the polar head groups of octadecylamine and octadecanoic acid molecules tend to form hydrogen bonds with the A8 residue, which also contribute to the stabilization of the aggregates. Overall, it may also explain for the thermal stabilization of the overall membrane structure which has been observed experimentally by pulsed-field-gradient NMR [22].

In the same experiments, a distinct increase of the membrane permeability for water was observed. This phenomenon is reproduced by the given results from the molecular dynamics calculations. During the simulated aggregation process, it was possible to follow the peptide-induced penetration of water molecules into the membrane. Obviously, it is caused by the formation of central pores within peptide clusters in the membrane. In addition, the influence of residue side chain orientation on pore formation was analysed. It was observed that the charged lysine side chain is stretched, interacting with the membrane head groups and establishing contact with the outside of the membrane [39]. In this way, the lysine side chain enhances the introduction of water molecules into the pore. The hydroxyl side chain of the serine residue is oriented toward the inner side of the pore, allowing water molecules to penetrate and flow into the pore region inside the peptide cluster. The opposite tendency is observed for nonpolar residues, so that the cyclic side chain of proline and the phenyl group of the phenylalanine are oriented toward the hydrophobic tails of octadecylamine and octadecanoic acid. During the time window of the simulation, the water content inside the pore fluctuates dynamically. This observation may be linked to the actual water permeation process which was induced by the peptide in a corresponding experiment [22]. Here, pulsed-field-gradient NMR reveal an increased degree of water permeation through the vesicle membranes with relatively low permeability when the octapeptide is added.

As the most important target for our present experimental studies, we are looking for additional peptides which are developed during long term evolution processes. From these, we expect additional functions which are linked to the survival of the corresponding vesicles. All in all, the simulation results clearly demonstrate why the octapeptide KSPFPFAA has a clear advantage against other peptide structures and therefore is accumulated in the experimental selection process. It forms a well-defined integration product with the bilayer membrane which is stabilizing the resulting composite structure thermodynamically. It enhances the membrane permeability for water, hereby relaxing the destructive osmotic pressure and leading to an increased survival rate of the corresponding vesicles. In order to further verify the results obtained in this study by the formation of peptide aggregates and pores, one could use other experimental techniques such as X-ray diffraction or atomic force microscopy, which provide additional information about the structural effects caused in the membrane. The most exciting result is the fact that relatively complex structures, such as shown in Figures 3 and 7, are formed spontaneously from quite simple educts such as amino acids and basic amphiphiles.

Author Contributions: Conceptualization, M.J.D. and C.M.; methodology, M.J.D.; writing—original draft preparation, M.J.D.; writing—review and editing, M.J.D. and C.M. All authors have read and agreed to the published version of the manuscript.

Funding: This research received no external funding.

Institutional Review Board Statement: Not applicable.

Informed Consent Statement: Not applicable.

Data Availability Statement: All data generated during this study are available from the corresponding author upon request.

Conflicts of Interest: The authors declare no conflict of interest.

References

1. Schrum, J.P.; Zhu, T.F.; Szostak, J.W. The origins of cellular life. *Cold Spring Harb. Perspect. Biol.* **2010**, *2*, a002212. [CrossRef] [PubMed]
2. Deamer, D. The role of lipid membranes in life's origin. *Life* **2017**, *7*, 5. [CrossRef] [PubMed]
3. Chen, I.A.; Szostak, J.W. A kinetic study of the growth of fatty acid vesicles. *Biophys. J.* **2004**, *87*, 988–998. [CrossRef] [PubMed]
4. Mansy, S.S. Membrane transport in primitive cells. *Cold Spring Harb. Perspect. Biol.* **2010**, *2*, a002188. [CrossRef]
5. Toparlak, Ö.D.; Wang, A.; Mansy, S.S. Population-level membrane diversity triggers growth and division of protocells. *JACS Au* **2021**, *1*, 560–568. [CrossRef] [PubMed]
6. Monnard, P.-A.; Deamer, D.W. Membrane self-assembly processes: Steps toward the first cellular life. *Anat. Rec.* **2002**, *268*, 196–207. [CrossRef]
7. Sacerdote, M.G.; Szostak, J.W. Semipermeable lipid bilayers exhibit diastereoselectivity favoring ribose. *Proc. Natl. Acad. Sci. USA* **2005**, *102*, 6004–6008. [CrossRef]
8. Deamer, D.W. Boundary structures are formed by organic components of the Murchison carbonaceous chondrite. *Nature* **1985**, *317*, 792–794. [CrossRef]
9. Rushdi, A.I.; Simoneit, B.R. Lipid formation by aqueous Fischer-Tropsch-type synthesis over a temperature range of 100 to 400 degrees C. *Orig. Life Evol. Biosph.* **2001**, *31*, 103–118. [CrossRef]
10. Schreiber, U.; Mayer, C.; Schmitz, O.J.; Rosendahl, P.; Bronja, A.; Greule, M.; Keppler, F.; Mulder, I.; Sattler, T.; Schöler, H.F. Organic compounds in fluid inclusions of Archean quartz—Analogues of prebiotic chemistry on early Earth. *PLoS ONE* **2017**, *12*, e0177570. [CrossRef] [PubMed]
11. Apel, C.; Mautner, M.; Deamer, D.W. Self-assembled vesicles of monocarboxylic acids and alcohols: Conditions for stability and for the encapsulation of biopolymers. *Biochim. Biophys. Acta Biomembr.* **2002**, *1559*, 1–9. [CrossRef]
12. Jordan, S.F.; Rammu, H.; Zheludev, I.N.; Hartley, A.M.; Marechal, A.; Lane, N. Promotion of protocell self-assembly from mixed amphiphiles at the origin of life. *Nat. Ecol. Evol.* **2019**, *3*, 1705–1714. [CrossRef]
13. Maurer, S.E.; Deamer, D.W.; Boncella, J.M.; Monnard, P.-A. Chemical evolution of amphiphiles: Glycerol monoacyl derivatives stabilize plausible prebiotic membranes. *Astrobiology* **2009**, *9*, 979–987. [CrossRef]
14. Groen, J.; Deamer, D.W.; Kros, A.; Ehrenfreund, P. Polycyclic aromatic hydrocarbons as plausible prebiotic membrane components. *Orig. Life Evol. Biosph.* **2012**, *42*, 295–306. [CrossRef]
15. Namani, T.; Deamer, D.W. Stability of model membranes in extreme environments. *Orig. Life Evol. Biosph.* **2008**, *38*, 329–341. [CrossRef]
16. Damer, B.; Deamer, D. The hot spring hypothesis for an origin of life. *Astrobiology* **2020**, *20*, 429–452. [CrossRef] [PubMed]
17. Longo, A.; Damer, B. Factoring origin of life hypotheses into the search for life in the solar system and beyond. *Life* **2020**, *10*, 52. [CrossRef] [PubMed]
18. Schreiber, U.; Locker-Grütjen, O.; Mayer, C. Hypothesis: Origin of life in deep-reaching tectonic faults. *Orig. Life Evol. Biosph.* **2012**, *42*, 47–54. [CrossRef] [PubMed]
19. Mayer, C.; Schreiber, U.; Dávila, M.J. Periodic vesicle formation in tectonic fault zones: An ideal scenario for molecular evolution. *Orig. Life Evol. Biosph.* **2015**, *45*, 139–148. [CrossRef] [PubMed]
20. Marshall, W.L. Hydrothermal synthesis of amino acids. *Geochim. Cosmochim. Acta* **1994**, *58*, 2099–2106. [CrossRef]
21. Mayer, C.; Schreiber, U.; Dávila, M.J. Selection of prebiotic molecules in amphiphilic environments. *Life* **2017**, *7*, 3. [CrossRef]
22. Mayer, C.; Schreiber, U.; Dávila, M.J.; Schmitz, O.J.; Bronja, A.; Meyer, M.; Klein, J.; Meckelmann, S.W. Molecular evolution in a peptide-vesicle system. *Life* **2018**, *8*, 16. [CrossRef]
23. Dennington, R.; Keith, T.A.; Millam, J.M. *GaussView*; Version 6; Semichem Inc.: Shawnee Mission, KS, USA, 2003.
24. Abraham, J.M.; Murtola, T.; Schulz, R.; Páll, S.; Smith, J.C.; Hess, B.; Lindahl, E. GROMACS: High performance molecular simulations through multi-level parallelism from laptops to supercomputers. *SoftwareX* **2015**, *1*, 19–25. [CrossRef]
25. Huang, J.; MacKerell, A.D., Jr. CHARMM36 all-atom additive protein force field: Validation based on comparison to NMR data. *J. Comput. Chem.* **2013**, *34*, 2135–2145. [CrossRef] [PubMed]
26. Berendsen, H.J.C.; Postma, J.P.M.; van Gunsteren, W.F.; Hermans, J. *Intermolecular Forces*; Pullmann, B., Ed.; Reidel: Dordrecht, The Netherlands, 1981; pp. 331–342.
27. Jorgensen, W.L.; Chandrasekhar, J.; Madura, J.D.; Impey, R.W.; Klein, M.L. Comparison of simple potential functions for simulating liquid water. *J. Chem. Phys.* **1983**, *79*, 926–935. [CrossRef]
28. Berendsen, H.J.C.; Postma, J.P.M.; DiNola, A.; Haak, J.R. Molecular dynamics with coupling to an external bath. *J. Chem. Phys.* **1984**, *81*, 3684–3690. [CrossRef]
29. Bussi, G.; Donadio, D.; Parrinello, M. Canonical sampling through velocity rescaling. *J. Chem. Phys.* **2007**, *126*, 014101. [CrossRef]
30. Parrinello, M.; Rahman, A. Polymorphic transitions in single crystals: A new molecular dynamics method. *J. Appl. Phys.* **1981**, *52*, 7182–7190. [CrossRef]
31. Hess, B.; Bekker, H.; Berendsen, H.J.C.; Fraaije, J.G.E.M. LINCS: A linear constraint solver for molecular simulations. *J. Comput. Chem.* **1997**, *18*, 1463–1472. [CrossRef]
32. Darden, T.; York, D.; Pedersen, L. Particle mesh Ewald: An N·log(N) method for Ewald sums in large systems. *J. Chem. Phys.* **1993**, *98*, 10089–10092. [CrossRef]
33. Humphrey, W.; Dalke, A.; Schulten, K. VMD: Visual molecular dynamics. *J. Mol. Graph.* **1996**, *14*, 33–38. [CrossRef]

34. Thøgersen, L.; Schiøt, B.; Vosegaard, T.; Nielsen, N.C.; Tajkhorshid, E. Peptide aggregation and pore formation in a lipid bilayer: A combined coarse-grained and all atom molecular dynamics study. *Biophys. J.* **2008**, *95*, 4337–4347. [CrossRef] [PubMed]
35. Chipot, C.; Pohorille, A. Folding and translocation of the undecamer of poly-L-leucine across the water-hexane interface. A molecular dynamics study. *J. Am. Chem. Soc.* **1998**, *120*, 11912–11924. [CrossRef] [PubMed]
36. Stockner, T.; Ash, W.L.; MacCallum, J.L.; Tieleman, D.P. Direct simulation of transmembrane helix association: Role of asparagines. *Biophys. J.* **2004**, *87*, 1650–1656. [CrossRef] [PubMed]
37. Hénin, J.; Pohorille, A.; Chipot, C. Insights into the recognition and association of transmembrane α-helices. The free energy of α-helix dimerization in glycophorin A. *J. Am. Chem. Soc.* **2005**, *127*, 8478–8484. [CrossRef]
38. Kulleperuma, K.; Smith, S.M.E.; Morgan, D.; Musset, B.; Holyoake, J.; Chakrabarti, N.; Cherny, V.V.; DeCoursey, T.E.; Pomès, R. Construction and validation of a homology model of the human voltage-gated proton channel hHV1. *J. Gen. Physiol.* **2013**, *141*, 445–465. [CrossRef]
39. Strandberg, E.; Killian, J.A. Snorkeling of lysine side chains in transmembrane helices: How easy can it get? *FEBS Lett.* **2003**, *544*, 69–73. [CrossRef]

Article

Influence of Metal Ions on Model Protoamphiphilic Vesicular Systems: Insights from Laboratory and Analogue Studies

Manesh Prakash Joshi [1,*], Luke Steller [2], Martin J. Van Kranendonk [2] and Sudha Rajamani [1,*]

1. Department of Biology, Indian Institute of Science Education and Research, Dr. Homi Bhabha Road, Pune 411008, Maharashtra, India
2. Australian Centre for Astrobiology, and School of Biological, Earth and Environmental Sciences, University of New South Wales, Kensington, NSW 2052, Australia; l.steller@unsw.edu.au (L.S.); m.vankranendonk@unsw.edu.au (M.J.V.K.)
* Correspondence: manesh.joshi@students.iiserpune.ac.in (M.P.J.); srajamani@iiserpune.ac.in (S.R.); Tel.: +91-20-2590-8061 (S.R.)

Citation: Joshi, M.P.; Steller, L.; Van Kranendonk, M.J.; Rajamani, S. Influence of Metal Ions on Model Protoamphiphilic Vesicular Systems: Insights from Laboratory and Analogue Studies. *Life* **2021**, *11*, 1413. https://doi.org/10.3390/life11121413

Academic Editors: Michele Fiore and Emiliano Altamura

Received: 24 November 2021
Accepted: 13 December 2021
Published: 16 December 2021

Publisher's Note: MDPI stays neutral with regard to jurisdictional claims in published maps and institutional affiliations.

Copyright: © 2021 by the authors. Licensee MDPI, Basel, Switzerland. This article is an open access article distributed under the terms and conditions of the Creative Commons Attribution (CC BY) license (https://creativecommons.org/licenses/by/4.0/).

Abstract: Metal ions strongly affect the self-assembly and stability of membranes composed of prebiotically relevant amphiphiles (protoamphiphiles). Therefore, evaluating the behavior of such amphiphiles in the presence of ions is a crucial step towards assessing their potential as model protocell compartments. We have recently reported vesicle formation by N-acyl amino acids (NAAs), an interesting class of protoamphiphiles containing an amino acid linked to a fatty acid via an amide linkage. Herein, we explore the effect of ions on the self-assembly and stability of model N-oleoyl glycine (NOG)-based membranes. Microscopic analysis showed that the blended membranes of NOG and Glycerol 1-monooleate (GMO) were more stable than pure NOG vesicles, both in the presence of monovalent and divalent cations, with the overall vesicle stability being 100-fold higher in the presence of a monovalent cation. Furthermore, both pure NOG and NOG + GMO mixed systems were able to self-assemble into vesicles in natural water samples containing multiple ions that were collected from active hot spring sites. Our study reveals that several aspects of the metal ion stability of NAA-based membranes are comparable to those of fatty acid-based systems, while also confirming the robustness of compositionally heterogeneous membranes towards high metal ion concentrations. Pertinently, the vesicle formation by NAA-based systems in terrestrial hot spring samples indicates the conduciveness of these low ionic strength freshwater systems for facilitating prebiotic membrane-assembly processes. This further highlights their potential to serve as a plausible niche for the emergence of cellular life on the early Earth.

Keywords: protocell; origins of life; prebiotic membranes; protoamphiphiles; metal ions; hot springs; N-acyl amino acid; analogue conditions

1. Introduction

Membrane compartmentalization is a key feature of cellular life and is also thought to have been a crucial step towards the formation of first protocells on the prebiotic Earth. Contemporary cell membranes are primarily made of phospholipids, which are diacyl lipids. There have been few attempts to show prebiotically plausible chemical routes for phospholipid synthesis on the early Earth [1–5]. Nonetheless, phospholipids and other such structurally complex diacyl lipids are mostly considered as a product of membrane evolution that potentially occurred over protracted timescales of life's evolution [6–8]. On the contrary, primitive membranes are thought to have been mainly composed of chemically simpler single chain amphiphiles (SCAs), such as fatty acids and their derivatives [7,9]. Given their plausible availability on the prebiotic Earth [10,11] and ability to self-assemble into membranes, these SCAs which presumably served as membrane components of early protocells can be collectively called as protoamphiphiles [12]. In general, the membrane-assembly of such protoamphiphiles is affected by several parameters like temperature, pH, and ionic strength of a medium [13–16]. Particularly, fatty acid vesicles are extremely sensitive to metal ions (especially divalent ions like Mg^{2+}), which interact with their negatively charged carboxylate head groups and induce the formation of aggregates. However, metal ions have been shown to be required at a relatively high concentration for other prebiotically pertinent processes such as protometabolic reactions [17,18], ribozyme activity [19–22], and nonenzymatic template-directed replication of RNA [23,24]. Therefore, it is necessary to systematically evaluate the metal ion stability of plausible prebiotic compartments to characterize their compatibility with the functioning of other prebiotic processes.

The influence of the overall ionic strength of an aqueous medium on prebiotic membrane-assembly and stability also has implications for discerning the plausible niche(s) where life might have originated on the early Earth [15]. Oceanic hydrothermal vents and terrestrial hot springs are the two widely considered niches for the origins of life on Earth [25,26]. However, the high salinity of seawater would have been detrimental for the prebiotic compartmentalization process [27,28]. In contrast, terrestrial fresh water bodies, such as hot springs and lakes, with their overall low ionic strengths, would have provided a more conducive environment for the membrane-assembly of protoamphiphiles [6,29].

Several decades of laboratory experiments with fatty acid vesicles using low ionic strength buffers provide a proof of concept to support this notion. Importantly, some of the recent advances in the field, wherein fatty acid-based systems have been shown to form vesicles in hot spring water samples collected from early Earth analogue sites, have also strongly supported the above mentioned hypothesis [27,30]. Moreover, these analogue experiments demonstrate the ability of these protoamphiphiles to assemble into vesicles in a natural environment, such as a hot spring, which contains a diverse set of ions in varying concentrations. These natural settings would likely be a closer mimic of a complex prebiotic soup than the stringently controlled laboratory buffered conditions [31]. Therefore, the membrane-assembly of protoamphiphiles under such natural analogue conditions further underscores their relevance and plausibility of constituting protocellular systems on the prebiotic Earth.

Nonetheless, fatty acids might not have been the only option that early protocellular systems would have explored to form compartments. Both meteoritic delivery and endogenous synthesis reactions, along with other amphiphile-co-solute interactions in a prebiotic soup, would have generated a diverse set of protoamphiphiles on the early Earth. Importantly, vesicles made of a heterogeneous pool of amphiphiles have been shown to be highly stable in the presence of metal ions. For example, mixed membranes composed of decanoic acid with other amphiphiles likes decanol, glycerol 1-monodecanoate, and decylamine are stable both in the presence of metal ions as well as under sea salt conditions [28,32–34]. Blended vesicles of fatty acids and phospholipids are also shown to be more robust than pure fatty acid vesicles in the presence of magnesium, hinting at the intermediate step towards protocell membrane evolution [35,36]. In addition to fatty acids, other admixtures of SCAs like cyclophospholipids and fatty alcohols also result in robust vesicles that can tolerate high concentrations of divalent cations [37].

Recently, we reported the formation of N-acyl amino acids in a lipid-amino acid mixture under prebiotically relevant wet-dry cycling conditions, and also showed that some of these NAAs readily self-assemble into vesicles at acidic pH [12]. Their vesicle formation behavior was also found to be temperature-dependent. Therefore, a high temperature of 60 °C should to be maintained to keep these amphiphiles in a vesicle form (see Section 2.2.1. for further details). NAA is an interesting class of protoamphiphiles containing an amino acid covalently linked to a fatty acid via an amide linkage (Figure S1). However, despite the undeniable potential of this unique amphiphile to serve as a protocell compartment, the self-assembly behavior and different physicochemical properties of NAA-based membranes are still largely unexplored. Furthermore, the effect of metal ions on these membranes has not been studied yet. Moreover, it is not known whether NAA-based amphiphile systems can self-assemble into membranes under natural aqueous conditions, such as in hot springs, which are characterized by a diversity of ions and ionic strengths.

Here, we systematically evaluated the effect of both divalent and monovalent metal ions on the stability of NAA-based vesicular systems. For this, we used N-oleoyl glycine-based model vesicles, and Mg^{2+} and Na^+ as the representative divalent and monovalent metal ions, respectively. We show that pure NOG vesicles can sustain Mg^{2+} concentrations in the order of up to 1:1 molar ratio, beyond which large magnesium-induced aggregates are formed. Interestingly, the addition of surfactants like Glycerol 1-monooleate to the NOG system results in a heterogeneous population of vesicles and droplets, which tolerates even higher concentrations of Mg^{2+}. These NOG-based systems are even more resistant to Na^+ ions and typically form aggregates only at a 100-fold higher concentration of Na^+ as compared to that of Mg^{2+}. Both pure and mixed NOG systems are able to assemble into vesicles in hot spring water samples containing multiple ions in varying concentrations. This highlights their potential to form membrane compartments under natural, fresh water hydrothermal conditions, underscoring the relevance of these terrestrial niches in facilitating prebiotic reactions pertinent to the emergence of early cellular life.

2. Materials and Methods

2.1. Materials

All the amphiphiles (N-oleoyl glycine, Glycerol 1-monooleate, and oleyl alcohol (OOH)) were purchased from Avanti polar lipids (Alabaster, AL, USA). EDTA disodium salt dihydrate and tri-sodium citrate (Qualigens) were procured from HiMedia (Mumbai, India) and Thermo Fisher Scientific (Mumbai, India), respectively. The rest of the chemicals and reagents, including acetic acid and sodium acetate (components of acetate buffer), as well as salts like magnesium chloride ($MgCl_2$) and sodium chloride (NaCl), were from Sigma-Aldrich (Bangalore, India). All the reagents were of analytical grade and used without further purification.

2.2. Methods

2.2.1. Vesicle Preparation

For the vesicle formation by NOG-based amphiphilic systems, a previously standardized protocol was used [12]. Typically, in metal ion stability reactions containing the pure NOG system, a master mix solution was prepared by taking an appropriate volume of NOG methanol stock (10 mg/mL) in a microcentrifuge tube. For NOG-surfactant mixed systems, appropriate volumes of 10 mg/mL methanol stocks of NOG and GMO (or OOH in some reactions) were mixed in a microcentrifuge tube. The methanol was evaporated under vacuum to form a dry lipid film, which was then hydrated with an appropriate volume of 200 mM acetate buffer pH 5 (prepared in ultrapure water (18.2 MΩ-cm)) to get a final total amphiphile concentration of 7.5 mM in the master mix reaction. This contained either NOG (pure system) or NOG + surfactant in 2:1 ratio (mixed system), unless and otherwise mentioned. This solution was then incubated at 60 °C for 1 h with constant shaking at 500 rpm, with intermediate mixing by vortexing and pipetting. Note that the heating step is very crucial for the vesicle formation by NOG-based amphiphilic systems, as the chain-melting transition temperature of NOG seems to be higher than room temperature [12]. Moreover, this high temperature should be maintained throughout the course of the experiment, in order to keep these systems in a vesicular form and to avoid NOG crystal formation [12]. This was achieved by incubating NOG-based amphiphile solutions in a thermomixer (ThermoMixer C; Eppendorf India) equipped with a temperature maintaining lid (ThermoTop; Eppendorf India) that also avoids volume changes due to evaporation-condensation during the incubation period.

2.2.2. Setting Up the Metal Ion Stability Experiments

Briefly, 40 μL aliquots of the NOG master mix vesicular solution were taken into separate microcentrifuge tubes. To these individual tubes, appropriate volumes of metal ion stock solution were externally added to get increasing concentrations of metal ion (usually in the range of 0 to 11 mM for Mg^{2+} and 0 to 600 mM for Na^+), while keeping the

NOG concentration constant. After adding the metal ion stock solution into individual reaction vials, the final volume was adjusted to 50 µL using 200 mM acetate buffer pH 5, to get a 6 mM working concentration of NOG. Control reaction contained only 6 mM NOG with no externally added metal ions. A similar procedure was followed for the experiments with the NOG + GMO mixed system. For the Mg^{2+} stability experiments, 100 mM $MgCl_2$ stock solution was used, while for the Na^+ stability experiments, 4 M NaCl stock solution was used. Both these stock solutions were prepared in 200 mM acetate buffer pH 5. After the addition of metal ions to vesicular solutions, the reactions were incubated for 30 min at 60 °C to allow the metal ion to interact with vesicles and these samples were then subjected to microscopy.

2.2.3. Microscopic Analysis

All the amphiphile-based higher order structures like vesicles, droplets, and metal ion-induced aggregates that could result in a solution, were visualized using differential interference contrast (DIC) microscopy (Axio Imager Z1, Carl Zeiss, Germany) under 40× objective (NA = 0.75). For the experiments characterizing vesicle formation by NOG-based amphiphile systems in hot spring water samples, the vesicular structures were further confirmed by staining vesicles with an amphiphilic dye and visualizing using epifluorescence microscopy. For this, vesicles were stained with 10 µM of Octadecyl Rhodamine B Chloride (R18) (Invitrogen, Thermo Fisher Scientific; India), which was generously donated by Pucadyil lab at IISER Pune. The staining was performed by adding an appropriate volume of R18 dye methanol stock (50 µM) while making the dry lipid film of the NOG-based amphiphiles. The fluorescent vesicles were visualized using filter set 43 HE (Ex: 550/25 nm, Em: 605/70 nm, Beamsplitter: FT 570), with rest of the settings being same as mentioned above for the DIC microscopy. Image acquisition was done using AxioVision software. Imaging of all the heated reaction solutions was performed as quickly as possible to avoid NOG crystal formation.

2.2.4. Vesicle Re-Formation by the Addition of Magnesium Chelators

Briefly, 6 mM pure NOG and NOG + GMO (2:1 ratio) vesicular solutions containing externally added 12 mM Mg^{2+} ions were prepared as mentioned above. This Mg^{2+} concentration was selected to ensure the formation of magnesium-induced aggregates in both the pure NOG and the NOG + GMO mixed systems. Firstly, the presence of aggregates was confirmed by DIC microscopy. Then, a chelator (EDTA or citrate) was added in 1:1 mole equivalents to that of Mg^{2+} to the same aggregate containing solution, followed by the incubation at 60 °C for 30 min. The dissociation of aggregates and the re-appearance of vesicles was monitored using DIC microscopy.

2.2.5. Vesicle Formation by NOG-Based Amphiphile Systems in Hot Spring Water Samples

I. The collection of water samples from early Earth analogue hot spring sites: Water samples were directly collected from two hot spring pools at Hells Gate Geothermal Reserve, Tikitere (abbreviated as TIKB and TIKC, respectively). These samples were filtered with a 0.22 µm polyethersulfone membrane filter (rinsed with 20 mL of sample) and stored in acid-washed high-density polyethylene bottles until further use. Samples were untreated and unpreserved in the field.

II. Geochemical analysis of hot spring water samples: The collected water samples were analysed for the identification and quantification of different ions present in them. These measurements were performed by following standard protocols as reported earlier [38]. Briefly, the concentrations of the major cations (Na^+, K^+, Ca^{2+}, Mg^{2+}, and NH_4^+) and anions (Cl^-, F^-, SO_4^{2-}, and NO_3^-) were measured using ion chromatography instrument Compact IC plus 882 (Metrohm, Herisau, Switzerland). Moreover, the alkalinity of the samples was measured using an auto-titrator Eco Titrator (Metrohm, Switzerland). The dissolved Silica concentrations were measured by conventional molybdenum-blue method using double beam UV-VIS Spectrophotometer (M.D.T.

INTERNATIONAL, Ambala, Haryana, India). The accuracy and precision of these analyses were monitored regularly, which had an average value of ±4%. The net inorganic charge balance (NICB) for these samples was around 1, which indicates good data quality.

III. Vesicle formation in hot spring water samples: Dry lipid films of pure NOG and mixed systems containing NOG + surfactant (GMO or OOH) in 2:1 ratio were prepared as previously described. These lipid films were hydrated with 50 µL of hot spring water sample (TIKB/TIKC) to get a total amphiphile concentration of 6 mM. The solution was further incubated at 60 °C for 1 h with constant shaking at 500 rpm, with intermittent mixing by vortexing and pipetting to facilitate the vesicle formation process. The initial pH of the hot spring water samples was 7–7.5, which decreased to 5–5.5 after the addition of the amphiphiles, likely because of the acidic nature of NOG. The formation of vesicles was checked by both DIC and epifluorescence microscopy as mentioned above.

All the experiments in this entire study were performed in at least three independent replicates.

3. Results

3.1. Effect of Mg^{2+} on the Stability of Pure NOG and NOG + GMO Mixed Systems

In our previous study, we showed that 6 mM NOG in 200 mM acetate buffer pH 5 results in the formation of a large number of vesicles [12]. Therefore, these conditions were used to evaluate the effect of different metal ions on the stability of NOG vesicles. We started this characterization with divalent metal ions using magnesium (Mg^{2+}) as a representative ion, because of its direct relevance to several prebiotic processes pertinent to the RNA world. As mentioned earlier, it is not only crucial for RNA folding and ribozyme activity [19,20], but also required for the nonenzymatic template-directed replication of RNA [39]. To test the effect of Mg^{2+} on NOG vesicles, NOG concentration was kept constant at 6 mM and the Mg^{2+} concentration was varied incrementally up to 11 mM. For the pure NOG system, it was observed that NOG vesicles could tolerate Mg^{2+} concentrations up to 1:1 molar ratio, beyond which magnesium ions induced a clumping of vesicles resulting in the formation of large-sized aggregates (Figures 1a and S2).

The morphology of these metal ion-induced aggregates was different from the NOG crystals that form at a lower temperature (Figure S3). Overall, the behavior of the pure NOG system in the presence of increasing concentrations of Mg^{2+} followed three distinct phases (Figure 1c). A vesicular phase was observed up to 5 mM Mg^{2+}, which contained only vesicles with no visible aggregates. It was followed by a transition phase at 6 mM Mg^{2+}, where small aggregates started forming in the solution, and both free vesicles and aggregates were present simultaneously. Finally, an aggregate phase was observed from 7 mM Mg^{2+} onwards, where the vesicular system formed large aggregates, with no free vesicles present in the solution.

Surfactants like monoglyceride and fatty alcohol are known to increase the stability of fatty acid vesicles towards metal ions [28,34]. Therefore, we sought to explore whether a similar effect is observed for NOG-surfactant mixed systems also. Firstly, the optimum NOG-surfactant mixed system to perform this experiment was narrowed down by systematically evaluating the vesicle formation behavior of different combinations of NOG-surfactant binary systems by varying their ratios. Surfactants used for these experiments were GMO and OOH with different NOG to surfactant ratios, including 1:1, 2:1, and 4:1, while the total amphiphile concentration was kept constant at 6 mM. In NOG + GMO mixed systems, it was observed that, irrespective of the ratios, a mixture of NOG and GMO always generated a heterogeneous population of vesicles and droplets (Figure S4; appendix A of our earlier study [30] details the criteria that we used to distinguish between vesicles and droplets using DIC microscopy). Similar results were observed even at a lower total amphiphile concentration of 3 mM (Figure S5). Nonetheless, a mixed system of NOG + GMO with 2:1 ratio was found to be optimum for studying the metal ion effect. This allowed the formation of large-sized vesicles as compared to the 1:1 ratio system, making it easy to observe them under the microscope (Figure S4). Furthermore, this 2:1 ratio also conferred greater stability in the presence of higher Mg^{2+} concentrations in comparison to the 4:1 ratio (Figure S6). In the case of the NOG + OOH system, we observed that OOH had a destabilizing effect on the NOG system, which predominantly led to droplet formation at all the three NOG to OOH ratios (Figure S7). Therefore, this system was not evaluated further for its metal ion stability.

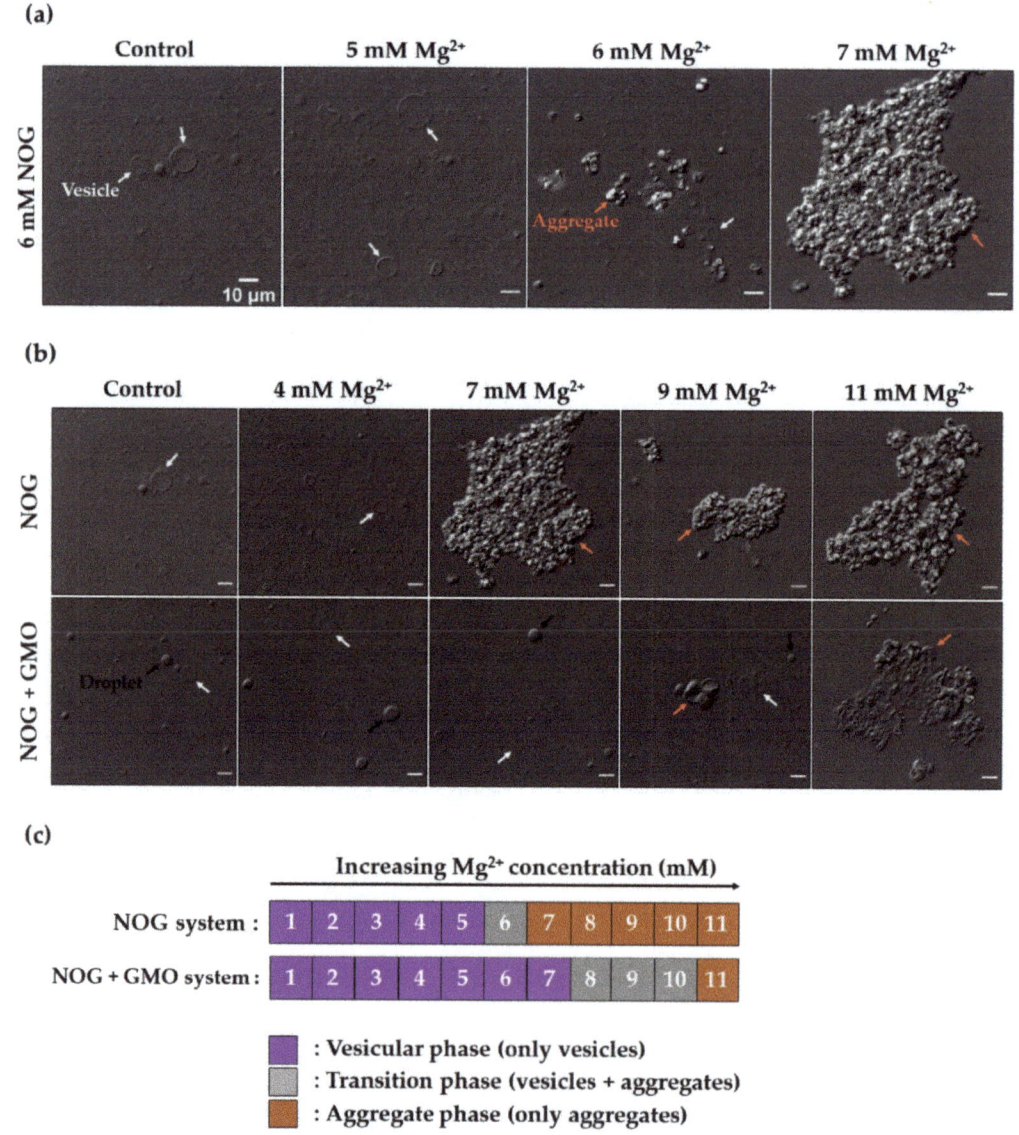

Figure 1. Effect of Mg^{2+} on pure NOG and NOG + GMO mixed vesicles. (**a**) NOG vesicles are stable in the control reaction

(no externally added Mg^{2+}) and in the presence of 5 mM Mg^{2+}. However, metal ion-induced aggregates appear in the reaction containing 6 mM Mg^{2+}, along with free vesicles. Finally, in the presence of 7 mM Mg^{2+}, vesicles completely collapse into large aggregates and no free vesicles are observed. (b) Addition of GMO to NOG system increases the stability of vesicles towards Mg^{2+} ion. 6 mM NOG system completely collapses in the presence of 7 mM Mg^{2+} and above. However, it takes 11 mM Mg^{2+} to induce large aggregates in the case of NOG + GMO (6 mM; 2:1 ratio) mixed system. Notably, NOG + GMO mixture results in a heterogeneous population of vesicles and droplets unlike pure NOG system, which predominantly forms vesicles. Vesicles, droplets and aggregates are indicated by white, black, and red arrows respectively. Imaging was done using DIC microscopy. Scale bar is 10 μm for all microscopy images, unless mentioned otherwise. (c) A distinct three-phase behavior of NOG-based vesicles is depicted with increasing concentrations of Mg^{2+}. The vesicular phase (purple) contains only vesicles and no metal ion-induced aggregates. It is followed by a transition phase (grey) where both free vesicles and small metal ion-induced aggregates are simultaneously present. Finally, the aggregate phase (orange) contains only large-sized aggregates. Although, the color code of the three phases has been described in terms of vesicles and aggregates, the NOG + GMO mixed system also contains droplets in addition to vesicles.

With NOG + GMO (6 mM; 2:1 ratio) mixed system, it was observed that the addition of GMO indeed increased the stability of NOG towards magnesium ions as this mixed system was able to tolerate Mg^{2+} concentrations till 11 mM (Figure 1b). Consistent with the pure NOG system, this mixed system also showed a three-phase behavior in the presence of increasing Mg^{2+} concentrations (Figure 1c). A vesicular phase containing a mixture of vesicles and droplets was observed up to 7 mM Mg^{2+}. It was followed by a transition phase from 8 to 10 mM Mg^{2+}, where small metal ion-induced aggregates were also present along with vesicles and droplets. Finally, the aggregate phase appeared from 11 mM Mg^{2+} onwards, where the system completely flocculated into large-sized aggregates. It is pertinent to note that the span of both the vesicular and transition phases of the NOG system increased after the addition of GMO to the system, which clearly indicated a stabilizing effect that was being conferred by the surfactant towards the metal ions. We also note that some variability in the aggregation behavior is expected to be observed in the transition phase, especially for the NOG + GMO mixed system, as it is difficult to achieve the exact same ratio of NOG to GMO in every resultant vesicle. This would essentially render vesicular systems with a lower concentration of GMO more vulnerable to metal ion-induced aggregation (Figure S6).

After evaluating the effect of Mg^{2+} on the stability of pure NOG and the NOG + GMO mixed systems, we tested whether the aggregates observed in the presence of Mg^{2+} were indeed magnesium-induced aggregates, and whether this aggregation phenomenon is reversible. To discern this, we added EDTA to the solution in 1:1 mole ratio to Mg^{2+}, after the formation of the aggregates. EDTA chelated Mg^{2+} ions in the solution, which resulted in the disruption of the large-sized aggregates and reappearance of vesicles in both the pure NOG and NOG + GMO mixed systems (Figure 2a). EDTA is an artificially synthesized chelating agent that is used majorly in commercial applications. However, there are many naturally occurring chelating agents that can also effectively chelate magnesium ions.

Given the scope of this study, we asked whether there are any prebiotically plausible chelating agents that might have provided further stability to protocell membranes in the presence of metal ions. Citrate is considered as an important metabolite both in extant as well as in proto-metabolic pathways [40], and is known for its chelation effect [41,42]. Therefore, we tested the effect of citrate on magnesium-induced aggregates and observed similar results as those with EDTA, where vesicles emerged from large aggregates (Figure 2b).

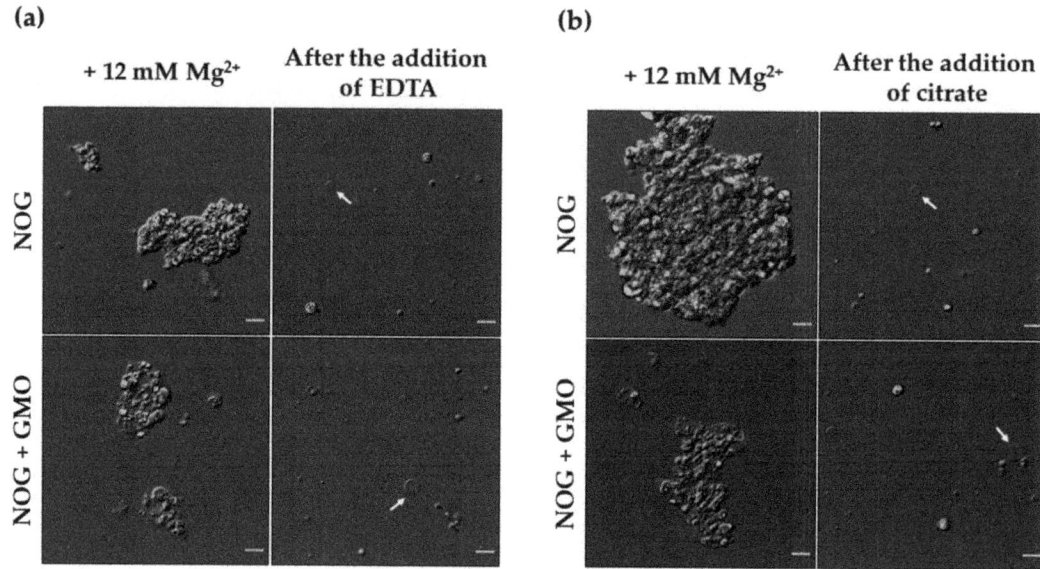

Figure 2. Re-formation of vesicles from magnesium-induced aggregates after the addition of a chelator. (**a**) Both pure NOG (6 mM) and NOG + GMO (6 mM; 2:1 ratio) mixed systems (top and bottom panels respectively) completely collapse into large aggregates in the presence of 12 mM Mg^{2+}. However, the addition of EDTA in 1:1 mole equivalents to that of Mg^{2+} results in the disassembly of aggregates with a concurrent reappearance of free vesicles (indicated by white arrows). (**b**) Similar effect is observed in the presence of citrate as a chelator, which was also added in 1:1 mole equivalents to that of Mg^{2+}. Imaging was done using DIC microscopy. Scale bar is 10 μm.

3.2. Effect of Na^+ on the Stability of Pure NOG and NOG + GMO Mixed Systems

After testing the effect of a divalent metal ion on the stability NOG-based systems, we aimed to characterize the behavior of these systems in the presence of a monovalent cation. For this, we used sodium (Na^+) as the representative ion. Previous studies have shown that fatty acid-based systems are more stable in the presence of monovalent cations than divalent ones [28]. We observed a similar behavior for NOG-based systems as well, where pure NOG vesicles formed aggregates only at and above 400 mM Na^+ (Figure 3), which is about two orders of magnitude higher than that for Mg^{2+} (7 mM). The NOG + GMO mixed system was even more stable and resulted in aggregate formation only at 500 mM Na^+ (Figure 3).

In addition to aggregate formation, another intriguing phenomenon was observed in the case of the pure NOG system, wherein the addition of 100 mM Na^+ resulted in vesicle shrinkage and also the formation of small-sized droplets in solution (Figures 3 and S8). This effect was consistent even for higher concentrations of Na^+ (i.e., at 200 mM and 300 mM). We hypothesized two plausible reasons for this effect. (1) The vesicle shrinkage might have occurred because of an osmotic shock faced by vesicles upon the external addition of NaCl to the solution. NOG vesicles were prepared in an acetate buffer that already contained around 143 mM Na^+ that was added during the buffer preparation process. Given that these vesicles are not readily permeable to charged species, the external addition of 100 mM Na^+ to a solution containing preformed vesicles would have generated a higher concentration of Na^+ in the exterior as compared to that of the vesicle lumen. This would have resulted in the outflow of water from vesicles causing the overall shrinkage of these vesicles, a phenomenon similar to plasmolysis. To validate this hypothesis, we performed an experiment where the Na^+ concentration in the acetate buffer was increased by 100 mM and this buffer with higher [Na^+] was used for vesicle preparation. The vesicle shrinkage in this reaction was not as dramatic as what was observed in the reaction with externally added Na$^+$ (Figure S9). This was likely because the Na$^+$ levels in the vesicle exterior and in the lumen would have gotten equilibrated during the vesicle formation process itself. Pertinently, it has been previously observed in the case of fatty acid systems that the vesicles undergo an osmotic shock upon the external addition of salts to a solution containing preformed vesicle [28]. Notably, this overall vesicle shrinkage phenomenon was less apparent in the NOG + GMO mixed system (Figure 3). It is known that the mixed membranes made of fatty acids and monoglycerides have a higher permeability than the pure fatty acid vesicles [7,34]. It is very possible that the presence of GMO increases the permeability of NOG + GMO mixed vesicles when compared to pure NOG vesicles. This probably would have allowed the faster equilibration of Na$^+$ between vesicle lumen and the exterior, thereby reducing the exosmosis of water and the resultant vesicle shrinkage. (2) The formation of droplets in the pure NOG system after the addition of Na$^+$ could be because of the charge screening of the deprotonated terminal carboxyl moiety (COO$^-$) of the NOG head group by Na$^+$ ions. This would make the overall system more hydrophobic, which could have resulted in the droplet formation. In such a case, a similar effect should also be observed with other monovalent cations. In order to test this hypothesis, we externally added 100 mM K$^+$ ions to the solution containing preformed NOG vesicles, which also resulted in droplet formation along with vesicle shrinkage similar to what was observed in the Na$^+$ reaction (Figure S10).

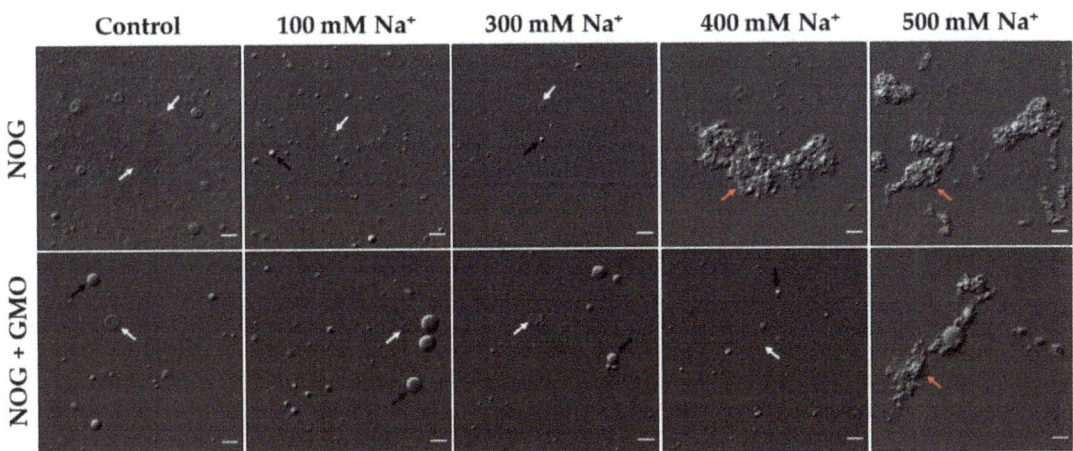

Figure 3. Effect of Na$^+$ on the stability of NOG-based vesicles. 6 mM pure NOG vesicles tolerate Na$^+$ concentrations up to 300 mM, beyond which large metal ion-induced aggregates are formed (top panel). Also, the external addition of Na$^+$ to pure NOG vesicular system results in vesicle shrinkage and the formation of small droplets as observed in the 100 mM and 300 mM Na$^+$ reactions. The NOG + GMO (6 mM; 2:1 ratio) mixed system (bottom panel) forms a heterogeneous population of vesicles and droplets, which survives Na$^+$ concentrations up to 400 mM, beyond which metal ion-induced aggregates are formed. Vesicles, droplets and aggregates are indicated by white, black and red arrows respectively. Imaging was done using DIC microscopy. Scale bar is 10 μm.

3.3. NOG-Based Amphiphile Systems Readily Form Vesicles in Hot Spring Water Samples

After systematically evaluating the stability of NOG-based systems in the presence of both monovalent and divalent metal ions, we set out to explore the self-assembly behavior of these systems in the presence of multiple cations and anions of different concentrations. This was to get a realistic sense of how NAA-based amphiphilic systems would behave in natural water bodies that often possess multiple ions in varying concentrations. To test this, we used water samples that were collected from hot spring sites, and we did this for two reasons. Firstly, mimicking such a system with diverse ionic strength in the laboratory is difficult, as these ions are added in the form of salts, which invariably results in the excess addition of some counter ions while preparing such ionic mimics. Secondly, and pertinently, terrestrial hot springs are considered as one of the plausible niches where life would have potentially originated on early Earth and Mars [26]. Therefore, studying prebiotic membrane-assembly processes by using water samples from contemporary terrestrial hot spring sites, which have topological features analogous to early Earth and Martian conditions, allows one to glean the feasibility of prebiotic compartmentalization under more realistic, diverse, and extreme environments. We have previously shown that fatty acid-based systems can assemble into vesicles in hot spring water [30], which motivated us to check whether a similar phenomenon could be observed in case of NAA-based systems too.

To perform these analogue experiments, we used water samples collected from two hot spring sites in Tikitere, New Zealand, which are abbreviated as TIKB and TIKC, and analysed for their ionic content (Table 1). Both TIKB and TIKC samples had a dominance of NH_4^+ and SO_4^{2-} ions. Several other cations (Na^+, K^+, Ca^{2+}, Mg^{2+}) and anions (Cl^-, F^-, NO_3^-, HCO_3^-) were also detected in varying concentrations in these water samples, highlighting the ionic diversity that such natural aqueous systems possess. However, it is important to note that the concentrations of both monovalent and divalent cations were far below the level at which they generally induce vesicle aggregation in laboratory settings.

Table 1. Geochemical analysis of the hot spring water samples that were used in this study. TZ^+ and TZ^- indicate sum of cations and sum of anions respectively in microequivalent units. TZ^+/TZ^- is net inorganic charge balance (NICB).

Hot Spring Sample	Major Cations					TZ^+	Major Anions					TZ^-		TZ^+/TZ^-
	NH_4^+	Na^+	K^+	Ca^{2+}	Mg^{2+}		NO_3^-	F^-	Cl^-	HCO_3^-	SO_4^{2-}		SiO_2	
	All Values in μM					μE	All Values in μM					μE	μM	
TIKB [†]	2756	743	285	98	31	4043	24	45	122	29	2550	5319	1128	0.76
TIKC [†]	3296	1007	301	167	27	4991	14	48	150	1458	2312	6294	1019	0.79

[†] The initial pH of TIKB and TIKC was 7–7.5, which decreased to 5–5.5 after the addition of the amphiphiles (likely because of the acidic nature of NOG).

It was observed that both pure NOG (6 mM) and NOG + GMO (6 mM; 2:1 ratio) mixed systems readily assembled into vesicles in both TIKB and TIKC samples (Figure 4). The resultant vesicle population was heterogeneous in terms of the size, shape and lamellarity of vesicles. Therefore, for better visualization of such higher order structures, these vesicles were stained using octadecyl rhodamine B chloride (R18) (an amphiphilic dye that readily partitions into membranes) and observed under epifluorescence microscopy, which further delineated the lamellarity of these vesicles. We further observed that a mixture of NOG + OOH (6 mM; 2:1 ratio) also resulted in vesicle formation in TIKC (Figure S11). Although the vesicle formation by lipid film hydration method is known to generate a morphologically diverse set of vesicles, the variety of ions and their relative concentrations in these hot spring samples could further affect the self-assembly and the morphology of these vesicles.

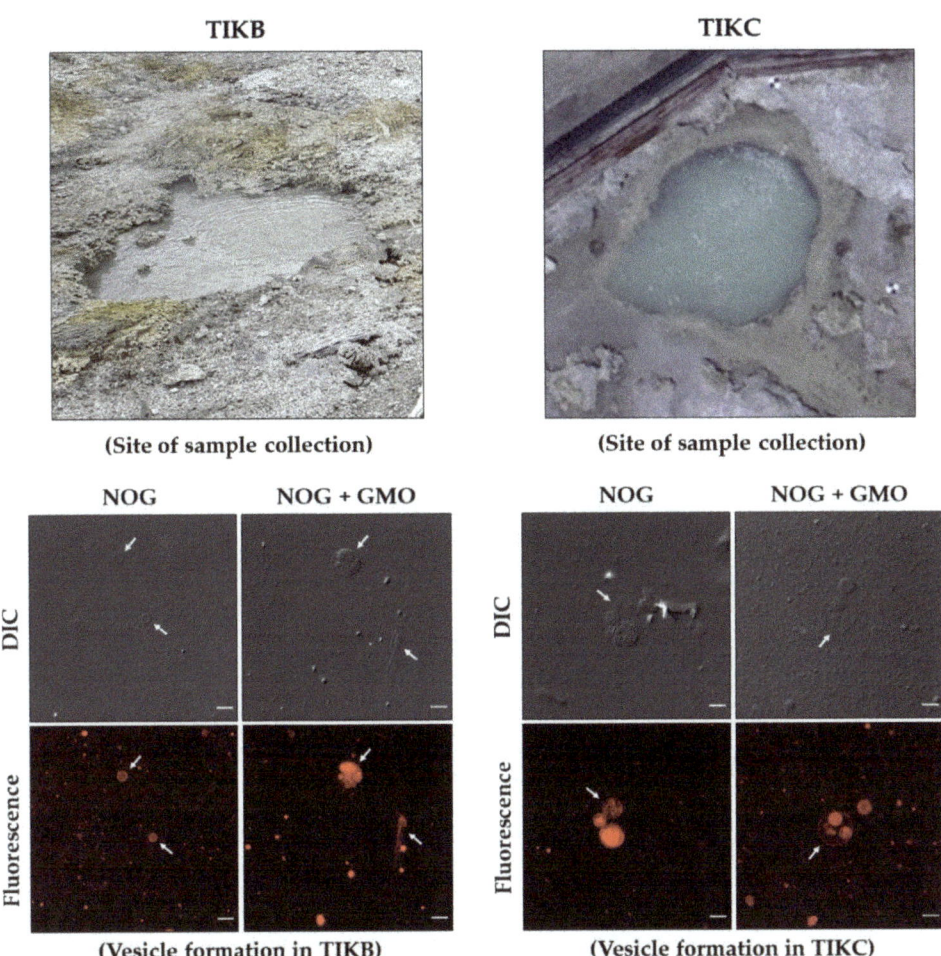

Figure 4. Vesicle formation by NOG-based amphiphile systems in hot spring water samples. Both pure NOG (6 mM) and NOG + GMO (6 mM; 2:1 ratio) mixed systems readily form vesicles (indicated by white arrows) in two different hot spring water samples with acronyms TIKB (**left panel**) and TIKC (**right panel**). Actual sites of sample collection are shown in top panels, while microscopy images for the vesicle formation in respective hot spring samples are shown in bottom panels. Vesicle morphology varies from unilamellar to multivesicular vesicles, which were visualized under DIC as well as fluorescence microscopy. For fluorescence imaging, vesicles were stained with an amphiphilic dye named octadecyl rhodamine-B chloride (R18). Fluorescence images are pseudocolored for a better visualization. Scale bar is 10 µm. TIKC sample collection site image was adapted from [43].

4. Discussion

N-acyl amino acid is a hybrid molecule comprising both fatty acid and amino acid moieties within the same structure. This amalgamation provides this amphiphile with great flexibility in terms of its being able to acquire different properties based on the varying nature of the amino acid head group, to better adapt to the fluctuating and extreme environmental conditions thought to have been prevalent on the prebiotic Earth. For example, the addition of glycine as a head group to the carboxyl terminal of oleic acid (OA), generates N-oleoyl glycine that can form vesicles at acidic pH, whereas OA itself is unable to do so. Thus, NAAs might have played a pivotal role towards bringing in the necessary diversity and adaptability in protocellular compartments, which is one of the fundamental requisites for the advent of Darwinian evolution. There have been few attempts to functionalize fatty acid membranes by doping them with small amounts of NAAs [44,45]. However, to the best of our knowledge, NAA itself as an amphiphilic system has not been systematically investigated in a prebiotic context, and the effect of ions on their self-assembly and stability is not yet known.

In this study, we evaluated the behavior of NOG-based model vesicular systems in the presence of metal ions, given the central importance of metal ions towards facilitating several pertinent processes in the RNA world [19,20,39]. Although the addition of an amino acid as a head group could potentially alter the basic physicochemical properties of fatty acids as mentioned earlier, especially in terms of the optimum pH for their vesicle formation, some of the features of NAAs and fatty acids may still be comparable. Herein, we demonstrate that the metal ion stability of these amphiphiles is one such feature. Previous studies exploring the effect of Mg^{2+} on fatty acid membranes show that vesicles made of only OA form large metal ion-induced aggregates when the Mg^{2+} concentration exceeds the fatty acid concentration [34,35]. We report a similar phenomenon with vesicles made of NOG (an NAA analogue of oleic acid), wherein these vesicles form large aggregates when the Mg^{2+} concentration just surpasses the NOG concentration. Furthermore, the NOG system was found to be much more stable in the presence of Na^+ than of Mg^{2+}, comparable to what has been reported for fatty acids [28]. Moreover, NOG vesicles undergo an osmotic shock upon the external addition of NaCl to the solution, which is consistent with previous observations with fatty acid vesicles [28].

This significant similarity between NAA and fatty acid vesicles in terms of their stability towards metal ions may likely stem from the nature of their terminal ionizable moiety. The behavior of an amphiphile that has an anionic head group in the presence of metal ions is primarily governed by the electrostatic interaction of its head group with the positively charged metal ion. Given that both fatty acids and NAAs have a carboxyl terminus (-COOH), which can deprotonate to form a carboxylate ion (-COO$^-$) depending on the pH of the solution, they show a comparable behavior in terms of their metal ion stability.

The incompatibility of SCA vesicles with high concentrations of metal ions (particularly divalent cations), which are required for nonenzymatic RNA strand copying and for ribozyme catalysis, has been considered a major obstacle towards the emergence of protocells [39]. One approach to overcome this difficulty is to systematically investigate different membrane compositions of prebiotically pertinent amphiphiles to discern which compositions generate stable membranes at high metal ion concentrations. This approach is also logical and realistic since a prebiotic soup would have likely harbored a plethora of amphiphiles, which inevitably would have generated compositionally heterogeneous membranes on the early Earth [9]. Interestingly, prebiotically plausible amphiphiles with a polar non-ionic head group such as monoglycerides have been shown to stabilize fatty acid vesicles in the presence of both divalent and monovalent cations [28,34]. Our study shows that such a stabilizing effect can also be achieved in the case of NAAs by generating blended membranes of NAAs and monoglycerides. Particularly, mixed vesicles composed of NOG and GMO were highly stable in the presence of both Mg^{2+} and Na^+ as compared to those made of only NOG.

Another possibility to tackle the above mentioned problem is to look for prebiotically plausible chelators that could form a complex with a metal ion in such a way that the coordination complex still remains functionally active to facilitate previously mentioned RNA-related reactions [39]. In such a scenario, it is possible to achieve low levels of free metal ions in the solution, which allow SCAs to sustain their vesicular form. In this context, citrate is an important molecule as it has an excellent chelating capacity and also a key metabolite in both extant as well as proto-metabolic pathways. Adamala et al. showed that citrate can chelate magnesium while still allowing nonenzymatic RNA template copying inside intact fatty acid vesicles at alkaline pH regime (pH 8) [42]. Given the influence of pH on the chelation ability of a chelator, our study demonstrates that citrate can effectively chelate magnesium even in the acidic regime of pH 5. The addition of citrate reversed the magnesium-induced aggregation phenomenon and allowed for the re-formation of vesicles, both in the pure NOG and the NOG + GMO mixed systems. It will be worthwhile to further evaluate whether NAA-based vesicular systems containing magnesium-citrate complex are compatible with RNA formation and ribozyme catalysis reactions.

Thus far, we discussed the adverse effect of metal ions on the stability of SCA vesicles. However, an equally important factor that is often neglected is that metal ions can actually facilitate, and are important for, vesicle self-assembly processes, albeit at a low concentration [46]. They can do so by salting-in the hydrophobic tails and stabilizing the negatively charged carboxylate moieties of the head group via electrostatic interactions. Therefore, from a prebiotic perspective, natural fresh water systems with their overall low ionic strengths would have been more favorable for a protocell compartmentalization process as compared to marine niches [47]. In this study, we showed that NOG-based model vesicular systems readily assemble into vesicles in hot spring water samples. These results have two-fold implications. Firstly, they demonstrate the ability of NAAs to assemble into vesicles in natural aqueous systems with varying ionic strengths, which further substantiates their candidature as model protocell membranes. Secondly, these results provide yet another piece of evidence for the conduciveness of terrestrial freshwater systems for the self-assembly of prebiotic compartments, in addition to supporting nonenzymatic oligomerization processes [31,48]. This, in turn, underscores the likelihood of these niches as plausible hatcheries for the origin and early evolution of life on Earth.

Overall, we discerned the influence of ions on the self-assembly and stability of NAA-based vesicular systems. This is an important parameter that needed to be systematically evaluated as researchers in the "Origins of life" field continue to work with this intriguing amphiphile system as a model protocell membrane and its plausible role in facilitating other prebiotically pertinent processes. It is worthwhile to mention in this context that several important aspects of this unique amphiphile, such as its stability under fluctuating conditions of temperature and pH, or its potential to catalyze different chemical reactions on the membrane surface via the side chain of its amino acid head group, are yet to be systematically explored. Future research in this direction will enable shedding greater light on the adaptability and functionality of early protocellular compartments in a prebioloical era.

Supplementary Materials: The following are available online at https://www.mdpi.com/article/10.3390/life11121413/s1, Figure S1: Chemical structures of different amphiphiles used in this study along with their abbreviations, Figure S2: Effect of Mg2+ on the stability of pure NOG vesicles, Figure S3: The morphology of metal ion-induced aggregates of NOG is different from that of NOG crystals, Figure S4: A mixed system of NOG and GMO forms a heterogeneous population of vesicles and droplets irrespective of the NOG to GMO ratio used, Figure S5: NOG + GMO (2:1) mixed system behaves similarly at different total amphiphile concentrations, Figure S6: Decreasing the GMO concentration in the NOG + GMO mixed system also decreases the overall Mg2+ stability of the system, Figure S7: NOG + OOH mixed system predominantly forms droplets irrespective of the NOG to OOH ratio, Figure S8: Effect of NaCl addition on pure NOG vesicles, Figure S9: NOG vesicle formation with increased Na+ concentration in the buffer, Figure S10: Effect of K+ on the stability of pure NOG vesicles, Figure S11: Vesicle formation behavior of NOG + OOH mixed system in hot spring water samples.

Author Contributions: M.P.J. and S.R. conceived and designed the experiments; M.P.J. performed the experiments; M.J.V.K. and L.S. contributed towards the collection of hot spring water samples and the analysis of geochemical data; M.P.J. and S.R. wrote the paper with crucial inputs from M.J.V.K. and L.S. All authors have read and agreed to the published version of the manuscript.

Funding: This research was supported by Department of Biotechnology, Govt. of India [BT/PR19201/BRB/10/1532/2016] and IISER Pune. M.J.V.K. and L.S. were supported by Australian Research Council Discovery Project funding DP180103204. M.P.J. was financially supported by UGC, government of India in the form of graduate fellowship (SRF).

Institutional Review Board Statement: Not Applicable.

Informed Consent Statement: Not Applicable.

Data Availability Statement: Not Applicable.

Acknowledgments: The authors wish to deeply acknowledge Rakesh Kumar Rout and Gyana Ranjan Tripathy for their help with the geochemical analysis of the hot spring samples. The authors also wish to thank the microscopy facility at IISER Pune. We also wish to acknowledge Hell's Gate Ltd. as managers of the Tourist venue, and the Whakapoungakau 24 Block—Tikitere Trust as the body entrusted with the kaitiakitanga (guardianship) of the region.

Conflicts of Interest: The authors declare no conflict of interest.

References

1. Liu, L.; Zou, Y.; Bhattacharya, A.; Zhang, D.; Lang, S.Q.; Houk, K.N.; Devaraj, N.K. Enzyme-free synthesis of natural phospholipids in water. *Nat. Chem.* **2020**, *12*, 1029–1034. [CrossRef]
2. Gibard, C.; Bhowmik, S.; Karki, M.; Kim, E.K.; Krishnamurthy, R. Phosphorylation, oligomerization and self-assembly in water under potential prebiotic conditions. *Nat. Chem.* **2018**, *10*, 212–217. [CrossRef] [PubMed]
3. Rao, M.; Eichberg, J.; Oró, J. Synthesis of phosphatidylethanolamine under possible primitive earth conditions. *J. Mol. Evol.* **1987**, *25*, 1–6. [CrossRef] [PubMed]
4. Hargreaves, W.R.; Mulvihill, S.J.; Deamer, D.W. Synthesis of phospholipids and membranes in prebiotic conditions. *Nature* **1977**, *266*, 78–80. [CrossRef] [PubMed]
5. Bonfio, C.; Caumes, C.; Duffy, C.D.; Patel, B.H.; Percivalle, C.; Tsanakopoulou, M.; Sutherland, J.D. Length-Selective Synthesis of Acylglycerol-Phosphates through Energy-Dissipative Cycling. *J. Am. Chem. Soc.* **2019**, *141*, 3934–3939. [CrossRef] [PubMed]
6. Deamer, D. The Role of Lipid Membranes in Life's Origin. *Life* **2017**, *7*, 5. [CrossRef]
7. Mansy, S.S. Model protocells from single-chain lipids. *Int. J. Mol. Sci.* **2009**, *10*, 835–843. [CrossRef]
8. Mansy, S.S. Membrane transport in primitive cells. *Cold Spring Harb. Perspect. Biol.* **2010**, *2*, 1–15. [CrossRef]
9. Sarkar, S.; Das, S.; Dagar, S.; Joshi, M.P.; Mungi, C.V.; Sawant, A.A.; Patki, G.M.; Rajamani, S. Prebiological Membranes and Their Role in the Emergence of Early Cellular Life. *J. Membr. Biol.* **2020**, *253*, 589–608. [CrossRef]
10. Lawless, J.G.; Yuen, G.U. Quantification of monocarboxylic acids in the Murchison carbonaceous meteorite. *Nature* **1979**, *282*, 396–398. [CrossRef]
11. McCollom, T.M.; Ritter, G.; Simoneit, B.R. Lipid synthesis under hydrothermal conditions by Fischer-Tropsch-type reactions. *Orig. Life Evol. Biosph.* **1999**, *29*, 153–166. [CrossRef]
12. Joshi, M.P.; Sawant, A.A.; Rajamani, S. Spontaneous emergence of membrane-forming protoamphiphiles from a lipid–amino acid mixture under wet–dry cycles. *Chem. Sci.* **2021**, *12*, 2970–2978. [CrossRef] [PubMed]
13. Mansy, S.S.; Szostak, J.W. Thermostability of model protocell membranes. *Proc. Natl. Acad. Sci. USA* **2008**, *105*, 13351–13355. [CrossRef] [PubMed]
14. Rubio-Sánchez, R.; O'Flaherty, D.K.; Wang, A.; Coscia, F.; Petris, G.; Di Michele, L.; Cicuta, P.; Bonfio, C. Thermally Driven Membrane Phase Transitions Enable Content Reshuffling in Primitive Cells. *J. Am. Chem. Soc.* **2021**, *143*, 16589–16598. [CrossRef]
15. Maurer, S. The Impact of Salts on Single Chain Amphiphile Membranes and Implications for the Location of the Origin of Life. *Life* **2017**, *7*, 44. [CrossRef] [PubMed]
16. Walde, P.; Namani, T.; Morigaki, K.; Hauser, H.; Namani, T.; Morigaki, K.; Hauser, H. Formation and Properties of Fatty Acid Vesicles (Liposomes). In *Liposome Technology*; CRC Press: Boca Raton, FL, USA, 2018; pp. 23–42.
17. Bonfio, C.; Godino, E.; Corsini, M.; Fabrizi de Biani, F.; Guella, G.; Mansy, S.S. Prebiotic iron–sulfur peptide catalysts generate a pH gradient across model membranes of late protocells. *Nat. Catal.* **2018**, *1*, 616–623. [CrossRef]
18. Muchowska, K.B.; Varma, S.J.; Chevallot-Beroux, E.; Lethuillier-Karl, L.; Li, G.; Moran, J. Metals promote sequences of the reverse Krebs cycle. *Nat. Ecol. Evol.* **2017**, *1*, 1716–1721. [CrossRef]
19. Bowman, J.C.; Lenz, T.K.; Hud, N.V.; Williams, L.D. Cations in charge: Magnesium ions in RNA folding and catalysis. *Curr. Opin. Struct. Biol.* **2012**, *22*, 262–272. [CrossRef]
20. Inoue, A.; Takagi, Y.; Taira, K. Importance of magnesium ions in the mechanism of catalysis by a hammerhead ribozyme: Strictly linear relationship between the ribozyme activity and the concentration of magnesium ions. *Magnes. Res.* **2003**, *16*, 210–217.
21. Attwater, J.; Wochner, A.; Holliger, P. In-ice evolution of RNA polymerase ribozyme activity. *Nat. Chem.* **2013**, *5*, 1011. [CrossRef]
22. Horning, D.P.; Joyce, G.F. Amplification of RNA by an RNA polymerase ribozyme. *Proc. Natl. Acad. Sci. USA* **2016**, *113*, 9786–9791. [CrossRef] [PubMed]
23. Bapat, N.V.; Rajamani, S. Templated replication (or lack thereof) under prebiotically pertinent conditions. *Sci. Rep.* **2018**, *8*, 15032. [CrossRef] [PubMed]
24. Bowler, F.R.; Chan, C.K.W.; Duffy, C.D.; Gerland, B.; Islam, S.; Powner, M.W.; Sutherland, J.D.; Xu, J. Prebiotically plausible oligoribonucleotide ligation facilitated by chemoselective acetylation. *Nat. Chem.* **2013**, *5*, 383–389. [CrossRef] [PubMed]
25. Baross, J.A.; Hoffman, S.E. Submarine hydrothermal vents and associated gradient environments as sites for the origin and evolution of life. *Orig. Life Evol. Biosph.* **1985**, *15*, 327–345. [CrossRef]

26. Damer, B.; Deamer, D. The hot spring hypothesis for an origin of life. *Astrobiology* **2020**, *20*, 429–452. [CrossRef]
27. Milshteyn, D.; Damer, B.; Havig, J.; Deamer, D. Amphiphilic Compounds Assemble into Membranous Vesicles in Hydrothermal Hot Spring Water but Not in Seawater. *Life* **2018**, *8*, 11. [CrossRef]
28. Monnard, P.-A.; Apel, C.L.; Kanavarioti, A.; Deamer, D.W. Influence of Ionic Inorganic Solutes on Self-Assembly and Polymerization Processes Related to Early Forms of Life: Implications for a Prebiotic Aqueous Medium. *Astrobiology* **2002**, *2*, 139–152. [CrossRef]
29. Deamer, D.W.; Georgiou, C.D. Hydrothermal Conditions and the Origin of Cellular Life. *Astrobiology* **2015**, *15*, 1091–1095. [CrossRef] [PubMed]
30. Joshi, M.P.; Samanta, A.; Tripathy, G.R.; Rajamani, S. Formation and Stability of Prebiotically Relevant Vesicular Systems in Terrestrial Geothermal Environments. *Life* **2017**, *7*, 51. [CrossRef]
31. Deamer, D. Where Did Life Begin? Testing Ideas in Prebiotic Analogue Conditions. *Life* **2021**, *11*, 134. [CrossRef] [PubMed]
32. Maurer, S.E.; Tølbøl Sørensen, K.; Iqbal, Z.; Nicholas, J.; Quirion, K.; Gioia, M.; Monnard, P.A.; Hanczyc, M.M. Vesicle Self-Assembly of Monoalkyl Amphiphiles under the Effects of High Ionic Strength, Extreme pH, and High Temperature Environments. *Langmuir* **2018**, *34*, 15560–15568. [CrossRef]
33. Namani, T.; Deamer, D.W. Stability of Model Membranes in Extreme Environments. *Orig. Life Evol. Biosph.* **2008**, *38*, 329–341. [CrossRef] [PubMed]
34. Sarkar, S.; Dagar, S.; Verma, A.; Rajamani, S. Compositional heterogeneity confers selective advantage to model protocellular membranes during the origins of cellular life. *Sci. Rep.* **2020**, *10*, 1–11.
35. Dalai, P.; Ustriyana, P.; Sahai, N. Aqueous magnesium as an environmental selection pressure in the evolution of phospholipid membranes on early earth. *Geochim. Cosmochim. Acta* **2018**, *223*, 216–228. [CrossRef]
36. Jin, L.; Kamat, N.P.; Jena, S.; Szostak, J.W. Fatty Acid/Phospholipid Blended Membranes: A Potential Intermediate State in Protocellular Evolution. *Small* **2018**, *14*. [CrossRef]
37. Toparlak, Ö.D.; Karki, M.; Egas Ortuno, V.; Krishnamurthy, R.; Mansy, S.S. Cyclophospholipids Increase Protocellular Stability to Metal Ions. *Small* **2019**, 1903381. [CrossRef] [PubMed]
38. Tripathy, G.R.; Goswami, V.; Singh, S.K.; Chakrapani, G.J. Temporal variations in Sr and ^{87}Sr/^{86}Sr of the Ganga headwaters: Estimates of dissolved Sr flux to the mainstream. *Hydrol. Process.* **2010**, *24*, 1159–1171. [CrossRef]
39. Szostak, J.W. The eightfold path to non-enzymatic RNA replication. *J. Syst. Chem.* **2012**, *3*, 2. [CrossRef]
40. Davey, S.G. Origin of Life: Cycling citrate sans enzymes. *Nat. Rev. Chem.* **2017**, *1*, 1. [CrossRef]
41. Glusker, J.P. Citrate conformation and chelation: Enzymic implications. *Acc. Chem. Res.* **2002**, *13*, 345–352. [CrossRef]
42. Adamala, K.; Szostak, J.W. Nonenzymatic template-directed RNA synthesis inside model protocells. *Science (80-.)* **2013**, *342*, 1098–1100. [CrossRef] [PubMed]
43. Dobson, M. Facies Mapping and Analysis of Diverse Hydrothermal Sedimentary Facies and Siliceous Spicular Sinter at Hell's Gate, Tikitere Geothermal Field, Taupō Volcanic Zone, New Zealand. Honours Dissertation, University of Auckland, Auckland, New Zealand, 2018.
44. Bonfio, C.; Russell, D.A.; Green, N.J.; Mariani, A.; Sutherland, J.D. Activation chemistry drives the emergence of functionalised protocells. *Chem. Sci.* **2020**, *11*, 10688–10697. [CrossRef] [PubMed]
45. Izgu, E.C.; Björkbom, A.; Kamat, N.P.; Lelyveld, V.S.; Zhang, W.; Jia, T.Z.; Szostak, J.W. N-Carboxyanhydride-Mediated Fatty Acylation of Amino Acids and Peptides for Functionalization of Protocell Membranes. *J. Am. Chem. Soc.* **2016**, *138*, 16669–16676. [CrossRef]
46. Maurer, S.E.; Nguyen, G. Prebiotic Vesicle Formation and the Necessity of Salts. *Orig. Life Evol. Biosph.* **2016**, *46*, 215–222. [CrossRef] [PubMed]
47. Mulkidjanian, A.Y.; Bychkov, A.Y.; Dibrova, D.V.; Galperin, M.Y.; Koonin, E.V. Origin of first cells at terrestrial, anoxic geothermal fields. *Proc. Natl. Acad. Sci. USA* **2012**, *109*, E821–E830. [CrossRef] [PubMed]
48. Dagar, S.; Sarkar, S.; Rajamani, S. Geochemical influences on nonenzymatic oligomerization of prebiotically relevant cyclic nucleotides. *RNA* **2020**, *26*, 756–769. [CrossRef]

Article

When Is a Reaction Network a Metabolism? Criteria for Simple Metabolisms That Support Growth and Division of Protocells

Paul G. Higgs

Department of Physics and Astronomy, Origins Institute, McMaster University, Hamilton, ON L8S 4M1, Canada; higgsp@mcmaster.ca

Abstract: With the aim of better understanding the nature of metabolism in the first cells and the relationship between the origin of life and the origin of metabolism, we propose three criteria that a chemical reaction system must satisfy in order to constitute a metabolism that would be capable of sustaining growth and division of a protocell. (1) Biomolecules produced by the reaction system must be maintained at high concentration inside the cell while they remain at low or zero concentration outside. (2) The total solute concentration inside the cell must be higher than outside, so there is a positive osmotic pressure that drives cell growth. (3) The metabolic rate (i.e., the rate of mass throughput) must be higher inside the cell than outside. We give examples of small-molecule reaction systems that satisfy these criteria, and others which do not, firstly considering fixed-volume compartments, and secondly, lipid vesicles that can grow and divide. If the criteria are satisfied, and if a supply of lipid is available outside the cell, then continued growth of membrane surface area occurs alongside the increase in volume of the cell. If the metabolism synthesizes more lipid inside the cell, then the membrane surface area can increase proportionately faster than the cell volume, in which case cell division is possible. The three criteria can be satisfied if the reaction system is bistable, because different concentrations can exist inside and out while the rate constants of all the reactions are the same. If the reaction system is monostable, the criteria can only be satisfied if there is a reason why the rate constants are different inside and out (for example, the decay rates of biomolecules are faster outside, or the formation rates of biomolecules are slower outside). If this difference between inside and outside does not exist, a monostable reaction system cannot sustain cell growth and division. We show that a reaction system for template-directed RNA polymerization can satisfy the requirements for a metabolism, even if the small-molecule reactions that make the single nucleotides do not.

Keywords: metabolism; autocatalytic set; origin of life; osmotic pressure; cell division; lipid membrane; bistable reaction system; template-directed RNA synthesis

Citation: Higgs, P.G. When Is a Reaction Network a Metabolism? Criteria for Simple Metabolisms That Support Growth and Division of Protocells. *Life* **2021**, *11*, 966. https://doi.org/10.3390/life11090966

Academic Editors: Michele Fiore and Emiliano Altamura

Received: 19 August 2021
Accepted: 8 September 2021
Published: 14 September 2021

Publisher's Note: MDPI stays neutral with regard to jurisdictional claims in published maps and institutional affiliations.

Copyright: © 2021 by the author. Licensee MDPI, Basel, Switzerland. This article is an open access article distributed under the terms and conditions of the Creative Commons Attribution (CC BY) license (https://creativecommons.org/licenses/by/4.0/).

1. Introduction

One of the long-standing questions in the area of the origin of life is whether the origin of replication of information-carrying polymers occurred before or after the origin of metabolism [1–3]. On the replication-first side, there is considerable experimental progress with non-enzymatic RNA replication [4,5], with polymerase ribozymes [6,7], and with the encapsulation of RNAs inside artificial protocells [8–13]. On the metabolism-first side, there is very little experimental evidence that small-molecule reactions can support cell maintenance, growth and division in the absence of genetic polymers. Although computational models of small-molecule assemblies can show compositional inheritance without having genetic sequences [14], there are persistent doubts as to whether such systems could evolve [15]. Since the capacity for Darwinian evolution is usually considered to be a defining feature of life [16], this would preclude any small-molecule autocatalytic network from being defined as life. However, whatever scenario we favour for the origin of life, we must acknowledge that metabolism is a key part of the puzzle.

There is no generally agreed definition of metabolism that is appropriate for research work in the origin of life. For the present purposes, we will simply define a metabolism as a chemical reaction system that can sustain growth and division of a cell. In our view, progress in understanding the origin of metabolism has been hampered by lack of a clear understanding of what properties a reaction system must have in order to constitute a metabolism. The aim of this paper is therefore to define criteria that a reaction system must satisfy in order to be able to support growth and division of a protocell, and to give examples illustrating the difference between systems that do and do not satisfy these criteria.

Theoretical models of metabolism go back to chemotons [17,18], which emphasize that the metabolic system is inside a membrane whose growth must also be sustained by the metabolism. Another important conceptual step was to show that an autocatalytic set of molecules is likely to exist in models of random reaction networks [19]. This led to the model of reflexively autocatalytic and food-generated (RAF) sets [20–23], which has been widely used recently to describe metabolic networks. An RAF set is a set of molecules and reactions such that every molecule is either a food molecule present in the environment or a biomolecule formed by the reactions, and such that every reaction is catalyzed by a molecule in the set. The expectation is that this set of molecules can maintain itself, given a continued supply of the food molecules. However, in our opinion there are several problems with the RAF model that make it incomplete as a theory. The requirement that every reaction be catalyzed by another molecule seems too strong when we are dealing with simple metabolisms in the earliest forms of life. In modern organisms, almost every reaction of small-molecule metabolites is catalyzed by enzymes. However, prior to the existence of RNA and proteins, there were no enzymes, so we need a theory that deals with the reactions of small molecules without insisting that the reactions be catalyzed. A further issue with the RAF model is that most of the work with this theory deals only with the structure of the reaction network, and not with rates and concentrations. The criteria for metabolisms that we use in this paper cannot be assessed without knowledge of rates and concentrations.

Another type of model that has been developed recently treats metabolic networks as cyclic processes [24–30]. In this framework, catalysts are treated on an equal footing with other molecules in the system, and all the steps in a catalytic process are specified. For example, suppose the reaction $A + B \rightarrow AB$ is catalyzed by another molecule C. In the RAF framework, this reaction would be treated as a single reaction that is assumed to occur if C is present, but the mechanism by which C controls the reaction would not be specified. The influences of catalysts on reactions are drawn as dashed arrows in the diagrams that represent RAF sets [20–23]. The way these influences operate is left unspecified. In contrast, in the cyclic process framework, the mechanism by which the catalyst participates in the process is specified. For example, the formation of AB could occur via two steps that involve C: $C + A \rightarrow CA; CA + B \rightarrow C + AB$. This shows that the catalyst C is regenerated unchanged at the end, and it also shows that the mechanism requires formation of another molecule CA, whose existence would not be known if the reaction were written as a single step.

If C were an enzyme catalyst, this mechanism would resemble the Michaelis–Menten theory of enzyme kinetics. The enzyme–substrate complex in the Michaelis–Menten theory (analogous to CA in this example) is usually considered to be a temporary physical association of the substrate to the enzyme surface. However, if we are dealing with small-molecule reactions, then physical associations seem less likely, and the formation of CA is more likely a chemical reaction involving chemical bond formation. The reaction $C + A \rightarrow CA$ is a chemical reaction that should be treated on the same footing as the original reaction $A + B \rightarrow AB$. If we say that the formation of AB is catalyzed, it means that the direct formation reaction is slow or does not occur, but the steps involving C are faster. In this framework, it does not make sense to insist that every reaction is catalyzed, otherwise we would need two additional catalysts to catalyze two the steps involving C, and further catalysts to catalyze these catalysts, and so on ad infinitum. Note that there is more than one way in which C could catalyze the first reaction. It could react with B before A: $B + C \to BC; A + BC \to AB + C$. Or the process could involve three steps instead of two: $C + A \to CA; CA + B \to CAB; CAB \to C + AB$. The dynamics of the reactions and the concentrations of the molecules will depend on the mechanism by which C participates in the process.

In this paper, we consider examples of reaction networks inside a cellular compartment that are described in the cyclic process framework. Transport of some types of molecule into and out of the cell is possible by diffusion through a semipermeable membrane, but the membrane may be impermeable to other kinds of molecules. We will divide molecules involved into three types: food molecules, biomolecules and waste molecules. Food molecules are present in the external environment and are imported into to the cell. Biomolecules are formed inside the cell and they remain at low or zero concentration outside the cell. The presence of these biomolecules at a high concentration inside the cell is one factor that distinguishes the living state inside from the non-living state outside. Waste molecules are those produced inside the cell which do not play a constructive role in further reactions in the cell and are exported from the cell.

The state with low biomolecule concentration outside the cell must be stable to perturbations that add small amounts of these molecules to the outside. Although it is possible in principle to begin with zero concentration of biomolecules outside and allow biomolecules to be present only inside, there will always be some possibility of leakage of small amounts of biomolecules from the inside to the outside, either through a small rate of permeability through the membrane or by occasional bursting of a cell. When this occurs, the external concentration of biomolecules must return to its original low or zero concentration. If the non-living state were not stable to small perturbations of this kind, the external environment would shift spontaneously to the living state, and there would no longer be any difference between the inside and outside.

When the reaction system maintains a positive osmotic pressure inside the cell, i.e., the total solute concentration is higher inside than outside, the osmotic pressure can drive growth of the system if the boundary is flexible. For a system enclosed in a lipid membrane, growth of the membrane surface area is possible if there is a supply of lipid available outside the cell or if new lipid is synthesized inside the cell by the reaction system itself.

The reaction system cannot be in thermodynamic equilibrium. Inside the cell, there must be a steady state with continual turnover, consisting of import of food molecules, conversion of food to biomolecules, decay of biomolecules to waste, and export of waste. If the inside were in equilibrium, all reactions would be balanced by equal and opposite reactions, and there would be no net flow of material through the system. In order to drive flow through the system inside the cell, the system outside must also be maintained out of equilibrium, i.e., the food molecules must have a higher concentration and the waste molecules must have a lower concentration that they would do if all reactions were in equilibrium. This is possible if some of the reactions that convert food molecules to biomolecules cannot occur outside the cell, or occur very slowly. If they do not occur at all, then food molecules can remain out of equilibrium at high concentration with no flow of material though the system. If the reactions occur slowly, then there will be a steady-state flow through the external reaction system that must be maintained by an environmental recycling or replenishment process (supply of food and removal of waste). The rate of the turnover due to environmental recycling outside can be compared with rate of the turnover due to metabolism inside the cell. We expect the turnover rate inside the cell to be higher than outside, otherwise the outside is "more alive" than the inside. In what follows, we define the metabolic rate as the rate of mass transfer per unit volume from food molecules to biomolecules. The volume of the external environment will be much larger than that of the cell. The per unit volume consumption rate of the food molecules outside the cell must be small, otherwise a massive supply rate of food molecules would be required that could not be maintained by an environmental recycling process.

We summarize the above considerations by the following three criteria which must be satisfied if the reaction network is to be considered as a metabolism that can support growth and division of a cell.

1. The reaction system should maintain a state of high biomolecule concentration inside the cell and low biomolecule concentration outside, and these states should be stable at the same time. Small perturbations should not cause either the death of the internal living system or the spontaneous springing to life of the external non-living system.
2. The total concentration inside the cell should be higher than outside, so that a positive osmotic pressure is maintained in the cell. This can drive growth of the cell if the boundary is flexible and lipid molecules are available to form new membrane.
3. The metabolic rate of the reaction system can be defined as the rate of mass flow from food molecules to biomolecules. The metabolic rate inside the cell should be higher than outside the cell.

These requirements can be satisfied if the reaction system is bistable. In this case, the reactions have the same rate constants inside and outside the cell, but there are two different steady-state concentrations of the molecules. The requirements can also be satisfied if the reaction system is monostable and if some of the reactions have different rate constants inside and out. For example, the decay rate of biomolecules to waste might be high on the outside and low on the inside, or the formation rates of biomolecules might be high on the inside and low on the outside. In this case, the difference in concentrations between the inside and outside can be maintained, but it is necessary to propose some additional ad hoc reason why the rate constants inside and outside the cell are different. This additional difference is not necessary in the case where the reaction system is bistable. We now introduce several different cases of reaction networks and show the difference between cases which satisfy the requirements to be a metabolism and cases which do not.

2. Materials and Methods

We begin with a set of nine reactions involving six kinds of molecule. The molecules are labelled A_n (n = 1, 2, 3, 4 or 5) and W. Molecules A_1 and A_2 are designated as food molecules (equivalent to the food set in the RAF model). These are present at high concentration in the external medium as well as inside the cell. Molecules A_3, A_4 and A_5 are designated as biomolecules. They are present at a significant concentration inside the cell, and in low (or zero) concentration outside. Molecule W is a waste molecule formed by breakdown of the biomolecules. Molecule A_n has mass n units and standard free energy of formation $\Delta G^0_{form} = (n-1) \times 0.2RT$. Molecule W has mass 1 and standard free energy of formation $\Delta G^0_{form} = -RT$.

The nine reactions are shown in Table 1. Reactions 1 to 4 are formation reactions of molecules A_2 to A_5 by progressive addition of A_1. Reactions 5 and 6 make cyclic autocatalytic processes involving A_3, A_4 and A_5. Reactions 7 to 9 are breakdown reactions in which the catalysts decay to W. All reactions are reversible. The free energy change for the reaction, ΔG, (shown in Table 1) is the difference between the free energy of formation of the products and the reactants. The net reaction rate of reaction i is r_i (where C_i denotes the concentration of A_i). In each case, the downhill reaction has a rate constant u_i and the uphill reaction has a rate constant $u_i K$, with an equilibrium constant K that is less than 1 (except for reactions 5 and 6, which have K = 1).

Table 1. Reactions and reaction rates in the metabolic system model.

Reaction Number	Reaction	ΔG	Reaction Rate	Equilibrium Constant
1	$A_1 + A_1 \rightleftarrows A_2$	+0.2RT	$r_1 = u_1\left(KC_1^2 - C_2\right)$	$K = \exp(-0.2)$
2	$A_1 + A_2 \rightleftarrows A_3$	+0.2RT	$r_2 = u_2(KC_1C_2 - C_3)$	$K = \exp(-0.2)$
3	$A_1 + A_3 \rightleftarrows A_4$	+0.2RT	$r_3 = u_3(KC_1C_3 - C_4)$	$K = \exp(-0.2)$
4	$A_1 + A_4 \rightleftarrows A_5$	+0.2RT	$r_4 = u_4(KC_1C_4 - C_5)$	$K = \exp(-0.2)$
5	$A_2 + A_4 \rightleftarrows 2A_3$	0	$r_5 = u_5\left(C_2C_4 - C_3^2\right)$	
6	$A_3 + A_5 \rightleftarrows 2A_4$	0	$r_6 = u_6\left(C_3C_5 - C_4^2\right)$	
7	$A_3 \rightleftarrows 3W$	−3.4RT	$r_7 = u_7\left(C_3 - K_7W^3\right)$	$K_7 = \exp(-3.4)$
8	$A_4 \rightleftarrows 4W$	−4.6RT	$r_8 = u_8\left(C_4 - K_8W^4\right)$	$K_8 = \exp(-4.6)$
9	$A_5 \rightleftarrows 5W$	−5.8RT	$r_9 = u_9\left(C_5 - K_9W^5\right)$	$K_9 = \exp(-5.8)$

There are several autocatalytic processes in this reaction network. The reason for including autocatalytic processes is the expectation that these can create a bistable reaction system that can maintain a difference between the concentrations inside and outside. Summing reactions 3 and 5 together gives process P3 + 5, which is autocatalytic formation of A_3 from A_1 and A_2. This is shown in Figure 1a, and can be written $A_1 + A_2 + A_3 + A_4 \rightarrow 2A_3 + A_4$. If A_3 is already present, then this process can form more, even if the direct formation of A_3 without catalysis (reaction 2) does not occur. Similarly, summing reactions 4 and 6 together gives process P4 + 6, which is autocatalytic formation of A_4 from A_1 and A_3. This is shown in Figure 1b, and can be written $A_1 + A_3 + A_4 + A_5 \rightarrow 2A_4 + A_5$. Summing reactions 4, 5 and 6 together gives process P4 + 5 + 6, which is autocatalytic formation of A_3. This is shown in Figure 1c, and can be written $A_1 + A_2 + A_3 + 2A_4 + A_5 \rightarrow 2A_3 + 2A_4 + A_5$. Processes P4 + 5 and P4 + 5 + 6 work in absence of direct synthesis of A_3 and A_4 (reactions 2 and 3). Molecule A_5 is a participant in these cyclic processes; thus, it is automatically formed if A_3 and A_4 are formed.

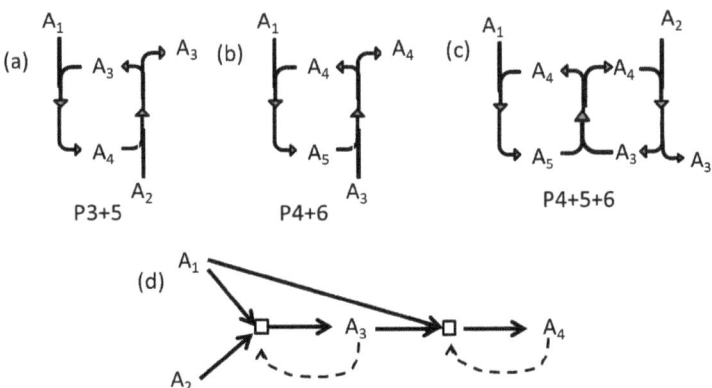

Figure 1. Autocatalytic processes can be formed by combining the reactions in Table 1, as described in the text. (a) Process P3 + 5 gives autocatalytic synthesis of A_3 from A_1 and A_2. (b) Process P4 + 6 gives autocatalytic synthesis of A_4 from A_1 and A_3. (c) Process P4 + 5 + 6 gives autocatalytic synthesis of A_3 from A_1 and A_2. (d) The reaction network drawn in the RAF manner, where squares indicate reactions and dashed arrows indicate catalytic effects.

Figure 1d shows the reaction network drawn in the RAF manner, where squares indicate reactions and dashed lines indicate catalytic effects. However, the mechanism of the reactions is not clear from Figure 1d, and the existence of A_5 is not apparent.

Autocatalytic processes become relevant when the corresponding direct synthesis reaction does not occur. Therefore, we consider the three cases shown in Table 2. Case 1 relies on P4 + 6 and P4 + 5 + 6 for formation of A_3 and A_4. The direct synthesis rate constants

u_2 and u_3 are 0. Case 2 relies on P3 + 5. Processes P4 + 6 and P4 + 5 + 6 cannot occur in Case 2 because u_6 = 0. We therefore need to include the direct synthesis of A_4 (u_3 = 1), but the direct synthesis of A_3 is still excluded (u_2 = 0). In Case 3, all the autocatalytic processes are eliminated ($u_5 = u_6 = 0$) and the direct synthesis reactions 1–4 are all included. In all three cases, the rate constants for the breakdown reactions 7–9 are equal to v, and we calculate the behaviour of the reaction systems as a function of the variable rate v. In each of the cases, we have set the rate of the reactions that sustain the metabolism to 1. Reactions must also occur sometimes that break down catalytic molecules. By setting the rates of these reactions to v we have a simple way of controlling the breakdown rates relative to the rates of the constructive reactions. Cases 1, 2 and 3 correspond to reaction systems that are autocatalytic and bistable, autocatalytic and monostable, and not autocatalytic, respectively. The results below demonstrate the differences between these cases. We show that Case 1 satisfies all the desired criteria for a metabolism, but some of these criteria are no longer satisfied in Cases 2 and 3.

Table 2. Values of the reaction rate constants in the different cases analyzed.

Rate Constant	Case 1	Case 2	Case 3
u_1	1	1	1
u_2	0	0	1
u_3	0	1	1
u_4	1	1	1
u_5	1	1	0
u_6	1	0	0
$u_7 = u_8 = u_9$	v	v	v

3. Results

3.1. Case 1

Firstly, we calculate the concentrations outside the cell. We assume that C_1 is maintained at C_1^{out} = 1M, and that reaction 1 can occur freely, so that C_2 is maintained in equilibrium with C_1; hence, $C_2^{out} = K(C_1^{out})^2$ = 0.819 M. W is maintained at a fixed low concentration W^{out} = 0.01 M. If W were in equilibrium with C_1 then its concentration would be $W^{eq} = C_1^{out} \exp(+1)$ = 2.718 M. Thus, by fixing W^{out} to be much less than W^{eq}, we ensure that the external environment is maintained out of equilibrium and that there is a driving force for the metabolic system inside the cell. The steady-state concentrations C_3, C_4 and C_5 are found by solution of the following differential equations using the fourth-order Runge-Kutta method until convergence is reached.

$$\frac{dC_3}{dt} = r_2 - r_3 + 2r_5 - r_6 - r_7 \quad (1)$$

$$\frac{dC_4}{dt} = r_3 - r_4 - r_5 + 2r_6 - r_8 \quad (2)$$

$$\frac{dC_5}{dt} = r_4 - r_6 - r_9 \quad (3)$$

The rates r_i are given in Table 1. As a measure of the metabolic rate, we use the rate of mass flow from food molecules (1 and 2) to biomolecules (3–5), which is

$$R_{met} = 3r_2 + r_3 + r_4 + 2r_5. \quad (4)$$

In the steady state, this rate is balanced by an equal rate of mass flow of biomolecules decaying to waste (Equation (5)):

$$R_{dec} = 3r_7 + 4r_8 + 5r_9. \quad (5)$$

In order to keep the concentrations of food molecules and waste fixed at the external concentrations, there must be a supply rate of food molecules and a removal rate of W by an environmental recycling or replenishment process. In the steady state, all four rates are equal (Equation (6)).

$$R_{supply} = R_{met} = R_{dec} = R_{remove} \qquad (6)$$

As this reaction system is bistable, the steady-state concentrations depend on the initial conditions. We consider two initial conditions, called low-concentration (LC) and high-concentration (HC) initial conditions. For the LC conditions, we start with $C_3 = C_4 = C_5 = 10^{-4}$ M. For the HC conditions, we set the concentrations to $C_n = \left(C_1^{out}\right)^n K^{n-1}$, which is the concentration that would occur if the molecules were in equilibrium with C_1 and there were no decay to W (i.e., if v were zero).

Figure 2 shows the steady-state concentrations that arise as a function of v, starting from the two initial conditions. There is a transition point at $v_1 = 0.016$. For $v < v_1$, there are two stable solutions—the HC solution with high concentrations of the three biomolecules and the LC solutions where the biomolecule concentrations are very small. Starting from HC initial conditions gives the HC solution, and starting from LC initial conditions gives the LC solution. For $v > v_1$, only the LC solution is stable, and this solution is reached starting from both initial conditions.

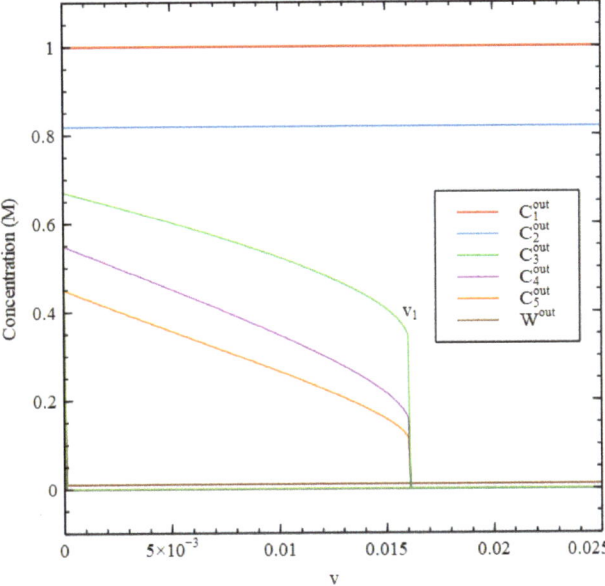

Figure 2. Steady-state concentrations C_n^{out} outside the cell as a function of the decay rate constant v for Case 1. For $v > v_1$ only the LC solution is stable. For $v < v_1$, both HC and LC solutions are stable.

We now consider concentrations inside the cell. Initially, we consider the cell to be a small compartment of fixed volume which does not grow. Later, we will consider cells with membranes that can grow and divide. For, the fixed-volume compartment, the cell boundary is permeable to A_1, A_2 and W, with inflow and outflow proportional to rate q, and it is impermeable to A_3, A_4 and A_5. Since A_1, A_2 and W have variable concentration inside the cell, we need three additional differential equations for these molecules.

$$\frac{dC_1}{dt} = q(C_1^{out} - C_1) - 2r_1 - r_2 - r_3 - r_4 \qquad (7)$$

$$\frac{dC_2}{dt} = q(C_2^{out} - C_2) + r_1 - r_2 - r_5 \tag{8}$$

$$\frac{dW}{dt} = q(W^{out} - W) + 3r_7 + 4r_8 + 5r_9 \tag{9}$$

The Equations (1)–(3) for C_3, C_4 and C_5 are the same as outside the cell (although the concentrations will be different). R_{met} and R_{dec} are calculated as before. Instead of the supply of food and removal of waste via environmental recycling, we now have diffusion of food into the cell and diffusion of waste out.

$$R_{in} = q(C_1^{out} + 2C_2^{out} - C_1 - 2C_2) \tag{10}$$

$$R_{out} = q(W - W^{out}) \tag{11}$$

In the steady state, the mass flow rate at all the steps is equal.

$$R_{in} = R_{met} = R_{dec} = R_{out} \tag{12}$$

Figure 3 compares steady-state concentrations inside and outside the cell. We first determine the external concentrations, starting from the LC initial conditions. Then, keeping the external concentrations fixed in their steady state, we determine the steady state inside the cell by numerical solution of the differential equations for the inside, starting from the HC initial conditions. We want to show that the HC solution inside the cell can be stable when the outside is in the LC state.

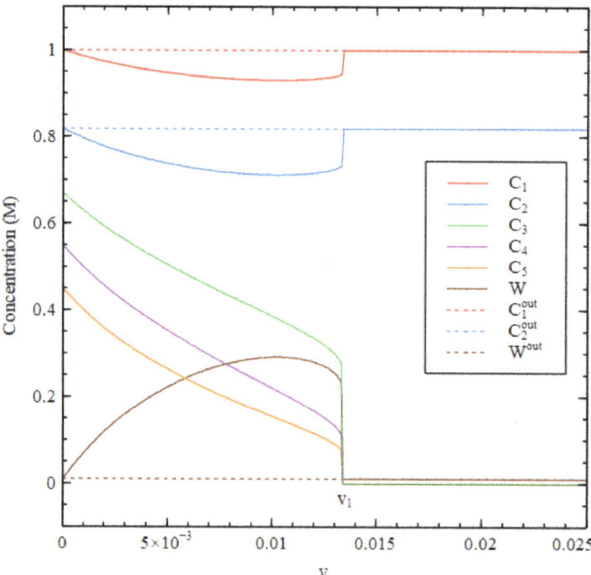

Figure 3. Steady-state concentrations inside (C_n) and outside (C_n^{out}) the cell as a function of decay rate v for Case 1.

There is an HC solution inside the cell that is stable up to a transition value v_1, which is slightly lower than the v_1 in Figure 2. It is apparent that the waste concentration W inside the cell is fairly high, whereas it is fixed at a low concentration outside. When the inside is in the HC state, C_1 and C_2 are lower inside than out because there has to be a net flow of food molecules inwards. The external concentrations C_3^{out}, C_4^{out} and C_5^{out} are very close to zero and are not shown in Figure 3.

It is useful to look at the total concentration, shown in Figure 4. For $v < v_1$, the total concentration is higher inside than out, hence there is a positive osmotic pressure inside the cell, as we require for Criterion 2. We will show in a later section of this paper that when this is the case, the osmotic pressure can drive growth and division of lipid vesicles. However, firstly, we want to establish under what conditions the reaction system can maintain a positive osmotic pressure in a compartment of fixed volume. For $v > v_1$, the total concentration inside collapses to the concentration outside, so the osmotic pressure is lost.

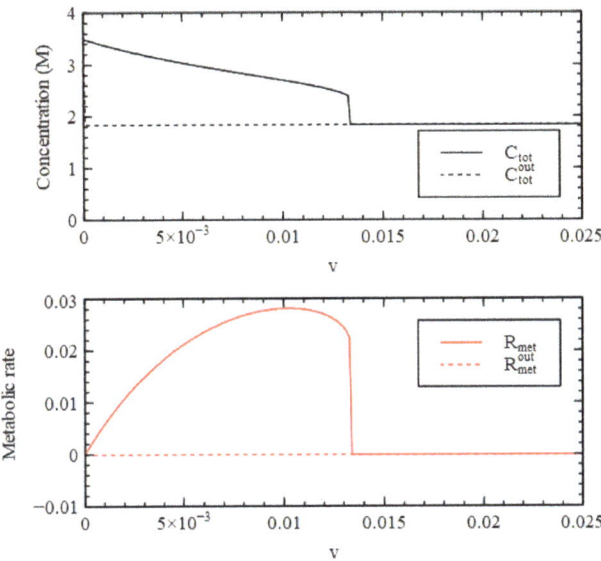

Figure 4. Total concentration inside (C_{tot}) and outside (C_{tot}^{out}) the cell and metabolic rate inside (R_{met}) and outside (R_{met}^{out}) as a function of decay rate v for Case 1.

A similar result is seen in Figure 4 for the metabolic rate R_{met} (Equation (4)). The external metabolic rate is very close to zero when the outside is in the LC state because the rate constants u_2 and u_3 are zero, and because the concentrations of the catalysts are very low, so the rates of the reactions in the catalytic processes are also very low. Thus, when $v < v_1$, R_{met} in the HC state inside the cell is much higher than R_{met}^{out} outside the cell, as we require for Criterion 3. Thus, all three criteria are satisfied for $v < v_1$, when the reaction system is bistable.

3.2. Case 2

We now consider Case 2, in which process P3 + 5 is permitted (by setting $u_3 = 1$), and where processes P4 + 6 and P4 + 5 + 6 are prevented (by setting $u_6 = 0$). This case still relies on an autocatalytic cycle to synthesize A_3, because the direct reaction 2 is not possible ($u_2 = 0$). Case 2 is monostable, therefore it does not matter whether we start with LC or HC initial conditions. Figure 5 shows the stable state concentrations inside and out. There is a transition point at $v_1 = 0.192$. For $v < v_1$, there is an HC solution with high C_3, C_4 and C_5, whereas for $v > v_1$, these concentrations fall very close to zero and the concentrations C_1, C_2 and W become equal to their external values.

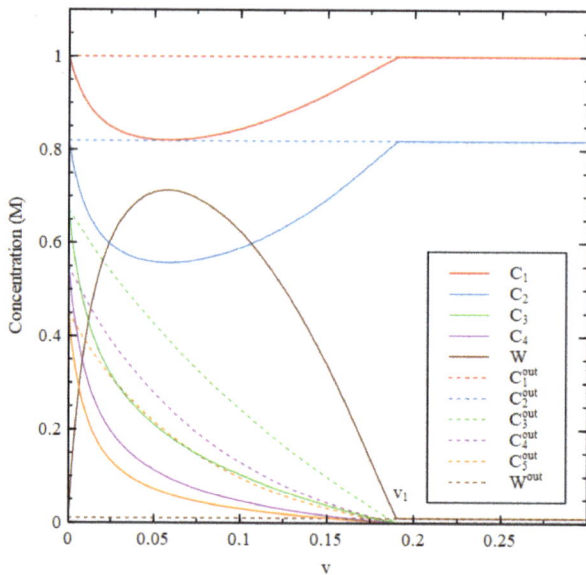

Figure 5. Steady-state concentrations inside (C_n) and outside (C_n^{out}) the cell as a function of decay rate v for Case 2.

In Case 2, biomolecules can survive at high concentration for much larger decay rates than in Case 1. Nevertheless, Case 2 is not a good model for a metabolism as it stands because all the molecules with the exception of W have a higher concentration outside than inside. Figure 6 shows that the total concentration inside is less than outside and the metabolic rate inside is less than outside. Thus, Case 2 does not satisfy Criteria 2 and 3. This shows that the existence of an autocatalytic process in the reaction system (P3 + 5 in this case) and the presence of biomolecules that are maintained at a high concentration by this process is not sufficient for the reaction system to satisfy the requirements for a metabolism.

The reason for the failure of Case 2 is that there is no stable LC state on the outside. We could in principle begin with with zero concentration of A_3 outside, and we could argue that A_3 would never form because $u_2 = 0$. However, in a real case, there would always be some possibility of leakage of A_3 from the inside, either via a small rate of permeability through the membrane or due to occasional bursting of cells. There is also a very small rate of formation of A_3 from W, via the reverse of reaction 7. Thus, we are bound to have a small initial concentration of these molecules by one means or another, and if the LC state is not stable, then the concentrations of A_3, A_4 and A_5 will rise to high values. This rules out Case 2 as a model for metabolism as it stands.

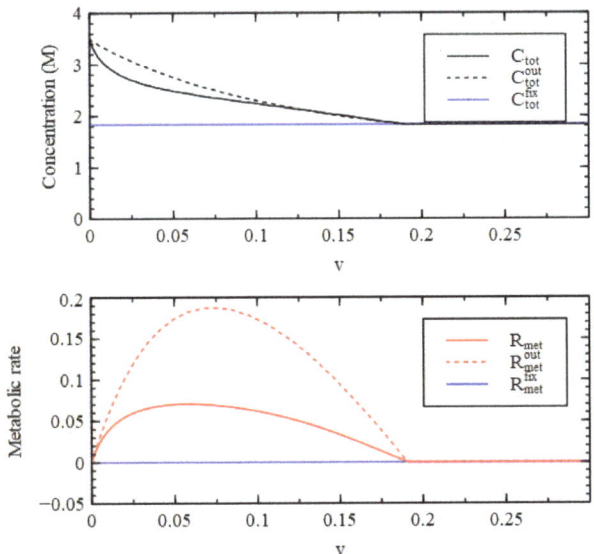

Figure 6. Total concentration and metabolic rate for Case 2, shown inside the cell (C_{tot}, R_{met}), outside the cell when v is the same inside and out (C_{tot}^{out}, R_{met}^{out}), and outside the cell when the external v is fixed at 0.3 (C_{tot}^{fix}, R_{met}^{fix}).

However, we can rescue Case 2 if we are prepared to make the assumption that the decay rate v of the biomolecules is larger on the outside than the inside. (We will discuss further below why this might be the case). We fixed the external decay rate to be $v_{out} = 0.3$, which is higher than v_1, and the decay rate v inside was allowed to vary as before. The concentrations inside the cell are the same as before, but C_3, C_4 and C_5 are now close to zero outside the cell because $v_{out} > v_1$. Figure 6 shows that when v_{out} is fixed at a high rate, the total external concentration C_{tot}^{fix} and the external metabolic rate R_{met}^{fix} are both lower than the internal values (C_{tot} and R_{met}).

Thus, with the additional assumption that the decay rate is higher outside than in, Case 2 now satisfies the required criteria for a metabolism. It is possible to think of reasons why this might be true in a real case. The decay reactions 7–9 might be driven by UV light, and being inside the compartment might provide some protection from this, so v would be larger outside than inside. Or possibly, the breakdown reactions could be driven by interaction with some additional molecule that is not permeable to the membrane, so the breakdown would occur outside but not inside. Although these reasons seem possible, they are somewhat ad hoc, in that an extra condition must be added that was not part of the definition of the reaction network. The actual degree of UV protection afforded by a lipid membrane may not be very large, and what impermeable molecule might be that would cause breakdown outside is unclear. Other reasons for the difference between inside and outside could be proposed if desired. However, if any of these mechanisms were true, we would have the rather odd situation that life would exist inside the cell because of the absence of an inhibiting effect that occurs outside the cell, rather than because of the presence of something constructive inside the cell. Life would be just waiting to pop up as soon as the inhibiting factor is removed! This does not fit with our intuition that the origin of life is an unlikely event that requires the coming together of favourable circumstances.

An alternative way to rescue Case 2, rather than making v larger on the outside, would be to make the reactions of the autocatalytic processes (u_3, u_4, u_5) small or zero on the outside. This would obviously make the model work, because the C_3, C_4 and C_5 would remain low outside, but again it is ad hoc. It is not obvious why the chemistry should be different outside the cell from inside. Piedrafita et al. [30] have studied two reaction networks which they call protometabolism 1 and 2—PM1 and PM2. PM2 is bistable, with similar properties to our Case 1, and PM1 is monostable, with similar properties to our Case 2. These authors did not look at the reactions outside the cell, and simply assumed that these reactions did not occur. This requires a reason why the reaction rates for the autocatalytic processes (analogous to our u_3, u_4, u_5) should be zero outside. This point was not considered by Piedrafita et al. [30]. We consider this issue to be important, because if the rate constants are equal inside and out and if the system has only one stable solution with high biomolecule concentration, the reaction system will not satisfy the criteria for a metabolism.

3.3. Case 3

So far, we have shown that in bistable reaction systems, there is a natural explanation of why the difference in total concentration and metabolic rate between the inside and the outside is maintained, whereas in monostable systems, these differences are only maintained if there is an extra ad hoc reason why reaction rate constants are different on the inside and outside. We now take this argument one step further, by showing that if we do allow rate constants to be different inside and out, then we no longer need there to be any autocatalytic processes at all.

In Case 3, all the direct reactions 1–4 occur with $u = 1$, and none of the autocatalytic processes occur, because $u_5 = u_6 = 0$. In this case, there is only one solution. The concentrations C_3, C_4 and C_5 reach high values both inside and outside, that decrease slowly with v, and there is no transition point. The total concentration is lower inside than out at small v, but becomes higher inside for high v, mostly because of the accumulation of W (Figure 7). However, the external metabolic rate is always much higher than the internal rate. It should be remembered that these metabolic rates are the rates of mass conversion from food molecules to biomolecules per unit volume. Given that the external environment has a much larger volume than the cell, this means huge amounts of food molecules must be supplied, which seems unreasonable.

Now we try to rescue Case 3 by fixing the external decay rate to be a higher rate than inside. If v_{out} is fixed at 1.0, then the total exterior concentration becomes C_{tot}^{fix}, shown in Figure 7, which is now lower than the interior concentration C_{tot} for the whole range of v. However, the external metabolic rate becomes R_{met}^{fix}, which is very high. So this still does not satisfy Criterion 3.

We can force Case 3 to satisfy all the criteria, if we simply say that the synthesis reactions of the biomolecules do not occur outside the cell ($u_2 = u_3 = u_4 = 0$). In this case only C_1, C_2 and W exist outside the cell, and R_{met}^{out} is fixed at zero, trivially, so Criterion 3 is also satisfied. Our point here is two-fold. Firstly, Case 3 seems to be too simple to be a good model for a metabolism, as it only works if we prohibit the direct synthesis reactions from occuring outside while allowing them to happen inside, which seems artificial. Secondly, however, if there really is a good reason why these reactions should be different inside and out, then this case shows that all three requirements for the existence of a metabolism can be satisfied with a reaction network having no autocatalytic processes at all.

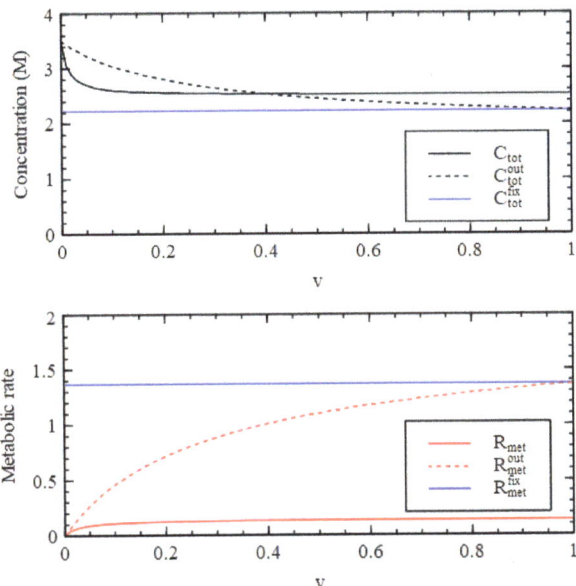

Figure 7. Total concentration and metabolic rate for Case 3, shown inside the cell (C_{tot}, R_{met}), outside the cell when v is the same inside and out (C_{tot}^{out}, R_{met}^{out}), and outside the cell when the external v is fixed at 1.0 (C_{tot}^{fix}, R_{met}^{fix}).

3.4. Cell Growth and Division

So far, we considered only compartments of fixed volume. In this section, we will consider lipid vesicle compartments in which growth and division are possible. The mechanism for growth and division used here is very similar to that introduced in references [28–30].

Experiments often use vesicles with a radius of approximately 50 nm [31,32]. We will therefore define a standard vesicle of this size, and measure sizes of growing vesicles relative to this standard size. The radius, suface area and volume of a standard vesicle are $r_{st} = 5 \times 10^{-8} m$, $S_{st} = 4\pi r_{st}^2$, and $V_{st} = 4\pi r_{st}^3/3$.

If a vesicle has a volume V, then its surface area, if it were a sphere, would be $S_{sph} = (36\pi V^2)^{1/3}$. However, the actual surface area of the vesicle also depends on the amount of lipid in the membrane. If L is the number of lipid molecules in the membrane, and a_L is the surface area per lipid in a relaxed membrane, then the natural surface area, if it is a relaxed state, is $S_L = a_L L/2$, where the factor of 2 occurs because of the two sides of the lipid bilayer. The actual surface area S is the larger of S_{sph} and S_L. When $S = S_L$ the vesicle is an irregular, non-spherical shape because its surface area is larger than the minimum area needed to enclose a sphere of volume V. When $S = S_{sph}$, the vesicle is spherical, and the membrane is under tension, because its surface area is larger than the natural surface area of a relaxed membrane with L lipid molecules.

The tension is proportional to the area strain ε, which is the relative increase in surface area above the natural area: $\varepsilon = (S/S_L - 1)$. This results in a pressure inside the vesicle proportional to ε/r. If the concentrations of solutes inside and outside the vesicle are different, then there is an osmotic pressure difference $\Delta\Pi = RT(C_{tot} - C_{tot}^{out})$. For simplicity, we have assumed that all the solutes behave like ideal gases, so the osmotic pressure is directly proportional to the concentration. The balance of osmotic pressure and membrane tension results in the following equation for the growth of the volume.

$$\frac{dV}{dt} = gV_{st}\frac{S}{S_{st}}\left(C_{tot} - C_{tot}^{out} - \frac{\tau r_{st}}{r}\left(\frac{S}{S_L} - 1\right)\right) \quad (13)$$

For convenience, we have incorporated the RT factor into the definition of constants τ and g. A factor of r_{st} has been added, so r is measured relative to the standard vesicle, and the constant τ for the membrane tension has units of concentration (molar). The volume increase is due to water passing through the membrane, therefor the rate scales in proportion to the surface area. A factor of V_{st} has been added to give a volume scale appropriate to the standard size vesicle. The time scale for growth is determined by g, which has units of time^{-1}molar^{-1}. In the examples below, we have $g = 1$, and $\tau = 10$.

If there is a fixed amount of lipid in the membrane, and a positive osmotic pressure, the volume will grow until the osmotic pressure is balanced by the membrane tension. There is a maximum area strain ε_c of approximately 0.1 that can be maintained by the membrane. A vesicle will burst and release some of its contents if ε exceeds ε_c [33,34]. Repeated cycles of swelling and bursting can be followed in experiments [35]. However, if the osmotic pressure is not too large, bursting will not occur. In this paper, we assume that the membrane is strong enough to maintain the pressure generated by the internal metabolic reactions without bursting.

If lipid is available in the solution outside and inside the membrane, then transfer of lipid molecules into and out of the membrane is possible, so the membrane surface area S_L can grow at the same time the volume is increasing. Lipids are only sparingly soluble, and above a critical concentration C^*, lipid bilayers will spontaneously form, until the concentration falls to C^*. The number of lipid molecules on each surface of the bilayer is S_L/a_L. The rate of motion of lipid molecules into and out of the membrane on the two sides is $\frac{S_L}{a_L}(k_{in}(C_L + C_L^{out}) - 2k_{out})$, where C_L and C_L^{out} are the lipid concentrations inside and outside the vesicle, and k_{in} and k_{out} are the rate constants for molecules entering and leaving the membrane. When the solution concentration is C^*, the membrane is in equilibrium with the solution. This means that $k_{in} = k_{out}/C^*$. We assume in what follows that the external lipid concentration is fixed at C^*, and the internal concentration will also tend to C^* if the membrane is relaxed. A relaxed vesicle will not increase in membrane area if the solution on both sides of the membrane has concentration C^*. However, it is observed experimentally [31] that transfer of lipids from relaxed vesicles to osmotically swollen vesicles occurs. This means that swollen vesicles gain lipids at solution concentrations where relaxed vesicles do not grow. To account for this, we assume that the rate of lipids entering the membrane increases when the membrane is under tension with input rate proportional to S/S_L, whereas the rate of lipids leaving the membrane remains unchanged when the membrane is under tension. The rate of change of the number of membrane lipids for a membrane under tension can therefore be written $\frac{S_L}{a_L}\left(\frac{k_{out}}{C^*}(C_L + C_L^{out})\frac{S}{S_L} - 2k_{out}\right)$. The rate of change of lipid surface area, assuming lipids flip between bilayers rapidly to keep equal numbers of molecules on both surfaces, is $a_L/2$ times the rate of exchange of molecules. Hence, finally:

$$\frac{dS_L}{dt} = k_{out}S_L\left(\frac{(C_L + C_L^{out})}{2C^*}\frac{S}{S_L} - 1\right) \quad (14)$$

The rate constant k_{out} is set to 1 in the examples below, which sets the timescale for membrane area growth comparable to that for vesicle volume growth.

Equation (14) says that the membrane area will grow either if the vesicle is under tension, or if the vesicle is relaxed but the internal concentration is higher than C^*. We therefore introduce further reactions into the system to allow for the possibility that the metabolism drives lipid synthesis inside the vesicle. A precursor molecule L_1 is assumed to be present in the external environment, and combination of L_1 with A_1 yields the lipid L. However, this direct reaction does not occur, but it is driven in two steps by a catalytic process involving A_3, as shown in Table 3. Thus, lipid synthesis will only occur when the metabolism is operating and maintaining a high concentration of A_3. We assume that the free energies of formation of the precursor L_1, the intermediate L_2 and the final lipid L are $0.4RT$, RT, and $0.8RT$, respectively. The free energies of the reactions are given in Table 3.

Table 3. Reactions for lipid synthesis.

Reaction Number	Reaction	ΔG	Reaction Rate	Equilibrium Constant
10	$L_1 + A_3 \rightleftharpoons L_2$	$+0.2RT$	$r_{10} = u_{10}(KL_1C_3 - L_2)$	$K = \exp(-0.2)$
11	$L_2 + A_1 \rightleftharpoons L + A_3$	$+0.2RT$	$r_{11} = u_{11}(KL_2C_1 - LC_3)$	$K = \exp(-0.2)$

As the volume is changing with time, it is easier to deal with the number of molecules in the vesicle, M_i, and then to determine the concentrations as $C_i = M_i/(N_A V)$, where V is in litres, C_i is in moles/litre and N_A is Avogadro's number $= 6.02 \times 10^{23}$. Equations (15)–(23) for the molecule numbers need to be solved concurrently with 13 and 14 for the volume and lipid surface area.

$$\frac{dM_1}{dt} = N_A V_{st} \frac{qS}{S_{st}} \left(C_1^{out} - \frac{M_1}{N_A V} \right) - N_A V (2r_1 + r_2 + r_3 + r_4 + r_{11}) \quad (15)$$

$$\frac{dM_2}{dt} = N_A V_{st} \frac{qS}{S_{st}} \left(C_2^{out} - \frac{M_2}{N_A V} \right) + N_A V (r_1 - r_2 - r_5) \quad (16)$$

$$\frac{dM_3}{dt} = N_A V (r_2 - r_3 + 2r_5 - r_6 - r_7 - r_{10} + r_{11}) \quad (17)$$

$$\frac{dM_4}{dt} = N_A V (r_3 - r_4 - r_5 + 2r_6 - r_8) \quad (18)$$

$$\frac{dM_5}{dt} = N_A V (r_4 - r_6 - r_9) \quad (19)$$

$$\frac{dM_W}{dt} = N_A V_{st} \frac{qS}{S_{st}} \left(W^{out} - \frac{M_W}{N_A V} \right) + N_A V (3r_7 + 4r_8 + 5r_9) \quad (20)$$

$$\frac{dM_{L1}}{dt} = N_A V_{st} \frac{qS}{S_{st}} \left(L_1^{out} - \frac{M_{L1}}{N_A V} \right) - N_A V r_{10} \quad (21)$$

$$\frac{dM_{L2}}{dt} = N_A V (r_{10} - r_{11}) \quad (22)$$

$$\frac{dM_L}{dt} = \frac{k_{out} S_L}{a_L} \left(\frac{M_L}{N_A V C^*} - 1 \right) + N_A V r_{11} \quad (23)$$

We have assumed that the membrane is permeable to the precursor L_1, but not to the intermediate L_2. The membrane is not permeable to the lipid L itself, but L can cross between the two sides by entering and leaving the membrane. In calculating the permeability terms above, we assumed that the constant q in Equations (7)–(9) for the fixed-volume compartment applies to a compartment with volume V_{st} and surface area S_{st}. If the rate of concentration change in the standard vesicle is q, the rate of change of the number of molecules is $N_A V_{st} q$. The rate of change in the vesicle of variable size increases in proportion to S, which gives a rate constant $N_A V_{st} \frac{qS}{S_{st}}$ in the equations above.

We suppose that the critical lipid concentration is $C^* = 4 \times 10^{-3}$ M, which is appropriate for fatty acids, and the external concentration is fixed at $C_L^{out} = C^*$. The external concentrations of the precursor and intermediate are $C_{L1}^{out} = 0.1$ M, and $C_{L2}^{out} = 0.0$ M. Figure 8 shows results using Case 1 reaction rates, assuming that L and L_1 are present, but the rate constants for reaction 10 and 11 are $u_L = 0$, so there is no lipid synthesis by the metabolism. The inside begins with HC initial conditions while the outside is in the LC state. Therefore, there is a positive osmotic pressure that drives increase in volume. The membrane is then under tension ($S > S_L$), so S_L increases due to addition of lipid molecules from the outside. Continued growth is seen at $v = 0.003$, but at $v = 0.005$ the internal concentrations collapse to the external concentration after a time, so there is no further growth.

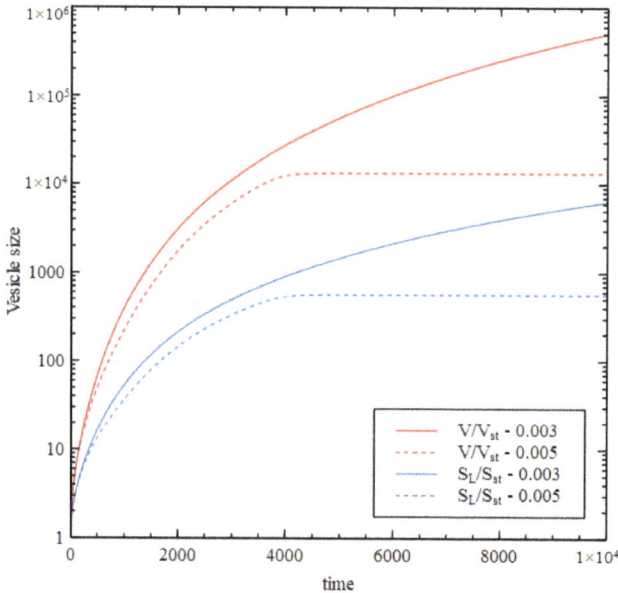

Figure 8. Ratio of the vesicle volume to standard vesicle volume (V/V_{st}) and ratio of lipid surface area to standard vesicle surface (S_L/S_{st}) as a function of time for Case 1 reaction network with no internal lipid synthesis ($u_L = 0$), for two values of the decay rate, $v = 0.003$ and 0.005.

Figure 9 shows the concentrations inside the cell at the end of the simulation run (time 10,000). There is a transition point at $v_1 \approx 0.004$. For $v < v_1$ the osmotic pressure difference is maintained by the metabolism, and the volume and lipid surface area grow indefinitely. If $v > v_1$ the osmotic pressure difference is not maintained and the vesicle stops growing. For $v < v_1$, all three criteria for a metabolism are satisfied while the vesicle is growing continuously.

When lipid synthesis occurs inside the vesicle, we need to consider the possibility of vesicle division. As has been shown previously [27–30], a spherical vesicle cannot divide without losing contents because the membrane area needed to enclose a sphere of volume V is less than the membrane area needed to enclose two spheres of volume $\frac{V}{2}$. Thus, if a vesicle is growing with a membrane under tension, as in Figure 8, then it cannot divide. For division to occur we need the membrane to be relaxed, i.e., the vesicle must be non-spherical, with a lipid surface area S_L greater than the minimal surface area required for the sphere of the current volume $S_{sph}(V)$. After division, the minimal surface area for the two smaller spheres is $2S_{sph}\left(\frac{V}{2}\right) = 2^{\frac{1}{3}} S_{sph}(V)$ Sph ivision, the minimal surface area for the two smaller spheres isrevious e area of a sphere of volume. Thus, division of the original vesicle can occur only if $S_L > 2^{1/3} S_{sph}(V)$.

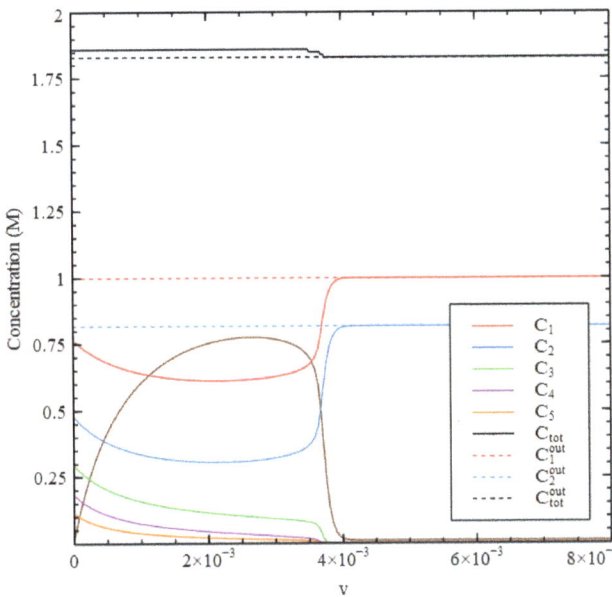

Figure 9. Concentrations inside the cell (C_n) after time 10^4 compared with stationary external concentrations (C_n^{out}), shown as a function of decay rate v for Case 1 reaction network with no internal lipid synthesis.

To achieve vesicle division at the same time as the osmotic pressure is driving volume increase, we need the surface area to grow proportionally faster than the volume. This means that the lipid concentration must be higher than C^*, so that there is a net addition of molecules to the relaxed membrane. The exterior concentration can be temporarily higher than C^* if there is a sudden addition of lipid to the external medium. It is known that sudden addition of lipid micelles to vesicles causes vesicle division [32,36,37]. On the other hand, if the external concentration remains at C^*, the internal concentration can become higher than C^* if there is lipid synthesis inside the vesicle.

We now consider the case of internal lipid synthesis driven by the metabolism. We consider Case 1 with $v = 0.003$, which is able to support continued vesicle growth when there is no internal lipid synthesis (as in Figure 8). We then add the lipid synthesis reactions 10 and 11, with rate constant u_L. We begin with one vesicle of standard size with internal concentrations equal to the HC initial conditions. Whenever the lipid surface area becomes large enough to satisfy the division condition, $S_L > 2^{1/3} S_{sph}(V)$, the vesicle is divided into two half-sized vesicles, and we continue to follow the growth of one of these. Figure 10 shows the volume as a function of time for five values of u_L. For $u_L = 0$, we have continued growth without division, as before. For $u_L > 0$, we have repeated growth and division. The time required for division and the size at which division occurs decrease with increasing lipid synthesis rate. Interestingly, it is possible for lipid synthesis to be too fast. For the highest u_L considered, the vesicle gets smaller on each division, because the division condition is reached before the size has doubled. Smaller vesicles have a larger surface area to volume ratio, so when the vesicle gets smaller, the permeability terms become larger relative to the internal reaction terms, and the difference between the internal and external concentrations cannot be maintained. Hence, growth and division stop after a few cycles.

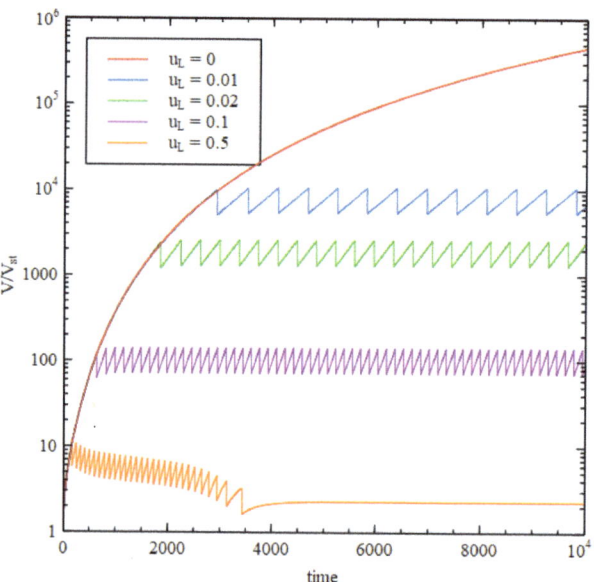

Figure 10. Case 1 reaction system in vesicle compartments with internal synthesis of lipid. Volume reltive to standard volume (V/V_{st}) is shown as a function of time for five different values of the lipid synthesis rate u_L. The decay rate is $v = 0.003$ in all cases.

In the case where continued growth occurs without division, as the growth is driven by autocatalytic reactions, we might expect that volume would grow exponentially with time. However, this is not so, because the rate of growth is limited by the rate of supply of food molecules. The vesicle settles into a state of steady growth where the internal concentrations are approximately constant in time and there is an approximately constant osmotic pressure. The membrane is only slightly swollen, and the tension term in Equation (13) is small in comparison to the osmotic pressure. In this case the rate of volume increase is roughly proportional to the surface area. The rate of import of food molecules is also proportional to surface area, as is required if the internal concentration is constant. We thus have $\frac{dV}{dt} \sim V^{2/3}$, hence V increases in proportion to t^3, not exponentially with time. However, in the case where repeated cell division occurs, the time required for each division is constant. The number of vesicles doubles at each division; hence, the number of vesicles increases exponentially with time. The time per division is only a constant when there is balanced growth of the membrane and the vesicle contents, i.e., the time for doubling of the vesicle volume is equal to the time for doubling of the membrane area (more details in [28]). These doubling times are dependent on vesicle size, and it can be seen that the vesicle naturally tends to a size when they are equal. In this self-reproducing state, each subsequent generation has the same size and composition as the previous one.

In this section, we have considered the Case 1 reaction network coupled to lipid synthesis and shown that it can drive vesicle growth and division. Any of the other models could also be coupled to lipid synthesis in the same way, so that in all the cases where the criteria for a metabolism are satisfied in a fixed-volume compartment, the addition of internal lipid synthesis will lead to growth and division in a lipid vesicle.

3.5. RNA Synthesis

The assumption of metabolism-first theories is that it is possible to have a self-sustaining small-molecule metabolic system without the presence of information-carrying polymers such as RNA. The previous examples have all focused on this case, showing that it is possible in theory, provided the reactions system satisfies the three criteria. If a small-molecule metabolism did get going in a protocell, it could later begin to synthesize nucleotides by some process that is coupled to the metabolism (in the way that the lipid synthesis was coupled to metabolism in the previous section). Polymerization of nucleotides could then arise in a protocell that already had an established metabolism.

However, there is still little experimental evidence for a self-sustaining small-molecule metabolism, so it is still possible to argue for replication-first. In this section, we make the point that template-directed synthesis of RNA is inherently autocatalytic because RNA strands are required to synthesize more strands. This means that the RNA polymerization process itself can produce a reaction system that satisfies the criteria for a metabolism. RNA World theories obviously require a means of synthesis of nucleotides before polymerization of nucleotides into RNA strands can occur. However, there is no requirement that the reaction pathways that generate nucleotides should be autocatalytic or that they should satisfy the criteria for a metabolism. This section will show a clear example where nucleotides are synthesized by a direct reaction pathway that is not autocatalytic and which does not by itself satisfy the metabolism criteria. Polymerization then occurs autocatalytically, and the metabolism requirements are satisfied due to the polymerization reactions, not the small-molecule reactions.

In this section, we identify molecule A_5 as a nucleotide, and we call it N hereafter. Reactions 1–4 represent steps in the pathway synthesizing nucleotides. We assume that $u_i = 1$ for all these direct reactions, and that the small-molecule autocatalytic processes do not occur (as in Case 3). We already showed above that Case 3 does not satisfy the metabolism criteria, unless we set u_i to be zero for the formation reactions outside the cell. Here, on the contrary, reactions 1–4 occur freely both inside and out. So the nucleotide-synthesis pathway does not constitute a metabolism by itself, and cannot support cell growth and division. We then suppose that nucleotides can polymerize by a reaction $N \to P$, where P represents a nucleotide that is part of an RNA polymer. Polymerization will generate many different sequences of many different lengths, and incorporation of all these things is beyond the scope of the present model. However, the simple $N \to P$ reaction captures the essence of the process. The polymerization rate is

$$r_{pol} = \left(s_P + r_P P + k_P P^2\right) N \left(1 - \frac{P}{P^*}\right), \qquad (24)$$

where s_P, r_P and k_P are rates of spontaneous polymerization (independent of current strand concentration), non-enzymatic template-directed synthesis (proportional to template concentration P), and ribozyme-catalyzed synthesis (proportional to both template and catalyst concentration—hence P^2). All three synthesis processes are proportional to nucleotide concentration N, and they are limited by some saturation of resources or space when P reaches a maximum concentration P^*. Inclusion of P^* is necessary to prevent a runaway increase in polymer concentration. Cleavage reactions convert polymeric nucleotides back to single nucleotides $P \to X$. We assume that the nucleotide X produced by cleavage is not the same as the nucleotide N prior to polymerization. For example, N could be an activated nucleotide, whereas X is not. We assume X cannot repolymerize, hence it is a waste molecule. The cleavage reaction occurs at rate aP. We ignore the other waste molecule W, because it is not important for the point we are making in this section. The decay reactions 7–9 do not occur in this example. Treating X as a waste molecule is the least favourable case for establishing polymerization, in comparison to cases where cleavage directly reforms N, or where recycling of X back to N is possible.

When k_P is high and s_P and r_P are both low, RNA polymerization is bistable. In earlier papers [38–41] we have shown that this bistability results in two states that we called living (high polymer concentration with RNA synthesis dominated by the catalyzed process), and non-living (low polymer concentration with RNA synthesis dominated by the spontaneous process). These earlier papers focused on RNA replication without considering metabolism. Here, we make a link to metabolism.

Figures 11 and 12 consider a reaction network for RNA synthesis consisting of reactions 1–4 plus polymerization and cleavage reactions. We set $s_P = 0.001$, $r_P = 0$, $k_P = 1$, and $P^* = 2$, so that the reaction system is bistable and RNA synthesis is predominantly autocatalytic. We consider a fixed-volume compartment. Molecules A_1 and A_2 are maintained at constant concentration outside, as before. Molecules A_3, A_4 and N are all formed outside as well as inside, and are considered as food molecules. The biomolecule is P. The waste molecule is X, which is fixed at a low concentration $X^{out} = 0.01$ M outside. Figure 11 shows the stationary concentrations as a function of cleavage rate. Outside the cell there is an HC and an LC solution, as with the Case 1 small-molecule network, but 'low' and 'high' refer to the concentration of P in this example because P is the molecule whose presence denotes the living state. There are two transition values in a. For $a > a_1$, only the LC concentration is stable. For $a_2 < a < a_1$, both states are stable and the state reached depends on the initial conditions. For $a < a_2$, only the HC state is stable. We start the outside in LC initial conditions and the inside in HC initial conditions. Thus, for $a_2 < a < a_1$, the inside goes to the HC state while outside goes to the LC state. For $a > a_1$, both inside and outside go to the LC state; for $a < a_2$, both inside and outside go to the HC state. Note that C_2, C_3, C_4 are also present in this example, but for clarity, they are not shown in Figure 11.

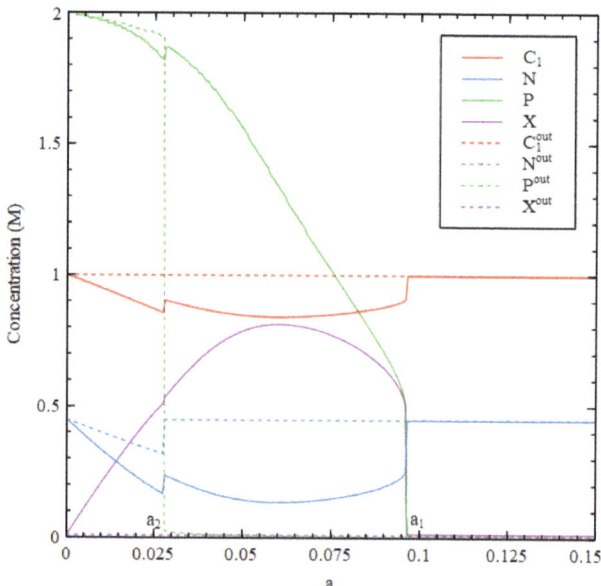

Figure 11. Stationary values of concentrations inside the cell (C_n, N, P, X) and outside the cell (C_n^{out}, N^{out}, P^{out}, X^{out}) in the RNA synthesis reaction nework as a function of the cleavage rate a.

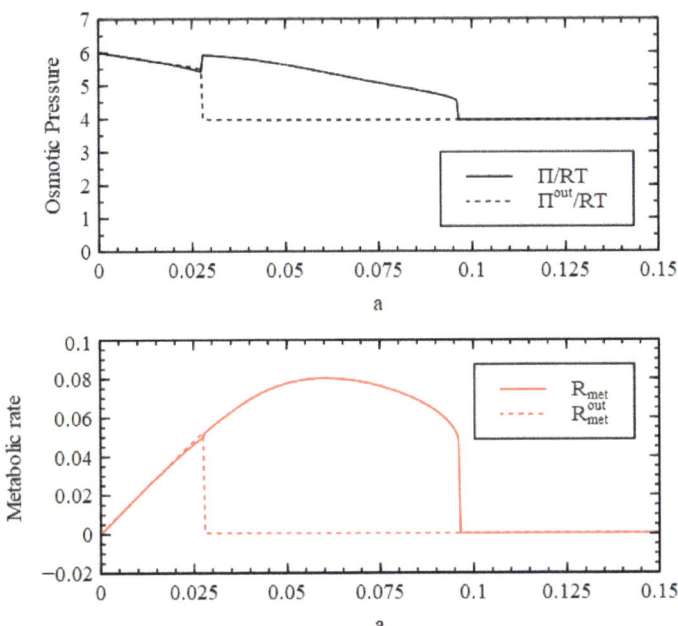

Figure 12. Osmotic pressure and metabolic rate inside $\left(\frac{\Pi}{RT}, R_{met}\right)$ and outside the cell $\left(\frac{\Pi^{out}}{RT}, R_{met}^{out}\right)$ in the RNA synthesis reaction nework as a function of the cleavage rate a.

The metabolic rate is the polymerization rate in this case $R_{met} = r_{pol}$. We also assume that the single nucleotides are ionized, so they make a contribution to the osmotic pressure which is twice the number of molecules. Under the ideal gas approximation, a polymer contributes only once per chain rather than once per monomer, so the contribution per polymeric nucleotide P is $1 + 1/l$ (1 for the ion and $1/l$ for the polymer). We set the polymer length to be $l = 50$. The resulting osmotic pressure is

$$\Pi/RT = C_1 + C_2 + C_3 + C_4 + 2(N + X) + (1 + 1/l)P. \tag{25}$$

There are theories giving the osmotic pressure of nucleic acids that are much more acurate than the ideal gas theory [42], but these are more complicated than we require here. The values of these osmotic coefficients do not affect the conclusions of this paper. The main point is that, even though the osmotic coefficient for P is less than that for N, since P is impermeable, the conversion of N to P creates a positive osmotic pressure difference that can drive cell growth. Figure 12 shows that both the osmotic pressure and the metabolic rate are higher inside than out when $a_2 < a < a_1$. Thus, all three criteria for a metabolism are satisfied by the autocatalytic process of RNA polymerization, even though they are not satisfied by the small-molecule reactions that generate the nucleotides. The origin of metabolism occurs only because of the origin of replication in this case.

4. Discussion

The three criteria for a metabolism that we proposed here emphasize the maintenance of a difference between the concentration of molecules inside and outside the cell. This difference has been largely ignored previously in both the RAF framework [20–23], which does not consider concentrations and reaction rates, and in the cyclic process framework [24–30], because the conditions outside the cell were not considered. The reaction systems studied above have similar properties to networks PM1 and PM2 studied by

Piedrafita et al. [30], although the specifications of the reaction networks are different. We now have two examples of bistable networks (PM2 and Case 1) and two examples of monostable networks (PM1 and Case 2). The theory given in this paper makes it clear why the difference between bistable and monostable networks is important. We are assuming that food molecules are present outside and inside the cell, and we are also assuming that the reactions that synthesize biomolecules from food molecules are possible inside the cell. Therefore, it is essential to ask why these same reactions do not occur outside as well. The most natural explanation is that the system is bistable. In that case, there is a stable solution where the biomolecules remain at low concentration outside at the same time as they are at high concentration inside. If the reaction system is monostable, then we need the synthesis of biomolecules to be slow (or prevented) outside the cell or we need the breakdown of biomolecules to be faster outside. While it is possible that these additional differences between inside and outside could exist, we see no compelling reason why they should always exist; hence, they seem like extra conditions that need to be added on *ad hoc*.

One interesting possibility that we have not considered is that physical interactions between molecules could produce some kind of clustering or phase separation leading to the formation of membraneless compartments [43]. We can envisage some kind of polymer with weak attractive interactions that would separate into moderate sized domains without being complete insoluble. If the small-molecule metabolites are attracted to the phase-separated polymer, then there would naturally be high concentrations and rapid reactions in the high-density regions and low concentrations in the dilute regions. Growth and division would then also require the synthesis of the polymer, so this case is not necessarily simpler than the cases of encapsulation in vesicles that we have already considered. Another similar possibility in the vesicle case is that the small-molecule metabolites are associated with the surface of the lipid membranes. This would also lead to a increased concentration of metabolites inside the vesicles (or close to the membranes on both sides). Modifications to lipid molecules can affect their ability to assemble into membranes, so if lipid molecules are involved in metabolic network, then the reactions in a compartment can be strongly coupled to formation, growth and division of the compartment (see [44] for examples).

Bistability has also come up in our previous papers related to the origin of RNA replication [38–41] in which RNA strands can be synthesized by random polymerization, by non-enzymatic template-directed replication, and by ribozyme-catalyzed replication. When the ribozyme-catalyzed rate is high, there is a bistable system, with an LC and an HC state which we referred to as living and non-living states. We pointed out that the origin of life must occur in a region of parameter space where both states are stable. Obviously life is not possible for rate parameters where only the non-living state is stable. However, the range where only the living state is stable also does not make sense because in that case life would spring up continuously and repeatedly. If both states are stable, then it is possible for life to originate by a chance event that brings together a high concentration of the right molecules in one place. In our previous work [38–40] we looked at how long it would take for a stochastic event to cause a compartment in which random RNAs were being synthesized by spontaneous polymerization to jump from the non-living state to the living state, in which ribozyme-catalyzed replication is predominant. Our point was that this jump can occur in a small system with finite volume but not in a well-mixed system with infinite volume. Thus, if the external environment begins in the LC non-living state, then it will remain in that state because concentration fluctuations will be small, whereas in a compartment of small volume, it is possible for a rare stochastic event to initiate a switch to the HC living state. This kind of stochastic transition to the living state could occur with the bistable Case 1 reaction network and with the RNA synthesis network discussed above. A similar effect has also been seen in [26].

There are a lot of similarities between the behaviour of the reaction network in Case 1, and the system for RNA synthesis. At first it appears that there are two transitions in the RNA synthesis model (Figure 11) and only one transition in Case 1 (Figure 3). However, the two transitions in the RNA case occur because we included a small non-zero rate of spontaneous polymerization ($s_P = 0.001$) as well as a high rate of RNA-catalyzed RNA synthesis ($k_P = 1$). If the spontaneous rate is set to zero, the transition at a_2 disappears, and the reaction system is bistable for all $a < a_1$. Similarly, in Case 1, we found that there was only one transition at v_1 when the direct synthesis rates, u_2 and u_3 were zero. However, if we set these rates to a small non-zero rate, then we find that a second transition appears at a decay rate $v_2 < v_1$. Case 1 is then bistable for $v_2 < v < v_1$, while for $v < v_2$ only the HC state is stable. This is the same behaviour seen in the RNA model. Another variation of the RNA model not shown here is to allow the linear catalysis term in the polymerization rate $r_P P$ to be high, but the quadratic catalysis term $k_P P^2$ to be zero, which would be true for the simplest descriptions of non-enzymatic template-directed replication in the absence of polymerase ribozymes. In this case the RNA network is autocatalytic but monostable, like Case 2. The reaction network can satisfy the criteria for a metabolism if there is an additional reason why the rate r_P is high inside and low outside, or why the cleavage rate a is low inside and high outside.

The analysis of Cases 1, 2 and 3 given above reveals the peculiarity that the biomolecules A_3, A_4 and A_5 can most easily sustain a metabolism only when their direct synthesis is not possible, as in Case 1. It is only when reactions 2 and 3 do not occur, that the autocatalytic processes involving reactions 4, 5 and 6 become important, and it is these processes that allow the difference between inside and outside the cell to be maintained. This is possibly a reason why small molecules alone cannot easily sustain a metabolism, and why clear experimental examples of small-molecule metabolisms have not been found. It would be necessary to find molecules for which the most obvious pathways of synthesis are blocked for some reason, but where a more complicated autocatalytic synthesis route exists.

The scenario that we have proposed for the RNA synthesis network in Section 3.4 gets us out of this difficulty. For the RNA World to originate, we need a plentiful supply of single nucleotides. The challenge to organic chemists is to find a route for nucleotide synthesis that is simple and direct and prebiotically plausible. This is already quite difficult, although there has been significant progress [45–48]. However, chemists do not need to solve a problem that would be much more difficult still, namely to synthesize nucleotides via a small-molecule reaction network that is autocatalytic and satisfies the requirements of a metabolism. As we have shown, since template-directed polymerization of RNA (or some similar polymer) is inherently autocatalytic, if the polymerization process gets going inside a protocell, this can sustain the metabolism without requiring prior existence of a small-molecule metabolism. In this scenario, the metabolism would originate only with the establishment of RNA replication, and this could naturally be coupled to growth and division of the cell. In parallel with this suggestion, we note that experimental studies aimed at creating artificial cells usually focus on polymerization of RNA within vesicles [8–13], assuming that nucleotides are supplied from the outside. They do not try to develop a small-molecule autocatalytic process for nucleotide synthesis.

The discussion in the previous paragraph is similar to that which arises when considering the origin of homochirality in biomolecules [49]. As nucleic acids and proteins are built from homochiral monomers, and as the synthesis of these monomers and polymers is catalyzed by the homochiral polymers, it is clear that a homochiral polymer system can maintain itself in the homochiral state. It is also possible that a small-molecule system could sustain a homochiral state via asymmetric autocatalysis, but less clear whether this actually occurred in practice prior to the origin of biopolymers. We therefore considered scenarios in which homochirality evolves at the same time as the origin of biopolymer replication, and compared these to scenarios in which the monomers were homochiral prior to the origin of replication [49]. We have since shown that in a model for templating of nucleotides, a symmetry-breaking phase transition can lead to the formation of homochiral RNAs from a racemic mixture of single nucleotides [50]. Thus, homochirality and a self-sustaining metabolism can both arise with the origin of template-directed RNA polymerization, even if neither of these exists at the small-molecule level.

Another general question in models for the origin of life is whether life should begin with a simple autocatalytic process where one kind of molecule makes more of the same thing, or whether it should begin with a collective process in which many kinds of molecules together catalyze the formation of the whole set. If we are discussing small-molecule reaction networks, then there seems to be no reason why the network should not be quite complex. The RAF theories show that autocatalytic sets involving large numbers of molecules are possible. Whether this would arise in practice depends on the details of the real chemical reaction rates, and it is difficult to make concrete predictions from theory. On the other hand, if we are dealing with genetic polymers, then observations of how sequence replication works in modern organisms would suggest that we need something simple first. In modern cells, we observe a simple mechanism for RNA and DNA synthesis that works with all sequences. Similarly, there is a single mechanism for protein synthesis that works with all sequences. We do not see collective mechanisms for replication involving many different mutually-dependent sequences, each of which synthesizes only a small number of other sequences. In our opinion, for the onset of replication, we need a general mechanism that works with all sequences, rather than a complex network of specific reactions that are sequence-specific.

It seems a long way from the simple reaction networks considered in this paper to the complex networks of metabolites of modern bacteria. The origin of metabolism needs to be linked to the possible chemistry in the prebiotic environment [51–53], and we also need to understand how to relate theoretical models of metabolism to the networks in modern bacteria [54,55]. However, it seems odd to use the RAF framework, in which every reaction must be catalyzed, to analyze a reaction system that might have existed prior to the origin of catalysts. In the analysis of bacterial metabolism [54,55], enzyme catalysts were excluded, and when the reactions involved cofactors, the cofactors were treated as the catalysts. For reactions with no cofactors, the authors needed to introduce an imaginary molecule called 'protein' to catalyze the reaction, because the existence of a catalyst for every reaction is a requirement of the RAF framework. This seems to be forcing the data to fit the RAF model when it does not naturally do so. We note that cofactors in modern biochemistry participate in reactions by being changed from one form to another (e.g., Coenzyme A gains and loses an acetyl group, or ADP/ATP gains and loses a phosphate), as occurs in the cyclic process framework for reaction dynamics. The cyclic process framework may be a better starting point for analysis of metabolic networks in early organisms. In any case, we would expect that complex networks in which every reaction is controlled by a specific catalyst may have evolved long after the origin of life, and only after the origin of RNA replication and protein synthesis. If some kind of metabolism did evolve early, then it must have been a simple one. The criteria proposed here define what is necessary for a reaction network to support growth and division of a protocell. Satisfying these criteria does not require that every reaction should be catalyzed.

Funding: This research was supported by the Natural Sciences and Engineering Research Council of Canada grant number 2017-05911.

Institutional Review Board Statement: Not applicable.

Informed Consent Statement: Not applicable.

Acknowledgments: We thank Kepa Ruiz-Mirazo and Gabriel Piedrafita for useful discussions.

Conflicts of Interest: The author declares no conflict of interest. The funders had no role in the design of the study; in the collection, analyses, or interpretation of data; in the writing of the manuscript, or in the decision to publish the results.

References

1. Anet, F.A.L. The place of metabolism in the origin of life. *Curr. Opin. Chem. Biol.* **2004**, *8*, 654–659. [CrossRef] [PubMed]
2. Pross, A. Causation and the origin of life. Metabolism or replication first? *Origins Life Evol. Biosp.* **2004**, *34*, 307–321. [CrossRef] [PubMed]

3. Shapiro, R. Small-molecule interactions were central to the origin of life. *Q. Rev. Biol.* **2006**, *81*, 105–125. [CrossRef] [PubMed]
4. Leu, K.; Kervio, E.; Obermeyer, B.; Turk-MacCleod, R.M.; Yuan, C.; Luevano, J.M., Jr.; Chen, E.; Gerland, U.; Richert, C.; Chen, I.A. Cascade of reduced speed and accuracy after errors in enzyme-free copying of nucleic acid sequences. *J. Am. Chem. Soc.* **2013**, *135*, 354–366. [PubMed]
5. Walton, T.; Pazienza, L.; Szostak, J.W. Template-directed catalysis of a multistep reaction pathway for nonenzymatic RNA primer extension. *Biochemistry* **2019**, *58*, 755–762. [CrossRef] [PubMed]
6. Attwater, J.; Raguram, A.; Morgunov, A.S.; Gianni, E.; Holliger, P. Ribozyme-catalyse RNA synthesis using triplet building blocks. *eLife* **2018**, *7*, e35255. [CrossRef] [PubMed]
7. Cojocaru, R.; Unrau, P.J. Processive RNA polymerization and promoter recognition in an RNA World. *Science* **2021**, *371*, 1225–1232. [CrossRef] [PubMed]
8. Chen, I.A.; Salehi-Ashtiani, K.; Szostak, J.W. RNA catalysis in model protocell vesicles. *J. Am. Chem. Soc.* **2005**, *127*, 13213–13219. [CrossRef] [PubMed]
9. Schrum, J.P.; Zhu, T.F.; Szostak, J.W. The origins of cellular life. *Cold Spring Harb. Perspect. Biol.* **2010**, *2*, a002212. [CrossRef] [PubMed]
10. Kurihara, K.; Okura, Y.; Matsuo, M.; Toyota, T.; Suzuki, K.; Sugawara, T. A recursive vesicle-based model protocell with a primitive model cell cycle. *Nat. Commun.* **2015**, *6*, 8352. [CrossRef] [PubMed]
11. Saha, R.; Verbanic, S.; Chen, I.A. Lipid vesicles chaperone and encapsulated RNA aptamer. *Nat. Commun.* **2018**, *9*, 2313. [CrossRef] [PubMed]
12. O'Flaherty, D.K.; Kamat, N.P.; Mirza, F.N.; Li, L.; Prywes, N.; Szostak, J.W. Copying of mixed-sequence RNA templates inside model protocells. *J. Am. Chem. Soc.* **2018**, *140*, 5171–5178. [CrossRef]
13. Lopez, A.; Fiore, M. Investigating prebiotic protocells for a comprehensive understanding of the origin of life: A prebiotic systems chemistry perspective. *Life* **2019**, *9*, 49. [CrossRef] [PubMed]
14. Segré, D.; Shenhav, B.; Kafri, R.; Lancet, D. The molecular roots of compositional inheritance. *J. Theor. Biol.* **2001**, *213*, 481–491. [CrossRef] [PubMed]
15. Vasas, V.; Szathmary, E.; Santos, M. Lack of evolvability in self-sustaining autocatalytic networks constrains metabolism-first theories for the origin of life. *Proc. Natl. Acad. Sci. USA* **2010**, *107*, 1470–1475. [CrossRef] [PubMed]
16. Higgs, P.G. Chemical evolution and the evolutionary definition of life. *J. Mol. Evol.* **2017**, *84*, 225–235. [CrossRef] [PubMed]
17. Ganti, T. Organization of chemical reactions into dividing and metabolizing units: The chemotons. *Biosystems* **1975**, *7*, 15–21. [CrossRef]
18. Ganti, T. *A Theory of Biochemical Supersystems*; University Park Press: Baltimore, MD, USA, 1979.
19. Kauffman, S.A. Autocatalytic sets of proteins. *J. Theor. Biol.* **1986**, *119*, 1–24. [CrossRef]
20. Mossel, E.; Steel, M. Random biochemical networks: The probability of self-sustaining autocatalysis. *J. Theor. Biol.* **2005**, *233*, 327–336. [CrossRef]
21. Hordijk, W.; Steel, M. Autocatalytic sets and boundaries. *J. Syst. Chem.* **2015**, *6*, 1. [CrossRef]
22. Hordijk, W.; Steel, M. Chasing the tail: The emergence of autocatalytic networks. *Biosystems* **2017**, *152*, 1–10. [CrossRef] [PubMed]
23. Hordijk, W.; Steel, M. Autocatalytic networks at the basis of life's origin and organization. *Life* **2018**, *8*, 62. [CrossRef] [PubMed]
24. Mavelli, F.; Ruiz-Mirazo, K. Stochastic simulations of minimal self-reproducing cellular systems. *Philos. Trans. R. Soc. B* **2007**, *362*, 1789–1802. [CrossRef] [PubMed]
25. Piedrafita, G.; Montero, F.; Moran, F.; Cardenas, M.L.; Cornish-Bowden, A. A Simple Self-Maintaining Metabolic System: Robustness, Autocatalysis, Bistability. *PLoS Comput. Biol.* **2010**, *6*, e1000872. [CrossRef]
26. Piedrafita, G.; Cornish-Bowden, A.; Moran, F.; Montero, F. Size matters: Influence of stochasticity of the maintenance of a simple model of metabolic closure. *J. Theor. Biol.* **2012**, *300*, 143–151. [CrossRef]
27. Piedrafita, G.; Ruiz-Mirazo, K.; Monnard, P.A.; Cornish-Bowden, A.; Montero, F. Viability conditions for a compartmentalized protometabolic system: A semi-empirical approach. *PLoS ONE* **2012**, *7*, e39480. [CrossRef]
28. Mavelli, F.; Ruiz-Mirazo, K. Theoretical conditions for the stationary reproduction of model protocells. *Integr. Biol.* **2013**, *5*, 324. [CrossRef]
29. Shirt-Ediss, B.; Ruiz-Mirazo, K.; Mavelli, F.; Solé, R.V. Modelling lipid competition dynamics in heterogeneous protocell populations. *Sci. Rep.* **2014**, *4*, 5675. [CrossRef]
30. Piedrafita, G.; Monnard, P.A.; Mavelli, F.; Ruiz-Mirazo, K. Permeability-driven selection in a semi-empirical protocell model: The roots of prebiotic systems evolution. *Sci. Rep.* **2017**, *7*, 3141. [CrossRef]
31. Chen, I.A.; Roberts, R.W.; Szostak, J.W. The emergence of competition between model protocells. *Science* **2004**, *305*, 1474–1476. [CrossRef]
32. Berclaz, N.; Blöchliger, E.; Müller, M.; Luisi, P.L. Marix effect of vesicle formation as investigated by cryotransmission electron microscopy. *J. Phys. Chem. B* **2001**, *105*, 1065–1071. [CrossRef]
33. Koslov, M.M.; Markin, V.S. Theory of osmotic lysis of lipid vesicles. *J. Theor. Biol.* **1984**, *109*, 17–39. [CrossRef]
34. Hallett, F.R.; Marsh, J.; Nickel, B.G.; Wood, J.M. Mechanical properties of vesicles. II A model for osmotic swelling and lysis. *Biophys. J.* **1993**, *64*, 435–442. [CrossRef]
35. Chabanon, M.; Ho, J.C.S.; Liedberg, B.; Parikh, A.N.; Rangamani, P. Pulsatile lipid vesicles under osmotic stress. *Biophys. J.* **2017**, *112*, 1682–1691. [CrossRef]

36. Zhu, T.F.; Szostak, J.W. Coupled growth and division of model protocell membranes. *J. Am. Chem. Soc.* **2009**, *131*, 5705–5713. [CrossRef]
37. Toparlak, O.D.; Wang, A.; Mansy, S.S. Population-level membrane diversity triggers growth and division of protocells. *J. Am. Chem. Soc.* **2021**, *1*, 560–568.
38. Wu, M.; Higgs, P.G. Origin of Self-replicating Biopolymers: Autocatalytic Feedback can Jump-start the RNA World. *J. Mol. Evol.* **2009**, *69*, 541–554. [CrossRef]
39. Higgs, P.G.; Wu, M. The importance of stochastic transitions for the origin of life. *Orig. Life. Evol. Biosph.* **2012**, *42*, 453–457. [CrossRef]
40. Wu, M.; Higgs, P.G. The origin of life is a spatially localized stochastic transition. *Biol. Direct* **2012**, *7*, 42. [CrossRef]
41. Shay, J.A.; Huynh, C.; Higgs, P.G. The origin and spread of a cooperative replicase in a prebiotic chemical system. *J. Theor. Biol.* **2015**, *364*, 249–259. [CrossRef]
42. Hansen, P.L.; Podgornik, R.; Parsegian, V.A. Osmotic properties of DNA: Critical evaluation of counterion condensation theory. *Phys. Rev. E* **2001**, *64*, 21907. [CrossRef]
43. Poudyal, R.R.; Cakmak, F.P.; Keating, C.D.; Bevilacqua, P.C. Physical principles and extant biology reveal roles for RNA-containing membraneless compartments in origins of life chemistry. *Biochemistry* **2018**, *57*, 2509–2519. [CrossRef]
44. Bonfia, C.; Russell, D.A.; Green, N.J.; Mariani, A.; Sutherland, J.D. Activation chemistry drives the emergence of functionalised protocells. *Chem. Sci.* **2020**, *11*, 10688. [CrossRef]
45. Powner, M.W.; Sutherland, J.D.; Szostak, J.W. The origins of nucleotides. *Synlett* **2011**, *14*, 1956–1964. [CrossRef]
46. Islam, S.; Powner, M.W. Prebiotic Systems Chemistry: Complexity Overcoming Clutter. *Chem* **2017**, *2*, 470–501. [CrossRef]
47. Stairs, S.; Nikmal, A.; Bučar, D.K.; Zheng, S.L.; Szostak, J.W.; Powner, M.W. Divergent prebiotic synthesis of pyrimidine and 8-oxo-purine ribonucleotides. *Nat. Commun.* **2017**, *8*, 15270. [CrossRef]
48. Fialho, D.M.; Roche, T.P.; Hud, N.V. Prebiotic Syntheses of Noncanonical Nucleosides and Nucleotides. *Chem. Rev.* **2020**, *120*, 4806–4830.
49. Wu, M.; Walker, S.I.; Higgs, P.G. Autocatalytic Replication and Homochirality in Biopolymers: Is Homochirality a Requirement for Life or a Result of It? *Astrobiology* **2012**, *12*, 818–829. [CrossRef]
50. Tupper, A.S.; Shi, K.; Higgs, P.G. The role of templating in the emergence of RNA from the prebiotic chemical mixture. *Life* **2017**, *7*, 41. [CrossRef]
51. Keller, M.A.; Turchyn, A.V.; Ralser, M. Non-enzymatic glycolysis and pentose phosphate pathway-like reactions in a plausible Archaean ocean. *Mol. Syst. Biol.* **2014**, *10*, 725. [CrossRef] [PubMed]
52. Stubbs, R.T.; Yadav, M.; Krishnamurthy, R.; Springsteen, G. A plausible metal-free ancestral analogue of the Krebs cycle composed entirely of α-ketoacids. *Nat. Chem.* **2020**, *12*, 1016–1022. [CrossRef] [PubMed]
53. Muchowska, K.B.; Varma, S.J.; Moran, J. Non-enzymatic metabolic reactions and life's origins. *Chem. Rev.* **2020**, *120*, 7708–7744. [CrossRef] [PubMed]
54. Sousa, F.L.; Hordijk, W.; Steel, M.; Martin, W.F. Autocatalytic sets in *E. coli* metabolism. *J. Syst. Chem.* **2015**, *6*, 4. [CrossRef] [PubMed]
55. Xavier, J.C.; Hordijk, W.; Kauffman, S.; Steel, M.; Martin, W.F. Autocatalytic chemical networks at the origin of metabolism. *Proc. R. Soc. B* **2020**, *287*, 20192377. [CrossRef]

Article

Plausibility of Early Life in a Relatively Wide Temperature Range: Clues from Simulated Metabolic Network Expansion

Xin-Yi Chu [†], Si-Ming Chen [†], Ke-Wei Zhao, Tian Tian, Jun Gao and Hong-Yu Zhang *

Hubei Key Laboratory of Agricultural Bioinformatics, College of Informatics, Huazhong Agricultural University, Wuhan 430070, China; chuxy@webmail.hzau.edu.cn (X.-Y.C.); csm930@webmail.hzau.edu.cn (S.-M.C.); zkw000320@webmail.hzau.edu.cn (K.-W.Z.); tiantian@whu.edu.cn (T.T.); gaojun@mail.hzau.edu.cn (J.G.)
* Correspondence: zhy630@mail.hzau.edu.cn; Tel.: +86-27-87285085
† These authors contributed equally.

Abstract: The debate on the temperature of the environment where life originated is still inconclusive. Metabolic reactions constitute the basis of life, and may be a window to the world where early life was born. Temperature is an important parameter of reaction thermodynamics, which determines whether metabolic reactions can proceed. In this study, the scale of the prebiotic metabolic network at different temperatures was examined by a thermodynamically constrained network expansion simulation. It was found that temperature has limited influence on the scale of the simulated metabolic networks, implying that early life may have occurred in a relatively wide temperature range.

Keywords: origin of life; metabolism; network expansion simulation; temperature; thermodynamics

1. Introduction

The temperature of the environment where life originated has elicited a long-term debate. Previous genome sequence-based studies on this issue reached inconsistent results (detailed in the Discussion) [1–7]. New perspectives are needed to address this issue. Metabolism-first world theory suggests that the origin of metabolism was earlier than the appearance of the genetic system and organic catalysts such as enzymes [8–11]. The feasibility (thermodynamics) and rates (kinetics) of the reactions that make up metabolism networks are associated with temperature. Therefore, studying the effects of temperature on metabolic reactions and networks may shed more light on the environmental conditions for the origin of life.

The network expansion algorithm has been used to trace the evolutionary history hidden in modern metabolism networks [12–15]. Using this method, Goldford et al. extracted a phosphorus-independent subset from the full KEGG metabolism, which represented a biosphere-level primitive metabolic network [13]. This network exhibited ancient features, such as the enzymes and protein folds belonging to the speculative last universal common ancestor (LUCA) proteomes, and thus was considered as a "metabolic fossil". They also showed that thermodynamics is an important limiting factor for the scale of the metabolic network. Interestingly, Goldford et al. found that in standard conditions, a thioester panteheine instead of phosphates (pyrophosphate or acetyl-phosphate) can significantly alleviate the thermodynamic bottleneck of the expansion of the metabolic network, though the latter is a more common energy source in modern metabolism. Goldford et al. suggested that the phosphorus-independent network provides a solution to the "phosphorus problem", which was caused by the low solubility and reactivity of Earth's main phosphate source—apatite [16]. This network also supports the previously proposed thioester world theory, which proposed that in the very early stage of life's origin, thioesters played multiple important roles in prebiotic metabolism [17]. Recently, based on their phosphorus-independent network, Goldford et al. investigated the impact of several

environmental factors on metabolism and found that temperature has a relatively limited influence on the scale of the metabolism network [14].

Nevertheless, phosphorus is an essential element of life and several possible prebiotic phosphorus sources have been proposed [18]. Phosphite, which is more soluble and has higher reactivity than apatite [16], can be generated from volcanic activity [19] or extraterrestrial schreibersite [20]. Another effective phosphorylating agent, polyphosphate, can also be generated in volcanic regions [21]. Phosphite and polyphosphate can be directly used in the synthesis of biomolecules [20,21], or be first converted to orthophosphate in plausible prebiotic environments [22]. Based on these findings, in a recent study [15], we constructed a phosphate-dependent network using Goldford et al.'s method. The obtained phosphate-dependent network could be as ancient as the phosphorus-independent counterpart and could synthesize ribose, which implies a connection between the metabolism world and the RNA world. Moreover, several phosphorylated intermediates such as glucose 6-phosphate can play the same role as thioester, significantly alleviating the thermodynamic bottleneck of network expansion. It is thus intriguing to explore how temperature affects the scale of this simulated phosphorus-dependent network.

2. Materials and Methods

2.1. Data Sources

The metabolism reactions and compounds were downloaded from the KEGG reaction database (release: 84.0) [23]. The Gibbs free energy of reactions were calculated by eQuilibrator [24,25].

2.2. Thermodynamically Constrained Network Expansion Simulation

The thermodynamically constrained network expansion simulation method has been described in detail in previous studies [13,15]. In brief, firstly, all KEGG metabolic reactions were downloaded and vague and unbalanced reactions were filtered out. The remaining reactions and their compounds constituted the background metabolism pool, which included 7376 reactions and 6460 compounds (Table S1). Secondly, the network expansion was started with a seed set of compounds that are considered to have existed on the primitive Earth. The seeds used in the phosphate-dependent network expansion were the same as in our previous study [15], which included water, dinitrogen, carbon dioxide, hydrogen sulfide, ammonia, acetate, formate, and glucose 6-phosphate. The first eight compounds are considered to have been abundant on prebiotic Earth and should participate in primitive biochemical reactions [13–15]. Glucose 6-phosphate is a phosphorylated intermediate from glycolysis which was speculated to be prebiotically synthesized [26,27]. Our previous study showed that glucose 6-phosphate can significantly alleviate the thermodynamic bottleneck. Without this compound, the network obtained by thermodynamically constrained expansion contained only dozens of metabolites [15]. In the thioester-dependent network expansion, glucose 6-phosphate was replaced by pantetheine. Note that Goldford et al. proposed that pantetheine can function as CoA thioesters [13]. Therefore, CoA and acetyl-CoA were also added into the seed set. Thirdly, the products of thermodynamically reachable reactions from the background metabolism pool enabled by available substrates were iteratively added into the compounds set until no new reactions or compounds could be produced.

Previously, the thermodynamically accessible standard of a reaction at 25 °C was defined as Gibbs free energy below 30 kJ/mol [13,15]. However, this standard is not appropriate at other temperatures. In this work, we calculated the lowest reaction free energy at the boundary of the physiological range of metabolite concentrations, referring to Goldford et al.'s recent study [14]. Briefly, the free energy of every reaction at different temperatures ($\Delta G'$) was calculated using the following equation:

$$\Delta G' = \Delta G^{\circ\prime} + RTln \prod_i a_i^{s_i} \quad (1)$$

where $\Delta G^{\circ\prime}$ is the Gibbs free energy under standard molar conditions, which was estimated by the eQuilibrator program using a group contribution method [24,25]. $\Delta G^{\circ\prime}$ is affected by pH, ionic strength, and also by the concentration of free Mg^{2+} (pMg) for some reactions like ATP hydrolysis [25]. These conditions were set as follows: pH = 7.0, ionic strength = 0.1 M, and pMg = 0.0, which is the same as our previous study [15]. R is the ideal gas constant, T is the temperature, a_i is the activity of metabolite i (represented by the metabolite's concentration), and s_i is the stoichiometric coefficient for metabolite i in a certain reaction. For all reactions, the reactant and product concentrations were set to 0.1 M and 10^{-6} M, respectively. These two values are considered to be the upper and lower bounds of the concentration of metabolites under physiological conditions. The water activity was set to 1 M, which is an essential assumption of eQuilibrator [24,25]. Reactions with positive free energy were removed from the background metabolism pool. About 40% of the biosphere-level metabolic network reactions had no accurate free energy estimation (3122/7376). The background metabolism pools including and excluding these reactions were both analyzed in this study. The free energy data of the reactions can be found in Table S2.

It should be noted that the change in Gibbs free energy (ΔG) contains two components: entropy change (ΔS) and enthalpy change (ΔH). However, data of ΔS are not available for most reactions so it was not taken into account when using eQuilibrator to calculate ΔG in different temperatures [24,25]. This simplification may cause some reactions to be incorrectly defined as thermodynamically favored.

3. Results

All chemical reactions are constrained by the thermodynamic rules. A reaction with a positive standard change in Gibbs free energy ($\Delta G > 0$) cannot proceed spontaneously. In modern organisms, these uphill reactions are usually coupled with exergonic reactions such as ATP hydrolysis to become energetically favorable. However, such stably coupled reaction systems may not have been available on primitive earth [28]. Therefore, the thermodynamic constraints could be an important limiting factor for primitive metabolism [13–15].

Existing organisms were found to survive from $-25\ °C$ to over 120 °C. *Planococcus halocryophilus* Or1, a bacterium isolated from the salty water veins of high Arctic permafrost, can remain metabolically active at $-25\ °C$ [29]. Hyperthermophilic archaea *Methanopyrus kandleri* strain 116 can proliferate at 122 °C [30]. Therefore, we simulated the expansion of the phosphate-dependent and thioester-dependent networks at the temperatures from $-25\ °C$ to 150 °C. Using Goldford et al.'s method [14], we calculated the ΔG of the metabolic reactions at $-25\ °C$, 0 °C, 25 °C, 50 °C, 75 °C, 100 °C, 125 °C, and 150 °C, respectively. As shown in Figure 1, from $-25\ °C$ to 150 °C, the scales of both phosphorus- and thioester-dependent networks only exhibit slight increases, regardless of whether the reactions include accurate free energy estimation or not.

When including the reactions without accurate free energy estimation, the scale of the phosphate-dependent network obtained at $-25\ °C$ is close to the thermodynamically constrained network constructed in our previous study (360 vs. 338 metabolites) [15]. At this temperature, this network is mainly composed of reactions that participate in glycolysis, the tricarboxylic acid (TCA) cycle, and carbon fixation, which supply the basic carbohydrates and energy to living systems. They can also produce eight proteinogenic amino acids (Figure 2A, Table S3). The reactions that are only thermodynamically feasible at temperatures higher than $-25\ °C$ provide 17 new metabolites (Table S3). Seven of these metabolites are involved in the amino acid metabolism, but none of them are proteinogenic amino acids. The other metabolites generated at higher temperatures are scattered across different pathways. When excluding the reactions without accurate free energy estimation, the network keeps the functions mentioned above, and the temperature increase has no significant influence on these functions (Figure S1A, Table S3). The thioester-dependent networks can produce two more proteinogenic amino acids but lack glycolysis-related carbohydrates. These functions are also not greatly influenced by the change in temperature (Figure 2B and Figure S1B, and Table S3).

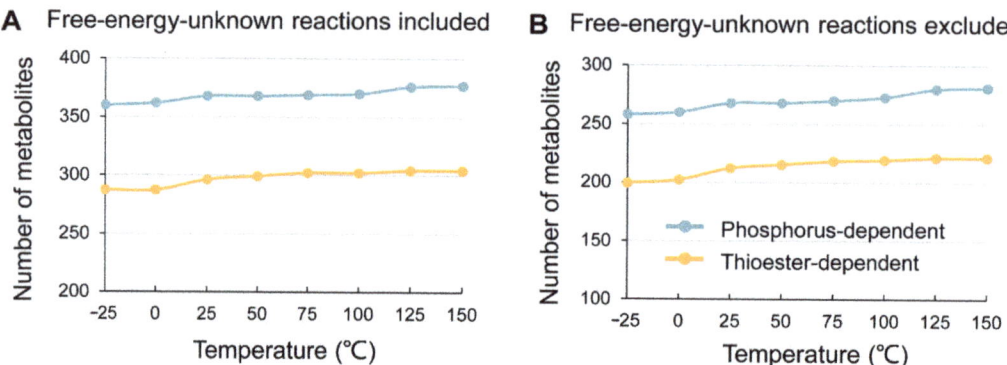

Figure 1. The scale of thermodynamically constrained networks at different temperatures. The figure shows the number of metabolites of the networks at different temperatures. (**A**) Including the reactions without accurate free energy estimation; (**B**) excluding the reactions without accurate free energy estimation. The green lines represent the phosphate-dependent networks, while the yellow lines represent the thioester-dependent networks. With the increase in temperature, the scales of the networks increase slightly.

Nucleotides are the basic components of RNA and constitute the cofactors of many proteins, so they are of great significance in the origin of life [31]. As there is no phosphorus in the thioester-dependent networks, it is obviously impossible for them to synthesize nucleotides (Figure 2B and Figure S1B). In fact, they cannot even synthesize ribose (Table S3). The phosphate-dependent networks can generate ribose 5-phosphate (KEGG ID C00117) at all tested temperatures, which is a precursor required for the synthesis of nucleotides (Table S3). Moreover, ribose 5-phosphate can be further phosphorylated to 5-phosphoribosyl diphosphate (PRPP) through KEGG reaction R01049, which is necessary for both purine and pyrimidine synthesis (Figure S2). Under a physiological condition, ATP provides the phosphate group and energy for the reaction. ATP is located at the end of one branch of the KEGG nucleotide synthesis pathway. Therefore, ATP cannot be used in a reaction that is located at the upstream end of the pathway, which prevented the phosphate-dependent networks from expanding along this path. However, there may be other ways to achieve the phosphorylation of ribose 5-phosphate. In fact, glucose-6-phosphate hydrolysis can provide phosphate and energy. We constructed such a reaction: ribose 5-phosphate + 2 glucose 6-phosphate => PRPP + 2 glucose. Under the substance concentration condition used in the network expansion simulation, this reaction is thermodynamically favorable even at -25 °C (with a free energy of -33.56 kJ/mol), implying that glucose-6-phosphate may play a role like ATP and facilitate the phosphate-dependent network to generate the nucleotides.

In modern life, ATP plays multiple roles in metabolism. In the background metabolism pool used in this study, ATP is the reactant or product of 471 reactions. These reactions cannot be reached by the network expansion because ATP is not available in the simulation. In primitive metabolism, other phosphorus-containing compounds with simpler structures may be more prevalent than ATP, and played similar roles to it in these reactions [32]. As shown in the fictional reaction to synthesize PRPP, their products may be obtained through reactions that do not rely on ATP. However, the possibility that ATP existed in the prebiotic world cannot be completely ruled out, because the abiotic synthesis of ATP from simple inorganic substances may be achieved in the prebiotic environment [33–35]. Therefore, we replaced the glucose-6-phosphate in the "seeds" with ATP and performed network expansions. The obtained metabolic networks have about 200 more metabolites than the phosphorus-dependent networks at the same temperatures. For these ATP-dependent networks, from −25 °C to 150 °C, only 24 metabolites were added (including the reactions without accurate free energy estimation, Table S3), showing a limited influence of temperature.

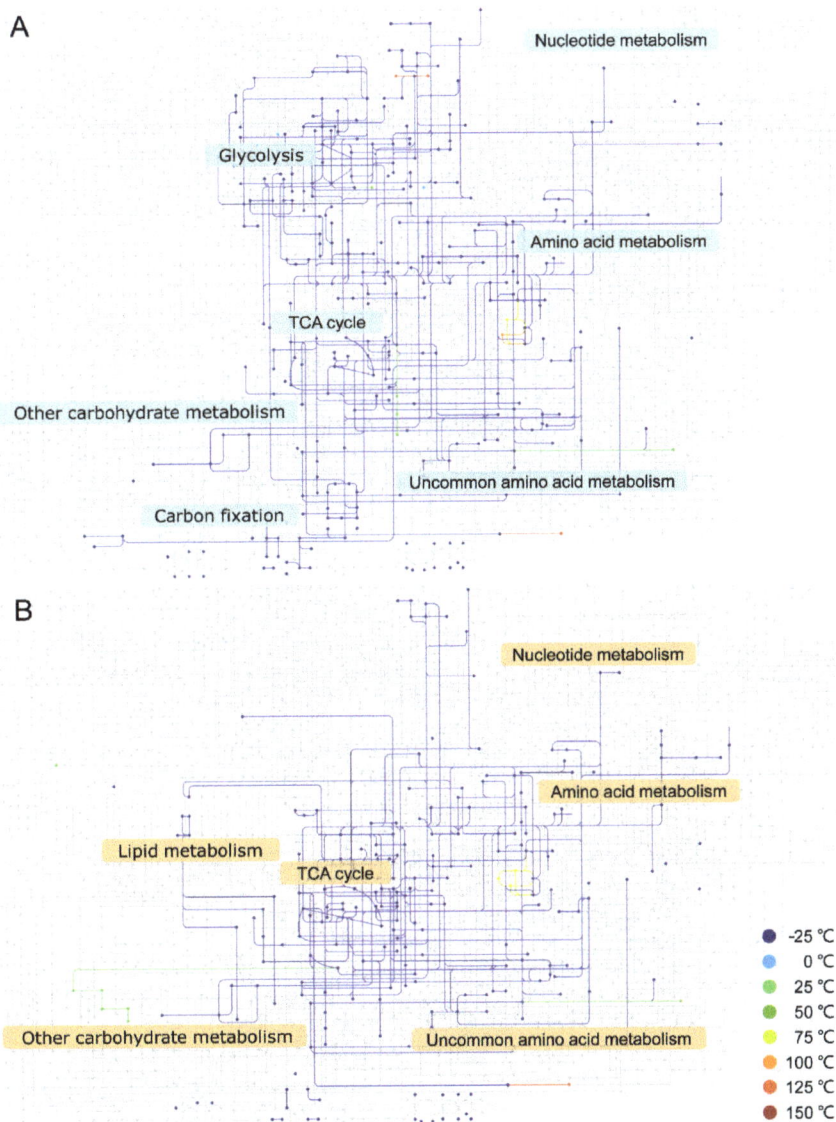

Figure 2. Metabolic networks at different temperatures. (**A**) Phosphate-dependent network; (**B**) thioester-dependent network. Reactions without accurate free energy estimation are included. The metabolites and reactions that appeared at different temperatures are represented by dots and lines of different colors, respectively. Light gray represents KEGG reference metabolites and reactions unavailable during the network expansions. Most reactions and metabolites can be achieved at −25 °C. One metabolite can be displayed as multiple nodes in the figure. The reactions and metabolites in different networks can be found in Table S3.

These findings showed that, at least in terms of thermodynamics, the scale and main functions of the simulated metabolic networks are slightly affected by temperature. The free energy characteristics (positive or negative) of most reactions do not change within the examined temperature range (Table S2), so the feasibility of these reactions is not largely influenced by temperature. These results suggest that, whether it is dependent on phosphate or thioester, metabolism may have originated in a relatively wide temperature range.

4. Discussion

There are many different opinions on the temperature at the origin of life. Darwin speculated that the first living organisms evolved in "warm little ponds" [36]. The Miller–Urey experiment showed that a warm, lightning-filled atmosphere that may have existed on primitive Earth can produce important biomolecules [37]. Hydrothermal environments were also "hot" in the area of the origin of life. Terrestrial hydrothermal fields and submarine hydrothermal vents can provide materials, energy, reducing power, and pH gradients, which may facilitate the synthesis of basic biomolecules [38–40]. Moreover, terrestrial hydrothermal fields can evaporate water to become dry and get wet by rain, which formed a natural dry–wet cycle [41]. Dry–wet cycles are very important for the synthesis of nucleotides under metal catalysis [42]. These effects are also significant for the polymerization of biological macromolecules such as peptides [43,44] and polynucleotides [41,45].

A criticism of the hydrothermal origin of life is that although the structures of most small molecule metabolites are stable, the high temperature can accelerate their reactions with other substances, such as water, and thus causes their thermal instability [46]. Miller and colleagues argued that important metabolites such as ribose and ATP are thermally unstable, thus life was not likely to originate at high temperatures [47,48]. However, the meaning of "stable" is relative. Some metabolites in metabolic pools have very short turnover times. Metabolites that can exist longer than their turnover time in organisms should have a chance to participate in metabolic reactions as reactants. Based on this principle, Bains et al. predicted the degradation rates of 63 metabolites at different temperatures and compared them with the intracellular half-life of these metabolites [46]. As a result, most of the metabolites were found to be unable to exist longer than their intracellular half-life at temperatures above 150–180 °C. Therefore, Bains et al. suggested that this temperature range is the upper limit of biochemistry.

Some basic materials for life also can be generated at cold temperatures. HCN polymerization may be an important procedure in biomolecule synthesis [49]. However, this process not likely to occur in warm environments due to the fast hydrolysis of HCN [50]. It has been found that the eutectic freezing of HCN and water at -21 °C can concentrate the former and promote its polymerization [50]. Moreover, eutectic freezing of NH_4CN solutions can generate nucleic acid bases and amino acids [51,52].

In addition to the conditions required for the synthesis of biomolecules, the living environment of modern organisms can also be used as a clue to explore the temperature for the origin of life. Previously, thermophilic bacteria and archaea were considered to be located near the roots of the phylogenetic tree, suggesting that life originated in hot temperatures [1–3]. However, in a newer rRNA-based phylogenetic tree of bacteria, mesophilic species rather than thermophiles are the closest branches to the root, showing that the thermophiles are latecomers in evolution [4]. Several studies inferred the temperature of the habitat for the LUCA based on the deduced rRNA GC contents or protein amino acid compositions, but no agreement has been reached [5–7].

The inconsistent results from gene sequence-based studies can be attributed to the uncertain early life evolution trajectory, which is difficult to properly characterize or experimentally test [53]. Compared with the genetic system, metabolic reactions may have an earlier origin and are more common among different species. Therefore, metabolism could be a window to detect the properties of the earliest life. In this study, the influence of temperature on metabolism was analyzed in terms of thermodynamics. We found that whether glucose 6-phosphate or thioester was used to alleviate the thermodynamic bottleneck of network expansion, there is no significant difference in the function and scale of the networks generated at different temperatures. These results imply that the origin of metabolism could have occurred in a relatively wide temperature range.

Although the thermodynamic data of most biochemical reactions at different temperatures are very scarce, several reactions that are crucial for network expansion may have this kind of information. To identify these network scale-limiting reactions, the network expansion was re-performed several times, wherein the reactions from the background metabolism pool were removed one by one. We found that the deletion of most reactions did not cause a significant reduction in the generated network, except for a few (Table S4). When excluding the reactions without accurate free energy estimation, reactions R00874, R01519, R01538, and R08570 are necessary for maintaining the scale of the phosphate-dependent network. Without any of these reactions, the network expansion ceased with up to eighty-nine metabolites. These four reactions are related to glucose metabolism. In R00874, glucose reacts with fructose to form gluconolactone and glucitol. The other three reactions are components of the pentose phosphate pathway, which starts with glucose and generates multiple important metabolites. For the thioester-dependent networks, reaction R00212 is necessary for network expansion. Without this reaction, the produced networks contained only dozens of metabolites. In this reaction, acetyl-CoA reacts with formate and generates CoA and pyruvate. These results showed that certain reactions can indeed deeply influence the expansion of metabolic networks. We tried to find out more about their feasibility at different temperatures, but no useful information was found. Further study on the thermodynamics of these key reactions may provide a better understanding of the temperature of the environment where life originated.

Thermodynamics is a very fundamental property for chemical reactions. Therefore, this study may also be meaningful for finding exoplanet life. Based on planetary physical parameters such as temperature, Lingam and Loeb formulated likelihood functions estimating the possibility of the existence of life on a certain planet [54]. In the calculation, they referred to the temperature range of the living creatures on Earth (262 to 395 K). Temperatures beyond this range will lead to a rapid decline in the possibility of life. The tested temperature range in this study is a bit wider (−25 to 150 °C, i.e., 248.15 to 423.15 K). The present analysis suggests that metabolic networks may originate in this range. In our simulation, the expansion of metabolic networks does not reject low temperature, which increases the possibility of life on cold celestial bodies. Primitive metabolism may occur in the saltwater lakes on Mars, the liquid methane of Titan, or a liquid ammonia ocean elsewhere.

In addition to the above discussion, the interplays between temperature and other factors such as the phase change of water (e.g., dry–wet cycle and eutectic freezing, as discussed above) and redox potential can promote metabolic reactions. In Goldford et al.'s study, redox potential had a decisive influence on the scale of the metabolism networks [14]. Temperature gradients can cause electrochemical potentials which can be used as a thermodynamic driving force. For example, the high temperature, pressure, and pH of deep sea alkaline hydrothermal vents elevate the redox potential of H_2 oxidation, which can be used as an electrochemical driving force for the abiotic reduction of CO_2 [55,56]. Intriguingly, temperature may also shape the structure of peptides [57], which could serve as catalysts in prebiotic metabolism. To date, most studies on the origin of metabolism have not considered the interactions between temperature and other environmental factors. Goldford et al.'s study investigated several environmental factors, but they are still independent of each other. The interplays between temperature and other factors and how they affect metabolism deserve more in-depth studies. Moreover, the feasibility of chemical reactions is also dependent on their kinetics. Thermodynamically favored reactions may have high energy barriers between the reactants and products, which make the reactions unfeasible in terms of kinetics. In future studies, we will explore the influence of temperature on the kinetics of key metabolic reactions by calculating the activation energies for the reactions.

Supplementary Materials: The following are available online at https://www.mdpi.com/article/10.3390/life11080738/s1, Figure S1: Metabolic networks at different temperatures. Figure S2: The link between phosphate-dependent networks and nucleotide metabolic pathways. Table S1: Background metabolism pool. Table S2: Gibbs free energy of the reactions at different temperatures. Table S3: Metabolites and reactions at different temperatures.

Author Contributions: Conceptualization, H.-Y.Z.; methodology, X.-Y.C., S.-M.C., K.-W.Z., T.T., and J.G.; validation, X.-Y.C. and K.-W.Z.; writing—original draft preparation, X.-Y.C. and T.T.; writing—review and editing, H.-Y.Z. and J.G.; project administration, X.-Y.C. and H.-Y.Z.; funding acquisition, H.-Y.Z. All authors have read and agreed to the published version of the manuscript.

Funding: This work has been supported by the National Natural Science Foundation of China (31870837).

Conflicts of Interest: The authors declare no conflict of interest.

References

1. Woese, C.R. Bacterial evolution. *Microbiol. Rev.* **1987**, *51*, 221–271. [CrossRef] [PubMed]
2. Achenbach-Richter, L.; Gupta, R.; Zillig, W.; Woese, C.R. Rooting the archaebacterial tree: The pivotal role of *Thermococcus celer* in archaebacterial evolution. *Syst. Appl. Microbiol.* **1988**, *10*, 231–240. [CrossRef]
3. Pace, N.R. Origin of life—Facing up to the physical setting. *Cell* **1991**, *65*, 531–533. [CrossRef]
4. Brochier, C.; Philippe, H. Phylogeny: A non-hyperthermophilic ancestor for bacteria. *Nature* **2002**, *417*, 244. [CrossRef] [PubMed]
5. Galtier, N.; Tourasse, N.; Gouy, M. A nonhyperthermophilic common ancestor to extant life forms. *Science* **1999**, *283*, 220–221. [CrossRef] [PubMed]
6. Boussau, B.; Blanquart, S.; Necsulea, A.; Lartillot, N.; Gouy, M. Parallel adaptations to high temperatures in the Archaean eon. *Nature* **2008**, *456*, 942–945. [CrossRef]
7. Groussin, M.; Boussau, B.; Charles, S.; Blanquart, S.; Gouy, M. The molecular signal for the adaptation to cold temperature during early life on Earth. *Biol. Lett.* **2013**, *9*, 20130608. [CrossRef] [PubMed]
8. Eck, R.V.; Dayhoff, M.O. Evolution of the structure of ferredoxin based on living relics of primitive amino acid sequences. *Science* **1966**, *152*, 363–366. [CrossRef]
9. Hartman, H. Speculations on the origin and evolution of metabolism. *J. Mol. Evol.* **1975**, *4*, 359–370. [CrossRef]
10. Morowitz, H.J.; Kostelnik, J.D.; Yang, J.; Cody, G.D. The origin of intermediary metabolism. *Proc. Natl. Acad. Sci. USA* **2000**, *97*, 7704–7708. [CrossRef]
11. Lanier, K.A.; Williams, L.D. The origin of life: Models and data. *J. Mol. Evol.* **2017**, *84*, 85–92. [CrossRef] [PubMed]
12. Ebenhöh, O.; Handorf, T.; Heinrich, R. Structural analysis of expanding metabolism networks. *Genome Inform.* **2004**, *15*, 35–45. [PubMed]
13. Goldford, J.E.; Hartman, H.; Smith, T.F.; Segrè, D. Remnants of an ancient metabolism without phosphate. *Cell* **2017**, *168*, 1126–1134. [CrossRef] [PubMed]
14. Goldford, J.E.; Hartman, H.; Marsland, R.; Segrè, D. Environmental boundary conditions for the origin of life converge to an organo-sulfur metabolism. *Nat. Ecol. Evol.* **2019**, *3*, 1715–1724. [CrossRef] [PubMed]
15. Tian, T.; Chu, X.-Y.; Yang, Y.; Zhang, X.; Liu, Y.-M.; Gao, J.; Ma, B.-G.; Zhang, H.-Y. Phosphates as energy sources to expand metabolism networks. *Life* **2019**, *9*, 43. [CrossRef]
16. Schwartz, A.W. Phosphorus in prebiotic chemistry. *Philos. Trans. R. Soc. B* **2006**, *361*, 1743–1749. [CrossRef]
17. De Duve, C. A research proposal on the origin of life. *Orig. Life Evol. Biosph.* **2003**, *33*, 559–574. [CrossRef]
18. Piast, R.W.; Wieczorek, R.M. Origin of life and the phosphate transfer catalyst. *Astrobiology* **2017**, *17*, 277–285. [CrossRef]
19. Glindemann, D.; De Graaf, R.M.; Schwartz, A.W. Chemical reduction of phosphate on the primitive earth. *Orig. Life Evol. Biosph.* **1999**, *29*, 555–561. [CrossRef]
20. Pasek, M.A. Schreibersite on the early Earth: Scenarios for prebiotic phosphorylation. *Geosci. Front.* **2017**, *8*, 329–335. [CrossRef]
21. Yamagata, Y.; Watanabe, H.; Saitoh, M.; Namba, T. Volcanic production of polyphosphates and its relevance to prebiotic evolution. *Nature* **1991**, *352*, 516–519. [CrossRef]
22. Benner, S.A.; Kim, H.J.; Carrigan, M.A. Asphalt, water, and the prebiotic synthesis of ribose, ribonucleosides, and RNA. *Acc. Chem. Res.* **2012**, *45*, 2025–2034. [CrossRef] [PubMed]
23. Ogata, H.; Goto, S.; Sato, K.; Fujibuchi, W.; Bono, H.; Kanehisa, M. KEGG: Kyoto Encyclopedia of Genes and Genomes. *Nucleic Acids Res.* **1999**, *27*, 29–34. [CrossRef] [PubMed]
24. Flamholz, A.; Noor, E.; Bar-Even, A.; Milo, R. eQuilibrator—The biochemical thermodynamics calculator. *Nucleic Acids Res.* **2012**, *40*, D770–D775. [CrossRef] [PubMed]
25. Noor, E.; Haraldsdóttir, H.S.; Milo, R.; Fleming, R.M.T. Consistent estimation of Gibbs energy using component contributions. *PLoS Comput. Biol.* **2013**, *9*, e1003098. [CrossRef]
26. Coggins, A.J.; Powner, M.W. Prebiotic synthesis of phosphoenol pyruvate by phosphorylation controlled triose glycolysis. *Nat. Chem.* **2017**, *9*, 310–317. [CrossRef]

27. Keller, M.A.; Turchyn, A.V.; Ralser, M. Non-enzymatic glycolysis and pentose phosphate pathway-like reactions in a plausible Archean ocean. *Mol. Syst. Biol.* **2014**, *10*, 725. [CrossRef]
28. Bar-Even, A.; Flamholz, A.; Noor, E.; Milo, R. Thermodynamic constraints shape the structure of carbon fixation pathways. *BBA Bioenergetics* **2012**, *1817*, 1646–1659. [CrossRef]
29. Mykytczuk, N.C.S.; Foote, S.J.; Omelon, C.R.; Southam, G.; Greer, C.W.; Whyte, L.G. Bacterial growth at −15 °C; molecular insights from the permafrost bacterium *Planococcus halocryophilus* Or1. *ISME J.* **2013**, *7*, 1211–1226. [CrossRef] [PubMed]
30. Takai, K.; Nakamura, K.; Toki, T.; Tsunogai, U.; Miyazaki, M.; Miyazaki, J.; Hirayama, H.; Nakagawa, S.; Nunoura, T.; Horikoshi, K. Cell proliferation at 122 °C and isotopically heavy CH_4 production by a hyperthermophilic methanogen under high-pressure cultivation. *Proc. Natl. Acad. Sci. USA* **2008**, *105*, 10949–10954. [CrossRef]
31. Chu, X.Y.; Zhang, H.Y. Cofactors as molecular fossils to trace the origin and evolution of proteins. *ChemBioChem* **2020**, *21*, 3161–3168. [CrossRef] [PubMed]
32. Whicher, A.; Camprubi, E.; Pinna, S.; Herschy, B.; Lane, N. Acetyl phosphate as a primordial energy currency at the origin of life. *Orig. Life Evol. Biosph.* **2018**, *48*, 159–179. [CrossRef] [PubMed]
33. Yamagata, Y. Prebiotic formation of ADP and ATP from AMP, calcium phosphates and cyanate in aqueous solution. *Orig. Life Evol. Biosph.* **1999**, *29*, 511–520. [CrossRef]
34. Akouche, M.; Jaber, M.; Maurel, M.C.; Lambert, J.F.; Georgelin, T. Phosphoribosyl pyrophosphate: A molecular vestige of the origin of life on minerals. *Angew. Chem. Int. Ed. Engl.* **2017**, *56*, 7920–7923. [CrossRef] [PubMed]
35. Yadav, M.; Kumar, R.; Krishnamurthy, R. Chemistry of abiotic nucleotide synthesis. *Chem. Rev.* **2020**, *120*, 4766–4805. [CrossRef] [PubMed]
36. Peretó, J.; Bada, J.L.; Lazcano, A. Charles Darwin and the origin of life. *Orig. Life Evol. Biosph.* **2009**, *39*, 395–406. [CrossRef]
37. Miller, S.L. A production of amino acids under possible primitive earth conditions. *Science* **1953**, *117*, 528–529. [CrossRef]
38. Deamer, D.; Damer, B.; Kompanichenko, V. Hydrothermal chemistry and the origin of cellular life. *Astrobiology* **2019**, *19*, 1523–1537. [CrossRef]
39. Damer, B.; Deamer, D. The hot spring hypothesis for an origin of life. *Astrobiology* **2020**, *20*, 429–452. [CrossRef]
40. Omran, A.; Pasek, M. A constructive way to think about different hydrothermal environments for the origins of life. *Life* **2020**, *10*, 36. [CrossRef]
41. Ross, D.; Deamer, D. Dry/wet cycling and the thermodynamics and kinetics of prebiotic polymer synthesis. *Life* **2016**, *6*, 28. [CrossRef] [PubMed]
42. Cheng, C.; Fan, C.; Wan, R.; Tong, C.; Miao, Z.; Zhao, Y. Phosphorylation of adenosine with trimetaphosphate under simulated prebiotic conditions. *Orig. Life. Evol. Biosph.* **2002**, *32*, 219–224. [CrossRef] [PubMed]
43. Fox, S.; Harada, K.; Kendrick, J. Production of spherules from proteinoids and hot water. *Science* **1959**, *129*, 1221–1223. [CrossRef]
44. Yu, S.-S.; Solano, M.; Blanchard, M.K.; Soper-Hopper, M.; Krishnamurthy, R.; Fernandez, F.M.; Hud, N.V.; Schork, F.J.; Grover, M.A. Elongation of model prebiotic proto-peptides by continuous monomer feeding. *Macromolecules* **2017**, *50*, 9286–9294. [CrossRef]
45. De Guzman, V.; Vercoutere, W.; Shenasa, H.; Deamer, D.W. Generation of oligonucleotides under hydrothermal conditions by non-enzymatic polymerization. *J. Mol. Evol.* **2014**, *78*, 251–262. [CrossRef] [PubMed]
46. Bains, W.; Xiao, Y.; Yu, C. Prediction of the maximum temperature for life based on the stability of metabolites to decomposition in water. *Life* **2015**, *5*, 1054–1100. [CrossRef]
47. Miller, S.L.; Bada, J.L. Submarine hot springs and the origin of life. *Nature* **1988**, *334*, 609–611. [CrossRef] [PubMed]
48. Miller, S.L.; Lazcano, A. The origin of life—Did it occur at high temperatures? *J. Mol. Evol.* **1995**, *41*, 689–692. [CrossRef] [PubMed]
49. Matthews, C.N.; Minard, R.D. Hydrogen cyanide polymers, comets and the origin of life. *Faraday Discuss.* **2006**, *133*, 393–401. [CrossRef]
50. Miyakawa, S.; Cleaves, H.J.; Miller, S.L. The cold origin of life: A. implications based on pyrimidines and purines produced from frozen ammonium cyanide solutions. *Orig. Life Evol. Biosph.* **2002**, *32*, 195–208. [CrossRef]
51. Miyakawa, S.; Cleaves, H.J.; Miller, S.L. The cold origin of life: B. implications based on pyrimidines and purines produced from frozen ammonium cyanide solutions. *Orig. Life Evol. Biosph.* **2002**, *32*, 209–218. [CrossRef] [PubMed]
52. Levy, M.; Miller, S.L.; Oró, J. Production of guanine from NH_4CN polymerizations. *J. Mol. Evol.* **1999**, *49*, 165–168. [CrossRef] [PubMed]
53. Doolittle, W.F. Phylogenetic classification and the universal tree. *Science* **1999**, *284*, 2124–2129. [CrossRef]
54. Lingam, M.; Loeb, A. Physical constraints on the likelihood of life on exoplanets. *Int. J. Astrobiol.* **2018**, *17*, 116–126. [CrossRef]
55. Ooka, H.; McGlynn, S.E.; Nakamura, R. Electrochemistry at deep-sea hydrothermal vents: Utilization of the thermodynamic driving force towards the autotrophic origin of life. *Chem. Electro. Chem.* **2019**, *6*, 1316–1323. [CrossRef]
56. Boyd, E.S.; Amenabar, M.J.; Poudel, S.; Templeton, A.S. Bioenergetic constraints on the origin of autotrophic metabolism. *Philos. Trans. A Math. Phys. Eng. Sci.* **2020**, *378*, 20190151. [CrossRef]
57. Brack, A.; Spach, G. Multiconformational synthetic polypeptides. *J. Amer. Chem. Soc.* **1981**, *103*, 6319–6323. [CrossRef]

Article

Thermodynamics of Potential CHO Metabolites in a Reducing Environment

Jeremy Kua *, Alexandra L. Hernandez and Danielle N. Velasquez

Department of Chemistry & Biochemistry, University of San Diego, San Diego, CA 92110, USA; alexandrahernandez@sandiego.edu (A.L.H.); dvelasquez@sandiego.edu (D.N.V.)
* Correspondence: jkua@sandiego.edu

Abstract: How did metabolism arise and evolve? What chemical compounds might be suitable to support and sustain a proto-metabolism before the advent of more complex co-factors? We explore these questions by using first-principles quantum chemistry to calculate the free energies of CHO compounds in aqueous solution, allowing us to probe the thermodynamics of core extant cycles and their closely related chemical cousins. By framing our analysis in terms of the simplest feasible cycle and its permutations, we analyze potentially favorable thermodynamic cycles for CO_2 fixation with H_2 as a reductant. We find that paying attention to redox states illuminates which reactions are endergonic or exergonic. Our results highlight the role of acetate in proto-metabolic cycles, and its connection to other prebiotic molecules such as glyoxalate, glycolaldehyde, and glycolic acid.

Keywords: origin of life; proto-metabolism; chemical evolution; thermodynamics; prebiotic chemistry

Citation: Kua, J.; Hernandez, A.L.; Velasquez, D.N. Thermodynamics of Potential CHO Metabolites in a Reducing Environment. *Life* **2021**, *11*, 1025. https://doi.org/10.3390/life11101025

Academic Editors: Michele Fiore and Emiliano Altamura

Received: 18 August 2021
Accepted: 28 September 2021
Published: 29 September 2021

Publisher's Note: MDPI stays neutral with regard to jurisdictional claims in published maps and institutional affiliations.

Copyright: © 2021 by the authors. Licensee MDPI, Basel, Switzerland. This article is an open access article distributed under the terms and conditions of the Creative Commons Attribution (CC BY) license (https://creativecommons.org/licenses/by/4.0/).

1. Introduction

The extant metabolism of living systems is complex. However, at its core [1], represented by the tricarboxylic acid (TCA) cycle and its chemical cousins, the metabolites consist of only a small subset of molecules containing the elements carbon, hydrogen, and oxygen. Present metabolism is highly regulated. Specific biochemical reactions are catalyzed by specialized enzymes with help from a variety of other co-factor molecules. However, at the dawn of life's origin, before the complex machinery of biochemical machinery evolved to support and sustain metabolism, what might a *proto*-metabolism look like?

As pointed out by Pross [2] in this Special Issue, "all material systems are driven towards more persistent forms." In the context of which specific molecules will persist (in equilibrium or at steady state) and thereby contribute to the small subset utilized by proto-metabolism, both thermodynamics and kinetics will play a role. Our present study will only focus quantitatively on the thermodynamics, while making qualitative reference to kinetics. This is partly due to limitations in our present methodology (see Methods section), partly to keep the problem at hand tractable, but also partly because we think a survey of the thermodynamics is an interesting tale in its own right.

Autocatalysis lies at the heart of persistence. If a molecule can catalyze its own production faster than its competitors, it will persist (maintaining a non-zero and possibly significant concentration as a function of time)—until its "food" runs out or a parasitic reaction shunts it out of the main network. For a broad overview of autocatalytic networks and their role at life's origin, we recommend the review by Hordijk and Steel [3] in this journal. Our present study focuses on just one class of reaction cycles: those that produce a net two-carbon molecule from one-carbon substrates. Extant life on our planet ultimately depends on fixing CO_2 as its carbon source to build biomass. While proto-life on the early Earth may have had a variety of carbon sources, the ability to incorporate C_1 molecules into larger structures remains fundamental for a self-sustaining system to emerge [4]. Thus, incorporating $C_1 + C_1 \rightarrow C_2$ into a sustaining cycle is fundamental.

In the absence of complex biomolecular enzymes, high-energy activation barriers are unlikely to be traversed at temperatures conducive to life. We can narrow down potential cycles by looking at thermodynamics: feasible chemistry will exclude overly endergonic reactions, while an overly exergonic reaction mid-cycle would require a subsequent endergonic step, thereby narrowing the possibilities further. We first examine the core TCA cycles and its chemical cousins building on work by Braakman and Smith [1], focusing on the reverse (i.e., reductive) direction in the absence of any enzymes or co-factors. (For recent reviews of non-enzymatic metabolic reactions including the potential role of a reverse TCA cycle at the origin of life, see Muchowska et al. [5] and Tran et al. [6]; for a review on the energetics of biomolecular synthesis, see Amend et al. [7])

We then expand the scope of potential metabolites beyond those used by extant life to a wider range of CHO compounds. We are not the first to examine this wider scope. Morowitz et al. suggested a set of 153 compounds ranging from C_1 to C_6 (70 in the C_1 to C_4 range) based on simple rules of thumb [8]. Meringer and Cleaves extended and refined this set using structure-generation methods from existing molecular databases to examine if the metabolites of the reverse TCA cycle are "optimal" [9]. (Their conclusion: maybe not.) Zubarev et al. examined the thermodynamics of a potential reverse TCA "supernetwork" (175 molecules, 444 reactions); they use computationally faster (but less accurate) semi-empirical methods to calculate the Gibbs free energies; and they conclude that there exist families of TCA-like reactions with similar energetic profiles [10].

Our present research complements these earlier "bird's eye view" studies by focusing on the details of the smaller C_1 to C_4 species. We calculate the (aqueous) Gibbs free energy of each species using first-principles quantum chemistry. Our method is more computationally-intensive (but not terribly so) and matches well with experimental data where available. To map the thermodynamic landscape, our set of 211 C_1 to C_4 compounds is much wider than those generated by the "Morowitz rules" [8] because we include highly reduced compounds that are unlikely to be metabolites. Our present study also limits the oxidizing and reducing agents to CO_2 and H_2, respectively. We do this to establish baseline data for future work.

While our data allow for a wide range of analyses, this paper will pick out a few key examples to illustrate how one might build a proto-metabolic cycle, why paying attention to redox states is important, and why particular molecules may be keystone species, thermodynamically-speaking. In particular, we will (1) explain why the $C_1 + C_1 \rightarrow C_2$ reaction must be incorporated into a cycle, (2) examine the close relationship between carbonyl compounds as potential metabolites, (3) suggest reasons for the centrality of acetate in core metabolic cycles, and (4) discuss specific reactions and molecules that could participate in proto-metabolic cycles in the absence of highly specific catalysts. In no way do we discount the importance of other compounds beyond the limited set we have calculated (compounds containing nitrogen, sulfur and phosphate; metal ions/clusters) that can act as potential catalysts and cofactors; nor will we discuss the important question of flows in non-equilibrium thermodynamic systems. Ongoing research in our laboratory explores these extensions but they are beyond the scope of the present work.

2. Materials and Methods

To construct our thermodynamic maps, we have established a protocol to calculate the *relative* aqueous free energy of each molecule using quantum chemistry. This protocol has been described in detail in our previous papers (most recently in [11]) and shows good agreement with available experimental results for CHO systems [12–14]. Herein, we briefly summarize the protocol and point out some of its limitations, reproducing some portions of text in our previous work [11] for clarity and reading ease.

The structure of each molecule is optimized and its electronic energy calculated at the B3LYP [15–18] flavor of density functional theory with the 6-311G** basis set. To maximize the probability of finding global minima, multiple conformers are generated using molecular mechanics (OPLS force field with water as a solvent) [19]. The optimized structures are embedded in a Poisson–Boltzmann continuum to calculate the aqueous solvation contribution to the free energy. While this does not provide a specific concentration, it assumes a dilute solution such that the electrostatic field generated by a neighboring solute molecule is effectively screened by the water solvent. One can consider all solutes to have the same relative concentrations in our calculations.

Zero-point energy corrections are included based on the calculated analytical Hessian. We apply the standard temperature-dependent enthalpy correction term (for 298.15 K) from statistical mechanics by assuming translational and rotational corrections are a constant times kT, and that low-frequency vibrational modes generally cancel out when calculating enthalpy differences. So far, this is standard fare.

Entropic corrections in aqueous solution are more problematic [20–22]. Changes in free energy terms for translation and rotation are poorly defined in solution due to restricted complex motion, particularly as the size of the molecule increases (thus increasing its conformational entropy). Free energy corrections come from two different sources: thermal corrections and implicit solvent. Neither of these parameters is easily separable, nor do they constitute all the required parts of the free energy. We follow the approach of Deubel and Lau [23], assigning the solvation entropy of each species as *half* its gas-phase entropy (calculated using standard statistical mechanics approximations similar to the enthalpy calculations described above), based on proposals by Wertz [24] and Abraham [25] that upon dissolving in water, molecules lose a constant fraction (~0.5) of their entropy.

When put to the test by first calculating the equilibrium concentrations in a self-oligomerizing solution of 1 M glycolaldehyde at 298 K, our protocol fared very well compared to subsequent NMR measurements [14]. That being said, our protocol does show systematic errors when calculating barriers, but since quantitative kinetics is not part of the present study, this is not an issue. When extended to include nitrogen-containing species, our protocol still does well but systematic errors do crop up for specific functional groups [26]. Going to a higher level of theory and/or including dispersion corrections does not improve the results (see Table S1); this may seem surprising but quantum chemistry is about error cancellation, and our protocol (with its foibles) seems to work well, at least for the compounds in this study.

The *relative* aqueous free energies, designated G_r, are calculated with respect to three reference molecules: H_2, H_2O, and H_2CO_3. We use H_2CO_3 instead of CO_2 because it allows a direct comparison with experimentally-derived thermodynamic data from Alberty [27]. Thus, if G is the free energy calculated from quantum chemistry, then the formation reaction of glycolaldehyde is $2\,H_2CO_3 + 4\,H_2 \rightarrow C_2H_4O_2 + H_2O$, and G_r(glycolaldehyde) = $G(C_2H_4O_2) + G(H_2O) - 2\,G(H_2CO_3) - 4\,G(H_2)$. In this study, free energy corrections for concentration differentials among species (to obtain chemical potentials) are not included.

We have calculated all compounds in their neutral form. Thus, acetic acid is calculated rather than acetate, because this gives much better results in comparison to experimental data as discussed in the Results section, where we compare our calculations of both the neutral and anionic forms for the TCA cycle. (Additionally, our protocol was established and validated only for neutral compounds.) Additional comparisons of neutral versus anions for other cycles are provided in Tables S2–S4. The protocol used for calculating anions is similar to the neutral, except for the addition of diffuse functions to the basis set of anions, a standard procedure in quantum chemical calculations.

Our set of calculated compounds includes all CHO molecules from C_1 to C_4 with carbon-carbon backbones (except alkynes). Select compounds with oxygen in the backbone (ethers, esters, anhydrides) are included, although non-exhaustively (they are less stable than their carbon-backbone isomers). For each unique molecular species, we report only the free energy of the most stable conformer. In a small number of cases, this may be the enol or the hydrate. For compounds with one chiral center, the biologically relevant one was calculated; for two chiral centers, the most stable diastereomer is reported. The list of compounds explicitly mentioned in the Results section and their corresponding G_r values can be found in Appendix A. The full list is provided in Table S5). A graphical comparison of our calculated G_r values to experimentally-derived ones is also shown in Appendix A.

Assigning H_2, H_2O, and H_2CO_3 as reference states allows us to quickly and easily visualize a map of the energy landscape for the myriad reactions that can take place. For a chemical reaction, the difference in free energies will be designated ΔG, calculated as G_r(products)–G_r(reactants). For example, ΔG for the conversion of acetic acid into pyruvic acid is simply G_r(pyruvic acid)–G_r(acetic acid), since the reference state molecules (H_2, H_2O, H_2CO_3) by definition have G_r values of zero.

3. Results

Our results and discussion are organized as follows: (1) First, we provide further validation of our methods by comparing our calculated ΔG values against experimental data for the TCA cycle. (2) Next, we examine the free energy profiles of four core CO_2-fixation cycles in Braakman and Smith [1]. (3) Then, we discuss why embedding $C_1 + C_1 \rightarrow C_2$ in a cycle is fundamental using the example of formaldehyde oligomerization. (4) This leads naturally to our exploring a wider scope of reactions and cycles that might have played a role in proto-metabolic systems. (5) Finally, in the Discussion section, we take a bird's eye view of the overall thermodynamic picture with an eye on redox states.

3.1. TCA Cycle: Comparison to Experimentally-Derived Data

Using experimentally-measured equilibrium constants, Alberty derived standard free energies of formation for a range of biochemical compounds [27]. Not surprisingly, there is a relatively good match when comparing the Gibbs free energies of reaction from Alberty's Table 4.10 (pH 7, 298.15 K, 0.25 M ionic strength, 1 M solute) to data provided from a standard biochemistry textbook [28], for the TCA cycle as shown in Table 1. The main discrepancy is Step 1, where Alberty has a more exergonic reaction by 3.2 kcal/mol. As a result, Alberty's net cycle is overall 3 kcal/mol more exergonic than the textbook value. Note that Alberty *assigns* $\Delta G = 0.0$ in Step 6 to match the experimental data when attempts to derive free energies for the enzyme-bound FAD proved unsatisfactory.

Table 1. ΔG in kcal/mol for reactions in the oxidative TCA cycle with co-factors.

Step	Reaction	Textbook	Alberty	Quantum
1	acetyl-CoA + oxaloacetate + H_2O → citrate + CoA	−7.5	−10.7	−4.5
2	citrate → cis-aconitate + H_2O	+2.0	+2.0	+3.8
3	cis-aconitate + H_2O → isocitrate	−0.5	−0.4	−3.0
4	isocitrate + NAD^+ → α-ketoglutarate + CO_2 + NADH	−2.0	−1.1	−0.3
5	α-ketoglutarate + NAD^+ + H_2O + P_i + ADP → succinate + CO_2 + NADH + ATP	−8.0	−8.6	−10.5
6	succinate + $FADenz_{ox}$ → fumarate + $FADenz_{red}$	~0	0.0	N/A
7	fumarate → malate	−0.9	−0.9	−1.2
8	malate + NAD^+ → oxaloacetate + NADH	+7.1	+6.9	+8.0
	Net Cycle:	−9.8	−12.7	−7.7

How do our quantum calculations compare? We do not explicitly calculate the co-factors CoA, NAD^+/NADH and ADP/ATP. Therefore, to correct for their inclusion, we recalculated Alberty's ΔG values for each reaction in the absence of these co-factors to find empirical corrections, which are: −7 kcal/mol for AcCoA → CoA + acetate; −9 kcal/mol for NAD^+ → NADH; and +7 kcal/mol for ADP + P_i → ATP. Both Alberty and the textbook split Step 5 into two separate reactions, but since we do not explicitly calculate succinyl-CoA as separate from succinate, we combined these into a single step.

Our results show some differences compared to the experimental data. We calculate larger energy changes for the hydration/dehydration sequence in Steps 2 and 3. For the $NAD^+/NADH$ couple, our Step 8 is slightly more endergonic, while our Steps 4 and 5 are more exergonic. Opposite to Alberty, our Step 1 ΔG value and net cycle are both less exergonic than the textbook data. However, we are more interested in the free energy profiles in the absence of these co-factors, and to that data we now turn.

Table 2 shows ΔG calculated from five sources of data:

- Alberty's Table 4.1 [27]: $\Delta_f G_i'^0$ values (298.15 K, pH 7, zero ionic strength, 1 M solute) as these correspond to our quantum calculations that do not include ionic strength.
- The larger experimentally-derived dataset from eQuilibrator [29,30] that built on and expanded the Alberty data: $\Delta_f G_i'^0$ values also at pH 7 and zero ionic strength.
- Our quantum calculated data for neutral molecules.
- Our quantum calculated data for anions (with the addition of diffuse functions in the basis set for all calculations, and adding bicarbonate as a reference compound).
- The group-additivity approach of Jankowski et al. [31]: $\Delta_f G'^0$ values (derived at zero ionic strength)

Table 2. ΔG in kcal/mol for reactions in the TCA cycle without co-factors.

Step	Reaction	Alberty	eQuil	Quantum (Neutral)	Quantum (Anions)	Jankowski
1	ACE + OXA → CIT	−0.81	−1.19	+2.49	+8.97	−0.71
2	CIT → CAC + H_2O	+2.01	+2.80	+3.80	−3.34	+4.59
3	CAC + H_2O → ISC	−0.43	+1.14	−3.02	+3.29	−3.17
4	ISC → OXS + H_2	+19.04	+11.09	+17.22	+11.30	+8.02
5	OXS + H_2O → AKG + H_2CO_3	−9.69	−2.31	−8.46	−7.51	−12.83
6	AKG + H_2O → SUC + H_2 + H_2CO_3	−6.75	−4.54	−7.98	−3.63	−4.16
7	SUC → FUM + H_2	+25.38	+23.78	+27.12	+21.53	+19.12
8	FUM + H_2O → MAL	−0.86	−0.74	−1.22	+4.74	−3.73
9	MAL → OXA + H_2	+16.01	+14.60	+15.08	+14.26	+8.02
	Net Cycle:	**+43.90**	**+44.63**	**+45.03**	**+49.61**	**+15.15**

The reaction cycle has nine steps because we explicitly include oxalosuccinate as a distinct species. For the metabolite names, we use the abbreviations of Braakman and Smith in their depiction of the four core-cycles (to be discussed in the next section); name–abbreviation connections are found in Appendix A. H_2 is the reductant in this system.

Not surprisingly, the Alberty and eQuilibrator data are very similar with the exception of OXS, where their $\Delta_f G_i'^0$ values differ significantly. This leads to the 7–8 kcal/mol uphill and downhill differences in Steps 4 and 5. However, the overall net cycles are similar at +44 kcal/mol. (With no co-factors the *oxidative* TCA cycle is rather endergonic!) The most endergonic step is the oxidation of SUC to FUM, and in the absence of the enzyme-bound-FAD complex, the raw cost of breaking two C–H bonds to form a π bond between the carbons and one H–H bond is expected to be approximately +25 kcal/mol.

Our quantum calculations for the neutral molecules are not too different (they match Alberty slightly better than eQuilbrator) and the overall cycle is +45 kcal/mol. This gives us confidence that our calculated ΔG values match reasonably well with experimental-derived data. In contrast, if we calculate all these compounds as anions (carboxylates) instead of the neutral COOH groups, there are significant differences in more than half the steps, although the overall cycle is not too different at +50 kcal/mol. (Using a higher level of theory does not improve the result; see Table S1.)

Experimentally-derived data are limited to compounds of biochemical interest. However, our goal is to expand the set of compounds to explore closely related compounds not utilized by extant biochemistry. Hence, we also used the group additivity method of Jankowski et al. [31] to calculate the reactions in this cycle. (Note that eQuilibrator also utilizes some group additivity in its calculations [30].) In this case, the Jankowski scheme does poorly when compared to Alberty. The overall cycle is +15 kcal/mol primarily due to underestimating the endergonicity of the oxidation steps (4, 7 and 9).

For the rest of the Results section, we will not be discussing the quantum calculations using anions since their ΔG values do not match well with experiment. (These, along with additional Jankowski and eQuilibrator data are included in Supplementary Materials for the interested reader.) Additionally, because the Alberty dataset is limited, we will compare our quantum calculations to eQuilibrator data where available.

3.2. Thermodynamics of Core Cycles for CO_2 Fixation

Since the oxidative TCA cycle is net endergonic, this provides an opportunity for exergonic CO_2 fixation by running the cycle in reverse. Figure 1 shows the four core cycles: TCA (in black), dicarboxylate/4-hydroxybutyrate (DC/4HB in red), 3-hydroxypropionate bicycle (3HP, in blue), and 3HP/4HB (in green). Figure 1 is adapted (under the Creative Commons license agreement) from Figure 6 in Braakman and Smith [1], where we have added the G_r values (in kcal/mol) from our quantum calculations for each metabolite.

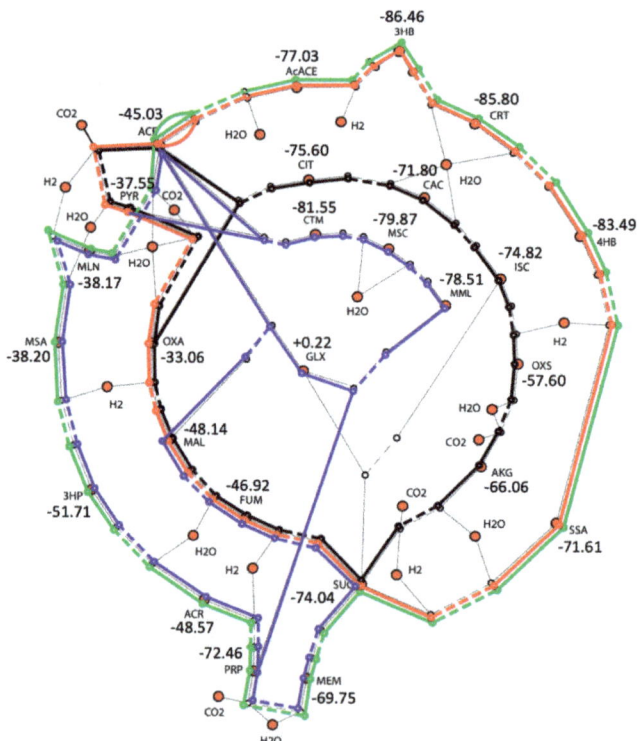

Figure 1. Metabolites and G_r values (in kcal/mol) for the four core metabolic cycles in Braakman and Smith.

As described in the Methods section, all G_r values are with respect to the reference molecules H_2, H_2O and H_2CO_3. This allows us to quickly calculate ΔG for any reaction of interest in these cycles. For example, oxaloacetatic acid (OXA) in the *reductive* TCA cycle is first reduced to malic acid (MAL). (We use the "acid" names rather than the anion names because our calculations are for the neutral form.) ΔG for this reaction is $-48.14 + 33.06 = -15.08$ kcal/mol. This reduction reaction is the exact opposite of its oxidative counterpart (Step 9 in Table 2). The overall cycle is exergonic with a net change of -45.03 kcal/mol (the exact opposite of $+45.03$ kcal/mol in Table 2). This value is also equal to the net chemical reaction of the cycle which converts two equivalents of CO_2 into acetic acid.

$$2 H_2CO_3 + 4 H_2 \rightarrow ACE + 4 H_2O \qquad (1)$$

Equation (1) is also the net reaction for the DC/4HB and 3HP/4HB cycles, and is a key example of incorporating the $C_1 + C_1 \rightarrow C_2$ reaction into a cycle.

In Figure 1, the G_r value for ACE is -45.03 kcal/mol since Equation (1) represents the formation of ACE with all other molecules being reference states. For eQuilibrator data, the equivalent G_r value for acetate is -44.63 kcal/mol (see Appendix A), which is exactly what you would expect given the net cycle in the oxidative direction from Table 2 is $+44.63$ kcal/mol.

3.2.1. The Reductive TCA Cycle

Let us examine the free energy changes for each step along the reductive TCA cycle by following the black lines in Figure 2. The reaction sequence is as follows:

1. OXA → MAL (reduction)—exergonic,
2. MAL → FUM (dehydration)—mildly endergonic,
3. FUM → SUC (reduction)—very exergonic,
4. SUC → AKG (formal addition of $CO_2 + H_2$)—endergonic,
5. AKG → OXS (formal addition of CO_2)—endergonic,
6. OXS → ISC (reduction)—exergonic,
7. ISC → CAC (dehydration)—mildly endergonic,
8. CAC → CIT (re-hydration)—mildly exergonic, and
9. CIT → OXA + ACE (splitting reaction)—mildly exergonic.

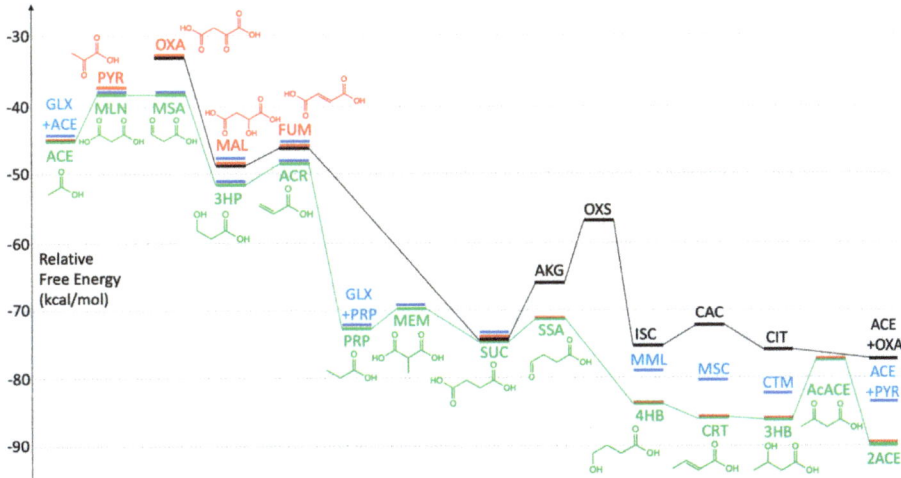

Figure 2. Reaction free energy profiles for the four core cycles.

The two mildly endergonic steps (2 and 7) are dehydration reactions. The two main endergonic steps (4 and 5) both involve oxidation because CO_2 (with a +4 oxidation state of carbon) is added. The first of these also adds an equivalent of the reductant H_2, i.e., a net change in oxidation of 2 units from SUC (+2) to AKG (+4). The second does not include H_2 and changes the oxidation state by 4 units: AKG (+4) to OXS (+8). Each step is ~8 kcal/mol uphill, and both steps occur in succession. Note that the first addition involves carboxylation at the terminus, while the second adds a branch to the backbone.

In contrast, the reduction steps (1, 3 and 6) are significantly downhill (ΔG of -15, -17 and -27 kcal/mol for OXA $\rightarrow$ MAL, OXS $\rightarrow$ ISC and FUM $\rightarrow$ SUC, respectively), and are the reason why the net cycle is overall exergonic. The final step splitting CIT (+6) to produce ACE (0) and regenerate OXA (+6) is mildly exergonic; OXA has the same overall oxidation state as CIT.

Could the reductive TCA cycle play a key role in prebiotic proto-metabolic systems? It is unclear. Having two successive endergonic reactions would require an oxidizing co-factor strong enough to compensate for the unfavorable uphill ~16 kcal/mol. The question of whether OXA could play the key role as the "recycled" molecule also poses problems given its kinetic instability to β-decarboxylation (and forming PYR). It is also unclear that the specific larger C_5 and C_6 compounds of the reductive TCA would be present in sufficient quantities to participate in a sustaining primordial cycle.

3.2.2. The 3HP/4HB Cycle

The 3HP/4HB cycle (green lines in Figures 1 and 2) mirrors the TCA cycle but only involves the smaller C_2 to C_4 metabolites. The net reaction is Equation (1) and the free energy change of the overall cycle is -45.03 kcal/mol. The "recycled" molecule is ACE itself. Let us consider the reaction steps in detail.

1. ACE $\rightarrow$ MLN (formal addition of CO_2)—endergonic,
2. MLN $\rightarrow$ MSA (acid-to-aldehyde reduction)—little change,
3. MSA $\rightarrow$ 3HP (reduction)—exergonic,
4. 3HP $\rightarrow$ ACR (dehydration)—mildly endergonic,
5. ACR $\rightarrow$ PRP (reduction)—very exergonic,
6. PRP $\rightarrow$ MEM (formal addition of CO_2)—endergonic,
7. MEM $\rightarrow$ SUC (methyl shift)—mildly exergonic,
8. SUC $\rightarrow$ SSA (acid-to-aldehyde reduction)—little change,
9. SSA $\rightarrow$ 4HB (reduction)—exergonic,
10. 4HB $\rightarrow$ CRT (dehydration)—mildly exergonic, which seems odd,
11. CRT $\rightarrow$ 3HB (re-hydration)—mildly exergonic,
12. 3HB $\rightarrow$ AcACE (oxidation)—endergonic, and
13. AcACE $\rightarrow$ 2 ACE (splitting reaction)—exergonic.

The first CO_2 addition takes place in the very first step at the α-carbon, converting ACE to MLN. This makes sense; highly oxidized CO_2 has a very positive carbon so we expect it to add to the more negative (or least positive) carbon on ACE, in this case the methyl group. For this oxidation step, $\Delta G = +7$ kcal/mol. Note, however, that in the absence of catalysts, the addition of CO_2 will have a high barrier.

The second step is a reduction of carboxylic acid to aldehyde to form MSA, and unlike reductions we have seen in the TCA cycle, it is energetically neutral. We will expand on this feature later, but for now, note that it involves a condensation (releasing H_2O) in concert with the addition of H_2. In contrast, the reactions we examined in the TCA cycle only involved H_2 addition (ketone to alcohol, or the very exergonic alkene to alkane).

The next several steps MSA $\rightarrow$ 3HP $\rightarrow$ ACR $\rightarrow$ PRP are similar to the OXA $\rightarrow$ MAL $\rightarrow$ FUM $\rightarrow$ SUC sequence. However, when PRP is oxidized by an equivalent of CO_2, branched MEM is produced rather than linear SUC. Once again, it makes sense to add CO_2 to the least positive carbon, in this case the α-carbon of PRP (in terms of alternant "plus" and "minus" carbons in the chain of a polar compound). Interestingly, this reaction is only mildly endergonic ($\Delta G = +3$ kcal/mol), noticeably less than previous CO_2 incorporations we have considered thus far. We will see this again in several cases when CO_2 is added to a carbon that is not bonded to oxygen (e.g., a methyl or methylene group).

Although exergonic, the isomerization of MEM to SUC whereby the methyl branch insinuates itself into the chain is unlikely to occur in a prebiotic milieu (not a problem for the evolved enzyme methylmalonyl-CoA mutase); and likely has a high barrier (uncatalyzed). Moving along, SUC $\rightarrow$ SSA $\rightarrow$ 4HB mirrors MLN $\rightarrow$ MSA $\rightarrow$ 3HP, although SUC $\rightarrow$ SSA is mildly exergonic instead of being energy neutral. You might expect dehydration of 4HB to be mildly endergonic, and it would be indeed if but-3-enoic acid was formed (see Section 3.4), but the formation of CRT also includes a double bond shift to the more stable isomer. Rehydration to 3HB then becomes mildly endergonic.

However, because 4HB/CRT/3HB have oxidation state −2, the cycle now requires an alcohol to ketone oxidation, uphill ~8 kcal/mol. Once zero-oxidation state AcACE is formed, it can split apart to two ACE molecules in a reaction resembling a reverse Claisen condensation; this reaction that recycles ACE is downhill (ΔG = −13 kcal/mol).

How does the 3HB/4HB cycle compare to TCA? It splits up the endergonic steps involving oxidative CO_2 addition, the second being only mildly endergonic, which might be advantageous. However, the cycle over-reduces, thus requiring a third endergonic oxidation step before splitting and recycling ACE. Furthermore, converting branched MEM to SUC, although thermodynamically favorable, is likely to be kinetically inaccessible without highly specific enzymes or co-factors.

Compared to eQuilibrator, our G_r values (in Figure 1) for the C_3 species are quite similar with the exception of PRP: eQuilibrator G_r values (in kcal/mol) are MLN (−39.42), MSA (−38.22), 3HP (−52.18), ACR (−49.53), PRP (−77.04). For PRP, our calculated G_r is −72.46, a difference of over 4 kcal/mol. For the C_4 species, our G_r values are consistently more negative by ~2 kcal/mol with the exception of 3HB and 4HB: eQuilibrator G_r values (in kcal/mol) are MEM (−67.76), SUC (−71.35), SSA (−69.51), 4HB (−77.78), CRT (−82.49), 3HB (−87.39), AcACE (−75.35). For 4HB, our calculated G_r is −83.49, a difference of over 5 kcal/mol. For 3HB, our calculated G_r is −86.46 which is less negative than eQuilibrator.

3.2.3. The DC/4HB Cycle

The DC/4HB cycle (red lines in Figures 1 and 2) utilizes the first half of the TCA cycle (OXA to SUC) and the second half of the 3HP/4HB cycle (SUC to AcACE). It also only involves the smaller C_2 to C_4 metabolites, and once again the net reaction is the same as Equation (1) and the overall cycle is −45.03 kcal/mol. However, it begins with ACE (and not OXA), and thus the recycled molecule is ACE itself.

The first two steps frontload the uphill oxidative incorporation of CO_2 equivalents. ACE → PYR is analogous to SUC → AKG, in that "$CO_2 + H_2$" are added to the terminal carboxylic acid, and the oxidation state changes by +2. The second step adds CO_2 to the least positive carbon, in this case the methyl of PYR (+2), to form OXA (+6) with an oxidation state change of +4 mirroring the AKG → OXS reaction. However, unlike SUC → AKG → OXS which is 16 kcal/mol uphill, ACE → PYR → OXA is only 12 kcal/mol uphill because the second step is less endergonic (ΔG = +5 kcal/mol). While this seems more feasible, the question is whether the first carboxylation of ACE to PYR is more or less favorable than to MLN. Thermodynamically, there is little difference (MLN is a mere 0.6 kcal/mol more stable), but kinetically things could be quite different depending on what primitive catalysts or co-factors might be present and whether H_2 can easily be incorporated as a reductant in this first step. Answering this question is beyond the scope of this article (we will tackle the kinetics in future work). As previously mentioned, CO_2 addition will have high barriers in the absence of catalysts, although our present work focuses quantitatively only on the thermodynamics.

The rest of the cycle was discussed earlier since it overlaps with the TCA and 3HP/4HB cycles. Once OXA is formed, it is mostly downhill all the way to 3HB except that over-reduction requires an uphill oxidation to AcACE prior to splitting apart into two ACE. Our calculated G_r for PYR (−37.55) is marginally less negative than eQuilibrator's value of −38.87 kcal/mol.

3.2.4. The 3HP Bicycle

Unlike the previous three cycles, the 3HP bicycle (blue lines in Figures 1 and 2) does not have the same net reaction. The first half of the cycle from ACE to PRP overlaps with the 3HP/4HB cycle, but then the path splits into two: (1) PRP can add GLX to access

C_5 compounds (MML, MSC, CTM) before splitting into ACE and PYR. (2) PRP can add CO_2 to access the C_4 compounds (MEM, SUC, FUM, MAL) before splitting into ACE and recovering GLX. Thus, the net reaction as shown in Equation (2) is

$$PRP + 2\,H_2CO_3 + 2\,H_2 \to ACE + PYR + 3\,H_2O \qquad (2)$$

and the net cycle is exergonic with a free energy change of -10.12 kcal/mol. While forming ACE from CO_2 contributes -45.03 kcal/mol as expected, oxidizing PRP to PYR is $-37.55 + 72.46 = +34.91$ kcal/mol. Hence, the net -10.12 kcal/mol.

GLX (glyoxylic acid / glyoxalate) is the key intermediate in the bicycle. It is a highly oxidized (+4) C_2 compound and a high-energy intermediate (G_r = +0.22). This means that when employed as a reactant, it can drive a downhill reaction to form thermodynamically more stable products. However, recycling GLX comes at a cost. The killer step, in this case, is the oxidation of SUC to FUM (ΔG = +27.12 kcal/mol), part of the sequence PRP $\to$ MEM $\to$ SUC $\to$ FUM $\to$ MAL $\to$ ACE + GLX which is oxidative and endothermic +27.65 kcal/mol. This is balanced by the downhill sequence starting from ACE and involving the C_5 compounds (ΔG = -37.77 kcal/mol), recycling ACE while converting GLX (+4) to PYR (+2), a net reductive reaction.

In the left half of Figure 2, there is a small gap separating the blue bars from the other bars, signifying the presence of GLX contributing G_r = + 0.22 kcal/mol to the relative free energy. We will revisit the potential wider role of GLX in Section 3.4.

3.3. Back to Basics: Incorporating $C_1 + C_1 \to C_2$ into a Cycle

While Stanley Miller's landmark experiment [32] demonstrated that the synthesis of amino acids was possible in a *highly* reducing atmosphere [33], the present scientific majority view as summarized by Kasting [34] is that CO_2 rather than CH_4 was the dominant C_1 building block available. Whatever the prebiotic milieu may have been, even if a range of carbon-containing compounds was available, the growth and sustenance of a protometabolic cycle requires "food" molecules—likely to be C_1, as this would deplete specific larger molecules in the cycle (via hydrolysis or other parasitic side-reactions) unless they are replenished by a C_1 source.

The problem, however, is that the direct $C_1 + C_1 \to C_2$, even if thermodynamically feasible, is kinetically challenging. In CO_2 (or HCO_2H, CH_2O, CH_3OH), carbon carries a partial positive charge because of the more electronegative oxygen. Making a new C–C bond between two such entities requires traversing high activation barriers. In the absence of co-factors or catalysts that could significantly lower the barrier (or using prohibitively high reaction temperatures), regular and constant production of C_2 (or larger) molecules would be very slow. While the carbon of CH_4 carries a small partial negative charge, CH_4 is generally unreactive to forming new C–C bonds (in the absence of metal-containing catalysts).

CO is an interesting case; its net dipole is close to zero although its carbon carries a small negative partial charge; but CO is much less stable (thermodynamically) than CO_2 and unlikely to exist in large quantities over a long period of time. It could, however, play an important role in situ, generated transiently from CO_2 in the presence of H_2. Using our reference states, this reaction as shown in Equation (3) is:

$$H_2CO_3 + H_2 \to CO + 2\,H_2O \qquad (3)$$

From our quantum calculations, $G_r(CO)$ = +3.33 kcal/mol, and thus this reaction is endergonic. (In contrast, formation of the other C_1 compounds from CO_2 and H_2 is exergonic, see the Discussion section.) This reaction could play a role in CO_2-fixation steps, as we have seen in the previous section, where several reactions involved the formal addition of "$CO_2 + H_2$" such as ACE (0) $\to$ PYR (+2).

The cycles we examined in the previous section all involve adding CO_2 to a species that is C_2 or larger. Why is this kinetically more feasible? A polar organic molecule, where one of the carbons carries a (significant) partial positive charge usually also contains neighboring carbons that are less positive or even have a (slight) partial negative charge. Addition of CO_2 can be "directed" towards the less positive carbon of this larger molecule with a lower barrier—an *umpolung*-like "strategy" if you will. This provides a feasible way to incorporate CO_2 without the cost of direct $C_1 + C_1$ addition.

However, how is the cycle created? At some point, a larger molecule would have to split into smaller entities (but not C_1 as this would simply be a reverse addition). Hence, the simplest cycle that can be generated involves a C_2, C_3, and a C_4 species, with the C_4 able to split into two C_2 species, as shown in Figure 3A.

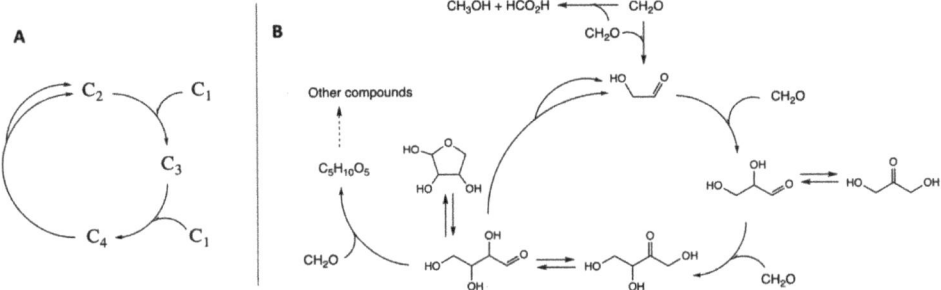

Figure 3. (**A**) Simplest cycle. (**B**) Key features of the CH_2O oligomerization cycle.

We illustrate this with a formaldehyde oligomerization cycle, part of the formose reaction [35], as shown in Figure 3B. In previous work [13], we calculated the free energy profiles for CH_2O oligomerization and explored both the thermodynamic and kinetic features of this cycle, which we now summarize briefly. Experimentally the reaction starting with solely CH_2O has an induction period because the direct dimerization of CH_2O to form glycolaldehyde (GA) has a very high barrier (recall our earlier discussion of why direct $C_1 + C_1 \rightarrow C_2$ is kinetically challenging). Once the C_2 species is formed, however, aldol addition of CH_2O proceeds rapidly (under appropriate experimental conditions) to generate a wide range of CHO-containing molecules, bypassing the direct $C_1 + C_1$.

The cycle has several interesting features. Aldol additions of CH_2O (hydrate) are exergonic (−6 kcal/mol, on average). The $C_2 + C_1 \rightarrow C_3$ kinetically favors glyceraldehyde (GLA), which may subsequently isomerize (via an enol intermediate) to thermodynamically more stable dihydroxyacetone (DHA). The kinetically favored $C_3 + C_1 \rightarrow C_4$ converts GLA into the ketone (erythrulose), which is thermodynamically more stable than its aldehyde isomer (erythrose). Erythrose can potentially cyclize, add further CH_2O units, or undergo a reverse aldol ($C_4 \rightarrow C_2 + C_2$) to regenerate GA. The splitting reaction is marginally endergonic (+2 kcal/mol). The overall cycle is exergonic by −7 kcal/mol.

We note three other interesting things about this system: (1) Active species in the cycle are in equilibrium with "off-cycle" compounds, which together form a "pool" of interconnected compounds—a feature that could stabilize the cycle and provide a point of regulatory control [36]. (2) Because the C_2 "recycling" species (GA) and the C_1 "food" species (CH_2O) are both at oxidation state zero, additional redox reactions are not required, thereby simplifying the situation. (3) Cannizzaro reactions (such as $2\ CH_2O + H_2O \rightarrow HCO_2H + CH_3OH$) can be parasitic on the cycle, but also provide access to compounds with a wider range of oxidation states; and experimentally the formose reaction generates a range of carboxylic acids [37], including those found in extant biochemistry.

This sets the stage for us to explore alternative cycles that are thermodynamically favorable, but possibly more feasible kinetically than the extant cycles we have analyzed in the previous section.

3.4. Alternative Cycles: A Brief Exploration

In extant biochemical cycles, ACE (overall zero oxidation state) is the key recycled C_2 species. The main goal of this section is to explore if this role can instead be played by one of its chemical cousins: ethanal (−4), glycolaldehyde (−2), glycolic acid (+2), or GLX (+4). We will also allude to CH_2O as a potential crossover C_1 food species, and we will see that intersecting cycles allow for a variety of possibilities. Before we embark, some words of caution. We remind the reader that our quantum calculations only provide thermodynamic data of the CHO metabolites; they do not take into account kinetics or even the presence of co-factors (such as sulfur-containing compounds) to tune the thermodynamics. We also chose to highlight a very limited set of specific examples that we think are interesting out of many possibilities. One should not over-interpret our results and fall into the trap of believing Kipling-esque *Just So* stories.

3.4.1. Minor Modifications to the 3HP/4HB Cycle

Before jumping into alternative C_2 species, one drawback to the extant 3HP/4HB cycle as discussed in Section 3.2.2 is over-reducing the carbon moiety, thereby requiring an endergonic step to generate AcACE before it splits into two ACE. A straightforward way to avoid this: If SSA is not reduced to 4HB but instead undergoes a thermodynamically favorable aldehyde-to-ketone isomerization (via the enol), AcACE can be formed directly from SSA, as shown in the right half of Figure 4 (long green line) We would expect this direct SSA → AcACE conversion in the absence of any reducing agents.

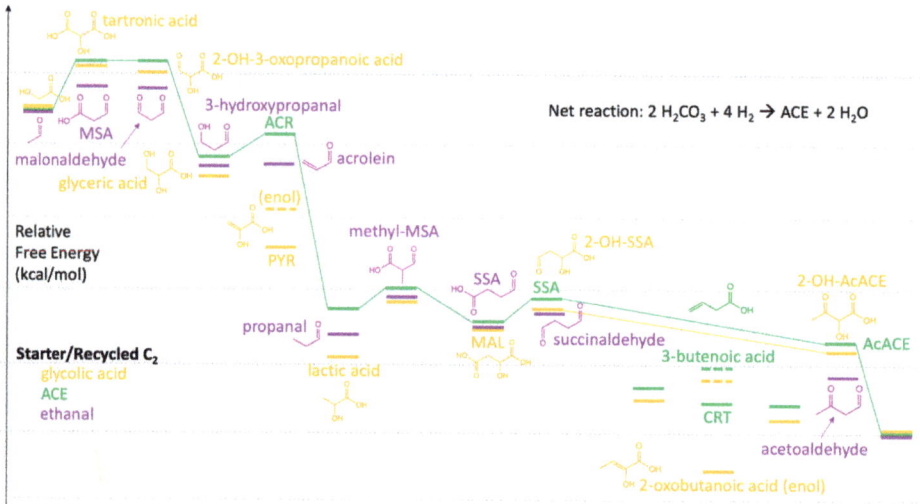

Figure 4. Relative energy profiles of modified 3HB/4HP cycles utilizing glycolic acid and ethanal.

However, the preceding reduction step, SUC → SSA, requires a reducing agent, and it may not be feasible to halt reduction of the carboxylic acid at the aldehyde depending on the reducing agent and the reaction conditions. If 4HB is formed, we would expect dehydration to result first in but-3-enoic acid (see Figure 4) followed by rehydration to 3HB, but now the endergonic oxidation must be carried out to form AcACE. Considering the pool of compounds in equilibrium, certainly CRT is in the mix as an isomer of but-3-enoic acid ($G_r = -80.19$); an isomer of 3HB and 4HB is 2-hydroxybutanoic acid ($G_r = -82.79$); and 2-oxobutanoic acid ($G_r = -68.75$) is an isomer of SSA and AcACE. The extant 3HP/4HB cycle utilizes the thermodynamically favorable isomers, but this may be less kinetically feasible for a proto-metabolic cycle that avoids or minimizes uphill steps.

3.4.2. Glycolic Acid as the Recycling C_2

Glycolic acid is produced in measurable quantities in formose reactions [37], Miller spark-discharge experiments [38], and plays a potential role as a scaffold in oligopeptide-forming chemistry [39]. Could it play a role as the recycling C_2 compound instead of ACE? In Figure 4, we plot the energy profile for glycolic acid (gold lines) relative to ACE (green lines) assuming the same reaction steps as the 3HP/4HB cycle. The net reaction is still similar to Equation (1) and the cycle overall exergonic at -45.03 kcal/mol.

The first CO_2 addition forming tartronic acid is endergonic (6 kcal/mol), marginally less (by 1 kcal/mol) compared to the 3HP/4HB cycle. Three exergonic reactions follow: Two reductions lead to glyceric acid, which when dehydrated, converts into the enol of PYR. This provides a connection to the DC/4HB cycle (red line in Figure 2), providing an alternative starting point to having two early successive endergonic reactions. However, a switch to DC/4HB would mean glycolic acid is depleted rather than recycled; the net reaction as shown in Equation (4) would then be:

$$\text{glycolic acid} + 2\,H_2CO_3 + 5\,H_2 \rightarrow 2\,\text{ACE} + 5\,H_2O \tag{4}$$

with an overall free energy of $-45.03 \times 2 + 18.08 = -71.98$ kcal/mol. (There is no cycle!)

However, let us keep following the gold line. Reduction of PYR leads to lactic acid where it may potentially accumulate as a thermodynamic sink. The second addition of CO_2 followed by a methyl shift leads to MAL. The analogous CO_2 addition in the green line was only 3 kcal/mol endergonic (see Section 3.2.2), but for the gold line this is back to the higher 7–8 kcal/mol range. From MAL, reduction of a carboxylic acid to an aldehyde is marginally endergonic and an aldehyde-to-ketone isomerization (long gold line) analogous to SSA → AcACE avoids the over-reduction and formation of the 2-oxobutanoic acid enol. The final split produces ACE and regenerates glycolic acid.

However, if 2-oxobutanoic acid was produced, recall in the previous section that it is a less stable structural isomer of AcACE and part of the same "pool". A keto-enol shift will convert it to AcACE which can split into two ACE. This would not recycle glycolic acid but provides a route to the 3HP/4HB cycle (green line in Figure 2), and illustrates one way such cycles crisscross and interconnect—a possible situation one might expect in a messier proto-metabolic milieu.

The kinetically challenging methyl-shift, however, remains problematic. Lactic acid could provide a way out. Unlike 3HP, ACR, or PRP, where the central carbon is the least positive and would be the site of CO_2 addition (forming branched C_4), in lactic acid the terminal methyl is the least positive carbon. If CO_2 were to be added there, it could provide a route to unbranched MAL ($G_r = -48.14$) and the DC/4HB cycle. This CO_2 addition would only be endergonic by 4 kcal/mol, given that G_r(lactic acid) is -52.03 kcal/mol, however it might still be kinetically challenging.

In fact, 3HP ($G_r = -51.71$) could convert to lactic acid through a dehydration-rehydration via ACR—and this provides a tantalizing cycle which modifies the 3HP/4HB cycle by patching in a small part of the DC/4HB: ACE → MLN → MSA → 3HP → lactic acid → MAL → FUM → SUC → SSA → AcACE → split. Glycolic acid is not needed in this case.

3.4.3. Ethanal or Glycolaldehyde (GA) as the Recycling C_2

Having investigated glycolic acid as an oxidized cousin of ACE, we now examine the free energy profile starting of its reduced counterparts. The purple lines in Figure 4 are for ethanal as the recycling C_2 species assuming the same reaction steps as the 3HP/4HB cycle; the net reaction is still Equation (1).

The first CO_2 addition converts ethanal to MSA (a connection to the 3HP route!) and this oxidation is mildly endergonic ($\Delta G = +3$ kcal/mol). Reduction of MSA to malonaldehyde is energy-neutral; reduction to 3-hydroxypropanal is exergonic as expected; dehydration to acrolein is energy-neutral; and then reduction to propanal is highly exergonic (analogous to ACR → PRP). The second CO_2 addition to a branched C_4 species, followed by the prebiotically questionable methyl shift, leads to SSA (in the 4HB route!). Isomerization to acetoaldehyde followed by splitting produces ACE and regenerates ethanal. The net reaction is Equation (1) and the net cycle is −45.03 kcal/mol. If instead, there is a switch to the 3HP/4HB cycle, ethanal is depleted rather than recycled in the net reaction as shown in Equation (5)

$$\text{ethanal} + 2\ H_2CO_3 + 3\ H_2 \rightarrow 2\ ACE + 3\ H_2O \tag{5}$$

with an overall free energy of $-45.03 \times 2 + 41.62 = -48.44$ kcal/mol, but not a cycle.

One might also imagine (not shown), a rehydration of acrolein to form lactaldehyde, which perhaps might lead directly to SSA via CO_2 addition to the terminal methyl. This might be possible if no reducing agent was present, but that is a challenge given the prior reduction steps.

What if we start from GA? The first addition of CO_2 leads to 2-OH-3-oxopropanoic acid (ΔG = +4 kcal/mol). Two reduction steps lead to glucic acid and glyceraldehyde (GLA), respectively. Dehydration leads to the enol form of methylglyoxal. If one were to add the second CO_2 to methylglyoxal, kinetically it would be most feasible to form 3,4-dioxobutyric acid (ΔG = +4 kcal/mol), thereby avoiding branched C_4 species (for the same reason as PYR → OXA). Reduction of the aldehyde and ketone to alcohols leads to 2-deoxythreonic acid, and then splitting produces ACE and regenerates GA. This last step is marginally endergonic (ΔG = +0.12 kcal/mol) but all other steps are exergonic except for the two CO_2 additions. This free energy profile is shown in Figure 5.

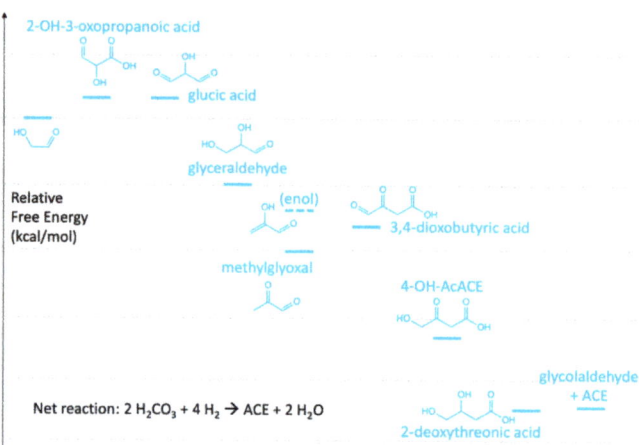

Figure 5. Potential cycle with glycolaldehyde as the C_2 recycling species.

This system has connections to CH_2O oligomerization. GA is the recycling C_2 species in the formose reaction, and GLA is the C_3 species. One could imagine coupled cycles where GLA in the formose cycle can undergo dehydration to methylglyoxal, thereby accessing the cycle in Figure 5. We could also imagine including CH_2O as an additional C_1 food source, which would allow aldol addition to ACE to form 3HP, skipping MLN and MSA in the (green-line) 3HP/4HB cycle; or better yet transformations that utilize an aldol addition to a C_3 to avoid forming branched C_4 species. Changing the food source would also change the net cycle reaction and thermodynamics.

However, before we get too excited at the tantalizing possibilities, keep in mind that our speculations of potential cycles are based on just the thermodynamics; we have not quantitatively accounted for the kinetics, or reducing agents appearing at the appropriate time, or hydrating/dehydrating conditions, just to name a few issues.

3.4.4. The Challenge of GLX

Earlier, we discussed the role of GLX as a high-energy intermediate, recycled in the 3HP bicycle pathway (Section 3.2.4). Could GLX play a role as the recycled C_2 species in a simpler cycle such as the ones we have just examined? One problem: Adding the first equivalent of CO_2 is significantly more endothermic, whether it produces oxomalonic acid (ΔG = +13 kcal/mol) with a +4 increase in oxidation state, or dioxopropanoic acid (ΔG = +14 kcal/mol) with concurrent addition of H_2 and an overall +2 increase in oxidation state. If instead GLX is reduced in an earlier step before addition of CO_2, this would be similar to the cases we have already examined.

Formation of GLX (G_r = +0.22) is marginally endergonic from the reference molecules. Alternatives include the reaction of CO_2 and H_2CO, the two "food" molecules we have considered, which is +5 kcal/mol uphill; or HCO_2H and CO which is +2 kcal/mol uphill (although CO would have to be generated prior). A prebiotic mixture might contain small amounts of GLX, but under reducing conditions, it would favorably convert into glycolic acid, ACE, or more reduced compounds.

However, as a high-energy reactant, GLX could drive thermodynamically favorable cycles. We previously saw (Section 3.2.4) that the downhill stretch of the 3HP bicycle has a free energy change of −37.77 kcal/mol. Stubbs et al. have experimentally designed a reverse TCA cycle analog that proceeds favorably in the absence of enzymes or metals as catalysts [40]. As shown in Table 3, our calculations show why this is thermodynamically favorable. The net chemical reaction as shown in Equation (6) is:

$$2 \text{ GLX} \rightarrow \text{ACE} + 2 \text{ H}_2\text{CO}_3 + \text{H}_2\text{O} \tag{6}$$

Table 3. ΔG in kcal/mol for the cycle designed by Stubbs et al.

Step	Reaction	Quantum	Jankowski
1	OXA + GLX + H_2O → maloylformate + H_2CO_3	−7.12	−15.39
2	maloylformate → fumaroylformate + H_2O	+0.78	+3.73
3	fumaroylformate + H_2 → AKG	−26.88	−22.33
4	AKG + GLX → isocitroylformate	+0.30	−2.56
5	isocitroylformate → aconitoylformate + H_2O	−0.30	+0.27
6	aconitoylformate + 2 H_2O → CAC + H_2CO_3 + H_2	−5.96	−4.46
7	CAC → CIT	−3.80	−5.49
8	CIT → OXA + ACE	−2.49	+0.71
	Net Cycle:	−45.47	−45.12

The overall free energy change is −45.47 kcal/mol. Interestingly, the Jankowski group-additivity method [31] yields a similar overall value for this cycle, although there are differences in the individual steps. (eQulibrator does not have $\Delta_f G_i'^0$ values for most of these "formates" so we are unable to make that comparison.)

This cycle does not fix carbon overall nor is it meant to. Instead, it essentially reduces GLX to ACE while releasing units of CO_2. GLX acts as an oxidizing agent, and this is its likely role in a proto-metabolic setting. With a G_r similar to the reference compounds, it is the C_2 equivalent of an "oxidizing" CO_2 unit.

As a second example, we examine a non-enzymatic cycle designed experimentally by Muchowska et al. [41]. This cycle marries parts of the oxidative TCA and glyoxalate cycles, but by utilizing GLX as the oxidizing agent, the net cycle is now marginally exergonic (−0.22 kcal/mol), instead of being +45.03 kcal/mol endergonic as we saw in Section 3.1. The net reaction as shown in Equation (7) is:

$$\text{GLX} + 3 \text{ H}_2\text{O} \rightarrow 2 \text{ H}_2\text{CO}_3 + 2 \text{ H}_2 \tag{7}$$

Individual steps in the cycle are shown in Table 4. The first half utilizes several compounds and reactions in common with *Stubbs* et al. [40], and we use the same names (from *Stubbs* et al.) in Table 3 rather than the different names used by Muchowska et al. (If comparing both papers, maloylformate = hydroxyketoglutarate; fumaroylformate = oxopentenedioate; isocitroylformate = oxalohydroxyglutarate.)

Table 4. ΔG in kcal/mol for the cycle designed by Muchowska et al.

Step	Reaction	eQuilibrator	Quantum	Jankowski
1	GLX + PYR → maloylformate	−3.09	−2.63	−0.75
2	maloylformate → fumaroylformate + H_2O	not available	+0.78	+3.73
3	fumaroylformate + H_2 → AKG	not available	−26.88	−22.33
4	AKG + GLX → isocitroylformate	not available	+0.30	−2.56
5	isocitroylformate + 2 H_2O → ISC + H_2CO_3 + H_2	not available	−9.28	−7.40
6	ISC → GLX + SUC	+1.39	+1.00	+5.08
7	SUC → FUM + H_2	+23.78	+27.12	+19.12
8	FUM + H_2O → MAL	−0.74	−1.22	−3.73
9	MAL → OXA + H_2	+14.60	+15.08	+8.02
10	OXA + H_2O → PYR + H_2CO_3	−5.16	−4.49	−14.64
	Net Cycle:	N/A	*−0.22*	*−15.46*

The cycle begins with PYR rather than OXA; Steps 2–4 are the same as Table 3; and Steps 7–9 are also found in the TCA cycle. While the main recycled compound is PYR, one equivalent of GLX is also recycled. (The two equivalents of GLX enter in Steps 1 and 4). From the quantum chemistry data, the very exergonic reductive Step 3 is balanced by the very endergonic oxidative Step 7. The Jankowski group additivity approach does not do as well here for the same reasons we saw in the TCA cycle. Where eQuilibrator values are available, our quantum calculations match up quite well. As in the previous example, GLX essentially acts a "high-energy" oxidizing agent allowing for exergonic steps.

4. Discussion

We are now ready to take a broader look at the overall thermodynamics of C_1 to C_4 species containing only carbon, hydrogen, and oxygen. We have arranged this map to emphasize redox states, inspired by Smith and Morowitz in their magisterial tome *The Origin and Nature of Life on Earth* [42]. We do not capture the intricacies they have investigated given the limited system size in our work, although our present analysis is influenced by Chapter 4 of their book.

Our overall thermodynamic map is displayed in Figure 6. Recall that the reference molecules (with G_r values of zero) are H_2, H_2O, H_2CO_3, and that our quantum chemical values are for neutral molecules in aqueous solution (zero ionic strength) under standard conditions at 298 K. The carbon "source" (CO_2, or H_2CO_3 in aqueous solution) has +4 oxidation state; and our present map represents a reducing environment with H_2 acting as the reductant. What features can we pick out under these conditions?

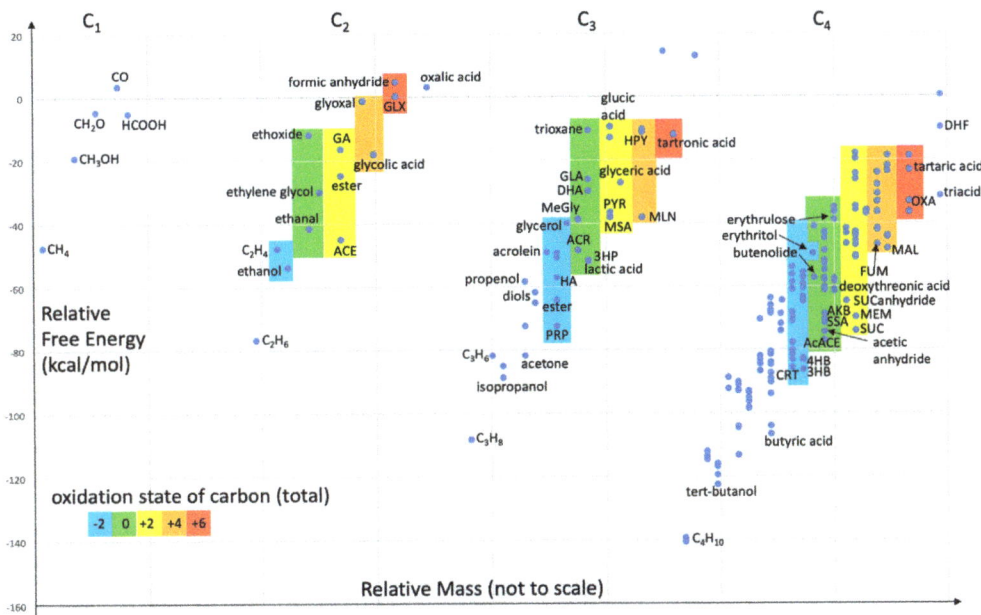

Figure 6. Overall thermodynamic redox map of C_1 to C_4 compounds (G_r values in kcal/mol).

For C_1 compounds, the most reduced compound CH_4 (−4) is the most stable, followed by CH_3OH (−2). Note that the aldehyde CH_2O (0) and the acid $HCOOH$ (+2) are of similar relative free energy. Carbon monoxide is the only compound higher in energy than the reference state. As size increases (towards the right in Figure 6), we see the same trends: alkanes, as the most reduced compounds, are the most stable; and relative free energies are less negative as oxidation state increases.

For C_2 compounds, ethanol is more stable than ethene, ethanal is more stable than ethylene glycol, ACE is more stable than GA, and glycolic acid is more stable than glyoxal. The same trends hold for larger (C_3 or C_4) compounds of the same oxidation state: an OH group is more stable than having a C=C double bond; COOH is more stable than separate C=O and OH groups; C=O is more stable than a diol; having acid and alcohol groups is more stable than two C=O groups (recall the Cannizzaro products in Section 3.3). Additionally, the ester methylformate is less stable than its acid isomer ACE, and formic anhydride is less stable than its isomer GLX.

Looking at the most stable compounds across oxidation states for the C_2 compounds, the series is ethane (−8), ethanol (−6), ethanal (−4), ACE (0), glycolic acid (+2), GLX (+4), oxalic acid (+6). Note that ACE is slightly more stable than ethanal, in the same way that HCO_2H is slightly more stable than CH_2O in the C_1 series. For the C_3 and C_4 compounds, simple ketones are more stable than their corresponding aldehydes and acids; although the acid is still more stable than the aldehyde (PRP is slightly more stable than propanal and butyric acid is slightly more stable than butanal). This similarity in energy between the acid and aldehyde may be an important feature in proto-metabolism by allowing essentially energy-neutral and thus thermodynamically reversible redox transformations, possibly aided by primitive redox catalysts. We previously alluded to this feature: acids, aldehydes and ketones potentially form an equilibrating pool of molecules that may exert some control and regulation [36,37]. To this, we might add esters and anhydrides that might allow for the sequestering of acid metabolites.

In a column of molecules sharing the same oxidation state, carboxylic acids (if they can be formed) are always the most thermodynamically stable. In the C_3 set for example, we have PRP (−2), lactic acid (0), MSA (+2), MLN (+4), and tartronic acid (+6). What are the most stable metabolites in the 3HP/4HB and the DC/4HB cycles? They are all acids and are among the most thermodynamically favored compounds in their respective columns. Once nature evolved specific enzymatic catalysts to reduce kinetic barriers, it seems to have maintained the primary use of thermodynamically stable compounds, the same ones you might expect to persist in a proto or pre-metabolic mixture.

An interesting metabolite not formally listed in the four core metabolic cycles is lactic acid, marginally more stable than its isomer 3HP (by 0.3 kcal/mol). We have speculated on its potential role in alternative (yet closely related) cycles in Section 3.4.4 but we do not yet know evolutionarily how it came to play its present role in extant metabolism. We should not discount it as an important player in a proto-metabolic system. Lactic acid's close cousins, 3HP and ACR are part of the 3HB/4HB cycle.

In an organic compound with only a single functional group, a ketone is noticeably more stable than its aldehyde isomer. However, the free energy difference is smaller in more oxidized molecules and there are potential reversals: MSA is marginally more stable than PYR. MSA also has a stable enol thanks to its carbonyl group in the β-position with respect to the acid group. That is also why AcACE (in the 4HB route) is the most stable zero oxidation state C_4 compound. Hence, if we imagined the simplest cycle utilizing the most stable acid molecules and CH_2O as "food" to avoid oxidation state changes, the C_2, C_3, C_4 candidates would be ACE, lactic acid (or 3HP), AcACE, respectively.

Could this be why ACE is the C_2 species recycled in extant core metabolisms? It has the same zero oxidation state as GA (in the formose reaction), but is significantly more stable. Glycolic acid and GLX are favorably reduced to ACE thermodynamically, and ACE is also more stable than its reduced counterpart ethanol. Perhaps ACE combines the twin features of stability and simplicity. Furthermore, having an anionic form (in contrast to ethanal) might provide an anchor to a proto-metabolic mineral surface as suggested by Wachterhauser in his origin-of-life proposals [43], potentially presaging the ubiquitous role of phosphates (perhaps later) in the evolution of metabolism. AcACE also plays a key role as the splitting C_4 species. Interestingly our calculations find that acetic anhydride is only marginally less stable than AcACE, and might play some role as a "pool" species if dehydrating conditions turn out to be important.

However, getting to AcACE is potentially tricky. We have proposed bypassing 4HB, CRT, 3HB, via a thermodynamically favorable aldehyde-to-ketone isomerization from SSA to avoid the late endergonic step. However, how do we get to SSA? At present, the most likely source is SUC, the only other molecule besides ACE to be present in all four core metabolisms shown in Figure 1. However, if SUC is being reduced to SSA, what stops reduction proceeding all the way to 4HB? It is unclear if there is a prebiotically plausible redox agent that will prevent reduction of the acid all the way to the alcohol, which may be why extant metabolism evolved this longer route. Additionally, how do you get to SUC? As previously discussed, routes involving MAL avoid the problem of converting a branched C_4 into an unbranched species, but there are uphill steps to get to MAL. As alternatives, we speculated on routes involving lactic acid or methylglyoxal that get to MAL with minimal endergonicity. However, just because a reaction is thermodynamically favorable does not mean it is feasible in a prebiotic milieu.

We have discussed in Section 3.3 why an overall $C_1 + C_1 \rightarrow C_2$ reaction, embedded in a cycle is a more feasible way of incorporating food molecules into a growing metabolic system compared to the direct simple addition. With ACE as the C_2 species, the net cycle is -45.03 kcal/mol. While two of its reduced cousins, ethanol and ethane, provide a more exergonic net reaction, both are less reactive and less likely to involve the rich carbonyl chemistry we have discussed. Ethanal could be a viable competitor (net cycle of -41.62 kcal/mol) but the lack of an anionic counterpart could be less robust for the anchoring advantages of selective surface chemistry. As to ACE's oxidative cousins, glycolic acid might work (net cycle of -18.08 kcal/mol) although it favorably converts to ACE under reducing conditions (the carboxylate could be "protected" by surface attachment), and we have already discussed the challenge and potential role of GLX. These considerations may cement the key role played by ACE.

Extant metabolites participating in core cycles occupy a particular range of oxidation states depending on the size range of compounds involved. In the C_1 to C_4 group, these are mostly in the -2 to $+4$ oxidation states with OXA ($+6$) being just outside that group. The C_5 and C_6 compounds are mostly in the $+2$ to $+6$ range, with transient OXS at $+8$. Why these ranges? We can only speculate for now that there is a balance between adding oxidizing food equivalents (CO_2 at $+4$) and reducing equivalents (H_2) to form stable compounds in some optimal "Goldilocks" zone. Over-reduction to rock-bottom (overly stable) alkanes may limit chemistry. However, some reduction is needed to drive a thermodynamically favorable cycle.

Most of this section has focused on discussing variable oxidation states, but we would be remiss if we did not briefly discuss variable protonation states. Our calculated free energies use only neutral species (because of the poorer anion results). This necessarily means we lump together multiple species into a single entity even if it exists in more than one protonation state at significant concentration depending on the pH. Until we devise a protocol that provides better results compared to experiment for anionic species, these are an undifferentiated part of the aforementioned "pool" of equilibrating species.

5. Conclusions

We have explored the thermodynamics of uncatalyzed reductive TCA cycle and its chemical cousins by calculating the relative free energies of potential CHO metabolites in aqueous solution under standard conditions in the hope that it may shed light on how proto-metabolic systems are constructed and sustained. By focusing on simple cycles that incorporate $C_1 + C_1 \rightarrow C_2$ as the net reaction for building biomass, we considered why acetate is uniquely positioned as the key recycled compound, and tracked the energy changes of chemical transformations with an eye on oxidation state changes.

Could other food molecules change the free energy map? Yes. We would expect systematic shifts in our calculated thermodynamics. Would introducing co-factors (such as thioesters) change the energy profiles? Definitely. Such coupling reactions may provide ways around otherwise unfavorable endergonic reactions. Could other reference molecules be used, say if we employed a different redox couple? Certainly. Considering the extreme case of strongly oxidizing conditions with available O_2 and no H_2, our map in Figure 6 would flip. CH_4 would be the least thermodynamically stable and CO_2 would be rock-bottom on the scale. Would varying concentrations and dynamic flows of material and energy modify the situation? Most certainly. These questions and many more could be asked, but their answers are beyond the scope of the present work. Our group is actively exploring a number of these interesting rabbit-holes, several of which have been hinted at in this article, that build on the present baseline results.

This baseline provides a preliminary step to exploring the effect of temperature, pH, and different effective concentrations (activities) of the various solutes including the reference species. One could recalculate the relative free energy changes using $\Delta G = \Delta G_0 + RT \ln Q$. Our baseline numbers provide the ΔG_0. Solute activities (not equal to unity) and the effect of pH (via a [H^+] term) in the reaction quotient Q would then result in different ΔG values. These only address equilibrium thermodynamics in a perturbative way. Temperature, pH, and solute activities are also expected to impact kinetics. Our group is working to calculate baseline activation barriers for reaction types in these cycles, although this requires the lengthy process of optimizing transition states.

Once baseline kinetics is established and we have made better estimates of the rate constants, we plan to incorporate both thermodynamic and kinetic parameters into a dynamic network model that tracks the fluxes of chemical species as they cycle through their myriad reactions. Our hope is that this will provide a preliminary model (with its attendant limitations) that represents the establishment of proto-metabolic cycles under non-equilibrium conditions, and we hope to provide a richer story in the future.

Supplementary Materials: The following are available online at https://www.mdpi.com/article/10.3390/life11101025/s1, Table S1: Comparison of ΔG (kcal/mol) for uncatalyzed oxidative TCA cycle reactions using M06-2X-D3 functional and the cc-pVDZ basis set; Table S2: Comparison of ΔG (kcal/mol) for uncatalyzed oxidative TCA cycle reactions using anions; Table S3: Comparison of ΔG (kcal/mol) for reductive 3HP/4HB Cycle; Table S4: Comparison of ΔG (kcal/mol) for additional reactions in DC/4HB cycle and 3HP bicycle; Table S5: Full set of compounds and Gr values (kcal/mol)

Author Contributions: Conceptualization, J.K.; methodology, J.K.; formal analysis, J.K. and A.L.H.; validation, J.K. and D.N.V.; investigation, J.K., A.L.H. and D.N.V.; writing—original draft preparation, J.K.; writing—review and editing, J.K.; visualization, J.K.; supervision, J.K. All authors have read and agreed to the published version of the manuscript.

Funding: This research received no external funding.

Institutional Review Board Statement: Not applicable.

Informed Consent Statement: Not applicable.

Data Availability Statement: The data presented in this study are available in Supplementary Material.

Acknowledgments: This research was supported by the University of San Diego. Shared computing facilities were provided by the saber2 and saber3 high-performance computing clusters at the University of San Diego.

Conflicts of Interest: The authors declare no conflict of interest.

Appendix A

Of the compounds in our full dataset of 223 compounds (see Table S5), 86 were found in the eQuilibrator database. Overall, from Figure A1, our quantum-calculated G_r values closely match the experimentally-derived data. There are two outliers: ethoxide, a three-membered ring (likely due to different determination of ring strain); and tartronic acid, for no apparent reason we can surmise although we think the eQuilibrator value is wildly incorrect. Our quantum values for similar di-acids such as malic and tartaric acid match up well.

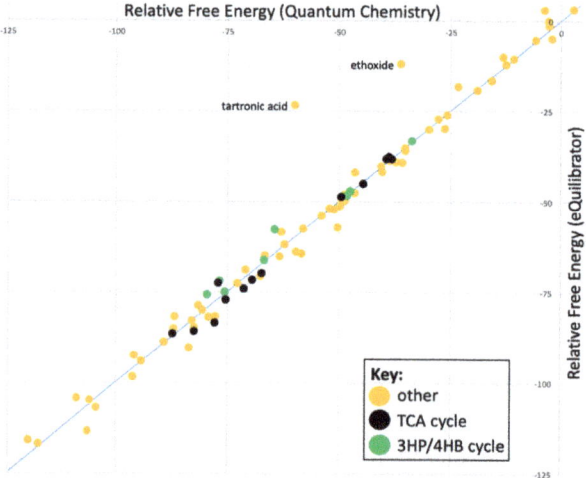

Figure A1. Comparison of quantum and available eQuilibrator G_r values (kcal/mol).

While a full list of all G_r values, compound names, abbreviations, molecular formulae, and overall carbon oxidation states are found in Table S5, the truncated list in Table A1 below provides an easy reference to the compounds explicitly mentioned in the Results section. If "(hyd)" is included in the name, it means the hydrate is the most stable form. If "(enol)" is included in the name, the enol is the most stable form.

Table A1. G_r values in kcal/mol for compounds in the Results section.

Name	Abbr.	Molecular Formula	Oxid. State	eQuilibrator G_r	Quantum G_r
formaldehyde (hyd)		CH2O	0	−2.08	−4.69
formic acid		CH4O2	2	−5.71	−5.23
ethanal		C2H4O	−2	−40.33	−41.62
acetic acid	ACE	C2H4O2	0	−44.63	−45.03
glylcolaldehyde	GA	C2H4O2	0	−15.68	−16.44
glycolic acid		C2H4O3	2	−23.33	−18.08
glyoxilic acid (hyd)	GLX	C2H4O4	4	−2.85	0.22
propanal		C3H6O	−4	−72.74	−72.43
acrolein		C3H4O	−2	−49.46	−49.18
propanoic acid	PRP	C3H6O2	−2	−77.04	−72.46
lactaldehyde (hyd)		C3H6O2	−2	−49.82	−51.06
3-OH-propanal		C3H6O2	−2	−48.77	−49.47
acrylic acid	ACR	C3H4O2	0	−49.53	−48.57
methylglyoxal (hyd)	MeGly	C3H4O2	0	−37.22	−38.93
malonaldehyde (enol)		C3H4O2	0	−38.63	−38.58
lactic acid		C3H6O3	0	−51.11	−52.03
3-OH-propanoic acid	3HP	C3H6O3	0	−52.18	−51.71
glyceraldehyde (hyd)	GLA	C3H6O3	0	−25.82	−25.99
di-OH-acetone	DHA	C3H6O3	0	−26.35	−29.71
pyruvic acid	PYR	C3H4O3	2	−38.87	−37.55
malonic semialdehyde (enol)	MSA	C3H4O3	2	−38.22	−38.20
glucic acid (enol)		C3H4O3	2		−12.87
glyceric acid		C3H6O4	2	−27.80	−27.12
malonic acid	MLN	C3H4O4	4	−39.42	−38.17
hydroxypyruvic acid	HPY	C3H4O4	4	−10.78	−10.33
2-OH-3-oxopropanoic acid (enol)		C3H4O4	4		−12.87
tartronic acid		C3H4O5	6	−12.53	−11.96
crotonic acid	CRT	C4H6O2	−2	−82.49	−85.80
3-butenoic acid		C4H6O2	−2		−80.19
succinaldehyde		C4H6O2	−2		−69.57
3-OH-butyric acid	3HB	C4H8O3	−2	−87.39	−86.46
4-OH-butyric acid	4HB	C4H8O3	−2	−77.78	−83.49
2-oxo-butanoic acid	AKB	C4H6O3	0	−70.92	−68.75
acetoacetic acid (enol)	AcACE	C4H6O3	0	−75.35	−77.03
succinic semialdehyde	SSA	C4H6O3	0	−69.51	−71.61
2-OH-but-3-enoic acid		C4H6O3	0		−54.93
erythrose		C4H8O4	0	−35.24	−35.00
erythrulose		C4H8O4	0	−35.84	−39.05

Table A1. Cont.

Name	Abbr.	Molecular Formula	Oxid. State	eQuilibrator G_r	Quantum G_r
2-deoxythreonic acid		C4H8O4	0		−61.59
2-oxobutenoicacid		C4H4O3	2		−41.86
succinic acid	SUC	C4H6O4	2	−71.35	−74.04
methylmalonic acid	MEM	C4H6O4	2	−67.36	−69.75
2-OH-3-oxobutyric acid		C4H6O4	2		−50.97
4-OH-3-oxobutyric acid		C4H6O4	2		−50.38
fumaric acid	FUM	C4H4O4	4	−47.57	−46.92
3,4-dioxobutyric acid (enol)		C4H4O4	4		−33.28
malic acid	MAL	C4H6O5	4	−48.31	−48.14
oxaloacetic acid (enol)	OXA	C4H4O5	6	−33.71	−33.06
mesaconic acid	MSC	C5H6O4	2	−80.63	−79.87
citramalic acid	CTM	C5H8O5	2	−77.66	−81.55
methylmalic acid	MML	C5H8O5	2	−81.51	−78.51
alpha-ketoglutaric acid	AKG	C5H6O5	4	−66.81	−66.06
fumaroylformic acid		C5H4O5	6		−39.18
maloylformic acid		C5H6O6	6	−38.63	−39.96
cis-aconitic acid	CAC	C6H6O6	6	−76.73	−71.80
citric acid	CIT	C6H8O7	6	−79.53	−75.60
isocitric acid	ISC	C6H8O7	6	−75.59	−74.82
oxalosuccinic acid	OXS	C6H6O7	8	−64.50	−57.60
isocitroylformic acid		C7H8O8	8		−65.54
aconitoylformic acid		C7H8O8	8		−65.84

References

1. Braakman, R.; Smith, E. The compositional and evolutionary logic of metabolism. *Phys. Biol.* **2013**, *10*, 011001. [CrossRef]
2. Pross, A. How Was Nature Able to Discover Its Own Laws-Twice? *Life* **2021**, *11*, 679. [CrossRef]
3. Hordijk, W.; Steel, M. Autocatalytic Networks at the Basis of Life's Origin and Organization. *Life* **2018**, *8*, 62. [CrossRef]
4. Russell, M.J.; Nitschke, W.; Branscomb, E. The inevitable journey to being. *Philos. Trans. R. Soc. B* **2013**, *368*, 2012054. [CrossRef]
5. Muchowska, K.B.; Varma, S.J.; Moran, J. Nonenzymatic Metabolic Reactions and Life's Origins. *Chem. Rev.* **2020**, *120*, 7708–7744. [CrossRef]
6. Tran, Q.P.; Adam, Z.R.; Fahrenbach, A.C. Prebiotic Reaction Networks in Water. *Life* **2020**, *10*, 352. [CrossRef]
7. Amend, J.P.; LaRowe, D.E.; McCollom, T.M.; Shock, E.L. The energetics of organic synthesis inside and outside the cell. *Philos. Trans. R. Soc. B* **2013**, *368*, 20120255. [CrossRef]
8. Morowitz, H.J.; Kostelnik, J.D.; Yang, J.; Cody, G.D. The origin of intermediary metabolism. *Proc. Natl. Acad. Sci. USA* **2000**, *97*, 7704–7708. [CrossRef]
9. Meringer, M.; Cleaves, H.J., 2nd. Computational exploration of the chemical structure space of possible reverse tricarboxylic acid cycle constituents. *Sci. Rep.* **2017**, *7*, 17540. [CrossRef]
10. Zubarev, D.Y.; Rappoport, D.; Aspuru-Guzik, A. Uncertainty of prebiotic scenarios: The case of the non-enzymatic reverse tricarboxylic acid cycle. *Sci. Rep.* **2015**, *5*, 8009. [CrossRef]
11. Kua, J.; Paradela, T.L. Early Steps of Glycolonitrile Oligomerization: A Free-Energy Map. *J. Phys. Chem. A* **2020**, *124*, 10019–10028. [CrossRef]
12. Kua, J.; Hanley, S.W.; De Haan, D.O. Thermodynamics and Kinetics of Glyoxal Dimer Formation: A Computational Study. *J. Phys. Chem. A* **2008**, *112*, 66–72. [CrossRef]
13. Kua, J.; Avila, J.E.; Lee, C.G.; Smith, W.D. Mapping the Kinetic and Thermodynamic Landscape of Formaldehyde Oligomerization under Neutral Conditions. *J. Phys. Chem. A* **2013**, *117*, 12658–12667. [CrossRef]

14. Kua, J.; Galloway, M.M.; Millage, K.D.; Avila, J.E.; De Haan, D.O. Glycolaldehyde Monomer and Oligomer Equilibria in Aqueous Solution: Comparing Computational Chemistry and NMR Data. *J. Phys. Chem. A* **2013**, *117*, 2997–3008. [CrossRef]
15. Vosko, S.H.; Wilk, L.; Nusair, M. Accurate spin-dependent electron liquid correlation energies for local spin density calculations: A critical analysis. *Can. J. Phys.* **1980**, *58*, 1200–1211. [CrossRef]
16. Becke, A.D. Density-functional exchange-energy approximation with correct asymptotic behavior. *Phys. Rev. A* **1988**, *38*, 3098–3100. [CrossRef] [PubMed]
17. Lee, C.; Yang, W.; Parr, R.G. Development of the Colle-Salvetti correlation-energy formula into a functional of the electron density. *Phys. Rev. B* **1988**, *37*, 785–789. [CrossRef] [PubMed]
18. Becke, A.D. Density-functional thermochemistry. III. The role of exact exchange. *J. Chem. Phys.* **1993**, *98*, 5648–5652. [CrossRef]
19. Banks, J.L.; Beard, H.S.; Cao, Y.; Cho, A.E.; Damm, W.; Farid, R.; Felts, A.K.; Halgren, T.A.; Mainz, D.T.; Maple, J.R.; et al. Integrated Modeling Program, Applied Chemical Theory (IMPACT). *J. Comput. Chem.* **2005**, *26*, 1752–1780. [CrossRef]
20. Warshel, A.; Florian, J. Computer simulations of enzyme catalysis: Finding out what has been optimized by evolution. *Proc. Natl. Acad. Sci. USA* **1998**, *95*, 5950–5955. [CrossRef] [PubMed]
21. Wiberg, K.B.; Bailey, W.F. Chiral diamines 4: A computational study of the enantioselective deprotonation of Boc-pyrrolidine with an alkyllithium in the presence of a chiral diamine. *J. Am. Chem. Soc.* **2001**, *123*, 8231–8238. [CrossRef] [PubMed]
22. Nielsen, R.J.; Keith, J.M.; Stoltz, B.M.; Goddard, W.A., 3rd. A computational model relating structure and reactivity in enantioselective oxidations of secondary alcohols by (-)-sparteine-Pd(II) complexes. *J. Am. Chem. Soc.* **2004**, *126*, 7967–7974. [CrossRef] [PubMed]
23. Deubel, D.V.; Lau, J.K. In silico evolution of substrate selectivity: Comparison of organometallic ruthenium complexes with the anticancer drug cisplatin. *ChemComm* **2006**, 2451–2453. [CrossRef] [PubMed]
24. Wertz, D.H. Relationship between the gas-phase entropies of molecules and their entropies of solvation in water and 1-octanol. *J. Am. Chem. Soc.* **1980**, *102*, 5316–5322. [CrossRef]
25. Abraham, M.H. Relationship between solution entropies and gas phase entropies of nonelectrolytes. *J. Am. Chem. Soc.* **1981**, *103*, 6742–6744. [CrossRef]
26. Thrush, K.L.; Kua, J. Reactions of Glycolonitrile with Ammonia and Water: A Free Energy Map. *J. Phys. Chem. A* **2018**, *122*, 6769–6779. [CrossRef]
27. Alberty, R.A. *Thermodynamics of Biochemical Reactions*; John Wiley & Sons, Inc.: Hoboken, NJ, USA, 2003.
28. Tymoczko, J.L.; Berg, J.M.; Stryer, L. *Biochemistry, A Short Course*, 3rd ed.; Table 19.11; W. H. Freeman and Co.: New York, NY, USA, 2015.
29. eQuilbrator. Available online: http://equilibrator.weizmann.ac.il/ (accessed on 1 March 2020).
30. Noor, E.; Bar-Even, A.; Flamholz, A.; Lubling, Y.; Davidi, D.; Milo, R. An integrated open framework for thermodynamics of reactions that combines accuracy and coverage. *Bioinformatics* **2012**, *28*, 2037–2044. [CrossRef]
31. Jankowski, M.D.; Henry, C.S.; Broadbelt, L.J.; Hatzimanikatis, V. Group contribution method for thermodynamic analysis of complex metabolic networks. *Biophys. J.* **2008**, *95*, 1487–1499. [CrossRef]
32. Miller, S.L. A production of amino acids under possible primitive earth conditions. *Science* **1953**, *117*, 528–529. [CrossRef] [PubMed]
33. Urey, H.C. On the Early Chemical History of the Earth and the Origin of Life. *Proc. Natl. Acad. Sci. USA* **1952**, *38*, 351–363. [CrossRef] [PubMed]
34. Kasting, J.F. Earth's early atmosphere. *Science* **1993**, *259*, 920–926. [CrossRef] [PubMed]
35. Breslow, R. On the Mechanism of the Formose Reaction. *Tetrahedron Lett.* **1959**, *1*, 22–26. [CrossRef]
36. Igamberdiev, A.U.; Eprintsev, A.T. Organic Acids: The Pools of Fixed Carbon Involved in Redox Regulation and Energy Balance in Higher Plants. *Front. Plant Sci.* **2016**, *7*, 1042. [CrossRef]
37. Omran, A.; Menor-Salvan, C.; Springsteen, G.; Pasek, M. The Messy Alkaline Formose Reaction and Its Link to Metabolism. *Life* **2020**, *10*, 125. [CrossRef] [PubMed]
38. Miller, S.L.; Urey, H.C. Organic Compound Synthesis on the Primitive Earth. *Science* **1959**, *130*, 245–251. [CrossRef]
39. Forsythe, J.G.; Yu, S.S.; Mamajanov, I.; Grover, M.A.; Krishnamurthy, R.; Fernandez, F.M.; Hud, N.V. Ester-Mediated Amide Bond Formation Driven by Wet-Dry Cycles: A Possible Path to Polypeptides on the Prebiotic Earth. *Angew. Chem. Int. Ed. Engl.* **2015**, *54*, 9871–9875. [CrossRef]
40. Stubbs, R.T.; Yadav, M.; Krishnamurthy, R.; Springsteen, G. A plausible metal-free ancestral analogue of the Krebs cycle composed entirely of alpha-ketoacids. *Nat. Chem.* **2020**, *12*, 1016–1022. [CrossRef]
41. Muchowska, K.B.; Varma, S.J.; Moran, J. Synthesis and breakdown of universal metabolic precursors promoted by iron. *Nature* **2019**, *569*, 104–107. [CrossRef]
42. Smith, E.; Morowitz, H.J. *The Origin and Nature of Life on Earth*; Cambridge University Press: Cambridge, UK, 2016.
43. Wachtershauser, G. The origin of life and its methodological challenge. *J. Theor. Biol.* **1997**, *187*, 483–494. [CrossRef]

Article

Oxidative Phosphorus Chemistry Perturbed by Minerals

Arthur Omran [1,2,*], Josh Abbatiello [2], Tian Feng [2] and Matthew A. Pasek [2]

1. Department of Chemistry, University of North Florida, 1 UNF Drive, Jacksonville, FL 32224, USA
2. Department of Geosciences, University of South Florida, Tampa, FL 33620, USA; jabbatiello@usf.edu (J.A.); tianfeng1@usf.edu (T.F.); mpasek@usf.edu (M.A.P.)
* Correspondence: N00431947@unf.edu

Abstract: Life is a complex, open chemical system that must be supported with energy inputs. If one fathoms how simple early life must have been, the complexity of modern-day life is staggering by comparison. A minimally complex system that could plausibly provide pyrophosphates for early life could be the oxidation of reduced phosphorus sources such as hypophosphite and phosphite. Like all plausible prebiotic chemistries, this system would have been altered by minerals and rocks in close contact with the evolving solutions. This study addresses the different types of perturbations that minerals might have on this chemical system. This study finds that minerals may inhibit the total production of oxidized phosphorus from reduced phosphorus species, they may facilitate the production of phosphate, or they may facilitate the production of pyrophosphate. This study concludes with the idea that mineral perturbations from the environment increase the chemical complexity of this system.

Keywords: Fenton chemistry; reduced phosphorus; pyrophosphate; chemical complexity; chemical evolution; minerals; schreibersite; olivine; serpentinite; ulexite

1. Introduction

Life is the ultimate example of a complex chemical system [1,2]. As such, it must be buttressed by energetic inputs to maintain homeostasis. Life requires energy, and present-day life uses phosphates such as adenosine triphosphate (ATP) or acetyl-phosphate as carrier molecules to transfer such energy [2]. In a completely simplistic system that one may fathom to "start from scratch," organophosphates such as nucleotide triphosphates are too sophisticated or evolved [3] and are likely products of an RNA world and the work of earlier metabolism [3]. A complex system may evolve organophosphates but perhaps not as the initial energy currency. Various inorganic phosphate species may have contributed to the prebiotic formation of protometabolic systems [4,5], at least if phosphorus is critical to life's origin [6]. Few simple phosphorus species exist in nature that may have supported life early on, such as reduced phosphorus species, phosphates (PO_4^{3-}), and polyphosphates [4].

A far-from-equilibrium system with environmental energy input could enable endergonic reactions. Life would need a system that readily liberates solvated electrons and provides a soluble phosphorus source from the environment. One such system that could provide electrons for redox chemistry may be the Fenton system. In this system, iron (II) ions in solution, or in a scaffolding, break down aqueous hydrogen peroxide into hydroxyl radicals (Scheme 1) [7–12]. Throughout this work, we represent this reaction using schemes of the imagined scaffolds in solution; this does not preclude simple, individual soluble ions of iron (II) from taking part in the reaction.

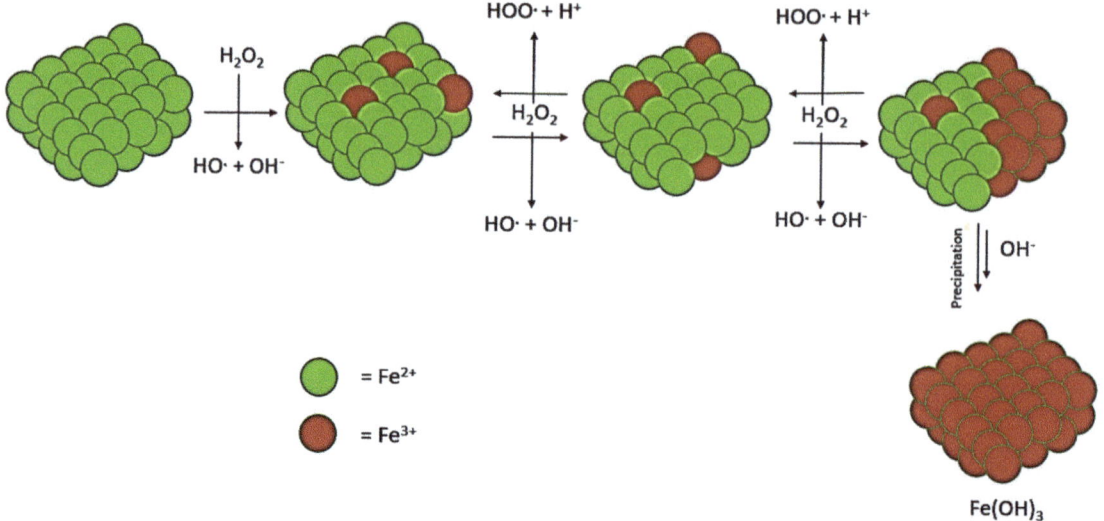

Scheme 1. Explanation of the Fenton reaction. Reaction of the iron complex is partially soluble and in solution, with iron (III) likely being solid and iron (II) more soluble.

When coupled to the oxidation of reduced phosphorus species, such as hypophosphite ($H_2PO_2^-$) and phosphite (HPO_3^{2-}), these hydroxyl radicals, liberated from the Fenton reaction, oxidized these reduced P compounds to phosphate and pyrophosphate ($P_2O_7^{4-}$) [13–15]. The reduced phosphorus species themselves are soluble compared to phosphate, which precipitates readily in the presence of divalent and trivalent metal cations [4,5,13,14]. Pyrophosphate may be evolutionarily or prebiotically important due to its being a potential precursor to ATP in the evolution of life [16–20]. Therefore, pyrophosphate may have been life's original energy currency, though this is debated [16–18].

This Fenton system, coupled with reduced phosphorus, may be a plausible system on rocky planets, such as the early Earth [5,14,15]. Firstly, the reduced P may have been provided by the meteoric input of iron meteorites, which are rich in the mineral schreibersite ($(Fe/Ni)_3P$), or from lightning strikes which may have formed schreibersite as well [13,21]. This mineral is a highly reduced P source that, when exposed to water, corrodes and forms reduced P species such as hypophosphite ($H_2PO_2^-$) and phosphite (HPO_3^{2-}), among other phosphate species (Figure 1) [13].

Secondly, hydrogen peroxide may have been plentiful on early rocky planets as one of the earliest oxidants. It may have formed via the photolysis of atmospheric water and rained down on the surface [22–24]. Additionally, peroxides form on the surface of icy moons, such as Europa or Enceladus, due to solar radiation of surface ice on the moons [25–27]. It is with these P, iron, and peroxide resources in mind that one may view the Fenton chemical system as plausible towards the emergence and chemical evolution of life.

A complex system is a system comprising many components with much interconnectedness and many relations between the system's components. This system has multiple components, iron species, P species, peroxides, etc., and the reactions affect the various species as well. As a complex system, this P-coupled Fenton chemistry is vulnerable to "dirty" chemistry, as are all plausible prebiotic chemistries. We therefore wish to understand the influence of minerals in the environment on this system.

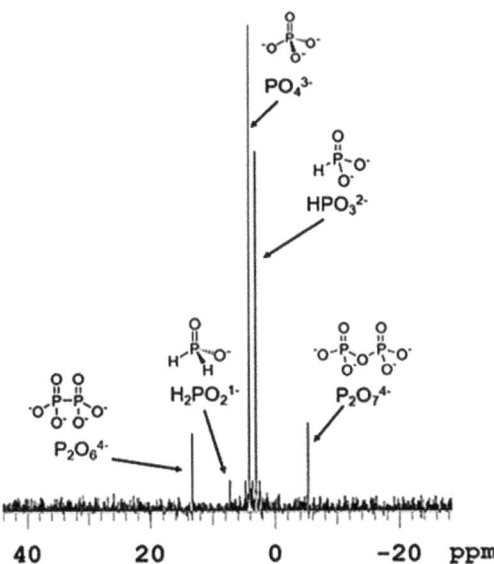

Figure 1. P^{31}-NMR spectrum of the corrosion of schreibersite. An amount of 0.5 Fe$_3$P powder was placed in 15 mL of water for 1 week under a 20% mixed CO$_2$/CH$_4$ and 80% N$_2$ atmosphere. The water contained multiple phosphorus species, including reduced phosphorus species, such as hypophosphite (H$_2$PO$_2^-$) and phosphite (HPO$_3^{2-}$).

2. Materials and Methods

Solutions of Fenton's reagent, 10 mL, were aliquoted into 20 mL borosilicate vials. Fenton's reagent was synthesized using 100 mM of FeCl$_2$ solution (Fischer, Waltham, MA, USA). The solutions contained a final concentration of 880 mM of H$_2$O$_2$ (Fischer); the peroxide was added very slowly. The solution also contained 100 mM of Na$_2$HPO$_3$ or 100 mM of NaH$_2$PO$_2$ (both from Sigma-Aldrich, St. Louis, MO, USA) as an initial phosphorus source. The H$_2$O$_2$ was added after the FeCl$_2$ and the P species.

In addition to the components of a Fenton solution and a phosphorus source, we added minerals individually to see if they influenced the product distributions. Each experiment contained a 0.05 g sample of a mineral or rock, including diopside, magnetite, serpentinite, quartz, kaolinite, siderite, newberyite, hematite, ulexite, schreibersite, orthoclase, olivine, hydroxyapatite, and struvite (Table S1). We selected these minerals based on the minerals' likely presence during the Hadean Eon or their prebiotic plausibility/utility (Table S1). The final experiment had no minerals as a control. Reactions took place at room temperature (25 °C) and were gently stirred for 48 h. Solutions were slowly titrated over a period of 5–10 min with 1 mL of 12 M NaOH. Fenton solutions were then filtered through 0.2 μm pore nylon syringe filters (Thermo Scientific, Waltham, MA, USA) using 10 mL Hamilton luer lock syringes.

An amount of 500 μL of each solution was aliquoted into NMR tubes, along with 10% by volume D$_2$O (Sigma-Aldrich, St. Louis, MO, USA) used to lock spectral measurements. Solutions were characterized using ^{31}P NMR on an INOVA Unity 400 MHz NMR. The INOVA 400 was run at 161.84 MHz, with 2052 normal scans per sample in H-coupled-NOE mode with a 1 s delay between acquisitions. The spectral width was 200 ppm, and the running temperature was set at 25 °C. The limit of detection for phosphorus species in solution on the INOVA 400 was 50–80 μg/g, well below our measurements and concentrations. All data analyses were run on SpinWorks 4.0 software and SigmaPlot 10.0 software.

Samples of the solution phases were also analyzed via inductively coupled plasma–optical emission spectrometry (ICP–OES) before undergoing a mineral extraction. Filtrates of the minerals in solution and the precipitated iron hydroxides were extracted using a 4:1 NaOH EDTA extraction solution. Filtrates were extracted for one week while stirring at room temperature, as described above. Extracts were collected and inspected via NMR, as stated above, and analyzed via ICP–OES.

Solution phase or mineral phase was analyzed via ICP–OES. The analysis was performed on the Avio 200 ICP–OES from PerkinElmer with a 1.5 mm loop. The internal standard was 100 ppm germanium. The calibration standards used were 0.02, 0.1, 1, 5, 50, and 100 ppm of phosphoric acid mixed with deionized water. A 10 ppm phosphorus acid standard, as well as a blank of deionized water, were used as quality controls throughout the run. The samples were all diluted 100-fold, and then run on the ICP–OES to measure P content.

3. Results and Discussion

The Fenton system generates radicals from peroxide. These hydroxyl radicals react with P-H bonds to generate P oxidation and polyphosphates (Scheme 2, Figure 2). Depending on which reduced P source is initially placed in the environment, this system will generate a complex mixture of P species at different redox states (Figure 2). If one starts with hypophosphite in solution, the products include phosphite, phosphate, pyrophosphate, and small amounts of hypophosphate, triphosphate, and pyrophosphite (rarely) (Figure S1, Table S2). Phosphite as a P source in Fenton reactions produced phosphates and pyrophosphate (with some triphosphate in small amounts) (Tables 1–3).

The Fenton chemistry requires soluble iron in solution to decompose hydrogen peroxide; it, too, can function as part of a heterogenous catalyst acting as a peroxide-decomposing scaffold in solution (Scheme 1) [7–12]. The preliminary work was carried out with hypophosphite solutional systems; increased oxidation of phosphorus species in solution was observed in the presence of hydroxyapatite and schreibersite (Table S2). After those experiments, this study mainly focused on phosphite undergoing oxidation due to the Fenton system. We verified that the use of insoluble iron mineral species without any soluble iron results in no reaction by adding powdered magnetite with no added soluble iron ($FeCl_2$) in solution, and we observed no to little (<5%) P oxidation to phosphate under such conditions.

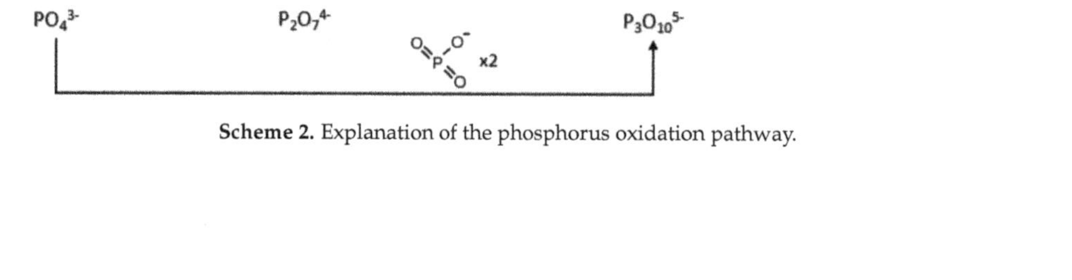

Scheme 2. Explanation of the phosphorus oxidation pathway.

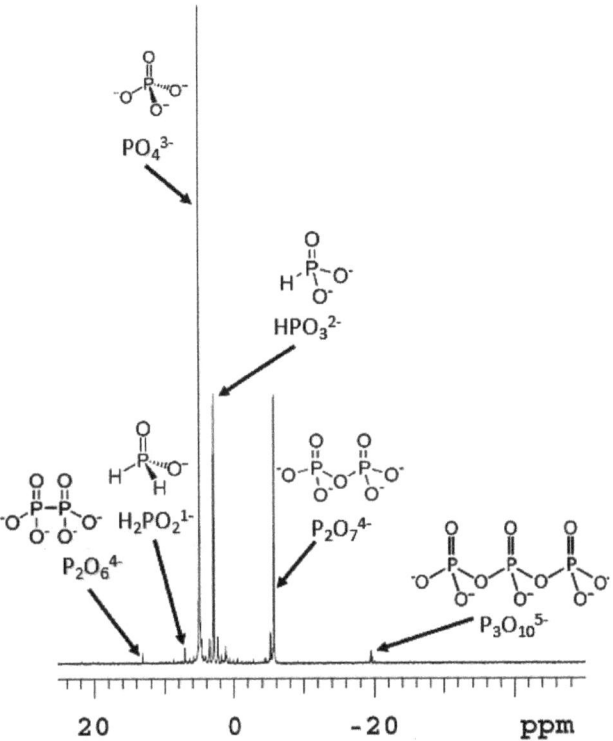

Figure 2. Complex mixture of different oxidation states of P species. Hypophosphite added to the Fenton reaction in the presence of schreibersite. Decoupled ^{31}P NMR.

Table 1. Mineral-added Fenton reactions with phosphite. Minerals inhibiting the Fenton reaction. All numeric values are integral percentages taken from NMR of single-batch simultaneous reactions.

Mineral	Phosphite	Phosphate	Pyrophosphate	Triphosphate
None	21.6	52.6	23.2	2.6
Magnetite	31.6	40.9	25	2.5
Orthoclase	39.7	47.8	11.5	1
Serpentinite	73.5	18.1	8.2	0.2

Table 2. Mineral-added Fenton reactions with phosphite. Minerals enhancing the Fenton reaction. All numeric values are integral percentages taken from NMR of single-batch simultaneous reactions.

Mineral	Phosphite	Phosphate	Pyrophosphate	Triphosphate
None	21.6	52.6	23.2	2.6
Calcite	4.5	75	19.7	
Diopside	5.2	64	27.3	2.6
Gypsum	21.7	63.7	14	0.6
Hydroxyapatite	3.8	77	18.5	0.6
Newberyite	3.1	78.7	18.2	
Ulexite	1.5	73.5	25	

Table 3. Mineral-added Fenton reactions with phosphite. Minerals facilitating phosphorus chemistry via the Fenton reaction. All numeric values are integral percentages taken from NMR of single-batch simultaneous reactions.

Mineral	Phosphite	Phosphate	Pyrophosphate	Triphosphate
Control (None)	21.6	52.6	23.2	2.6
Hematite	19.7	49.3	28	3
Kaolinite	15	53.5	28.3	3.2
Sand	14.2	54.7	28.4	2.7
Schreibersite	11.5	52	33.3	3.1
Siderite	17.6	48.7	30.7	3
Struvite	7.6	47.4	44.1	0.9

A criticism of the Fenton system is that it may be harmful to organic compounds and degrade them. Indeed, Fenton's reagent was developed by Henry Fenton in 1894 as an analytical agent to detect organics such as tartaric acid [7]. It is currently used as a green chemical solution to clean up and break down organics [10,11]. In these experiments, no organics were added as substrates; thus, no degradation occurred, but phosphorylation of organic substrates would likely require addition of the organic after the Fenton reaction has ceased. For example, surface pooling or small ponds that receive peroxide downpour could plausibly house these reactions, producing phosphates and pyrophosphates. Any organics introduced into these ponds after the reactions completed might then be phosphorylated due to other environmental factors, such as UV light.

3.1. Mineral Perturbation: Inhibition

The phosphite-added Fenton reactions had diverse outcomes. Firstly, the control group that had 100 mM of $NaHPO_3$ and 100 mM of $FeCl_2$ produced 53% phosphate, 23% pyrophosphate, and ~3% triphosphates, with 21% phosphite left. The no-mineral-added control was used as a benchmark to compare against the mineral-added reactions. Magnetite, serpentinite, and orthoclase seemed to inhibit the oxidation of the phosphite in the Fenton solution phase (Table 1). The least amount of P oxidation occurred with these minerals present, with about 30–40% of the phosphite left unreacted in solution.

It is possible that an antioxidant mechanism, such as scavenging the peroxide radicals, is involved in this result. As the Fenton system (Scheme 1) occurs and radicals are generated, a mineral in the environment may act to inhibit the system by using up the radicals before they are used to fuel P oxidation (Figure 3). Both serpentinite and magnetite are Fe^{2+}-containing rocks and minerals. Likely, stable products would include water and oxygen gas (Figure 3).

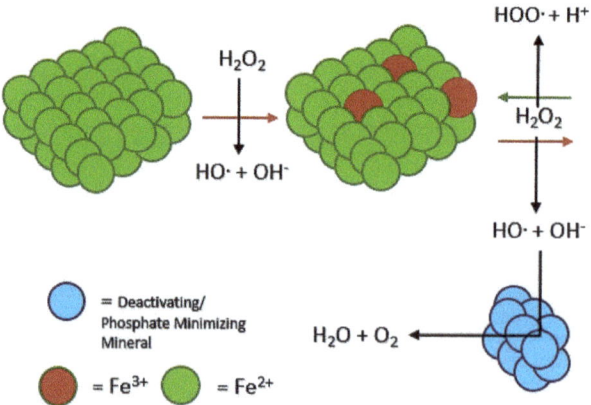

Figure 3. Diagram of how minerals may inhibit the Fenton reaction.

3.2. Mineral Perturbation: Phosphate Promotion

Another outcome observed was the generation of more phosphate in solution than phosphite (Table 2). Some minerals managed to perturb the system towards 60–78% phosphate, but not a great amount more; these included newberyite, calcite, diopside, gypsum, hydroxyapatite, and ulexite (Table 2). The gypsum mineral-added systems did not exhibit an increase in total oxidation, with unreacted phosphite present at a concentration like that of the control. One must then assume this phenomenon or mineral "influence" to be based on the minerals interacting with the phosphorus species, and not just the peroxide in solution (Figure 4). Interestingly, some minerals showed a similar influence as well as increased total oxidation of phosphite (Table 2); these minerals included calcite, newberyite, hydroxyapatite, diopside, and ulexite, with 90+% total oxidation of phosphite in solution. One may suspect better phosphite reactivity towards hydroxyl radicals in addition to the mineral working on the phosphite itself (Figure S2).

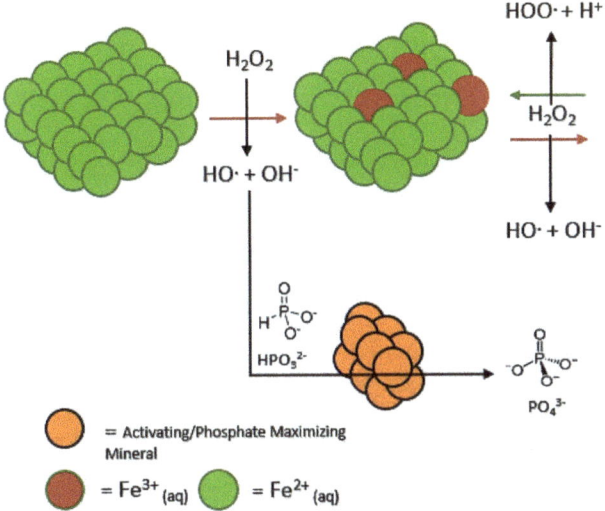

Figure 4. Diagram of how minerals may enhance the Fenton reaction.

3.3. Mineral Perturbation: Pyrophosphate Promotion

A third type of influence or perturbation was observed. Like the phosphate increase observed, some minerals appear to increase the amount of pyrophosphate (Table 3). Diopside, kaolinite, sand (quartz), and hematite showed slight increases of pyrophosphate in solution. Schreibersite, siderite, and struvite showed the greatest production of pyrophosphate at 30+%, with marked increases in total oxidation compared to the control. We observed the increased total oxidation of phosphorus for each of these mineral groups. Diopside, kaolinite, sand (quartz), and hematite all showed 27+% production of pyrophosphate in solution. It could be that these minerals directly bind the phosphorus in solution. As the Fenton system proceeds, the phosphorus species cluster together on the mineral surface, facilitating the production of pyrophosphate and other polyphosphates (Figure 5). Additionally, accompanying increases in total oxidation are slight increases in detected triphosphate species (Table 3). Overall, triphosphate does not seem to be affected like phosphate or pyrophosphate, potentially due to its breaking down into smaller P species or being at a low enough concentration that variations are hard to detect.

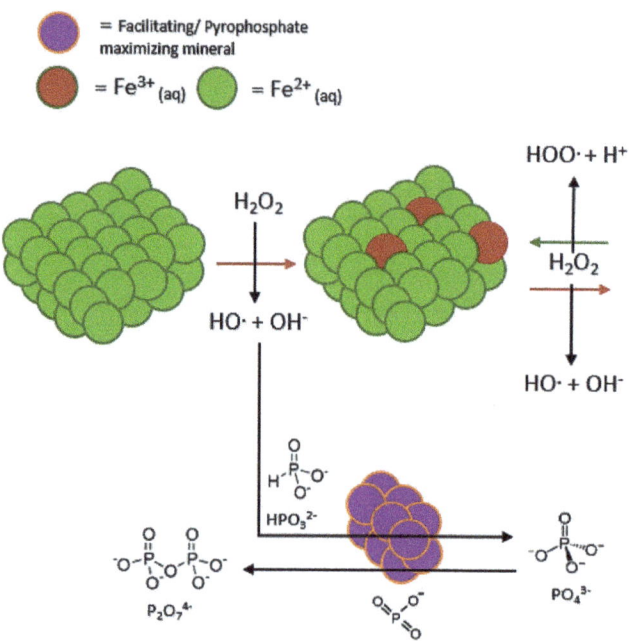

Figure 5. Diagram of how minerals may be facilitating phosphorus chemistry via the Fenton reaction.

The phosphorus studied in this system is specifically the phosphorus in aqueous solution that can be detected via NMR. The phosphorus adsorbed to the mineral phase was also inspected. Schreibersite had triphosphate and hypophosphate in the mineral phase, differing from the other minerals. This is likely due to the corrosion of the schreibersite itself while in aqueous solution, which is known to produce these anions [28]. Struvite had a noted increase in pyrophosphate. Struvite, being a phosphate-bearing mineral, likely added phosphate to the reactions. Struvite's mineral phase adsorbed about 50+% additional pyrophosphate, likely due to the reaction of phosphates with phosphite in the mineral phase.

3.4. Additional Verifications

When compared to standards of our phosphorus reagents or to no reaction, we saw no oxidation of phosphorus or oligomers of phosphorus, similarly to what we saw in the absence of soluble iron. It is important to mention as well that olivine did not seem to influence the product distributions, so some minerals do not influence the reactions (Table S4). In terms of reproducibility, these reactions are chaotic in that they rely on free radical formation from peroxide breakdown. These reactions in general can be compared to one another within datasets as long as the same peroxide solution was used at the same concentrations at roughly the same time. With that in mind, we can clearly see within the datasets the influence of our minerals on the Fenton system. A plethora of diverse oxidized P species was observed; this included pyrophosphite on rare occasions (Figure S1, Table S2).

Obviously, the chemical composition of the minerals matters in their effect on the system, in this study we focused on the product distributions mostly, looking for overarching trends on the system holistically. We did note additional trends based on mineral composition. Minerals have various compositions and absorptivity (Tables S3 and S4). The adsorptions of P in the mineral phase vs. the solution phase was inspected via ICP–OES (Table S3). Mineral stability and reactivity matter in terms of P recovery in solution and in the mineral phase itself. Very stable samples, such as our serpentinites, were shown to be less perturbing than minerals with greater solubility or reactivity, such as hydroxyapatite. We expected to see such variation based on the diversity of mineral and rock species selected (Table S1).

The majority of our starting phosphorus was accounted for in the solution and mineral phase using ICP–OES, ranging from ~70% to 120%. Samples with lower P detection were still in acceptable ranges for ICP–OES, allowing us to track where the P in the system was mostly located. The lower detections may be due to incomplete extraction with the mineral phase, so some P would cling to the mineral and not be accounted for. Additionally, some of the P might get lost in the iron oxides formed by the Fenton process. Most of the P was present in the original filtrate, as the filtrate had P contents of 51–117% of the initial added P (with the exception of hydroxyapatite, which bore sparingly little P). The extracted P in contrast was usually 1–18% of the total added P. Schreibersite, struvite, and newberyite had a higher total phosphorus content via ICP–OES due to the release of P from these minerals, accounting for the 100+% P detected (Table 3). In addition, the extracts tended to have a higher abundance of pyrophosphate vs. phosphate, suggesting that pyrophosphate binds to the mineral surface. In contrast, the phosphite to phosphate ratios are generally similar in both the filtrates and the extracts, although we note that, in general, this ratio is more varied in those samples that released more P upon extraction (e.g., hydroxyapatite). We note as well that serpentinite had more phosphite than phosphate in both the solution phase and the extraction, suggesting either a phosphite source in the serpentinite, or an impedance of phosphite oxidation via the serpentinite. It is possible the serpentine rock exchanges Mg for Fe during the Fenton process, especially if the rock is rich in brucites. Further study is needed for serpentinite regarding its relationships with reduced P species.

4. Conclusions

Phosphorus species pertain to a great many minerals, of various physical properties and reactivities, valuable to prebiotic chemistries [4,5,28–32]. Not only has reduced phosphorus been detected in meteorites, but also in rocks and minerals originating here on Earth. Other complex systems seem viable in multiple environments, [33–35] with some models incorporating phosphorus into self-assembling systems [36].

The Fenton system induces an oxidation of reduced P to phosphate and polyphosphates and is a complex chemical system. This chemical system is a plausible prebiotic system based on the simplicity and plausibility of its reagents. This chemical system may increase in complexity based on the minerals in its "dirty" environment. Those minerals are additional components, after all, contributing to systemic complexity. Complexity was measured by the increase in the distribution of types of P species in solution. These P species increased in their oxidation states and increased in their molecular weights as compared to the reduced P species we started with. There is a limit to the system's complexity in that complex organics are vulnerable to Fenton chemistry. Tolerance to peroxides may have been necessary for life originally [24].

Many types of perturbations were observed. The perturbations are based on the diversity of mineral species and compositions. These include promoting the formation of phosphate, promoting the formation of pyrophosphate, or stymieing the oxidative process altogether. The importance of the "dirty" environment and the "messy" product distributions may be fundamental in our understanding prebiotic chemistry in terms of chemical complexity. The minerals in the environment increase the chemical space of possibility for this open system and may even act as switches or controls for an open system and its environment. In a multimineral system or microenvironment, these switches may add together or oppose each other, totaling to a sum perturbation for the system, determining the systems' product distribution. Future work with multiple minerals per reaction system is needed to explore this.

Supplementary Materials: The following supporting information can be downloaded at: https://www.mdpi.com/article/10.3390/life12020198/s1. Figure S1: Fenton reaction with serpentinite and hypophosphite added, Figure S2: Additional possible Fenton reaction diagram, Table S1: List of plausibly prebiotic minerals, Table S2: Hypophosphite added Fenton reactions solutional product distribution table, Table S3: ICPOES data, Table S4: Mineral extraction NMR data.

Author Contributions: Conceptualization, A.O. and M.A.P.; methodology, A.O., T.F. and J.A.; resources, M.A.P.; writing—original draft preparation, A.O.; writing—review and editing, A.O.; visualization, A.O.; supervision, A.O.; project administration, M.A.P.; funding acquisition, M.A.P. All authors have read and agreed to the published version of the manuscript.

Funding: This work was jointly supported by the NSF and the NASA Astrobiology Program, under the NSF Center for Chemical Evolution (CHE-1504217). This work was supported the NASA Habitable Worlds Program (Grant 436 80NM0018F0612).

Institutional Review Board Statement: Not applicable.

Informed Consent Statement: Not applicable.

Data Availability Statement: Not applicable.

Acknowledgments: We would like to thank members of the Center for Chemical Evolution Phosphorus working group for helpful discussions and suggestions.

Conflicts of Interest: The authors declare no conflict of interest.

References

1. Whitesides, G.M.; Ismagilov, R.F. Complexity in chemistry. *Science* **1999**, *284*, 89–92. [CrossRef]
2. Mikhailov, A.; Ertl, G. Chemical complexity. In *The Frontiers Collection*; Springer: Berlin/Heidelberg, Germany, 2017.
3. White, H.B. Coenzymes as fossils of an earlier metabolic state. *J. Mol. Evol.* **1976**, *7*, 101–104. [CrossRef]
4. Pasek, M.A. Thermodynamics of Prebiotic Phosphorylation. *Chem. Rev.* **2019**, *12*, 4690–4706. [CrossRef] [PubMed]
5. Pasek, M.A.; Gull, M.; Herschy, B. Phosphorylation on the early earth. *Chem. Geol.* **2017**, *475*, 149–170. [CrossRef]
6. Goldford, J.E.; Hartman, H.; Smith, T.F.; Segrè, D. Remnants of an ancient metabolism without phosphate. *Cell* **2017**, *168*, 1126–1134. [CrossRef] [PubMed]
7. Fenton, H.J.H. Oxidation of tartaric acid in presence of iron. *J. Chem. Soc.* **1984**, *65*, 899–910. [CrossRef]
8. Mao, H.; Wang, P.; Zhang, D.; Xia, Y.; Li, Y.; Zeng, W.; Zhan, S.; Crittenden, J.C. Accelerating FeIII-Aqua Complex Reduction in an Efficient Solid–Liquid-Interfacial Fenton Reaction over the Mn–CNH Co-catalyst at Near-Neutral pH. *Environ. Sci. Technol.* **2021**, *55*, 13326–13334. [CrossRef]
9. Bakac, A. Oxygen Activation with Transition-Metal Complexes in Aqueous Solution. *Inorg. Chem.* **2010**, *49*, 3584–3593. [CrossRef]
10. Gokulakrishnan, N.; Pandurangan, A.; Sinha, P.K. Catalytic Wet Peroxide Oxidation Technique for the Removal of Decontaminating Agents Ethylenediaminetetraacetic Acid and Oxalic Acidfrom Aqueous Solution Using Efficient Fenton Type Fe-MCM-41 Mesoporous Materials. *Ind. Eng. Chem. Res.* **2009**, *48*, 1556–1561. [CrossRef]
11. Chang, M.-E.; Chen, T.-A.; Chern, J.-I. Initial Degradation Rate of p-Nitrophenol in Aqueous Solution by Fenton Reaction. *Ind. Eng. Chem. Res.* **2008**, *47*, 8533–8541. [CrossRef]
12. Duesterberg, C.K.; Mylon, S.E.; Waite, T.D. pH Effects on Iron-Catalyzed Oxidation using Fenton's Reagent. *Environ. Sci. Technol.* **2008**, *42*, 8522–8527. [CrossRef] [PubMed]
13. Pasek, M.A.; Lauretta, D.S. Aqueous corrosion of phosphide minerals from iron meteorites: A highly reactive source of prebiotic phosphorus on the surface of the early Earth. *Astrobiology* **2005**, *5*, 515–535. [CrossRef]
14. Herschy, B.; Chang, S.J.; Blake, R.; Lepland, A.; Abbott-Lyon, H.; Sampson, J.; Atlas, Z.; Kee, T.P.; Pasek, M.A. Archean phosphorus liberation induced by iron redox geochemistry. *Nat. Commun.* **2018**, *9*, 1–7. [CrossRef] [PubMed]
15. Pasek, M.A. Production of potentially prebiotic condensed phosphates by phosphorus redox chemistry. *Angew. Chem. Int. Ed.* **2008**, *47*, 7918–7920. [CrossRef] [PubMed]
16. Holm, N.G.; Baltscheffsky, H. Links between hydrothermal environments, pyrophosphate, Na+, and early evolution. *Orig. Life Evol. Biosph.* **2011**, *41*, 483–493. [CrossRef]
17. Rao, N.N.; Gómez-García, M.R.; Kornberg, A. Inorganic polyphosphate: Essential for growth and survival. *Annu. Rev. Biochem.* **2009**, *78*, 605–647. [CrossRef] [PubMed]
18. Brown, M.R.; Kornberg, A. Inorganic polyphosphate in the origin and survival of species. *Proc. Natl. Acad. Sci. USA* **2004**, *101*, 16085–16087. [CrossRef] [PubMed]
19. Zhang, H.; Ishige, K.; Kornberg, A. A polyphosphate kinase (PPK2) widely conserved in bacteria. *Proc. Natl. Acad. Sci. USA* **2002**, *99*, 16678–16683. [CrossRef]
20. Mulkidjanian, A.Y.; Bychkov, A.Y.; Dibrova, D.V.; Galperin, M.Y.; Koonin, E.V. Origin of first cells at terrestrial, anoxic geothermal fields. *Proc. Natl. Acad. Sci. USA* **2012**, *109*, E821–E830. [CrossRef]
21. Hess, B.L.; Piazolo, S.; Harvey, J. Lightning strikes as a major facilitator of prebiotic phosphorus reduction on early Earth. *Nat. Commun.* **2021**, *12*, 1–8. [CrossRef]

22. Levine, J.S.; Hays, P.B.; Walker, J.C.G. The evolution and variability of atmospheric ozone over geological time. *Icarus* **1979**, *39*, 295–309. [CrossRef]
23. Levine, J.S.; Graedel, T.E. Photochemistry in planetary atmospheres. *Eos Trans. AGU* **1981**, *62*, 1177–1181. [CrossRef]
24. Ślesak, I.; Ślesak, H.; Kruk, J. Oxygen and hydrogen peroxide in the early evolution of life on earth: In silico comparative analysis of biochemical pathways. *Astrobiology* **2012**, *12*, 775–784. [CrossRef]
25. Loeffler, M.J.; Baragiola, R.A. The state of hydrogen peroxide on Europa. *Geophys. Res. Lett.* **2005**, *32*. [CrossRef]
26. Carlsonm, R.W.; Andersonr, S.; Johnsonw, E.; Smythea, D.; Hendrixc, R.; Barthl, A.; Soderblomg, A.; Hansent, B.; Mccordj, B.; Daltonr, B.; et al. Hydrogen peroxide on the surface of Europa. *Science* **1999**, *283*, 2062–2064. [CrossRef]
27. Filacchione, G.; Adriani, A.; Mura, A.; Tosi, F.; Lunine, J.I.; Raponi, A.; Ciarniello, M.; Grassi, D.; Piccioni, G.; Moriconi, M.L.; et al. Serendipitous infrared observations of Europa by Juno/JIRAM. *Icarus* **2019**, *328*, 1–13. [CrossRef]
28. Pasek, M.A.; Dworkin, J.P.; Lauretta, D.S. A radical pathway for organic phosphorylation during schreibersite corrosion with implications for the origin of life. *Geochim. Cosmochim. Acta* **2007**, *71*, 1721–1736. [CrossRef]
29. Kaye, K.; Bryant, D.E.; Marriott, K.E.; Ohara, S.; Fishwick, C.W.; Kee, T.P. Selective Phosphonylation of 5′-Adenosine Monophosphate (5′-AMP) via Pyrophosphite [PPi(III)]. *Orig. Life Evol. Biosph.* **2016**, *46*, 425–434. [CrossRef]
30. Feng, T.; Gull, M.; Omran, A.; Abbott-Lyon, H.; Pasek, M.A. Evolution of Ephemeral Phosphate Minerals on Planetary Environments. *ACS Earth Space Chem.* **2021**, *5*, 1647–1656. [CrossRef]
31. Feng, T.; Lang, C.; Pasek, M.A. The origin of blue coloration in a fulgurite from Marquette, Michigan. *Lithos* **2019**, *342*, 288–294. [CrossRef]
32. Gull, M.; Omran, A.; Feng, T.; Pasek, M.A. Silicate-, magnesium ion-, and urea-induced prebiotic phosphorylation of uridine via pyrophosphate; revisiting the hot drying water pool scenario. *Life* **2020**, *10*, 122. [CrossRef] [PubMed]
33. Omran, A.; Pasek, M. A Constructive Way to Think about Different Hydrothermal Environments for the Origins of Life. *Life* **2020**, *10*, 36. [CrossRef] [PubMed]
34. Omran, A.; Menor-Salavan, C.; Springsteen, G.; Pasek, M.A. The Messy Alkaline Formose Reaction and its Link to Metabolism. *Life* **2020**, *10*, 125. [CrossRef] [PubMed]
35. Omran, A. Plausibility of the Formose Reaction in Alkaline Hydrothermal Vent Environments. *Orig. Life Evol. Biosph.* **2020**, 1–13. [CrossRef] [PubMed]
36. Michael, O.G.; Pere, M.; Bess, V.; Anuradha, S.K.A.; Laura, M.B.; Omran, A.; Videau, P.; Swenson, V.; Leinen, L.J.; Fitch, N.W.; et al. Plausible Emergence and Self Assembly of a Primitive Phospholipid from Reduced Phosphorus on the Primordial Earth. *Orig. Life Evol. Biosph.* **2021**, *53*, 185–213.

Article

The Combinatorial Fusion Cascade to Generate the Standard Genetic Code

Alexander Nesterov-Mueller * and Roman Popov

Institute of Microstructure Technology, Karlsruhe Institute of Technology (KIT), 76344 Eggenstein-Leopoldshafen, Germany; roman.popov@axxelera.com
* Correspondence: alexander.nesterov-mueller@kit.edu

Abstract: Combinatorial fusion cascade was proposed as a transition stage between prebiotic chemistry and early forms of life. The combinatorial fusion cascade consists of three stages: eight initial complimentary pairs of amino acids, four protocodes, and the standard genetic code. The initial complimentary pairs and the protocodes are divided into dominant and recessive entities. The transitions between these stages obey the same combinatorial fusion rules for all amino acids. The combinatorial fusion cascade mathematically describes the codon assignments in the standard genetic code. It explains the availability of amino acids with the even and odd numbers of codons, the appearance of stop codons, inclusion of novel canonical amino acids, exceptional high numbers of codons for amino acids arginine, leucine, and serine, and the temporal order of amino acid inclusion into the genetic code. The temporal order of amino acids within the cascade is congruent with the consensus temporal order previously derived from the similarities between the available hypotheses. The control over the combinatorial fusion cascades would open the road for a novel technology to develop artificial microorganisms.

Keywords: origin of genetic code; prebiotic chemistry; time order of canonical amino acids

Citation: Nesterov-Mueller, A.; Popov, R. The Combinatorial Fusion Cascade to Generate the Standard Genetic Code. *Life* **2021**, *11*, 975. https://doi.org/10.3390/life11090975

Academic Editors: Michele Fiore and Emiliano Altamura

Received: 2 September 2021
Accepted: 14 September 2021
Published: 16 September 2021

Publisher's Note: MDPI stays neutral with regard to jurisdictional claims in published maps and institutional affiliations.

Copyright: © 2021 by the authors. Licensee MDPI, Basel, Switzerland. This article is an open access article distributed under the terms and conditions of the Creative Commons Attribution (CC BY) license (https://creativecommons.org/licenses/by/4.0/).

1. Introduction

The origin of the standard genetic code (SGC), more specifically, the codon distribution over canonical amino acids is one of the fundamental scientific problems. Mastering the molecular apparatus for generating artificial code would enable novel efficient microorganisms for scientific, medical, and industrial applications.

Koonin distinguished three major theories—error minimization, coevolution, and stereochemical—that strive to explain the regularities in the standard genetic code [1]. If the origin of the genetic code is considered in the scope of protein evolution, then this "top-down" approach leads to the consensus about the gradually evolved SGC from some initial primordial code. For example, P. Higgs, using the advanced error minimization model, showed that his four-column theory with a primordial code consisting of non-biologically synthesized amino acids Gly, Ala, Asp, Glu, and Val fit perfectly with a row of predictions from the coevolution theory [2]. S.E. Massey proposed that error minimization of the SGC could arise via the genetic code expansion, facilitated by the duplication of genes encoding charging enzymes and adaptor molecules [3,4].

The theory of coevolution was proposed by Wong in 1975 [5]. It is based on the idea that the genetic code originally consisted of a few amino acids—precursors, which occupied all available coding triplets and were subsequently replaced by their products—late amino acids. Wong defined one major center of amino acids consisting of Glu, Asp, Ala, Ser, and Gly, from which 11 other amino acids evolved, and the two minor centers of Phe-Tyr and Val-Leu. This idea turned out to be very productive as it stimulated research on synthetic pathways from prebiotic chemistry to microbial metabolism, which have made significant progress to date [6,7]. A detailed description of the coevolution theory and its variations can be found, for example, in the works of Di Gulio [8] or Wong et al. [9]. A controversial issue in coevolution theory is the principle of codon transfer from precursors to products (see discussions between author Di Giulio and reviewers Koonin, Knight and Higgs [10]). Wong himself, faced with the fact that Leu and Arg each occupy as many as six coding triplets, but do not belong to the major center, assumed: "To acquire and retain a high plurality, like Leu and Arg, the amino acid had to be both early in arrival and inert in reactivity" [5].

Remaining within the framework of this "top-down" approach, other arguments can be made in favor of the gradually evolving genetic code. Bases guanine G and cytosine C prevail in the acceptor stem of modern tRNA, and therefore the amino acids associated with the codons from these bases appeared earlier in the code [11,12]. However, the very unstable canonical amino acid arginine does not fit into this concept. The late branching of Class I aaRS (aminoacyl-tRNA synthetases) implies an early origin of the amino acids coded by the Class II aaRS [13,14]. The gradual expansion of the coding space as GC–GCA–GCAU genetic code was proposed by Hartman and Smith [15]. Kubyshkin and Budisa supported this hypothesis, demonstrating the correlation of this scheme with the hierarchy of the protein folding [16].

The anticodon binding domain typically provides added specificity for tRNA substrate interactions with aaRS [17,18]. This feature is missing by LeuRS, SerRS, and AlaRS, which may point out an early origin of the respective amino acids in the genetic code.

The modern stereochemical theory profits from the progress in aptamer technology and does not need the assumption about the gradual entry of amino acids into the code. It focuses on identifying the mechanisms of selective interactions between amino acids and their codons, or anticodons [19–21]. Yarus et al. reported statistically significant affinity for the interactions between Arg, His, Ile, Phe, Trp, and their cognate triplets [22]. These results were interpreted as evidence of the existence of an early stereochemical era during the code evolution [23].

The presence of only a portion of the 20 canonical amino acids in abiotic synthesis experiments, for example, in Miller's experiment [24] implies that amino acids were added to the SGC during its evolution. Although experimental conditions of such experiments such as the composition of gases have been criticized [25] and there are no proven connections with the time when amino acids entered the code and their appearance on Earth, Miller's results are widely used as key evidence for the gradually evolving genetic code [12,26].

Today, the assumption about the gradually evolving SGC has taken on the status of a postulate. However, this postulate fulfills a constructive role in reconciling modern trends in the science of the code origin: while the stereochemical theory develops a molecular base for the primordial code, the error minimization, and coevolution theories should complementarily describe its evolution to the SGC.

Meanwhile, there are reasons to believe that the origin of the SGC may not relate to the evolution of proteins. Koonin and Novozhilov argued that "attempts to decipher the primordial stereochemical code by comparative analysis of modern translation system components are likely to be futile" [23]. Since the boundary between the primordial code and the code that should have developed adaptively could not be drawn, it is possible to extend this statement to most of the genetic code.

This assumption is not new. According to Woese et al., the evolution of aminoacyl-tRNA synthetases certainly influenced the formation of the modern translation mechanisms, but did not shape the codon assignments [27]. The pioneers of the quantitative metric to measure the robustness of codes to error, Haig and Hurst, believed that "the code could acquire its major features before the evolution of proteins [28]." These ideas were put forward over 20 years ago, but have not received much more attention. Against the background of the rapid development of DNA sequencers since 2000, it was first more important to understand how the genomes and proteomes of the most archaic microorganisms relate to the genetic code [29]. This trend has contributed to the intensive development of the theory of coevolution and error minimization in the scope of protein evolution.

The outstanding work of Vetsigian et al. has given a new perspective on the origin of the code, focusing on the issue of its universality [30]. The authors concluded that "horizontal transfer of genes and perhaps other complex elements among the evolving entities [a dynamic far more rampant and pervasive than our current perception of horizontal gene transfer], is required to bring the evolving translation apparatus, its code, and by implication the cell itself to their current condition." In other words, the emergence of a universal code would be possible as a result of the coexisting competing entities with a horizontal fusion of their molecular apparatus.

The recent discovery of the fusion rules integrated into the SGC led to a simple and mathematically exact description of the codon distribution over canonical amino acids [31,32]. According to these rules, the modern genetic code arose from the fusion of dominant and recessive protocodes, which initially competed for the same triplets. It was concluded that almost all canonical amino acids were already involved in the reproductive apparatus of coexisting protocodes. In this paper, we consider the combinatorial fusion of competing entities in the scope of the combinatorial fusion cascade. We write out combinatorial fusion rules that are objective properties of the genetic code, discuss their rationale, and focus on how combinatorial fusional cascade could exist.

2. Combinatorial Fusion Cascade and Its Formalism

Figure 1 suggests that the fusion of protocodes into standard genetic code was part of a larger event that we call the combinatorial fusion cascade. The combinatorial fusion cascade started with coexisting pairs of amino acids. Each of these pairs had complementary monobasic codons AAA/UUU or GGG/CCC. By analogy with protocodes, these pairs were also divided into dominant and recessive ones, depending on the efficiency of their ancient reproduction mechanism. The terms "dominant" and "recessive" are borrowed from classical genetics and refer to the fact that the dominant entities do not change their initial codon-amino acid assignments after the fusion. In contrast, the recessive entities acquire new triplets.

The transition from the initial pairs to the protocodes consisted of the expansion of the initial monobasic triplets with complementary bases (indicated with blue letters in Figure 1). Therefore, the dominant pair Gly/Pro received additional codons GGC and CCG, in which there were substitutions by C or G in the third position, respectively. In the case of the recessive pair Arg/Ala, the same substitutions occurred in the first or the first and the third positions. As a result, Arg received two codons CGG and CGC, and Ala two codons GCC and GCG. The recessive Arg/Ala pair gave up their original codons in favor of the dominant Gly/Pro pair. This example illustrates the formation of the dominant GC-protocode as an intermediate stage to the SGC.

This fusion of the initial pairs to the dominant and recessive protocodes is summarized by the following rules:

Rule 1: The second-position bases do not change in any code.
Rule 2: C and G as well as U and A are exchangeable only in the third position in the dominant initial pairs.
Rule 3: C and G as well as U and A are exchangeable either in the first position or simultaneously in the first and third positions in the recessive initial pairs.

Combinatorial fusion rules of the protocodes to the SGC have the same pattern: dominant codes retain their original codons and have a single mutation at the third position. Combinatorial fusion rules for codon triplets in the protocodes state (Figure 1, red letters) have the form:

Rule 1: The second-position bases do not change in any code.
Rule 2: A and G as well as U and C are exchangeable only in the third position in the dominant protocodes.
Rule 3: A and G as well as U and C are exchangeable either in the first position or simultaneously in the first and third positions in the recessive protocodes.

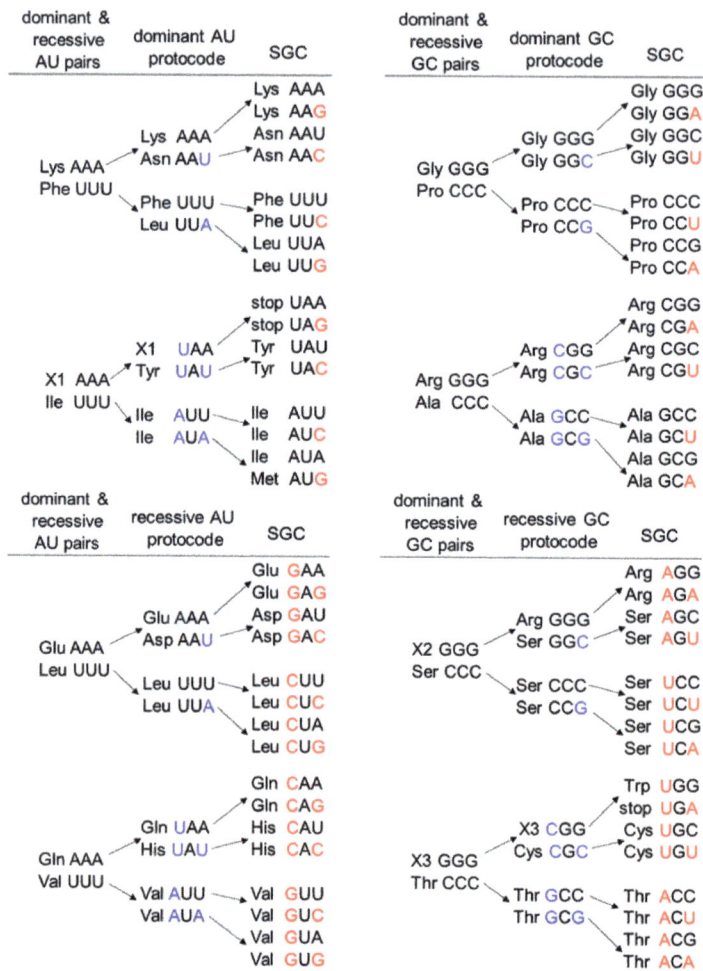

Figure 1. Combinatorial fusion cascade of the canonical amino acids leading to the codon assignments in the SGC. The blue letters indicate the fusion rules for the dominant and recessive AAA/UUU- and GGG/CCC-pairs to the protocodes. The red letters indicate the fusion rules for dominant and recessive AU- and GC-protocodes to the SGC. The fusion pattern is identical for all amino acids: The third position changes in the codons of the dominant entities. The first or the first and the third positions change in the codons of the recessive entities.

The fusion rules for the dominant protocodes correspond to the most spontaneously occurring mutation types (A -> G and G -> A as well as C -> U and U -> C) in the third codon (1st anticodon) position noted by Crick shortly after the publication of the code table [33]. This wobble position is occupied by a modified base that is part of the universal genetic code and was probably present in the last universal common ancestor [23]. In contrast, positions 1 and 2 are devoid of mutations in the case of the dominant protocodes. According to Copley et al., these regularities include strong correlations between the first base of codons and the precursor from which the encoded amino acid is synthesized and between the second base of codons and the hydrophobicity of the encoded amino acid [34].

The fusion rules for the recessive protocodes are based on the same mutations, but in the 1st position or in positions 1 and 3. Having identified this fact from the fusion cascade,

Nesterov-Mueller et al. suggested that such combinations could be induced within the framework of the kissing hairpin geometry by means of complementary codons [32]. Attempts to explain the emergence of the genetic code from complementary tRNA hairpins were also undertaken by Rodin and Ohno [35]. These kissing hairpins served as proto-tRNAs carrying two amino acids and were a structural element of the protocodes. Hairpins in recessive protocodes had a significantly lower concentration than hairpins in dominant ones. Therefore, they were inhibited by the dominant hairpins and occupied the remaining free combinatorial combinations after the fusion event, which correspond to the fusion rules for the recessive protocodes.

By analogy, we assume that the concentration of the dominant initial pairs in the cascade was higher than in recessive ones, which could be determined by different lengths of linear fragments of complementary RNAs. The initial pairs of triplets had a relatively simple organization including only one hydrophobic and one polar amino acid. At the same time, they could reproduce themselves and participate in the competition with other initial pairs. Most likely, the combinatorial fusion cascade was more extensive and contained other pairs with non-canonical hydrophobic and polar amino acids. However, the reconstruction of the combinatorial cascade from the standard genetic code, according to the fusion rules, led to exactly eight initial pairs of amino acids (total of 16 amino acids). This is equal to the number of amino acids in the doublet codon code proposed by Copley et al. based on the possible synthesis pathways of amino acids from α-keto acid precursors covalently attached to dinucleotides [34]. At the same time, the bases and phosphates of the dinucleotide are proposed to have enhanced the rates of synthetic reactions, leading to amino acids.

Individual parts of combinatorial fusion rules can be reformulated within the framework of group theory. However, we do not believe that group theory is a suitable formalism for describing the combinatorial fusion cascade. The idea of using group theory to explain the origin of the genetic code appeared in the 60s, immediately after the publication of the code table [36]. Afterward, several mathematical approaches to the genetic code in terms of symmetry properties have been developed [37–39]. According to D.L. Gonzalez et al. [40], the problem with such descriptions was the difficulty in providing a biological interpretation. A deeper reason for this is that symmetries derived from group theory describe conservation laws in conservative systems. However, the genetic code is not a conservative system, neither in terms of volume nor in terms of energy or mass transport. Consequently, the group theory approach, although it can exactly describe certainly not random symmetries of the SGC and their breaking, could not lead to its ultimate goal—conservation laws and the conclusion about the coexisting protocodes with identical triplets. In contrast, combinatorial fusion rules express this intrinsic fact of the SGC in an explicit and probably the simplest form.

3. Results and Discussion

3.1. Entering Amino Acids into the Combinatorial Fusion Cascade and Codon Assignment in the SGC

Using Figure 1, the development of the genetic code can be represented in two stages: The entrance of the bases U and A or G and C into the monobasic codons and then the fusion of the protocodes to the SGC. These processes obey a simple mathematical description that is uniform for the entire cascade.

This separation of SGC visually highlights the already known properties of the code and reveals features that were not visible in other representations of the code, for example, in the original tabular form. The formation of an even number of codons for canonical amino acids is a direct consequence of the combinatorial fusion cascade. Ideally, each amino acid from the initial pairs should have four codons. Deviations from four codons (two or six codons per amino acid as well as an odd number of codons) are associated with the entry of new amino acids or the disappearance of old ones.

The capture of new amino acids into the cascade led to a reduction in the number of codons to two per amino acids. At the same time, the new incoming amino acids had similar properties to those that they replaced. For example, aspartic acid (Asp) acquired the codon AAU derived from the AAA of glutamic acid (Glu). In the dominant AU-protocode, the hydrophobic amino acid phenylalanine (Phe) was substituted by the hydrophobic amino acid leucine (Leu). Thus, this transition of Leu from the recessive AU-protocode resulted in the six codons.

The amino acids arginine and serine (Ser) also have a maximum number of six codons. According to the combinatorial fusion cascade, these amino acids acquired an additional two codons after the exclusion of the amino acid X2 from the recessive protocode. It is noteworthy that the division into protocodes showed that each protocode used one positively charged amino acid: dominant AU—Lys, recessive AU—His, dominant GC—Arg, and recessive GC—Arg (X2). These positively charged amino acids may significantly contribute to the specific interactions between negatively charged RNAs and antient peptides. Most likely, the lost X2 was a positively charged amino acid and its substitution occurred with arginine, which carried a positive charge and had the same cognate codon GGG in the dominant protocode. A candidate for X2 could be the guanidinooxy analogue of L-arginine—the weaker base canavanine (Cav). Cav is found in some legumes. It easily integrates into proteins instead of arginine and is toxic to many organisms [41]. The role of Cav in microorganisms, especially in nitrogen fixation, is still far from being understood [42].

The appearance of an odd number of codons (three stop codons, one for methionine (Met), and one for tryptophan (Trp)) refers to the later stabilization of the SGC after the combinatorial fusion. Met and Trp are the most poorly represented amino acids in the genomes. Trp entered the code after the fusion of the protocodes occupying one of the initial stop codon UGG. Similar substitutions were also observed for the stop codons in the SGC. Non-canonical amino acid pyrrolysine (Pyl) occupied stop codon UAG in some prokaryotes, which was necessary to develop the methane metabolism [43,44]. The free stop codon UGA was adopted for selenocysteine (Sec) [45].

3.2. Temporal Order of Amino Acids

The combinatorial fusion cascade reveals the principle of the amino acid entry into the genetic code as well as the evolution of triplets. All canonical amino acids (except Trp and Met) were already assigned to their cognate codons in the protocodes' code before the modern-type translation appeared. In this context, one can compare the temporal order of amino acids from the combinatorial fusion cascade (Figure 2a) with the consensus chronology developed by Trifonov (Figure 2b). This approach is unique because forty different single-factor criteria and multi-factor hypotheses about the chronological order of the appearance of amino acids in the early evolution are summarized in consensus ranking. Such a compact representation is very convenient for our case.

Analyzing the consensus chronology, Trifonov made four conclusions:

1. The first amino acids to have been incorporated in early code were of abiotic origin, namely those that were obtained in classical imitation experiments by S. Miller.
2. In the development of the triplet code, a major role was played by the thermostability of codon–anticodon interactions.
3. New codons appeared in complementary pairs.
4. New codons were simple derivatives of chronologically earlier ones.

The chronology of amino acids in the combinatorial fusion cascade showed good congruence with the consensus chronology according to Trifonov. Exceptions are some cases related to the temporal order, which is based on Miller's experiment [24]. For example, the negatively charged aspartic acid (Asp) entered the code later than its negatively charged analog glutamic acid (Glu). In contrast, in consensus-time order, Asp is one of the first amino acids in the genetic code. Phenylalanine (Phe) and lysine (Lys) are the initial pair in the dominant protocode, but according to Trifonov, they entered the code much later just before the rare canonical amino acid methionine.

(a) Amino acid chronology↓	Dominant protocodes	Recessive protocodes
I. Stage of the initial pairs	Lys, Phe X1, Ile Gly, Pro Arg, Ala	Glu, Leu Gln, Val X2, Ser X3, Thr
II. Stage of the coexisting protocodes	Asn Leu (UUA/G) Tyr	Asp His Cys Arg (AGA/G) Ser (AGU/C)
III. After-fusion-stage	Met stopcodon (UAA/G)	Trp stopcodon (UGA/G)

(b)

Amino acid chronology↓
Gly
Ala
Val
Asp
Pro
Ser
Glu
Leu
Thr
Arg
Ser (AGU/C)
Arg (AGA/G)
Asn
Lys
Gln
Leu (UUA/G)
Ile
Cys
His
Phe
Met
Tyr, stop
Trp, stop

Figure 2. (a) Amino acid chronology of the combinatorial fusion cascade; (b) amino acid chronology according to the consensus temporal order after Trifonov [46]. Both the combinatorial fusion cascade and the consensus temporal order indicate a later acquisition of the codons UUA/UUG by the amino acid Leu, AGA/AGG by Arg, and AGU/AGC by Ser. As a result, each of these amino acids acquired six codons in the SGC. The color denotes the belonging of the amino acid to the stage of the combinatorial fusional cascade: red—the stage of the initial pairs, blue—the stage of coexisting protocodes, brown—after-fusion-stage.

Using only the temperature stability parameter without taking into consideration all the consensus numerous criteria, Trifonov obtained, in particular, the following chronological orders: Arg > Ser (AGU/C) > Cys; Lys > Gln > Leu (UUA/G); Arg (AGA/G) > Trp; His > Met (sign > means here "earlier"). These estimates completely coincide with the time order of the combinatorial fusion cascade. It is worth noticing that the consensus approach and the combinatorial fusion cascade are congruent in the late assignment of AGA and AGG codons to Arg as well UUA and UUG codons to Leu. Within the cascade, this is simply explained by the transition of Arg with the GGG codon from the dominant GC code to the recessive one as well as the transition of Leu with the UUA codon from the recessive AU code to the dominant one (Figure 1).

Trifonov's consensus principle 3 regarding the complementarity of new codons also finds a simple explanation within the combinatorial fusion cascade. Codon complementarity is a central element of the cascade (see Section 3.3).

Consensus principle 4 about the sequential entry of new codons into code as simple derivatives of chronologically earlier ones leaves room for different interpretations. If it is understood as a modification of a codon by one letter, then it contradicts the third fusion rule (for recessive entities), which allows for the appearance of a new codon with the replacement of bases in the first and third positions. This is where the fundamental difference between the combinatorial fusion cascade and the hypothesis about a gradually evolving code is manifested: The code arose not through progressive evolution, but because of the competition of dominant and recessive entities for the same codons.

3.3. Horizontal Transfer of "Complex Elements among the Evolving Entities"

The simulation model of Vetsigian, Woese, and Goldenfeld mentioned in the introduction cannot be directly applied for the combinatorial fusion cascade, since it does not consider the expansion of the code due to the appearance of new bases in triplets. However, it possesses a high degree of generalization, calculating the communal evolution of the competing initially random entities and their genomes in the form of a freely defined codon usage matrix with and without horizontal gene transfer (HGT) [30]. The authors demonstrated that when the HGT is present, the tendency to diversity between the competing entities is reduced and the code tends to achieve near universality.

In the combinatorial fusion cascade, this HGT principle is expressed in an extreme form of the fusion between the recessive and dominant entities. The combinatorial fusion shows how the competing protocodes can be upgraded without destroying their functionality: The original complementarity between small (most of them hydrophobic) and large (hydrophilic) amino acids in all competing protocodes is retained in the SGC. It is obvious for the dominant entities. For example, the codons of Lys and Phe (AAA and UUU) remain complimentary along the entire cascade. In the recessive AU-protocode, for example, Asp and Val lost their initial complementary codons AAU and AUU, but remained complementary in the SCG with new codons GAC and GUC, respectively. An added value from the combinatorial fusion consists of generating new amphiphilic combinations. For example, Lys becomes complementary not only to Phe, but also to hydrophobic Leu from the recessive AU-protocode at the end of the combinatorial fusion cascade.

Amphiphilic amino-acid–RNA complexes could contribute to the compartmentalization that was likely a crucial stage in the emergence of life [47–49] and primordial enzymatic functions where amino acids were brought together by non-covalent interactions [50]. Thus, the genetic code generated via the combinatorial fusion cascade could perform other functions before switching to the modern-type translation.

Compartmenting of the ancient protocodes could explain their spatial coexistence, similar to the coexistence of different microorganisms isolated by lipid bilayer. Stueken et al. proposed a study of the origin of life within the global context of the Hadean Earth as a global chemical reactor to benefit from identifying linkages between organic precursors, minerals, and fluids in various environmental contexts [51]. Assuming this approach, the protocodes could emerge independently in different geographic zones. For example, the recessive GC protocode containing Cys could arise in an area with a high concentration of sulfur atoms. An interesting fact of the combinatorial fusion cascade is that the canonic amino acids Lys and its non-canonic derivative Pyl, or structurally similar Sec and Cys are located in the same protocodes, respectively. One of the explanations for this fact is that non-canonical amino acids Pyl and Sec are X1 and X3, which were lost when the spreading across the Earth protocodes merged and then later rediscovered by microorganisms in conditions similar to those where the protocodes appeared [31]. Preiner et al. studying the ancient metabolic pathways of amino acids indicated the existence of the same synthesis pathway for Lys, Pyl, Met, Ile, and Asn [7]. Note that all these amino acids are associated with the dominant AU protocode.

To what extent the combinatorial fusion took place in reality depends on the principal issue: "Why did the combinatorial fusion cascade start with monobasic triplet codons, while only random polymerization of nucleotides could occur on the primitive Earth?". The answer to this question is not yet known, but may be in the special features of the oligonucleotide replication. P.W. Kudella et al. showed that linking short oligomers from a random sequence pool in a templated ligation reaction significantly reduces the sequence space of product strands [52]. The principles for reducing strand entropy identified in that article are also applicable to non-enzymatic replication.

4. Conclusions and Outlook

The concept of a gradually evolving genetic code dominates the science about the origin of the SCG. However, this view is only a hypothesis and cannot be used to negate other approaches such as the combinatorial fusion cascade.

Combinatorial fusion rules provide a simple description of codon assignments based on the fusion of the four protocodes. The corresponding protocodes competed for a limited number of codons. Therefore, dominant and recessive protocodes appeared. It turned out that these competition patterns can be followed up to individual coexisting amino acid pairs. They existed in the initial stage of the combinatorial fusion cascade—the explosion-like transition from prebiotic molecules to the first forms of life.

Along with the mathematical description of the codon assignments, combinatorial fusion cascade explained many features of the standard genetic codes as availability of amino acids with the even and odd numbers of codons, the appearance of stop codons, inclusion of novel canonical amino acids, exceptional high numbers of codons for amino acids arginine, leucinem and serine, and the temporal order of amino acid inclusion into the code.

The time order of amino acids within the cascade demonstrated a good congruence with the consensus time order calculated by Trifonov. The difference between both time orders arose only where the combinatorial fusion cascade contradicted the postulate of the progressive addition of amino acids into the genetic code.

An important property of the combinatorial fusional cascade is the preservation of complementarity between the codons of hydrophobic and hydrophilic canonical amino acids and the generation of new amphiphilic pairs after the fusion of dominant and recessive entities.

The combinatorial fusion cascade broadens the view on biotechnology. As noted at the end of §3, amino acid-modified amphiphilic hairpin pairs as well as replication processes were presumably the driving forces behind the combinatorial fusion cascade. Since the chemical foundations of the functionalization of RNA with amino acids as well as replication processes are well studied and can individually be reproduced in laboratory conditions, there is a possibility of laboratory combination of these processes for an artificial combinatorial fusional cascade. Depending on the set of artificial amino acids for the cascade (which cannot be integrated into proteins) as well as the environment that may be of technical interest, artificial self-replicating entities with the desired biochemical pathways could be developed. Further study of the principles of the combinatorial fusional cascade could help in the search for early life forms on Earth and beyond, and in the long-term, for developing extraterrestrial habitats.

Author Contributions: Conceptualization, methodology and writing, A.N.-M. and R.P. All authors have read and agreed to the published version of the manuscript.

Funding: This research was funded by DFG, grant number AOBJ655892.

Institutional Review Board Statement: Not applicable.

Informed Consent Statement: Not applicable.

Acknowledgments: We acknowledge support from the KIT-Publication Fund of the Karlsruhe Institute of Technology.

Conflicts of Interest: The authors declare no conflict of interest.

References

1. Koonin, E.V. Frozen Accident Pushing 50: Stereochemistry, Expansion, and Chance in the Evolution of the Genetic Code. *Life* **2017**, *7*, 22. [CrossRef]
2. Higgs, P.G. A four-column theory for the origin of the genetic code: Tracing the evolutionary pathways that gave rise to an optimized code. *Biol. Direct* **2009**, *4*, 16. [CrossRef] [PubMed]
3. Massey, S.E. The neutral emergence of error minimized genetic codes superior to the standard genetic code. *J. Theor. Biol.* **2016**, *408*, 237–242. [CrossRef]

4. Di Giulio, M. A Non-neutral Origin for Error Minimization in the Origin of the Genetic Code. *J. Mol. Evol.* **2018**, *86*, 593–597. [CrossRef] [PubMed]
5. Wong, J.T. A co-evolution theory of the genetic code. *Proc. Natl. Acad. Sci. USA* **1975**, *72*, 1909–1912. [CrossRef]
6. Muchowska, K.B.; Varma, S.J.; Moran, J. Synthesis and breakdown of universal metabolic precursors promoted by iron. *Nature* **2019**, *569*, 104. [CrossRef]
7. Preiner, M.; Xavier, J.C.; Vieira, A.D.; Kleinermanns, K.; Allen, J.F.; Martin, W.F. Catalysts, autocatalysis and the origin of metabolism. *Interface Focus* **2019**, *9*. [CrossRef]
8. Di Giulio, M. The origin of the genetic code: Theories and their relationships, a review. *Biosystems* **2005**, *80*, 175–184. [CrossRef]
9. Wong, J.T.F. Coevolution theory of the genetic code at age thirty. *Bioessays* **2005**, *27*, 416–425. [CrossRef] [PubMed]
10. Di Giulio, M. An extension of the coevolution theory of the origin of the genetic code. *Biol. Direct* **2008**, *3*, 37. [CrossRef] [PubMed]
11. Gospodinov, A.; Kunnev, D. Universal Codons with Enrichment from GC to AU Nucleotide Composition Reveal a Chronological Assignment from Early to Late Along with LUCA Formation. *Life* **2020**, *10*, 81. [CrossRef] [PubMed]
12. Higgs, P.G.; Pudritz, R.E. A Thermodynamic Basis for Prebiotic Amino Acid Synthesis and the Nature of the First Genetic Code. *Astrobiology* **2009**, *9*, 483–490. [CrossRef] [PubMed]
13. Carter, C.W.; Wills, P.R. Hierarchical groove discrimination by Class I and II aminoacyl-tRNA synthetases reveals a palimpsest of the operational RNA code in the tRNA acceptor-stem bases. *Nucleic Acids Res.* **2018**, *46*, 9667–9683. [CrossRef]
14. Caetano-Anolles, G.; Wang, M.L.; Caetano-Anolles, D. Structural Phylogenomics Retrodicts the Origin of the Genetic Code and Uncovers the Evolutionary Impact of Protein Flexibility. *PLoS ONE* **2013**, *8*, e72225. [CrossRef]
15. Hartman, H.; Smith, T.F. The evolution of the ribosome and the genetic code. *Life* **2014**, *4*, 227–249. [CrossRef] [PubMed]
16. Kubyshkin, V.; Budisa, N. The Alanine World Model for the Development of the Amino Acid Repertoire in Protein Biosynthesis. *Int. J. Mol. Sci.* **2019**, *20*, 5507. [CrossRef] [PubMed]
17. Giege, R.; Sissler, M.; Florentz, C. Universal rules and idiosyncratic features in tRNA identity. *Nucleic Acids Res.* **1998**, *26*, 5017–5035. [CrossRef]
18. Pang, Y.L.; Poruri, K.; Martinis, S.A. tRNA synthetase: tRNA aminoacylation and beyond. *Wiley Interdiscip Rev. RNA* **2014**, *5*, 461–480. [CrossRef] [PubMed]
19. Yarus, M. Amino acids as RNA ligands: A direct-RNA-template theory for the code's origin. *J. Mol. Evol.* **1998**, *47*, 109–117. [CrossRef]
20. Yarus, M. RNA-ligand chemistry: A testable source for the genetic code. *RNA* **2000**, *6*, 475–484. [CrossRef]
21. Yarus, M.; Caporaso, J.G.; Knight, R. Origins of the genetic code: The escaped triplet theory. *Annu. Rev. Biochem.* **2005**, *74*, 179–198. [CrossRef] [PubMed]
22. Yarus, M.; Widmann, J.J.; Knight, R. RNA-Amino Acid Binding: A Stereochemical Era for the Genetic Code. *J. Mol. Evol.* **2009**, *69*, 406–429. [CrossRef]
23. Koonin, E.V.; Novozhilov, A.S. Origin and Evolution of the Universal Genetic Code. *Annu. Rev. Genet.* **2017**, *51*, 45–62. [CrossRef]
24. Miller, S.L. A Production of Amino Acids under Possible Primitive Earth Conditions. *Science* **1953**, *117*, 528–529. [CrossRef]
25. Bada, J.L. New insights into prebiotic chemistry from Stanley Miller's spark discharge experiments. *Chem. Soc. Rev.* **2013**, *42*, 2186–2196. [CrossRef]
26. Brooks, D.J.; Fresco, J.R.; Lesk, A.M.; Singh, M. Evolution of amino acid frequencies in proteins over deep time: Inferred order of introduction of amino acids into the genetic code. *Mol. Biol. Evol.* **2002**, *19*, 1645–1655. [CrossRef] [PubMed]
27. Woese, C.R.; Olsen, G.J.; Ibba, M.; Soll, D. Aminoacyl-tRNA synthetases, the genetic code, and the evolutionary process. *Microbiol. Mol. Biol. Rev.* **2000**, *64*, 202–236. [CrossRef] [PubMed]
28. Haig, D.; Hurst, L.D. A Quantitative Measure of Error Minimization in the Genetic-Code. *J. Mol. Evol.* **1991**, *33*, 412–417. [CrossRef] [PubMed]
29. Weiss, M.C.; Preiner, M.; Xavier, J.C.; Zimorski, V.; Martin, W.F. The last universal common ancestor between ancient Earth chemistry and the onset of genetics. *PLoS Genet.* **2018**, *14*, e1007518. [CrossRef] [PubMed]
30. Vetsigian, K.; Woese, C.; Goldenfeld, N. Collective evolution and the genetic code. *Proc. Natl. Acad. Sci. USA* **2006**, *103*, 10696–10701. [CrossRef] [PubMed]
31. Nesterov-Müller, A.; Popov, R. Die Botschaft von LUCA—Der letzte universelle gemeinsame Vorfahre. *Biospektrum* **2020**, *26*, 488–489. [CrossRef]
32. Nesterov-Mueller, A.; Popov, R.; Seligmann, H. Combinatorial Fusion Rules to Describe Codon Assignment in the Standard Genetic Code. *Life* **2021**, *11*, 4. [CrossRef]
33. Lei, L.; Burton, Z.F. Evolution of Life on Earth: tRNA, Aminoacyl-tRNA Synthetases and the Genetic Code. *Life* **2020**, *10*, 21. [CrossRef]
34. Copley, S.D.; Smith, E.; Morowitz, H.J. A mechanism for the association of amino acids with their codons and the origin of the genetic code. *Proc. Natl. Acad. Sci. USA* **2005**, *102*, 4442–4447. [CrossRef]
35. Rodin, S.N.; Ohno, S. Four primordial modes of tRNA-synthetase recognition, determined by the (G,C) operational code. *Proc. Natl. Acad. Sci. USA* **1997**, *94*, 5183–5188. [CrossRef] [PubMed]
36. Rumer, Y.B. Translation of 'Systematization of Codons in the Genetic Code [I]' by Yu. B. Rumer (1966). *Philos. Trans. R. Soc. A* **2016**, *374*, 20150446. [CrossRef] [PubMed]

37. Antoneli, F.; Forger, M. Symmetry breaking in the genetic code: Finite groups. *Math. Comput. Model.* **2011**, *53*, 1469–1488. [CrossRef]
38. Lenstra, R. Evolution of the genetic code through progressive symmetry breaking. *J. Theor. Biol.* **2014**, *347*, 95–108. [CrossRef]
39. Hornos, J.E.M.; Hornos, Y.M.M. Algebraic Model for the Evolution of the Genetic-Code. *Phys. Rev. Lett.* **1993**, *71*, 4401–4404. [CrossRef]
40. Gonzalez, D.L.; Giannerini, S.; Rosa, R. On the origin of degeneracy in the genetic code. *Interface Focus* **2019**, *9*, 20190038. [CrossRef] [PubMed]
41. Emmert, E.A.B.; Milner, J.L.; Lee, J.C.; Pulvermacher, K.L.; Olivares, H.A.; Clardy, J.; Handelsman, J. Effect of canavanine from alfalfa seeds on the population biology of Bacillus cereus. *Appl. Environ. Microb.* **1998**, *64*, 4683–4688. [CrossRef] [PubMed]
42. Kamo, T.; Sakurai, S.; Yamanashi, T.; Todoroki, Y. Cyanamide is biosynthesized from L-canavanine in plants. *Sci. Rep.* **2015**, *5*, 10527. [CrossRef]
43. Srinivasan, G.; James, C.M.; Krzycki, J.A. Pyrrolysine encoded by UAG in Archaea: Charging of a UAG-decoding specialized tRNA. *Science* **2002**, *296*, 1459–1462. [CrossRef]
44. Hao, B.; Gong, W.M.; Ferguson, T.K.; James, C.M.; Krzycki, J.A.; Chan, M.K. A new UAG-encoded residue in the structure of a methanogen methyltransferase. *Science* **2002**, *296*, 1462–1466. [CrossRef] [PubMed]
45. Donovan, J.; Copeland, P.R. The Efficiency of Selenocysteine Incorporation Is Regulated by Translation Initiation Factors. *J. Mol. Biol.* **2010**, *400*, 659–664. [CrossRef] [PubMed]
46. Trifonov, E.N. Consensus temporal order of amino acids and evolution of the triplet code. *Gene* **2000**, *261*, 139–151. [CrossRef]
47. Jia, T.Z.; Chandru, K.; Hongo, Y.; Afrin, R.; Usui, T.; Myojo, K.; Cleaves, H.J. Membraneless polyester microdroplets as primordial compartments at the origins of life. *Proc. Natl. Acad. Sci. USA* **2019**, *116*, 15830–15835. [CrossRef] [PubMed]
48. Szostak, J.W.; Bartel, D.P.; Luisi, P.L. Synthesizing life. *Nature* **2001**, *409*, 387–390. [CrossRef] [PubMed]
49. Schreiber, U.; Locker-Grutjen, O.; Mayer, C. Hypothesis: Origin of Life in Deep-Reaching Tectonic Faults. *Orig. Life Evol. Biosph.* **2012**, *42*, 47–54. [CrossRef] [PubMed]
50. New, R.; Bansal, G.S.; Bogus, M.; Zajkowska, K.; Rickelt, S.; Toth, I. Use of Mixed Micelles for Presentation of Building Blocks in a New Combinatorial Discovery Methodology: Proof-of-Concept Studies. *Molecules* **2013**, *18*, 3427–3441. [CrossRef]
51. Stueken, E.E.; Anderson, R.E.; Bowman, J.S.; Brazelton, W.J.; Colangelo-Lillis, J.; Goldman, A.D.; Som, S.M.; Baross, J.A. Did life originate from a global chemical reactor? *Geobiology* **2013**, *11*, 101–126. [CrossRef] [PubMed]
52. Kudella, P.W.; Tkachenko, A.V.; Salditt, A.; Maslov, S.; Braun, D. Structured sequences emerge from random pool when replicated by templated ligation. *Proc. Natl. Acad. Sci. USA* **2021**, *118*, e2018830118. [CrossRef] [PubMed]

Review

Questions and Answers Related to the Prebiotic Production of Oligonucleotide Sequences from 3′,5′ Cyclic Nucleotide Precursors

Judit E. Šponer [1,*], Jiří Šponer [1], Aleš Kovařík [1], Ondrej Šedo [2], Zbyněk Zdráhal [2], Giovanna Costanzo [3] and Ernesto Di Mauro [3]

1. Institute of Biophysics of the Czech Academy of Sciences, Královopolská 135, 61265 Brno, Czech Republic; sponer@ncbr.muni.cz (J.Š.); kovarik@ibp.cz (A.K.)
2. Central European Institute of Technology, Masaryk University, Kamenice 5, 62500 Brno, Czech Republic; ondrej.sedo@ceitec.muni.cz (O.Š.); zdrahal@sci.muni.cz (Z.Z.)
3. Institute of Molecular Biology and Pathology, CNR, Piazzale A. Moro 5, 00185 Rome, Italy; giovannamaria.costanzo@cnr.it (G.C.); ernesto.dimauro@uniroma1.it (E.D.M.)
* Correspondence: judit@ncbr.muni.cz

Abstract: Template-free nonenzymatic polymerization of 3′,5′ cyclic nucleotides is an emerging topic of the origin of life research. In the last ten years, a number of papers have been published addressing various aspects of this process. These works evoked a vivid discussion among scientists working in the field of prebiotic chemistry. The aim of the current review is to answer the most frequently raised questions related to the detection and characterization of oligomeric products as well as to the geological context of this chemistry.

Keywords: polymerization; cyclic nucleotides; prebiotic chemistry

1. Introduction

Production of oligonucleotide sequences from nucleotide precursors in the absence of enzymes and template molecules is one of the most challenging problems of modern origin of life research. 2′,3′ cyclic nucleotides (see Figure 1) have long been considered as potent precursors of the most ancient oligonucleotide sequences formed on our planet [1–4]. This motivated numerous attempts dealing with the reconstruction of their synthesis under plausible prebiotic conditions [5–8]. Nonetheless, nucleotides containing a phosphodiester linkage confined in a strained 5-membered ring are unstable in an acidic environment and undergo a quick ring-opening reaction on the timescale of minutes [9].

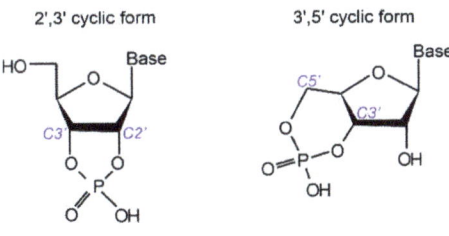

Figure 1. Structural formulas of 3′,5′ and 2′,3′ cyclic nucleotides.

In contrast, the 3′,5′ cyclic form possesses sufficient stability to withstand extreme conditions as conferred by low-pH environments, most likely ubiquitous on the early Earth [10,11]. Partly because of its significance played in modern biology (cGMP and cAMP are cofactors of key biochemical reactions [12]) and partly because of its unique aggregation properties, 3′,5′ cyclic guanosine monophosphate (hereafter 3′,5′ cGMP, the rest of the abbreviations used are explained at the end of the paper) was the target of a series of studies aimed at unraveling the process that could lead to the formation of the first oligonucleotide sequences on the early Earth [13–18].

The concept of polymerization studies using 3′,5′ cyclic nucleotides is strikingly different from that of preceding works published on prebiotic polymerization processes, which set a goal to find high-yielding chemical routes to oligonucleotides. These latter studies (for an overview see e.g., [19]) use highly activated monomers, because they enable achieving higher polymerization yields. Nevertheless, the lower stability of highly activated precursors (e.g., phosphorimidazolides, nucleotide triphosphates or 2′,3′ cyclic nucleotides) raises questions regarding their accumulation and long-term survival in the prebiotic pool.

Polymerization studies targeting 3′,5′ cyclic nucleotides place a special emphasis on identifying the conditions that are compatible with a sustainable oligonucleotide production on longer time scales in a geochemical context relevant to the early Earth. This requires not only sufficiently stable monomers, but also a robust chemistry. As Albert Eschenmoser suggested [20], the robustness of a prebiotic synthetic network lies in the multimodality of catalytically accessible chemical pathways between reactants and final products, because it enables adaptation of the chemical system to the quickly changing chemical environment.

In the current paper we provide a brief account of the available knowledge on the polymerization of 3′,5′ cyclic nucleotides. Since this research has been the focus of general attention for a long time, a number of questions and objections have been raised by the prebiotic chemistry community related to this topic. Our main goal with this paper is to provide an in-depth answer to the most frequent ones and to clarify those disputed points, which emerged after publishing the original studies.

2. How Could 3′,5′ Cyclic Nucleotides Accumulate in the Prebiotic Pool?

Synthesis of 2′,3′ cyclic nucleotides [5–8] was demonstrated by a number of studies both in aqueous and non-aqueous media. In contrast, no attention has so far been devoted to the prebiotic synthesis of 3′,5′ cyclic nucleotides. Nevertheless, based on historical studies by Khorana et al., a simple potential prebiotic route can be outlined to these compounds: Reference [21] illustrates that in the presence of carbodiimides nucleoside 5′-phosphates can be converted into the 3′,5′ cyclic form. Since carbodiimides form in large amounts upon thermal treatment of formamide in the presence of meteorites [22] and nucleoside-5′-phosphates belong to the main products of phosphorylation reactions conducted in formamide [7], accumulation of 3′,5′ cyclic nucleotides seems to be plausible in a formamide-rich medium under prebiotic conditions. Work on this topic is in progress.

3. Mechanism of Ring-Opening Polymerization Reactions Involving 3′,5′ Cyclic Nucleotides

Ring-opening polymerization reactions of cyclic phosphate and phosphonate esters in non-aqueous medium are well-known in the literature [23,24]. These reactions are commonly catalyzed by DBU suggesting that a base-catalyzed ring-opening polymerization operates in this case. In the first reports on polymerization of 3′,5′ cGMP the highest reaction yields were observed at pH 9 achieved with Tris-HCl buffering suggesting a base-catalyzed reaction mechanism [14]. Addition of DBU also promoted the reaction [14]. Recent studies on the pH-dependence of the reaction conducted in dry state indicated that the reaction proceeds by an acid- rather than a base-catalyzed mechanism [25]. We note, that this observation does not contradict the original observation made in a pH = 9 Tris-HCl buffer, since a recent study showed that the pH of this buffer exhibits an extraordinarily strong temperature dependence [26]. Thus, the actual pH of the Tris-HCl buffer could be 3–4 orders of magnitude lower at the temperature of the polymerization experiment (75–85 °C). At this temperature, a large part of the water present in the buffer (commonly only 15 µl is added to the dry material) evaporates and condenses in the tip of the test tube, thus, the pH-shift towards acidic becomes even more pronounced, making the acid-catalyzed mechanism more plausible [25].

A computational investigation has shown that a ladder-like stacked supramolecular architecture may host both the base- as well as the acid-catalyzed reaction routes (see Figure 2a) [16]. The only difference between the two pathways is that while in the base-catalyzed reaction pathway the nucleophile is activated by deprotonation of the ribose O3′ position, in the acid-catalyzed pathway the phosphate of the substrate is activated by protonation (see Figure 3). Let us note that use of both activation strategies is widespread in organic chemistry to facilitate addition-elimination-type substitution reactions [27].

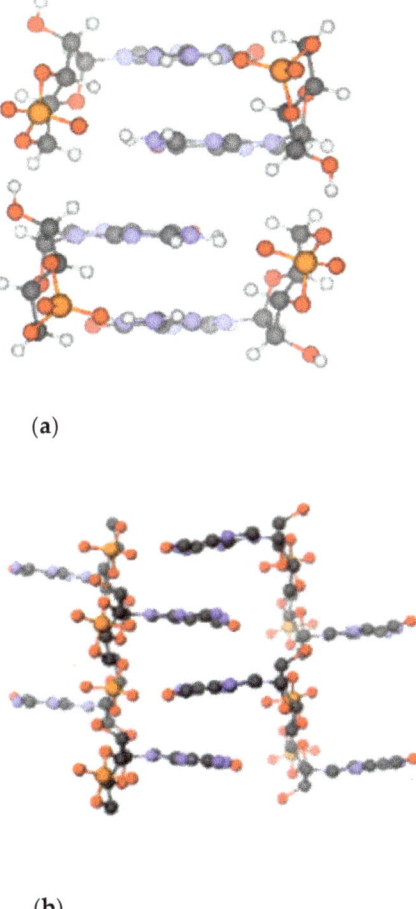

(a)

(b)

Figure 2. *Cont.*

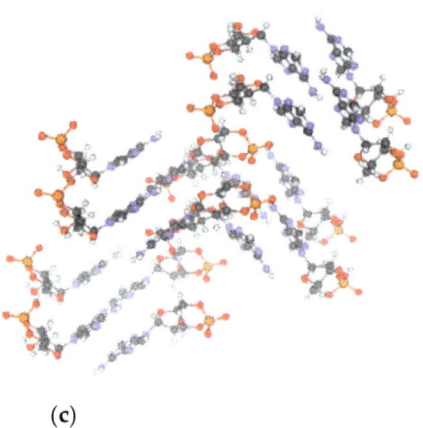

(c)

Figure 2. Theoretically predicted optimum model for the transphosphorylation of 3′,5′ cGMP [16] (**a**) as well as crystal structures of 3′,5′ cGMP [28] (**b**) and 3′,5′ cAMP [29] (**c**). Color coding: C—grey; O—red; N—blue; P—orange; H—white.

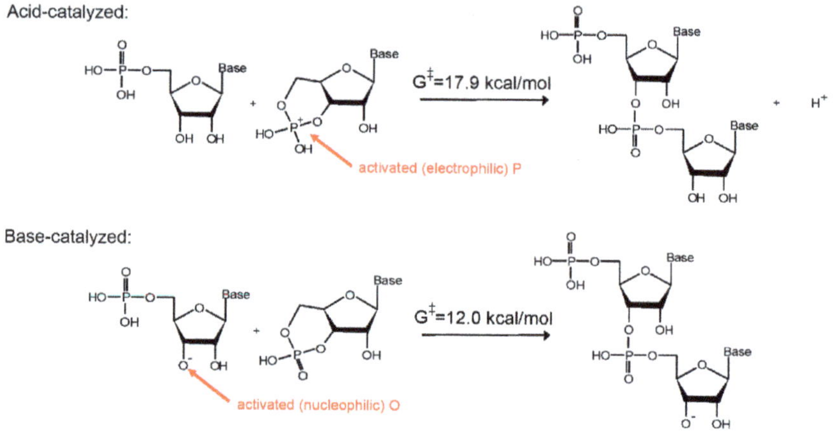

Figure 3. Mechanistic models for the chain-extension step of the ring-opening polymerization of 3′,5′ cGMP. $G^{\ddagger}$ refers to the computed activation free energy of the reaction step [16,25].

Amazingly, the stacked structure of 3′,5′ cGMP predicted by state-of-the art electronic structure (QM) calculations is essentially identical to the stacking pattern seen in the X-ray structure (see Figure 2b), reflecting the unique capability of 3′,5′ cGMP to self-associate by stacking. The electronic structure calculations in addition without doubt prove that this robust ground-state stacking arrangement is readily capable to support a reactive geometry for the oligomerization process. In view of this we conclude that all claims that 3′,5′ cGMP cannot adopt a reactive conformation, etc., should be dismissed as scientifically unfounded. Due to its stacking properties 3′,5′ cGMP can clearly approach a reactive conformation for polymerization on its own and does not need help from any other midwife molecules or activating agents. The ability of biopolymers and their monomers to self-assemble into reactive geometries could play a pivotal role at the emergence of life on our planet [30].

Our computations show that both mechanisms have similar activation energies and thermodynamics [16,25]. Thus, both mechanisms are chemically plausible per see. Nevertheless, the acid-catalyzed pathway seems to be more likely in a prebiotic context. This is because the reaction requires formation of a well-defined supramolecular architecture prior to the transphosphorylation reactions which is more easily achievable under acidic conditions.

It has been found that the crystal structure [28,31] of free acid- and Na-form cGMP provides optimum steric conditions for the transphosphorylation reactions necessary to form intermolecular phosphodiester linkages between the cyclic monomers (see Figure 2b) [16]. Notably, the crystal structure of cAMP [29] (Figure 2c) is less suited to support the polymerization chemistry. In line with this, only short (3–4 nt long) oligomers form from $3',5'$ cAMP in contrast to $3',5'$ cGMP, which polymerizes to a length up to 15–20 nucleotides. In addition, polymerization of $3',5'$ cAMP is considerably slower as compared to that of $3',5'$ cGMP [32].

While the computed thermodynamic and kinetic parameters are very similar for the acid- and base-catalyzed mechanisms, crystallization of the cGMP monomers under acidic and basic conditions is strikingly different [18,25]. While the H-form (i.e., acidic) monomers crystallize on the timescale of minutes, crystallization of the Na-form (i.e., basic salt) material requires several days [31] due to the apparent electrostatic repulsion between the negatively charged cGMP$^-$ monomers. Faster crystallization of the acid-form cGMP molecules enables faster formation of the stacked supramolecular architecture that is the prerequisite of the polymerization reaction. This is compatible only with a low pH environment and an acid-catalyzed mechanism [25].

4. Is Detection of Oligonucleotides Formed from $3',5'$ cGMP by Denaturing Gel Electrophoresis Biased by Noncovalent Adducts?

A number of methods have been used for detection of the various products formed in the oligomerization of $3',5'$ cGMP. Among them, electrophoresis on denaturing gels has been the most frequently applied technique mainly because of its high sensitivity. The method was used in combination with radioactive [13,14,16–18,32] as well as fluorescent labeling [15,25] (for a recent detailed description of methodological details see [18,25]).

The radioactive labeling has often been criticized because of the use of enzymes and the necessity of a precipitation step that is used to remove the excess ^{32}P-ATP prior to loading the labeled material on the gel. Potentially, this could lead to the formation of noncovalent aggregates. This possibility has been tested with a simple experiment, in which guanosine has been treated in the same way as $3',5'$ cGMP [18]. Aggregation properties of H-form $3',5'$ cGMP are very similar to those of guanosine. Thus, it is reasonable to expect, that if aggregates formed from guanosine withstand the precipitation and labeling procedure, they would have to produce multiple band structures on the denaturing gel, similar to those of the oligomerization products. As Figure 4 illustrates, this has not been observed. Further, this experiment excludes the possibility that the polynucleotide kinase used for phosphorylation could induce oligomer formation.

Denaturation PAGE is standardly run in the presence of 50% urea at elevated temperatures which effectively disrupts intermolecular bonds. Together, detection of non-covalent aggregates can be excluded. The fact that the oligomers are efficiently labeled in a kinase reaction suggests that a considerable portion of the polymerization products contains free $5'$-OH groups.

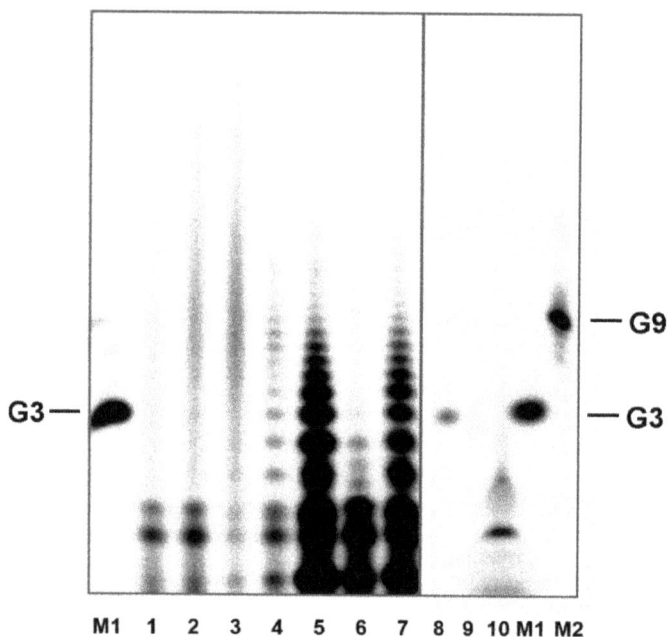

Figure 4. Control experiments with variously treated 3′,5′ cGMP as well as with guanosine and guanosine 5′-triphosphate (GTP) adopted from [18]. Lane 1: 150 μl 1 mM cGMP-H freeze-dried in an Eppendorf tube (for ~1 h) without subsequent thermal treatment at elevated temperatures. Lane 2: 150 μl 1 mM cGMP-H freeze-dried in an Eppendorf tube (for ~1 h) and treated at 80 °C for 5 min. Lanes 3 and 4: the same as lane 2, except that the thermal treatment at 80 °C was performed for 15 and 60 min, respectively. Lane 5: 66 × 3 μl of 1 mM cGMP-H solution deposited in a dropwise manner on the bottom of a preheated (80 °C) borosilicate beaker. The next drop was always deposited on the same place after drying of the previous one. The dry material was treated at 80 °C for additional 5 h. Lane 6: 6 × 7.5 μl of 1 mM cGMP-H solution was deposited in a dropwise manner into 5 spots on the bottom of a preheated (80 °C) borosilicate beaker. The next drop was always deposited on the same place after drying of the previous one. The dry material was treated at 100 °C for additional 5 h. Note that the optimum temperature for the polymerization reaction is ~75–85 °C, according to [14] and [16], i.e., this experiment was performed outside the optimum reaction temperature range. Lane 7: 30 × 6.6 μl of 1 mM cGMP-H solution deposited in a dropwise manner on the bottom of a preheated (80 °C) borosilicate beaker. The next drop was always deposited on the same place after drying of the previous one. The dry material was treated at 80 °C for additional 5 h. Lane 8: 150 μl of 2 mM (~300 ng) G3 oligomer (from Biomers) was freeze-dried in an Eppendorf tube, followed by a heat-treatment at 80 °C for 3 h. Lane 9 (almost empty): 150 μl of 1 mM guanosine freeze-dried in an Eppendorf tube and subsequently treated at 80 °C for 5 h. Since guanosine has the same aggregation properties as 3′,5′ cGMP-H, absence of oligomerization products shows that the kinase used for phosphorylation does not recognize non-covalent aggregates of monomers. Lane 10: 150 μl of 1 mM GTP prepared and treated as the guanosine sample (diffuse smeared bands towards the bottom are likely caused by the highly charged character of the molecule). Samples were precipitated using the procedure given in [18]. M1 = G3 (Biomers) marker. M2 = G9 (Biomers) marker.

5. Is Mass Spectrometric Detection of Oligonucleotides Formed from Cyclic Nucleotide Precursors Biased by Non-Covalent Adducts?

A frequently used methodology for oligonucleotide detection is MALDI-ToF mass spectrometry. The approach was used for detection of oligonucleotides formed from cyclic nucleotide precursors as well. Nevertheless, soon after its first use it has been shown that this method is often biased by non-covalent adduct formation, since the molar weight of adducts consisting of n cyclic monomers equals to that of an n-mer oligomer having a 2′,3′ cyclic end [33]. Likewise, molecular weight of an n-mer oligomer is the same as that of n-monomers plus a water molecule.

Studying the fragmentation spectra of the species detected with MALDI MS could be one of the potential remedies for this problem. For example, inspection of the polymerization products formed from 3′,5′ cAMP allowed a successful identification of pApA dimers. The fragmentation spectra clearly revealed signals corresponding to pA as well as pAp fragments, which exhibited roughly the same intensities as those found in the fragmentation spectra of pApA standards [32]. Unfortunately, the same method cannot be used for identifying pGpG products, since intensity of the pG signal, which also corresponds to the hydrolyzed monomer, is an order of magnitude higher than that of the more important pGp fragment.

In our studies dealing with the polymerization of 3′,5′ cCMP [34], we used ESI-MS detection, which (at least for this monomer) was less biased by formation of non-covalent adducts. In particular, in the spectrum of the heat-treated sample, we did not observe the peak shifted by +18 Da, corresponding to the adduct of the monomer with water or hydrolyzed monomer, whereas the signal corresponding to the pCpC covalent product was observed. Fragmentation of this signal produced the expected pCp and pC species, verifying the covalent character of the pCpC dimer.

To assess the extent of non-covalent adduct formation in the MALDI-ToF MS and ESI-MS spectra of polymerization products formed from guanine-containing monomers, we investigated the spectra of guanosine, a molecule with remarkably similar aggregation properties as that of 3′,5′ cGMP. Figure 5 shows that in contrast to the ESI-MS detection of oligoC sequences discussed in [34], non-covalent guanosine aggregates up to tetramer are observed both by MALDI-ToF as well as by ESI mass spectrometry. This is likely caused by the outstandingly high propensity of guanine-derivatives to form stacked aggregates independently on the medium. On the other hand, as expected, in lack of covalent oligomer formation, there is almost no difference between the spectra of the untreated and heat-treated (conducted at 80 °C for 5 h in dry form) materials.

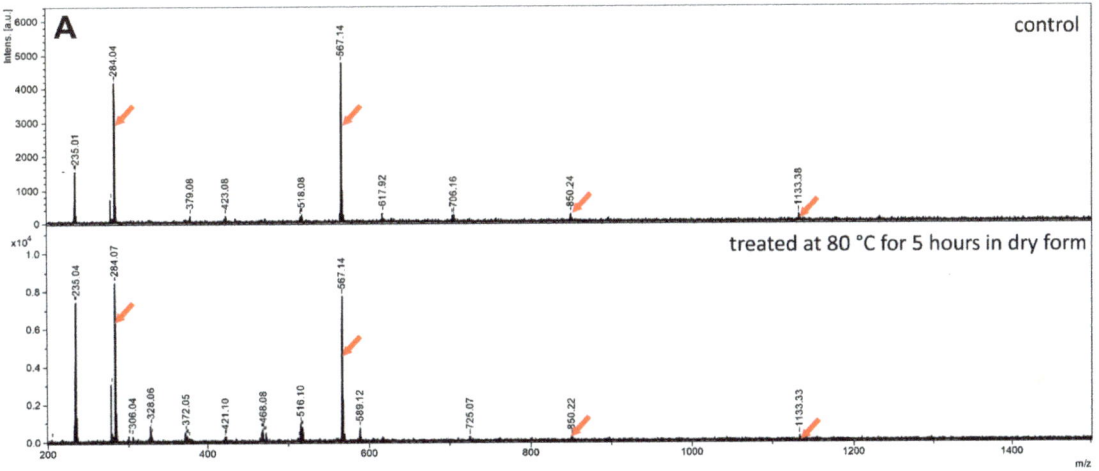

Figure 5. *Cont.*

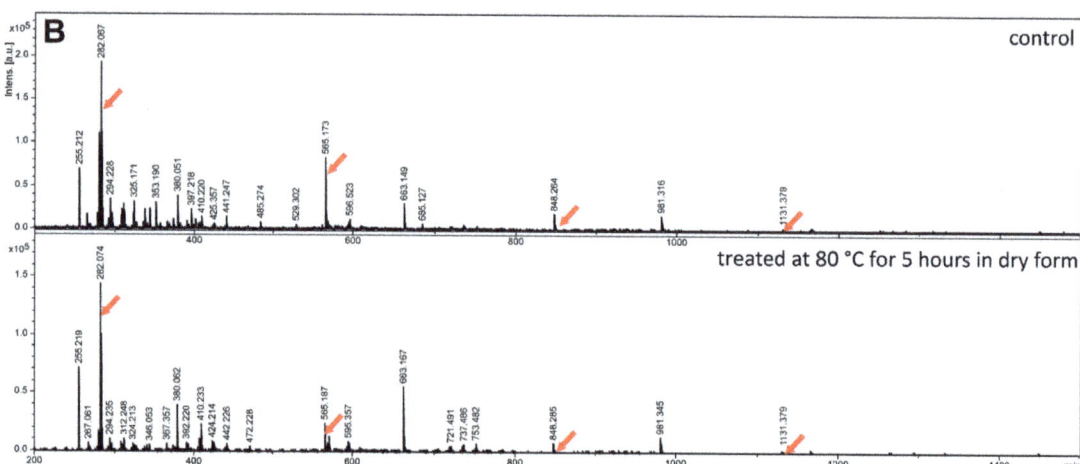

Figure 5. MALDI-ToF MS (measured in reflectron positive detection mode, panel (**A**)) and ESI-MS (measured in negative ion mode, panel (**B**)) spectra of untreated and heat-treated guanosine samples. The heat treatment was performed in dry state at 80 °C for 5 h. Signals of guanosine and its adducts are indicated with an arrow.

6. What Kind of Other Detection Methods Support Oligonucleotide Formation from 3′,5′ cGMP?

In addition to detection by electrophoresis and mass spectrometry, the covalent nature of the oligomers formed in the polymerization of 3′,5′ cGMP has been studied by a number of other methods (for a summary of methods applied for characterization of products formed from 3′,5′ cyclic nucleotides, see Table 1).

Table 1. Summary of the methodologies used for detection of oligonucleotides formed from 3′,5′ cyclic nucleotide precursors.

Method	Monomer	References
Denaturing gel-electrophoresis with ^{32}P-labelling	3′,5′ cGMP 3′,5′ cAMP	[13,14,16–18] [32]
Denaturing gel-electrophoresis with intercalating fluorescent labelling	3′,5′ cGMP	[15,25]
MALDI-ToF	3′,5′ cGMP 3′,5′ cAMP 3′,5′ cCMP	[14–16] [32] * [34]
ESI-MS	3′,5′ cCMP	[34] *
HPLC-ESI-MS	3′,5′ cGMP	[25]
^{31}P-NMR	3′,5′ cGMP	[14]
Infrared nanospectroscopy	3′,5′ cGMP	[17]

* Including MS/MS fragmentation analysis.

In [14] ^{31}P-NMR spectroscopy showed that upon heat treatment of 3′,5′ cGMP new signals appear in the 0 to −2.6 chemical shift range, characteristic to RNA oligonucleotides. Since the signal of phosphorus of acyclic or 2′,3′ cyclic monomers falls into the positive chemical shift range, the new signals resulting from sample treatment have been attributed to RNA-oligomers formed from 3′,5′ cyclic precursors.

Infrared nanospectroscopy has been used to follow the change of the IR spectrum of a 3′,5′ cGMP sample deposited on gold surface upon heat-treatment at 80 °C [17]. The analysis has shown that a new band appears at 970 cm^{-1}, i.e., in the fingerprint region characteristic to RNA-backbones, in the spectrum of the heat-treated material.

A recent study combines HPLC-separation and mass spectrometry to identify oligonucleotide products formed when polymerizing 3′,5′ cGMP [25]. This investigation has shown that besides non-covalent aggregates a noticeable amount of covalently bound oligomers is formed.

In summary, although the individual experimental methods used to monitor the oligomerization of 3′,5′ cNMP compounds are complicated by some uncertainties, on aggregate, the 3′,5′ cNMP polymerization reactions have been characterized by multiple independent methods and are thus more thoroughly scrutinized than many other oligomerization processes discussed in contemporary prebiotic chemistry literature.

7. Does the Oligomerization Reaction of 3′,5′ cGMP Proceed in the Presence of Water?

The reaction was studied under a multitude of experimental conditions. Its temperature optimum was found to be ca. 80 °C [13,14,16]. It was found that at least one drying step is necessary to get oligonucleotides longer than trimers [15,17]. Oligomer formation is regularly observed after ca. 1 h of reaction time. The reaction yields exhibit an oscillatory behavior on longer (several days) time scales, regardless of whether the reaction is conducted in dry state or in a nearly saturated solution [18]. The oscillatory behavior of the reaction has pointed at the importance of phase-separation processes in this chemistry. Nonetheless, oligomer (>trimer) formation is detectable on timescales of several weeks, assumed that at least partial drying of the monomers occurs in the course of sample treatment.

Thus, in summary, our investigations show that the presence of water in the reaction mixture is not poisonous for the polymer formation if part of the material is at least temporarily present in dry form during the sample treatment. Nonetheless, only short (≤3 nt long) oligomers form if drying of monomers is prevented [25,35].

8. Technical Requirements of the Oligomerization Experiment

When studying the polymerization of 3′,5′ cGMP, it was found that the drying process itself induces oligomer formation. Thus, conducting the polymerization experiment in a straightforward manner requires the use of a specially purified material that was never precipitated or dried during the sample preparation procedure. Using material purchased in dry form, due to its oligonucleotide content, produces false positive results.

When working with minerals, the specimens must be thoroughly cleaned by using H_2O_2, ethanol to remove their organic material content, which may degrade the oligomers formed in the polymerization reaction.

9. How Much is the Polymerization of 3′,5′ cGMP Sensitive to the Presence of Cations? May It Be Relevant in a Geological Context?

Earlier reports interpreted the absence of polymer formation from an Na-form salt of 3′,5′-cGMP by the fact that cations block the anionic ring-opening polymerization mechanism [16]. A recent study revisited this issue, and showed that presence of cations in the reaction mixture is not as restrictive as thought before [25]. As long as the polymerization is performed in an acidic environment, i.e., the monomers are neutral molecules, cations start inhibiting the reaction at very large (ca. 50×) excess. This means that basic pH rather than cations exert an inhibitory effect on the reaction [25].

The argument that the polymerization reaction is sensitive to cations logically suggested that such chemistry might be irrelevant to a real prebiotic environment, where free cations were likely available in large amounts. To project laboratory experiments into more geologically plausible models, we recently tried deposition of the cGMP monomers from a dropping solution on various heated mineral surfaces relevant to an early Earth scenario. The minerals chosen for our experiments were all included in Hazen's "preliminary species list" [36]. We found that minerals, like amorphous and crystalline silica forms, mica, amphibole and other silicates, like andalusite, may host the reaction. On the other hand, minerals that chemically react with the acid-form of 3′,5′ cGMP leading to its deprotonation (like carbonate- and serpentine-group minerals or olivine) are incompatible with the chemistry [18]. This way the free acid form of 3′,5′ cGMP is converted into a salt-form material, which does not crystallize on the time scales of the experiment. The lack of crystal order in the dry deposited material then interferes with the oligonucleotide formation (see Figure 6).

In summary, the ions per se do not prevent the oligomerization process.

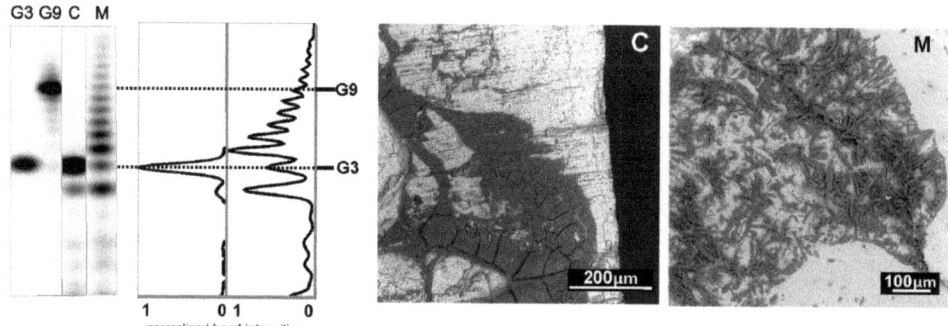

Figure 6. 3′,5′ cGMP crystallizes and polymerizes on muscovite mica (M). No crystallization and polymer formation are observed on calcite (C). From left to right: electrophoretic analysis of the polymerization products formed by heat-treatment (80 °C, 5 h) of the dry material deposited in a dropwise manner on the mineral surface; densitometric analysis of the electrophoretic images; electron microscopic images of the dry materials deposited on the mineral surfaces. Adopted from [18].

10. Does Depurination Hamper Formation of Short Oligonucleotide Sequences from 3′,5′ cGMP?

Higher propensity of guanine-containing nucleotides to depurinate in acidic conditions is well known [37]. Nevertheless, depurination reactions also require the presence of water in the reaction mixture. Under drying conditions, where oligonucleotide formation has been demonstrated, the depurination process is not significant and is unable to outcompete the polymerization reaction. It does not interfere with the oligomerization. Our recent mass spectrometric investigation (using both ESI- and MALDI-ToF MS methods) focused on the low molecular weight fraction products present in the reaction mixture and has shown that sample treatment in dry form at 80 °C for 5 h results only in a negligible depurination of the cyclic monomers [18]. In contrast, under dry-wet cycling conditions used by Rajamani et al. [38], half-life of acyclic 5′ AMP was estimated to be 6.5 h only. This study suggests that repeated dry-wet cycles may interfere with formation of oligonucleotides from purine nucleotide precursors.

We note, that depurinated short oligonucleotide sequences have been detected among the polymerization products by gel-electrophoresis [18]. Their origin is not necessarily associated with the polymerization process, since they could be the result of the sample treatment required by the radioactive end labeling and the denaturing gel electrophoresis. Signals of depurinated oligomer products were not observed in the MALDI-ToF MS analysis presented in [16].

11. Conclusions: What are the Potential Reasons for the Erratic Reproducibility of 3′,5′ cGMP Polymerization Experiments?

Lane et al. [39] rightfully designated reproducibility of the 3′,5′ cNMP oligomerization reactions as "erratic". This means that the reaction requires suitable physical and chemical conditions to proceed, and if these conditions are not fulfilled, the reaction is inhibited, which can lead to the erroneous claim that the experiments could not be reproduced. However, erratic behaviors of chemical processes are not uncommon (on example could be crystallization processes which often require many empirical trial and error experiments to find the right condition) and have nothing to do with reproducibility. In fact, some degree of erratic behavior may be advantageous in chemical evolution, as it indicates the reaction is variable and sensitive to external conditions. Therefore, it can be regulated and optimized by external factors, and coupled with other reactions. In other words, while we certainly found some suitable conditions upon which the 3′,5′ cGMP oligomerization occurs, we almost certainly did not find the optimal conditions. Importantly, this polymerization reaction does not seem to need a sophistically tailored chemical environment with other supportive chemical species; it appears to be responsive to straightforward changes in common physical-chemistry conditions.

Apart from trivial reasons (like not ensuring conditions in which relatively fragile short RNA sequences are capable of surviving; or using a salt form of the cyclic nucleotide monomer), over the years we identified several other reasons that might potentially hamper detection of oligonucleotides formed from 3′,5′ cGMP in specific experimental conditions.

We showed that the reaction itself exhibits an oscillatory behavior on longer time scales [17]. This feature originates in the highly non-equilibrium nature of this chemistry and manifests itself as the strong time dependence of the reaction yields. For this reason, if the reaction time is improperly selected for a given experiment, only low amounts of short (≤ 3 nt long) oligomers may be detected. Thus, it is advisable to study the reaction on longer time scales at several time points for a successful detection.

Another very important feature of the reaction is that at least one drying step is needed for a continued oligonucleotide production. This means that to a lesser extent, longer (>3 nt) oligonucleotides do form even in the presence of water, if the experimental setup allows for partial drying of the sample. In our previous studies we showed that this can be achieved by rewetting of the dry material, or by partial drying of the sample during the heat treatment [17,25].

The method used for drying may also influence the outcome of detection for two reasons. Prolonged (several hours long) drying in a low-performance vacuum evaporator may cause depurination of the nucleotide monomers. Further, too fast drying at elevated temperatures may occasionally lead to the formation of a glassy, non-crystalline material with randomly placed monomers even when using an H-form material. Obviously, in this case the lack of a properly organized supramolecular architecture interferes with the oligonucleotide formation. Many doubts about the reproducibility of the cGMP oligomerization reaction may stem from variability between experiments leading to different amounts and quality of crystalline materials.

Author Contributions: All authors contributed to writing—original draft preparation. All authors have read and agreed to the published version of the manuscript.

Funding: This research was funded by the Czech Science Foundation (grant 19-03442S). G.C. and E.D.M. were supported by the Italian Space Agency (ASI) project N. 2019-3-U.0, "Vita nello spazio—Origine, Presenza, Persistenza della vita nello Spazio, dalle molecole agli estremofili" (Space life—OPPS). This work was supported by the project SYMBIT reg. number: CZ.02.1.01/0.0/0.0/15_003/0000477 financed by the ERDF. CIISB, Instruct-CZ Centre of Instruct-ERIC EU consortium, funded by MEYS CR infrastructure project LM2018127, is gratefully acknowledged for the financial support of the measurements at the CEITEC Proteomics Core Facility.

Institutional Review Board Statement: Not applicable.

Informed Consent Statement: Not applicable.

Data Availability Statement: Data reported in this paper can be obtained from the authors upon request.

Acknowledgments: The authors thank Dieter Braun for his comments on the manuscript.

Conflicts of Interest: The authors declare no conflict of interest.

Abbreviations

3′,5′ cGMP	3′,5′ cyclic guanosine monophosphate
3′,5′ cGMP-H	free adic form of 3′,5′ cyclic guanosine monophosphate
3′,5′ cAMP	3′,5′ cyclic adenosine monophosphate
3′,5′ cCMP	3′,5′ cyclic cytosine monophosphate
5′ AMP	adenosine 5′ monophosphate
3′,5′ cNMP	3′,5′ cyclic nucleotide
nt	nucleotide
pG	5′ guanylic acid
pGpG	5′ diguanylic acid
pGp	5′ guanylic acid 3′ phosphate
pC	5′ cytidylic acid
pCpC	5′ dicytidylic acid
pCp	5′ cytidylic acid 3′ phosphate
DBU	1,8-diazabicyclo(5.4.0)undec-7-ene
MALDI	matrix-assisted laser desorption/ionization
ToF	time of flight
ESI	electrospray ionization
MS	mass spectrometry
HPLC	high-performance liquid chromatography

References

1. Verlander, M.S.; Lohrmann, R.; Orgel, L.E. Catalysts for self-polymerization of adenosine cyclic 2′, 3′-phosphate. *J. Mol. Evol.* **1973**, *2*, 303–316. [CrossRef]
2. Usher, D.A.; Yee, D. Geometry of the dry-state oligomerization of 2′,3′-cyclic phosphates. *J. Mol. Evol.* **1979**, *13*, 287–293. [CrossRef]
3. Tapiero, C.M.; Nagyvary, J. Prebiotic formation of cytidine nucleotides. *Nature* **1971**, *231*, 42–43. [CrossRef]
4. Morávek, J.; Kopecký, J.; Škoda, J. Thermic phosphorylations. 6. Formation of oligonucleotides from uridine 2′(3′)-phosphate. *Collect. Czech. Chem. Commun.* **1968**, *33*, 4120–4124. [CrossRef]
5. Powner, M.W.; Gerland, B.; Sutherland, J.D. Synthesis of activated pyrimidine ribonucleotides in prebiotically plausible conditions. *Nature* **2009**, *459*, 239–242. [CrossRef] [PubMed]
6. Powner, M.W.; Sutherland, J.D.; Szostak, J.W. Chemoselective multicomponent one-pot assembly of purine precursors in water. *J. Am. Chem. Soc.* **2010**, *132*, 16677–16688. [CrossRef] [PubMed]
7. Costanzo, G.; Saladino, R.; Crestini, C.; Ciciriello, F.; Di Mauro, E. Nucleoside phosphorylation by phosphate minerals. *J. Biol. Chem.* **2007**, *282*, 16729–16735. [CrossRef] [PubMed]
8. Crowe, M.A.; Sutherland, J.D. Reaction of cytidine nucleotides with cyanoacetylene: Support for the intermediacy of nucleoside-2′,3′-cyclic phosphates in the prebiotic synthesis of RNA. *Chembiochem* **2006**, *7*, 951–956. [CrossRef] [PubMed]
9. Shabarova, Z.; Bogdanov, A. *Advanced Organic Chemistry of Nucleic Acids*; VCH Verlagsgesellschaft mbH: Weinheim, Germany, 1994.
10. Mulkidjanian, A.Y.; Bychkov, A.Y.; Dibrova, D.V.; Galperin, M.Y.; Koonin, E.V. Origin of first cells at terrestrial, anoxic geothermal fields. *Proc. Natl. Acad. Sci. USA* **2012**, *109*, E821–E830. [CrossRef]
11. Kompanichenko, V.N. Exploring the Kamchatka geothermal region in the context of life's beginning. *Life* **2019**, *9*, 41. [CrossRef]
12. Newton, A.C.; Bootman, M.D.; Scott, J.D. Second messengers. *Cold Spring Harb. Perspect. Biol.* **2016**, *8*, a005926. [CrossRef] [PubMed]
13. Costanzo, G.; Pino, S.; Ciciriello, F.; Di Mauro, E. Generation of long RNA chains in water. *J. Biol. Chem.* **2009**, *284*, 33206–33216. [CrossRef]
14. Costanzo, G.; Saladino, R.; Botta, G.; Giorgi, A.; Scipioni, A.; Pino, S.; Di Mauro, E. Generation of RNA molecules by a base-catalysed click-like reaction. *ChemBioChem* **2012**, *13*, 999–1008. [CrossRef] [PubMed]
15. Morasch, M.; Mast, C.B.; Langer, J.K.; Schilcher, P.; Braun, D. Dry polymerization of 3′,5′-cyclic GMP to long strands of RNA. *ChemBioChem* **2014**, *15*, 879–883. [CrossRef] [PubMed]
16. Šponer, J.E.; Šponer, J.; Giorgi, A.; Di Mauro, E.; Pino, S.; Costanzo, G. Untemplated nonenzymatic polymerization of 3′,5′ cGMP: A plausible route to 3′,5′-linked oligonucleotides in primordia. *J. Phys. Chem. B* **2015**, *119*, 2979–2989. [CrossRef] [PubMed]

17. Costanzo, G.; Šponer, J.E.; Šponer, J.; Cirigliano, A.; Benedetti, P.; Giliberti, V.; Polito, R.; Di Mauro, E. Sustainability and chaos in the abiotic polymerization of 3′,5′ cyclic guanosine monophosphate: The role of aggregation. *ChemSystemsChem* **2021**, *3*, e2000056. [CrossRef]
18. Šponer, J.E.; Šponer, J.; Výravský, J.; Šedo, O.; Zdráhal, Z.; Costanzo, G.; Di Mauro, E.; Wunnava, S.; Braun, D.; Matyášek, R.; et al. Non-enzymatic, template-free polymerization of 3′,5′ cyclic guanosine monophosphate on mineral surfaces. *ChemSystemsChem* **2021**. Available online: https://chemistry-europe.onlinelibrary.wiley.com/doi/full/10.1002/syst.202100017 (accessed on 4 August 2021). [CrossRef]
19. Dorr, M.; Loffler, P.M.G.L.; Monnard, P.A. Non-enzymatic polymerization of nucleic acids from monomers: Monomer self-condensation and template-directed reactions. *Curr. Org. Synth.* **2012**, *9*, 735–763. [CrossRef]
20. Eschenmoser, A. Question 1: Commentary referring to the statement "The origin of life can be traced back to the origin of kinetic control" and the question "Do you agree with this statement; and how would you envisage the prebiotic evolutionary bridge between thermodynamic and kinetic control?" stated in section 1.1. *Orig. Life Evol. Biosph.* **2007**, *37*, 309–314.
21. Smith, M.; Drummond, G.I.; Khorana, H.G. Cyclic phosphates. IV.1 Ribonucleoside-3′,5′ cyclic phosphates. A general method of synthesis and some properties. *J. Am. Chem. Soc.* **1961**, *83*, 698–706. [CrossRef]
22. Saladino, R.; Botta, G.; Delfino, M.; Di Mauro, E. Meteorites as catalysts for prebiotic chemistry. *Chem.-Eur. J.* **2013**, *19*, 16916–16922. [CrossRef]
23. Iwasaki, Y.; Yamaguchi, E. Synthesis of well-defined thermoresponsive polyphosphoester macroinitiators using organocatalysts. *Macromolecules* **2010**, *43*, 2664–2666. [CrossRef]
24. Steinbach, T.; Ritz, S.; Wurm, F.R. Water-soluble poly(phosphonate)s via living ring-opening polymerization. *ACS Macro Lett.* **2014**, *3*, 244–248. [CrossRef]
25. Wunnava, S.; Dirscherl, C.F.; Výravský, J.; Kovařík, A.; Matyášek, R.; Šponer, J.; Braun, D.; Šponer, J.E. Acid-catalyzed RNA-oligomerization from 3,5-cGMP at an air-water interface. *Chem. Eur. J.* **2021**, in press.
26. Reineke, K.; Mathys, A.; Knorr, D. Shift of pH-value during thermal treatments in buffer solutions and selected foods. *Int. J. Food Prop.* **2011**, *14*, 870–881. [CrossRef]
27. Carey, F.A.; Sundberg, R.J. *Advanced Organic Chemistry. Part A: Structure and Mechanism*; Springer Science+Business Media, LLC: New York, NY, USA, 2007; p. 325.
28. Chwang, A.K.; Sundaralingam, M. The crystal and molecular structure of guanosine 3′,5′-cyclic monophosphate (cyclic GMP) sodium tetrahydrate. *Acta Crystallogr. B* **1974**, *30*, 1233–1240. [CrossRef]
29. Varughese, K.I.; Lu, C.T.; Kartha, G. Crystal and molecular structure of cyclic adenosine 3′,5′-monophosphate sodium salt, monoclinic form. *J. Am. Chem. Soc.* **1982**, *104*, 3398–3401. [CrossRef]
30. Scognamiglio, P.L.; Platella, C.; Napolitano, E.; Musumeci, D.; Roviello, G.N. From prebiotic chemistry to supramolecular biomedical materials: Exploring the properties of self-assembling nucleobase-containing peptides. *Molecules* **2021**, *26*, 3558. [CrossRef]
31. Slepokura, K. Purine 3′:5′-cyclic nucleotides with the nucleobase in a syn orientation: cAMP, cGMP and cIMP. *Acta Crystallogr. C* **2016**, *72*, 465–479. [CrossRef]
32. Costanzo, G.; Pino, S.; Timperio, A.M.; Šponer, J.E.; Šponer, J.; Nováková, O.; Šedo, O.; Zdráhal, Z.; Di Mauro, E. Non-enzymatic oligomerization of 3′,5′ cyclic AMP. *PLoS ONE* **2016**, *11*, e0165723. [CrossRef]
33. Burcar, B.T.; Cassidy, L.M.; Moriarty, E.M.; Joshi, P.C.; Coari, K.M.; McGown, L.B. Potential pitfalls in MALDI-TOF MS analysis of abiotically synthesized RNA oligonucleotides. *Orig. Life Evol. Biosph.* **2013**, *43*, 247–261. [CrossRef] [PubMed]
34. Costanzo, G.; Giorgi, A.; Scipioni, A.; Timperio, A.M.; Mancone, C.; Tripodi, M.; Kapralov, M.; Krasavin, E.; Kruse, H.; Šponer, J.; et al. Nonenzymatic oligomerization of 3′,5′-cyclic CMP induced by proton and UV irradiation hints at a nonfastidious origin of RNA. *ChemBioChem* **2017**, *18*, 1535–1543. [CrossRef] [PubMed]
35. Šponer, J.E.; Šponer, J.; Nováková, O.; Brabec, V.; Šedo, O.; Zdráhal, Z.; Costanzo, G.; Pino, S.; Saladino, R.; Di Mauro, E. Emergence of the first catalytic oligonucleotides in a formamide-based origin scenario. *Chem.-Eur. J.* **2016**, *22*, 3572–3586. [CrossRef] [PubMed]
36. Hazen, R.M. Paleomineralogy of the Hadean eon: A preliminary species list. *Am. J. Sci.* **2013**, *313*, 807–843. [CrossRef]
37. Kunkel, T.A. Mutational specificity of depurination. *Proc. Natl. Acad. Sci. USA* **1984**, *81*, 1494–1498. [CrossRef]
38. Mungi, C.V.; Bapat, N.V.; Hongo, Y.; Rajamani, S. Formation of abasic oligomers in nonenzymatic polymerization of canonical nucleotides. *Life* **2019**, *9*, 57. [CrossRef] [PubMed]
39. Whicher, A.; Camprubi, E.; Pinna, S.; Herschy, B.; Lane, N. Acetyl phosphate as a primordial energy currency at the origin of life. *Orig. Life Evol. Biosph.* **2018**, *48*, 159–179. [CrossRef] [PubMed]

Prebiotic Peptides Based on the Glycocodon Theory Analyzed with FRET

Jozef Nahalka [1,2,*] and Eva Hrabarova [1,2]

1. Institute of Chemistry, Centre for Glycomics, Slovak Academy of Sciences, Dubravska Cesta 9, SK-84538 Bratislava, Slovakia; chemhra@savba.sk
2. Institute of Chemistry, Centre of Excellence for White-Green Biotechnology, Slovak Academy of Sciences, Trieda Andreja Hlinku 2, SK-94976 Nitra, Slovakia
* Correspondence: nahalka@savba.sk

Abstract: In modern protein–carbohydrate interactions, carbohydrate–aromatic contact with CH–π interactions are used. Currently, they are considered driving forces of this complexation. In these contacts, tryptophan, tyrosine, and histidine are preferred. In this study, we focus on primary prebiotic chemistry when only glycine, alanine, aspartic acid, and valine are available in polypeptides. In this situation, when the aromatic acids are not available, hydrogen-bonding aspartic acid must be used for monosaccharide complexation. It is shown here that (DAA)n polypeptides play important roles in primary "protein"–glucose recognition, that (DGG)n plays an important role in "protein"–ribose recognition, and that (DGA)n plays an important role in "protein"–galactose recognition. Glucose oxidase from *Aspergillus niger*, which still has some ancient prebiotic sequences, is chosen here as an example for discussion.

Keywords: prebiotic chemistry; protein–monosaccharide recognition; protein–monosaccharide interactions; FRET analysis; glycocodon theory; glucose oxidase

1. Introduction

The origin and evolution of modern biochemistry and cellular life are a very interesting field of research; each human generation postulates new theories. Unfortunately, a top-down approach from the perspectives of modern biochemistry and molecular biology is easier to formulate than a bottom-up approach starting from the physicochemical properties of nucleic and amino acid polymers [1]. Nevertheless, 50 years of research exploring the origin and early evolution of life has created the concepts of an "RNA world" [2,3], a "protein world" [4,5], and a "protein–RNA coevolution world" [6,7] and many other hypotheses; however, the role of sugars is still relatively underestimated. After nucleotides and amino acids, sugars can be considered the third alphabet of life and are used to transfer information from the cellular environment via protein–glycan interactions [8]. Cells are covered by a "sugar coat" (glycocalyx) attached to membrane proteins and lipids; the carbohydrate moieties of membrane glycolipids and glycoproteins provide a first line of contact to the surrounding environment and are unique for different bacterial colonies or animal cells/tissues. The gel-like sugar layer has protective and sensing functions; for example, the glycocalyx of vascular endothelial cells serves as a negatively charged molecular sieve and acts as a signaling platform [9,10]. Eukaryotes, and some bacteria, adopt the "sugar code" to protein synthesis; N-linked protein glycosylation is performed in the endoplasmic reticulum (ER); and the N-linked glycosyl moiety controls protein folding and finalization. Different cells/tissues provide different N-glycosylation patterns/signatures; therefore, the protein expression is highly specific and controlled; when out of control, polypeptides aggregate and form inclusion bodies, which is the main problem in recombinant protein production. Oligosaccharides are assembled at the ER membrane on a lipid carrier by glycosyltransferases, and then transferred to asparagine

(N-X-S/T sequons) of the polypeptide chain by an oligosaccharyltransferase [8,11]. The activity and composition of glycosyltransferases are specific for each cell/tissue.

In early prebiotic chemistry, when the first peptide–monosaccharide interaction was selected and conserved, the number of different amino acids in short peptides and the number of three-dimensional structures were limited; in light of this, minimal amounts of amino acids and minimally ordered peptide chains were used for primary interactions and for primary glyco-code evolution. It seems that, today, N-linked protein glycosylation in the ER uses evolved glyco-codes to control protein folding and finalization. On the other hand, our previous work gave a number of examples from "modern" proteins in which disordered protein chains recognize/bind the monosaccharide. For example, in the Bacteriocin LLPA–mannose complex (3M7J), a disordered QGDGN175 sequence recognizes and binds mannose [12,13]. The glycocodon theory states that the monosaccharide units of a 3D glycan structure are recognized by the glycocodons, which are short amino acid sequences (sequons) composed of a polar/aromatic amino acid and an amino acid couple specific for the monosaccharide [12,13].

Ikehara proposed a prebiotic protein world enriched with peptides/proteins composed of glycine, alanine, aspartic acid, and valine (G, A, D, and V, respectively; also abbreviated as Gly, Ala, Asp, and Val) [4,5]. A simple BLAST search against the human genome for decapeptides with a combination of runs composed of two amino acids showed that the G, A, D, and V amino acids are extremely overrepresented [14]. In light of this, the first glycocodons were derived from ancient prebiotic conditions when only the G, A, D, and V amino acids were available in building primary proteins. For example, DAA (Asp-Ala-Ala) and DGV (Asp-Gly-Val) were proposed for glucose (Glc), DGG (Asp-Gly-Gly) was proposed for ribose (Rib), (G)DGD (Asp-Gly-Asp) was proposed for mannose (Man), and DAG (Asp-Ala-Gly) was proposed for galactose (Gal) [12,13]. These proposals were based on the measured frequencies of G, A, D, and V amino acid couples in monosaccharide-specific proteins (e.g., 6321 proteins specific for Glc) and were verified on the protein–monosaccharide structures available in the Protein Data Bank (PDB) [12]. Figure 1 depicts the measured frequencies for Glc, Gal, and Man. Glucose oxidase (GO) from *Aspergillus niger* is shown in the figure as an example of conserved ancient sequences.

The GO monomer has a deeply buried flavin adenine dinucleotide cofactor (FAD/FADH$_2$, ≈15 Å from the surface), and the open, deep pocket with a dimension of ≈10 Å × 10 Å provides access to the active site [15]. Access is covered by the glycocodons, which attract Glc towards the catalytic area (Figure 1A). In the catalytic area, Glc molecules form hydrogen bonds with R512, N514, H516, and Y68; the O1 hydroxyl group of Glc is located close to the key H516; and protons are transferred from C1-Glc to H516 [15]. In GO, H516GV(G) is an example glycocodon, which is a short, disordered chain, and helps organize the catalytic area in preparation for a reaction (Figure 1A).

In this study, we constructed FRET fusion proteins composed of a cyan fluorescent protein (CFP) and a yellow fluorescent protein (YFP), a common pair of fluorophores for the biological use of FRET (fluorescence resonance energy transfer). The mentioned color variants of the green fluorescent protein were fused using the peptide linker composed of a triple tandem of glycocodons for Rib, Glc, Gal, and Man. The results are discussed in the next sections; it is shown that the FRET signal changes in accordance with the glycocodon theory.

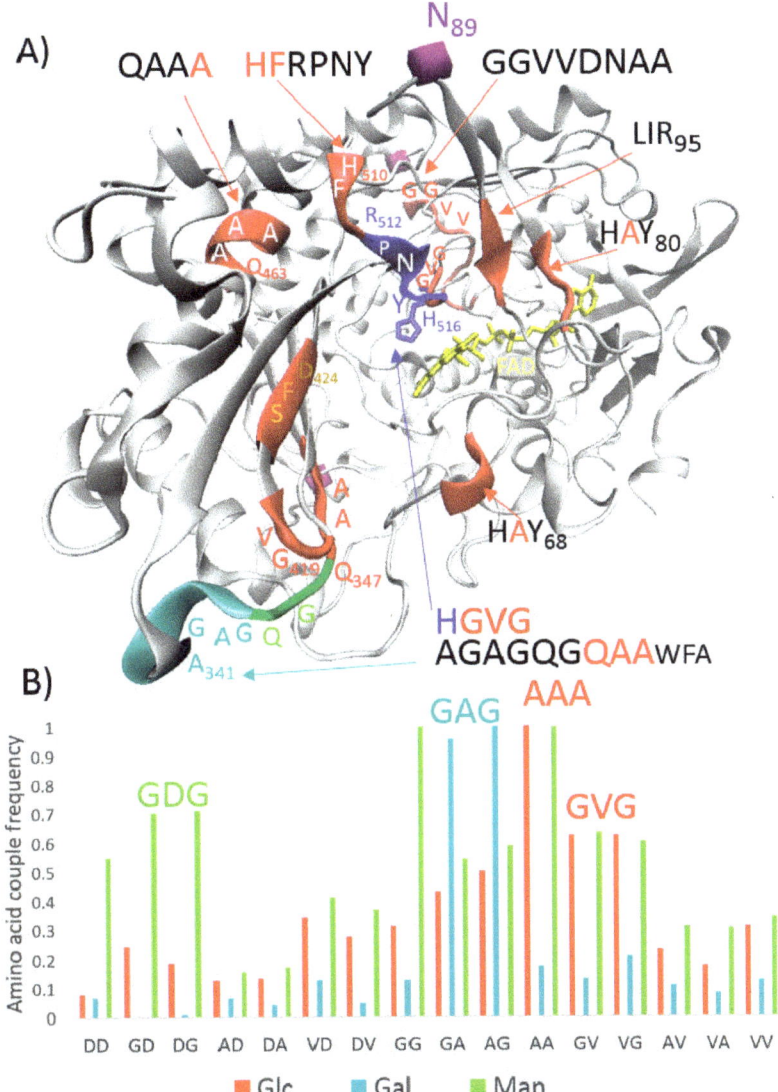

Figure 1. An explanation of the glycocodon theory [12]. (**A**) GO monomer—glucose oxidase from *Aspergillus niger*. N89—glycosylation site for an extended carbohydrate moiety forming a bridge between two monomers. GO is inactive without this N-linked glycosyl moiety. Red highlights—sequons/glycocodons for glucose (Glc); cyan highlights—ancient sequons/glycocodons for galactose (Gal); green highlights—ancient sequons/glycocodons for mannose (Man); and blue highlights—catalytic area with a key histidine. Protons are transferred from C1-Glc to H516 [15]. (**B**) Frequencies of amino acid couples, composed of G, A, D, and V amino acids, in monosaccharide-specific proteins, data from [12].

2. Materials and Methods

2.1. Cloning and Expression

For CFP-YFP-fused constructs, the mTurquoise2 and SYFP2 protein sequences were used (https://www.fpbase.org (accessed on 22 April 2021)). Codon optimization for *E. coli*, gene synthesis and cloning to the pET-51b (+) plasmid were performed using GenScript. Ten amino acids in the fusion linker were then mutated to the tested linkers (Figure 2C) using GenScript. Chemically competent *E. coli* BL21 (DE3)T1R cells were transformed with the plasmids and cultivated on ampicillin/LB/agar plates. The colonies were regrown in 30 mL LB medium supplemented with ampicillin (150 g/L) for approximately 20 h at 36 °C; then, 15 mL of the solution was transferred to 100 mL for 4 h at 37 °C and 150 rpm; and the recombinant expression was accomplished by adding an inductor (IPTG; 95.2 g/L) for 20 h at 20 °C and 150 rpm. After centrifugation (30 min, $4000\times g$ rpm, and +4 °C), the biomass was suspended in a minimal volume of water and immediately lyophilized. The yields of the dry cells varied from 120 to 150 mg per cultivation flask. The lyophilized cells were kept at -30 °C until further use.

2.2. Isolation of FRET Constructs

The FRET mutant variants—each of the two parallels (25 mg)—were lysed in Bug Buster Protein Extraction Reagent (Novagen, Darmstadt, Germany) detergent, cooled on ice, and centrifuged, and the cleared lysates were loaded on a 1 mL Strep-Tactin Superflow Plus Cartridge (QIAGEN, Hilden, Germany). According to the QIAGEN protocol, the proteins were washed and isolated in a volume of 4 mL of Strep-Tactin elution buffer (pH 8.0). The concentration of proteins was assayed using a Total Protein Kit, Micro Lowry, Onishi & Barr Modification (Sigma-Aldrich, St. Louis, MO, USA). The elution products were kept at -30 °C until further use.

2.3. The Protein–Monosaccharide Interactions and FRET Analysis

The protein–monosaccharide interactions were investigated using the measurement of fluorescence spectra using a TECAN Infinite M200 apparatus. Prior to this, all of the mutants were diluted with a Strep-Tactin washing buffer (pH 8.0) to achieve as comparable a starting fluorescence as possible. The sugars were weighed (5, 10, 15, and 20 mg) and dissolved in 100 µL diluted protein solutions. Afterwards, they were kept at +4 °C overnight to achieve an optimal steady state, and after centrifugation (1 min, $13,000\times g$ rpm, and +4 °C), 100 µL were pipetted onto a 96-well black microplate. Additionally, reference samples of all of the protein mutants were measured without sugar.

3. Results

The results are depicted in Figure 2. As can be seen, the studied peptide is connected to the C-end of CFP and to the N-terminus of YFP by three prolines (Figure 2A). The CFP-(QGG)$_3$Q-YFP construct and the influence of Rib on FRET is shown as an example (Figure 2B). The CFP-donor emits light at 480 nm wavelength, and the YFP-acceptor emits light at 530 nm wavelength. In the FRET concept, the FRET efficiency depends mostly on the distance between the CFP-donor and the YFP-acceptor; therefore, the acceptor emission is maximal at the shortest distance between the CFP and YFP domains. When the monosaccharide is in the contact with the linker and is inserted between the domains, the acceptor emission decreases and the donor emission increases. In our experiments, the constructs showed that relatively low FRET signals (530 nm) and relatively high concentrations of the monosaccharide had to be applied; however, the ratios between the emission peak highs (480/530) that are dependent on monosaccharide concentration were perfectly aligned on a graph, with a correlation coefficient close to 1 (Figure 2B). The slope coefficient ($\times 10^{-5}$) was chosen as a determinant of the interaction between a triple glycocodon and a monosaccharide (Figure 2C); in the text below, the numerals in parentheses denote the slope coefficient $\times 10^{-5}$.

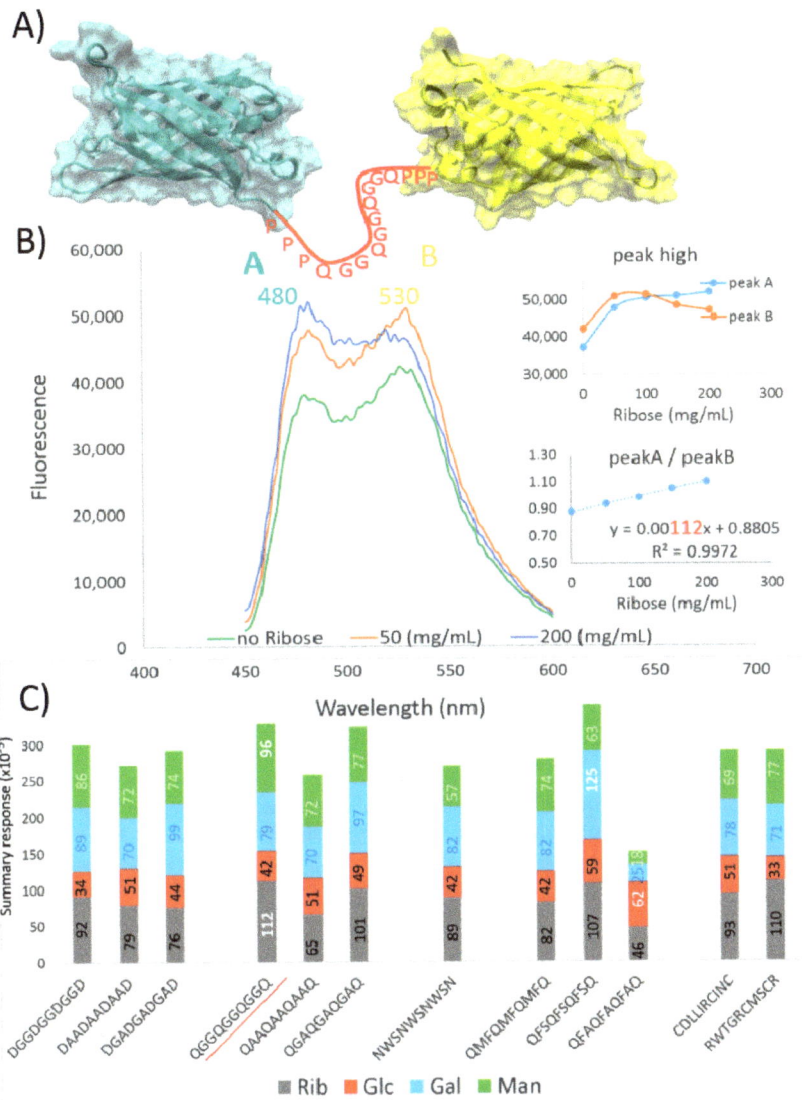

Figure 2. Peptides based on the glycocodon theory analyzed with FRET. (**A**) Illustration of the protein constructs used: cyan fluorescent proteins (CFP) and yellow fluorescent proteins (YFP) are linked by PPP-(glycocodon)$_3$-PPP linkers. (**B**) In the FRET constructs, the CFP-donor emits light at the 480 nm wavelength and the YFP-acceptor emits light at the 530 nm wavelength. The slope of change in the ratio between peak high A vs. peak high B, which are dependent on monosaccharide concentration, was used for the evaluation. (**C**) Responses of various linkers to increasing concentrations of the monosaccharides: red—glucose (Glc), cyan—galactose (Gal), green—mannose (Man), and grey—ribose (Rib). The summary response shows how the linker reacts to the monosaccharides tested overall; the monosaccharide-specific FRET responses ($\times 10^{-5}$) are color-coded within the columns.

At first, we tested ancient DGG, DAA, and DGA glycocodons; the (DGG)$_3$D sequence had the highest response to Rib (92), (DAA)$_3$D had the highest response to Rib (79), and (DGA)$_3$D had the highest response to Gal (99). In the case of Glc, the maximal response was in the (DAA)$_3$D construct (51). In other words, each construct responded similarly for the tested monosaccharides, and (DAA)$_3$D gave a slightly lower response. However, according to monosaccharide type, Rib gave the highest response in (DGG)$_3$D compared to (DAA)$_3$D and (DGA)$_3$D, Glc gave the highest response in (DAA)$_3$D compared to (DGG)$_3$D and (DGA)$_3$D, Gal gave the highest response in (DGA)$_3$D compared to (DGG)$_3$D and (DAA)$_3$D, and Man gave the highest response in (DGG)$_3$D. The results confirmed that, in the postulated prebiotic conditions, DGG is the best "choice" for Rib, DAA is the best for Glc, and DGA is the best for Gal; however, this does not mean that the sequons interact only with these best monosaccharides.

In the second experiment, we used the same linkers, but aspartic acid was substituted with glutamine (Q). The responses presented the same patterns as those in the previous experiment, but the responses to the monosaccharides were slightly higher in the cases of (QGG)$_3$Q and (QGA)$_3$Q and lower in the case of (QAA)$_3$Q (Figure 2C). The (QGG)$_3$Q construct provided maximal responses for Rib (112) and Man (96) when tested against all of the tested linkers assessed in this work.

In the third experiment, we tested some "modern" glycocodons: the amino acid couples from the library of twenty amino acids presently used in biology, which were found to have the best frequencies in the Gal- and Glc-specific proteins [12]. Asparagine (N) was used for the Gal glycocodon, and Q was used for Glc glycocodons; in other words, the NWS sequon was tested for Gal and QMF, QFS, and QFA were tested for Glc. Surprisingly, (QFS)$_3$Q was "the best choice" for Gal and (QFA)$_3$Q was the best for Glc. In fact, (QFA)$_3$Q had the lowest response to the monosaccharides used but had the highest response to Glc (62) amongst all of the tested linkers (Figure 2C), and (QFS)$_3$Q had the highest response to the monosaccharides used and had the highest response to Gal (125).

Lastly, we tested random linkers as a control experiment; the linkers were designed for other purposes and not for studying glycocodons. Figure 2C shows the results for CDLLIRCINC and RWTGRCMSCR. The linkers had equal summary responses, but CDLLIRCINC was a "good choice" for Glc and RWTGRCMSCR gave a better response to Rib.

4. Discussion

The quality of information about protein–glycan interactions available in the PDB is still limited; however, high-quality structures for a few sets of monosaccharide forms connected by a small diversity of glycoside linkages are available in sufficient quantity to study protein–carbohydrate interactions [16]. Hydrogen bonds between the carbohydrate hydroxyl groups and polar amino acids in the binding of carbohydrates by proteins were well recognized from the first structures; however, in the last decade, electrostatic and electronic complementarity between carbohydrates and aromatic residues have been shown to play key roles [17]. In monosaccharide binding sites, tryptophan (W) >> tyrosine (Y) > histidine (H) are greatly preferred, asparagine (N) > aspartic acid (D) >> glutamine (Q) are slightly preferred, and aliphatic residues are disfavored [17]. The CH–π stacking contribution to the overall binding energy ranges from −4 kcal/mol to −8 kcal/mol; currently, stacking CH–π interactions are considered driving forces of protein–carbohydrate complexation [18]. However, the calculated energies in the systems without CH–π interactions are in the range from −0.2 to −3.2 kcal/mol; hence, they can also be important for aromatic amino acid and carbohydrate binding processes [19]. In the enzymes used for glycan synthesis/transformations, weaker interactions that enable the release of small carbohydrate fragments after an enzymatic reaction are possibly preferred [19]. To illustrate this, H and Y residues are preferred in GO and only W426 is close to the catalytic area (Figure 1A). On the other hand, not only the frequencies of the residues in catalytic/binding sites are important, but directed evolution studies also usually identify residues far from the active site that have significant impacts on activity and function [20]. Conformational dynamics plus "something else" also play important roles. In the glycocodon theory, the frequency of amino acid couples in whole proteins has been studied in accordance with monosaccharide recognition [12]. In light of the bottom-up model, a study was rooted in "primary" proteins composed of G, A, D, and V amino acids; AA and GV couples were identified as the maximal frequencies for the recognition of Glc, GA and AG were identified as that for Gal, and GD and DG were identified as that for Man (Figure 1B) [12]. For example, in the case of GO, the ancient conserved AGAGQGQAA sequence can be found in front of the catalytic Glc-binding location, doubled GGVVDNAA sequon behind the catalytic Glc-binding location, and H516GV(G) inside of the catalytic Glc-binding location (Figure 1A). QAA and GVD can be considered the glycocodons for Glc, (G)QGQ can be considered the glycocodon for Man, and (AG)AGQ can be considered the glycocodon for Gal. In fact, GO from *Aspergillus niger* is specific for Glc, but the catalytic area also accepts Man (1% relative to Glc) and, at very low levels, Gal (0.08%) [21].

It seems that CH–π interactions and aromatic residues such as W, Y, and H are important and that aliphatic residues close/next to stacking amino acids can also be relevant in carbohydrate recognition; for example, in the case of GO, the ancient GVG sequence is precisely behind the key catalytic H516, which theoretically forms the HGV(G) glycocodon/sequon (Figure 1A). In light of this, we studied the DAA, DGA, and DGG sequons in response to Rib, Glc, Gal, and Man (Figure 2). Tripled glycocodons were used as linkers, and constructs such as CFP-(DGG)$_3$D-YFP were used to study the change in FRET in the presence of the mentioned monosaccharides. For Glc, to our satisfaction, DAA was confirmed as "the best choice" among the tested primary sequons and, similarly DGG was the best for Rib and DGA was the best for Gal (Figure 2c). GDGD was predicted for Man, and, in actuality, the highest response was found for the DGGDGGDGGD linker. The conditions were the same, and the studied linkers only differed slightly in sequence and 3D form. It seems that the primary "prebiotic" model was well designed. The measurements confirmed the controversial glycocodon theory, which was developed for these modelled primary "prebiotic" conditions, where the first polypeptide–monosaccharide interactions were evolved and conserved. The results (see the responses of the monosaccharides in Figure 2c) also correspond well with the known monosaccharide glycation efficiency: Glc << Gal ≤ Man << Rib. Glc is the least reactive, and Rib is the most reactive; therefore, Glc rather than other stereoisomers has emerged as the universal metabolic fuel and Rib has emerged as the RNA component [22]. As can be seen from the GO structure (Figure 1A), in conserved ancient sequons, aspartic acid is often exchanged for glutamine, histidine, or asparagine. When we changed D for Q in the GG, AA, and GA linkers, QGGQG-GQGGQ showed the maximal response for Rib and Man amongst all of the tested linkers (Figure 2c). As mentioned above, hydrogen-bonding Q has a much lower frequency in carbohydrate-binding sites than hydrogen-bonding D [17]. Despite Q being observed much less proximally to carbohydrates in the binding centers, it is functional in the glycocodons, and in our experiments, providing "stronger interactions" for Rib and Man, Glc response did not change. A (G)QGQ glycocodon blocks access to the catalytic center of GO. Perhaps Glc more readily "flows" to the catalytic center of GO than Man, and therefore there is strong preference for Glc. Logically, it seems that the substitution of prebiotic D for N, Q, H, Y, W, R, or K is specific for monosaccharide recognition and binding.

In the third set of experiments, we tested samples from the predicted "modern" glycocodons/sequons for Glc and Gal. For Glc, MF/FM, FS/SF, and AF/FA showed high frequencies in the Glc-binding proteins [12]; we used Q as the third amino acid and tested the QMF, QFS, and QFA sequons. For Gal, WN/NW and WS/SW were among the best measured frequencies [12]; therefore, we tested the NWS glycocodon. Surprisingly, the QFS glycocodon had the highest response for Gal and QFA had the highest response for Glc amongst all of the tested linkers (Figure 2C). (QFA)$_3$Q was remarkable because it provided the lowest "summary response" and the maximal response for Glc amongst all of the linkers. It seems that the substitution of one alanine with phenylalanine in the DAA/QAA ancient Glc glycocodon was convenient in the evolution of protein–Glc interactions; the response increased from 51 for DAA and QAA to 62 for QFA (Figure 2c). The QFS signal increased to 59 in our experiments; in fact, it was previously proposed as a sequon for GlcNAc. The GA/AG and AG/AA frequencies were measured as the maxima for Gal and GlcNAc, respectively [12,13]. This could explain why the QFS glycocodon is among the best glycocodons for Gal recognition in our experiments. QMF surprisingly did not increase but, contrarily, lowered the response to Glc compared to DAA/QAA. It seems that Q is not suitable as a third amino acid for this Glc glycocodon. In the case of (NWS)$_3$N, Gal was preferred, but the response decreased compared to DGA/QGA; however, the response to other monosaccharides decreased further. It seems that, in modern glycocodons, the preference is balanced between maximal specificity and maximal "response".

The random CDLLIRCINC linker, used as a control experiment, showed exactly the same response for Glc as the (QAA)$_3$Q linker. The IL/LI amino acid couple has been recognized among the best frequencies for Glc [12]; it seems that the aliphatic LI couple in combination with positively charged arginine (R) could be a regular modern glycocodon for Glc. For example, the LIR95 sequon can be found close to the catalytic center of GO (Figure 1A).

5. Conclusions

The stacking CH–π interactions provided by amino acids such as W, Y, or H are essential for protein–monosaccharide complexation in lectins or glyco-enzymes; however, aliphatic residues and other amino acids next to these key aromatic amino acids, currently considered "unfavorable", are very important for monosaccharide recognition, transport, and binding. Polar hydrogen-bonding residues such as D and N may partly substitute aromatic acid in short-sequence sequons, which recognize and respond to monosaccharide type. The sequons, called glycocodons, are composed of one aromatic (W, Y, or H) or polar (D, N, or Q) amino acid and one amino acid couple convenient for monosaccharide recognition. In prebiotic chemistry, when only the G, A, D, and V amino acids are available, the first glycocodons evolve: DAA(A) and DGV(G) for Glc, DGG(G) for Rib, DGA(G) for Gal, and (D)GDG for Man. Interestingly, these prebiotic sequons/glycocodons are still evolutionarily conserved and can be found in monosaccharide-specific proteins, for example, H516GV(G), Q463AA(A), and AGAGGQ347AA in GO from *Aspergillus niger* (Figure 1A).

The FRET analysis confirmed that, in prebiotic chemistry, "the best choices" were DAA for Glc, DGG for Rib, DGA for Gal, and DGDG for Man.

In future, it is necessary to study how these prebiotic glycocodons evolved into modern glycocodons. For example, in the case of QAA(A), the change into QFA markedly improved Glc recognition, as is shown in the presented results.

Author Contributions: J.N. designed the experiments, evaluated the data, wrote the manuscript, and prepared the figures. E.H. propagated the biomass-expressing recombinant proteins, isolated the protein constructs, and measured the interactions. All authors have read and agreed to the published version of the manuscript.

Funding: This research was funded by APVV, grant number APVV-18-0361.

Institutional Review Board Statement: Not applicable.

Informed Consent Statement: Not applicable.

Data Availability Statement: Not applicable.

Acknowledgments: Working at home office during the COVID-19 pandemic requires patience from relatives; in light of this, I (J.N.) gratefully acknowledge my wife.

Conflicts of Interest: The authors declare no conflict of interest.

References

1. Auboeuf, D. Physicochemical foundations of life that direct evolution: Chance and natural selection are not evolutionary driving forces. *Life* **2020**, *10*, 7. [CrossRef] [PubMed]
2. Robertson, M.P.; Joyce, G.F. The origins of the RNA world. *Cold Spring Harb. Perspect. Biol.* **2012**, *4*, a003608. [CrossRef]
3. Mizuuchi, R.; Lehman, N. Limited sequence diversity within a population supports prebiotic RNA reproduction. *Life* **2019**, *9*, 20. [CrossRef] [PubMed]
4. Ikehara, K. Pseudo-replication of [GADV]-proteins and origin of life. *Int. J. Mol. Sci.* **2009**, *10*, 1525. [CrossRef] [PubMed]
5. Ikehara, K. Evolutionary steps in the emergence of life deduced from the bottom-up approach and GADV hypothesis (Top-Down Approach). *Life* **2016**, *6*, 6. [CrossRef]

6. Caetano-Anollés, G.; Seufferheld, M.J. The coevolutionary roots of biochemistry and cellular organization challenge the RNA world paradigm. *J. Mol. Microbiol. Biotechnol.* **2013**, *23*, 152–177. [CrossRef] [PubMed]
7. Carter, C.W. What RNA world? Why a peptide/RNA partnership merits renewed experimental attention. *Life* **2015**, *5*, 294. [CrossRef]
8. Gabius, H. The sugar code: Why glycans are so important. *BioSystems* **2018**, *164*, 102–111. [CrossRef]
9. Uchimido, R.; Schmidt, E.P.; Shapiro, N.I. The glycocalyx: A novel diagnostic and therapeutic target in sepsis. *Crit. Care* **2019**, *23*, 16. [CrossRef] [PubMed]
10. Zeng, Y. Endothelial glycocalyx as a critical signalling platform integrating the extracellular haemodynamic forces and chemical signalling. *J. Cell. Mol. Med.* **2017**, *21*, 1457–1462. [CrossRef]
11. Aebi, M. N-linked protein glycosylation in the ER. *Biochim. Biophys. Acta* **2013**, *1833*, 2430–2437. [CrossRef] [PubMed]
12. Nahalka, J. Glycocodon theory-the first table of glycocodons. *J. Theor. Biol.* **2012**, *307*, 193–204. [CrossRef]
13. Nahalka, J.; Hrabarova, E.; Talafova, K. Protein-RNA and protein-glycan recognitions in light of amino acid codes. *Biochim. Biophys. Acta Gen. Subj.* **2015**, *1850*, 1942–1952. [CrossRef] [PubMed]
14. Nahalka, J. Quantification of peptide bond types in human proteome indicates how DNA codons were assembled at prebiotic conditions. *J. Proteom. Bioinform.* **2011**, *4*, 153–159. [CrossRef]
15. Mano, N. Engineering glucose oxidase for bioelectrochemical applications. *Bioelectrochemistry* **2019**, *128*, 218–240. [CrossRef]
16. De Meirelles, J.L.; Nepomuceno, F.C.; Peña-García, J.; Schmidt, R.R.; Pérez-Sánchez, H.; Verli, H. Current status of carbohydrates information in the protein data bank. *J. Chem. Inf. Model.* **2020**, *60*, 684–699. [CrossRef]
17. Hudson, K.L.; Bartlett, G.J.; Diehl, R.C.; Agirre, J.; Gallagher, T.; Kiessling, L.L.; Woolfson, D.N. Carbohydrate-aromatic interactions in proteins. *J. Am. Chem. Soc.* **2015**, *137*, 15152–15160. [CrossRef]
18. Houser, J.; Kozmon, S.; Mishra, D.; Hammerová, Z.; Wimmerová, M.; Koča, J. The CH–π interaction in Protein–Carbohydrate binding: Bioinformatics and in vitro quantification. *Chem. Eur. J.* **2020**, *26*, 10769–10780. [CrossRef]
19. Stanković, I.M.; Blagojević Filipović, J.P.; Zarić, S.D. Carbohydrate—Protein aromatic ring interactions beyond CH/π interactions: A protein data bank survey and quantum chemical calculations. *Int. J. Biol. Macromol.* **2020**, *157*, 1–9. [CrossRef]
20. Gardner, J.M.; Biler, M.; Risso, V.A.; Sanchez-Ruiz, J.M.; Kamerlin, S.C.L. Manipulating conformational dynamics to repurpose ancient proteins for modern catalytic functions. *ACS Catal.* **2020**, *10*, 4863–4870. [CrossRef]
21. Leskovac, V.; Trivić, S.; Wohlfahrt, G.; Kandrač, J.; Peričin, D. Glucose oxidase from Aspergillus niger: The mechanism of action with molecular oxygen, quinones, and one-electron acceptors. *Int. J. Biochem. Cell. Biol.* **2005**, *37*, 731–750. [CrossRef] [PubMed]
22. Bunn, H.F.; Higgins, P.J. Reaction of monosaccharides with proteins: Possible evolutionary significance. *Science* **1981**, *213*, 222–224. [CrossRef] [PubMed]

Article

Combinatorial Fusion Rules to Describe Codon Assignment in the Standard Genetic Code

Alexander Nesterov-Mueller [1,*], Roman Popov [1] and Hervé Seligmann [1,2,3]

1. Institute of Microstructure Technology, Karlsruhe Institute of Technology (KIT), 76344 Eggenstein-Leopoldshafen, Germany; roman.popov@kit.edu (R.P.); varanuseremius@gmail.com (H.S.)
2. The National Natural History Collections, The Hebrew University of Jerusalem, Jerusalem 91904, Israel
3. Laboratory AGEIS EA 7407, Team Tools for e-GnosisMedical & LabcomCNRS/UGA/OrangeLabs Telecoms4Health, Faculty of Medicine, Université Grenoble Alpes, F-38700 La Tronche, France
* Correspondence: Alexander.Nesterov-Mueller@KIT.edu

Abstract: We propose combinatorial fusion rules that describe the codon assignment in the standard genetic code simply and uniformly for all canonical amino acids. These rules become obvious if the origin of the standard genetic code is considered as a result of a fusion of four protocodes: Two dominant AU and GC protocodes and two recessive AU and GC protocodes. The biochemical meaning of the fusion rules consists of retaining the complementarity between cognate codons of the small hydrophobic amino acids and large charged or polar amino acids within the protocodes. The proto tRNAs were assembled in form of two kissing hairpins with 9-base and 10-base loops in the case of dominant protocodes and two 9-base loops in the case of recessive protocodes. The fusion rules reveal the connection between the stop codons, the non-canonical amino acids, pyrrolysine and selenocysteine, and deviations in the translation of mitochondria. Using fusion rules, we predicted the existence of additional amino acids that are essential for the development of the standard genetic code. The validity of the proposed partition of the genetic code into dominant and recessive protocodes is considered referring to state-of-the-art hypotheses. The formation of two aminoacyl-tRNA synthetase classes is compatible with four-protocode partition.

Keywords: standard genetic codes; codon assignment; tRNA; aminoacyl-tRNA synthetase classes

1. Introduction

Covering more than 50 years of the literature on the origin of the genetic code, renowned specialists E.V. Koonin et al. put a shortlist of widespread statements about the code properties and aspects of its evolution [1]:

1. "The code is effectively universal: Departures from code universality in extant organisms are minor and of secondary origin.
2. The code is non-randomly organized and is highly robust to errors, although it is far from being globally optimal.
3. Evolution of the code involved expansion from a limited set of primordial amino acids toward the modern canonical set."

We doubt that statement 3 from this list is a significant prerequisite for clarifying the codon assignment in the standard genetic code (SGC). The most commonly used arguments for the sequential entry of amino acids into SGC are differences in the prebiotic abundance of amino acids [2,3], indications of GC-rich content in the most archaic RNAs [4], or branching history of two aminoacyl-tRNA synthetase (aaRS) superfamilies [5]. Even though canonical amino acids and RNAs with a specific base content may have appeared on Earth in different ways and at different times, there is no evidence that the non-randomly organized SGC arose as a result of successive expansions with amino acids. R.D. Knight et al. critically reviewed hypotheses based on phylogenetic analysis and emphasized the absence of any evidence of code expansion during the evolution of synthetases [6].

Indeed, such a "progressively evolutionary" view is rather a common perception of historical processes. Maybe intuitive acceptance of statement 3 is the reason for criticisms (also by Koonin et al. [1]) of code origin hypotheses, such as the stereochemical hypothesis, that "failed to provide clear solutions" or "does not find general confirmation" [6,7]. Despite rapidly growing genomic databases, aptamer screening technology, and extensive computational efforts, the SGC origins remain unknown, even called the universal enigma [8].

The central issue of the genetic code origin is a rational explanation for the assignment of amino acids to different numbers of codons [9]. Several mathematical approaches of the genetic code in terms of symmetry properties and group theory have been developed [10–12]. The problem with such descriptions is the difficulty in providing a biological interpretation. The tessera hypothesis follows a different approach [13]. It is a unified mathematical framework that accounts for the degeneracy properties of both nuclear and mitochondrial genetic codes. According to this model, the early versions of the genetic code had codons of four base length. This assumption would solve the conundrum that regular triplet codon–anticodon duplexes are too unstable to allow primitive ribosome-free translation [14]. Indeed, predicted coding by quadruplet codons increases with mean body temperature in lizards [15] and coevolves with predicted tRNAs with expanded anticodons [16]. There is a row of observations originating from tRNAs with expanded eight-nucleotide anticodon loops [17–19] or mass spectrometry analyses of peptides corresponding to the translation of the human mitogenome according to codons with more than three nucleotides [20,21]. These observations could support the tessera hypothesis, but they are not necessarily evidencing a primitive genetic code with expanded codons. The tessera hypothesis proposes a rather indirect way to the SGC via transformation of the tessera code with codons of four nucleotides to the Juke's ancestral code with codons of three nucleotides (16 amino acids and two stop codons) [22]. Using further assumptions, Juke postulated two additional successive expansions of the ancestral code (via Ile and Thr, and later via Met and Trp) to finally fully describe the SGC.

Instead of introducing four-base codons, we describe the exact degeneracy of the genetic code with simple fusion rules. In this study, we consider four single base-pair protocodes as originally independent coexisting codes.

2. Results

2.1. Single Base-Pair Codes and Combinatorial Rules of Their Fusion

Tables 1 and 2 represent the standard genetic code but in a special form of four protocodes. Twenty proteinogenic amino acids are distributed over the two dominant and two recessive protocodes. The terms "dominant" and "recessive" are borrowed from classical genetics and refer to the fact that the dominant protocodes do not change their initial codon/amino acid assignments after the fusion. In contrast, the recessive codes acquire new triplets.

Table 1. Dominant AU and GC protocodes and their transformation to the standard genetic code (SGC) after the fusion. The codons of the SGC are obtained due to mutations A↔G or U↔C in the third position of the protocodes. The red letters illustrate the transformation.

AU Code			GC Code		
Amino Acid	Before Fusion	SGC	Amino Acid	Before Fusion	SGC
Lys	AAA	AAA, AAG	Gly	GGG	GGG, GGA
Asn	AAU	AAU, AAC	Gly	GGC	GGC, GGU
Ile	AUU	AUU, AUC	Ala	GCC	GCC, GCU
Phe	UUU	UUU, UUC	Ala	GCG	GCG, GCA
Leu	UUA	UUA, UUG	Pro	CCC	CCC, CCU
Tyr	UAU	UAU, UAC	Pro	CCG	CCG, CCA

Table 1. Cont.

| Met (Ile + Met) * | AUA | AUA, AUG | Arg | CGC | CGC, CGU |
| stop | UAA | UAA, UAG | Arg | CGG | CGG, CGA |

* The asterisk indicates the codon assignments of the SGC that we will discuss in the next sections.

Table 2. Recessive AU and GC protocodes and their transformation to the SGC after the fusion. The codons of the SGC are obtained due to mutations A↔G or U↔C in the first position or the first and the third positions of the protocodes. The red letters illustrate the transformation.

	AU Code			GC Code	
Amino Acid	before Fusion	SGC	Amino Acid	before Fusion	SGC
Gln	UAA	CAA, CAG	stop (Trp + stop) *	CGG	UGG, UGA
His	UAU	CAU, CAC	Cys	CGC	UGC, UGU
Leu	UUU	CUU, CUC	Ser	CCC	UCC, UCU
Val	AUU	GUU, GUC	Ser	CCG	UCG, UCA
Val	AUA	GUA, GUG	Thr	GCC	ACC, ACU
Asp	AAU	GAU, GAC	Thr	GCG	ACG, ACA
Leu	UUA	CUA, CUG	Ser *	GGC	AGC, AGU
Glu	AAA	GAA, GAG	Arg *	GGG	AGG, AGA

* The asterisks indicate the codon assignments of the SGC that we will discuss in the next sections.

The peculiarity of this construction is that the number and type of codons for each amino acid in the SGC are determined according to rules 1–3. The red letters in Table 1 illustrate the changes according to these rules. The asterisks indicate the codon assignments of the SGC that we will discuss in the next sections.

Rule 1: The second-position bases do not change in any code.

Rule 2: A and G, as well as U and C, are exchangeable only in the third-position base in the dominant protocodes.

Rule 3: A and C, as well as U and G, are exchangeable either in the first position or simultaneously in the first and third positions in the recessive protocodes.

The derivation of these rules occurs automatically as soon as the coexistence of dominant and recessive codes is accepted. The advantage of these rules is that they are uniform for all amino acids and reduce the problem of codon assignment to a simple mathematical function.

Following the fusion rules, one can calculate the number of codons for the stop codon and each amino acid. For example, the stop codon UAG of the SGC originates from the stop codon UAA by substitution of A for G. Amino acid Lys has only two codons AAA and AAG in the SGC, because Lys had only one codon AAA in the protocode.

Each protocode contains only one positively charged amino acid (dominant codes—Lys and Arg and recessive codes—His and Arg (X4)). These positively charged amino acids may significantly contribute to the specific interactions between negatively charged RNAs and protopeptides.

2.2. Combinatorial Fusion Rules Preserve Complementarity of Codons for Specific Clusters of Amino Acids

Table 3 shows the four protocodes in form of complementary clusters **AUa**, **GCa**, **AUā**, and **GCā**. These clusters emerged automatically by writing down the complementary codons within the protocodes. It turned out that the properties of the amino acids significantly differ between complementary clusters. Clusters **AUā** and **GCā** include all hydrophobic canonical amino acids Met, Leu, Ile, Val, Ala, Pro, Phe, and two small amino acids Ser and Thr. In contrast, clusters **AUa** and **GCa** include only charged and polar amino acids. These four clusters represent the well-known evolutionary columns [7] that are frequently used to demonstrate the hypotheses of the genetic code expansion from the single Gly code to the SGC [23].

Table 3. Distribution of the amino acids within the protocodes into the complementary clusters **AUa**, **GCa**, **AUā**, and **GCā**. Clusters **AUā** and **GCā** consist of only the "small" amino acids. **AUa** and **GCa** consist only of the charged or polar amino acids.

Amino Acids, Cluster a	AU Codons, Cluster a	AU Codons, Cluster ā	Amino Acids, Cluster ā	Amino Acids, Cluster a	GC Codons, Cluster a	GC Codons, Cluster ā	Amino Acids, Cluster ā
Lys	AAA	UUU	Phe	Gly	GGG	CCC	Pro
Asn	AAU	AUU	Ile	Gly	GGC	GCC	Ala
Tyr	UAU	AUA	Ile (Ile+ Met) *	Arg	CGC	GCG	Ala
stop	UAA	UUA	Leu	Arg	CGG	CCG	Pro
Gln	UAA	UUA	Leu	Trp + stop	CGG	CCG	Ser
His	UAU	AUA	Val	Cys	CGC	GCG	Thr
Asp	AAU	AUU	Val	Ser *	GGC	GCC	Thr
Glu	AAA	UUU	Leu	Arg *	GGG	CCC	Ser

* The asterisks indicate the codon assignments of the SGC that we will discuss in the next sections.

Table 4 shows the distribution of the complementary codons and corresponding amino acids after the fusion in the SGC. This complementarity has already been noticed by Rodin and Ohno [24]. The complementarity of the codons after the fusion changed in such a way that additional complementary codons appeared in the dominant protocodes. For example, a new pair Lys-Leu (AAG-CUU) is added to the Lys-Phe (AAA-UUU) from the protocode. The complementarity of amino acid codons in the recessive protocodes changed totally because the old coexisting codes disappeared. However, fusion rules preserved the original distribution of amino acids within the clusters **AUa**, **GCa**, **AUā**, and **GCā**.

Table 4. Distribution of the amino acids within the SGC into the complementary clusters **AUa**, **GCa**, **AUā**, and **GCā**. The distribution into the clusters remains after the fusion.

Amino Acids, Cluster a	AU Codons, Cluster a	AU Codons, Cluster ā	Amino Acids, Cluster ā	Amino Acids, Cluster a	GC Codons, Cluster a	GC Codons, Cluster ā	Amino Acids, Cluster ā
Lys	AAA	UUU	Phe	Gly	GGG	CCC	Pro
Lys	AAG	CUU	Leu	Gly	GGA	UCC	Ser
Asn	AAU	AUU	Ile	Gly	GGC	GCC	Ala
Asn	AAC	GUU	Val	Gly	GGU	ACC	Thr
Tyr	UAU	AUA	Met	Arg	CGC	GCG	Ala
Tyr	UAC	GUA	Val	Arg	CGU	ACG	Thr
stop	UAA	UUA	Leu	Arg	CGG	CCG	Pro
stop	UAG	CUA	Leu	Arg	CGA	UCG	Ser
Gln	CAA	UUG	Leu	Trp	UGG	CCA	Pro
Gln	CAG	CUG	Leu	stop	UGA	UCA	Ser
His	CAU	AUG	Met	Cys	UGC	GCA	Ala
His	CAC	GUG	Val	Cys	UGU	ACA	Thr
Asp	GAU	AUC	Ile	Ser *	AGC	GCU	Ala
Asp	GAC	GUC	Val	Ser *	AGU	ACU	Thr
Glu	GAA	UUC	Phe	Arg *	AGG	CCU	Pro
Glu	GAG	CUC	Leu	Arg *	AGA	UCU	Ser

* The asterisks indicate the codon assignments of the SGC that we will discuss in the next sections.

3. Discussion

3.1. Kissing Proto tRNAs

The fact of codon complementarity before and after the fusion indicates the importance of the specific loop-loop interactions (kissing) between the proto tRNAs for the ancient translation. The kissing contacts were experimentally detected in the case of bacterial and viral systems, where they are prevalent in regulatory complexes [25,26]. This is also in line with the self-referential hypothesis for genetic code origins that assumes "kissing" between complementary tRNA anticodons [27,28], forming a structure similar to the ribosomal peptide elongation core [29]. The complementary hairpin kissing complexes are relatively stable. They demonstrate dissociation constants in the low-to-medium nanomolar range [25,30–33].

Fusion rules represent the discrimination of A/G and U/C. This discrimination in codon recognition is known as a wobble position in the anticodon of the modern-type tRNA [23]. The wobble position is occupied by a modified base that is part of the universal genetic code and was probably present in Last Universal Common Ancestor (LUCA) [34]. Fusion rule 2 applies the A/G and U/C discrimination to the 3rd codon (1st anticodon) position. Thus, fusion rule 2 preserves the kissing contact between the 10-base loop of the proto tRNAs for the amino acids from the clusters **AUa** and **GCa** and the 9-base loop of the corresponding proto tRNAs for the amino acids from the clusters **AUā** and **GCā** (Figure 1, left).

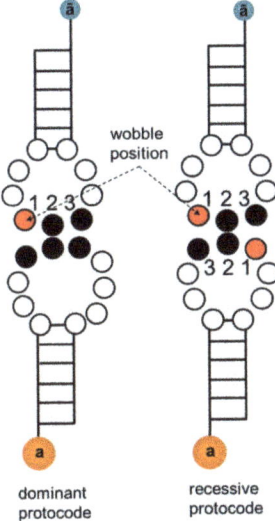

Figure 1. Schematic view of kissing proto tRNAs in form of hairpins. The red circle represents the wobble position. (**left**) Dominant protocode: kissing contact via a 9-base loop and a 10-base loop. The geometry of wobble positions corresponds to fusion rules 2. (**right**) Recessive protocode: kissing contact via two 9-base loops. The geometry of wobble positions corresponds to fusion rule 3.

The kissing contact in the recessive codes is represented by the two 9-base loops (Figure 1, right). This loop kissing allows for A/G and U/C discrimination both in the 1st and 3rd codon positions. Thus, fusion rule 3 has the same function as rule 2 to preserve the kissing contacts in the protocode after the fusion. The recessive protocodes lost their initial codon assignments after the fusion, because the new codons formed kissing loops with stronger affinity. For example, Gln-Leu tRNA kissing was initially formed by the complementary codons AAA-UUU (Table 3). After the fusion, the kissing Gln-Leu tRNA geometry was extended with codon pairs CAA-UUG und CAG-CUG (Table 4) with a greater affinity that made the initial assignment unnecessary.

The difference in the size of kissing loop geometries between the dominant and recessive protocodes caused the orthogonality of the ancient coexisting translation apparatuses.

The formation of proto tRNA pairs provided the advantage for their better recognition by the ancient aminoacyl-tRNA synthetases (proto-aaRS): (i) each pair had a more complex structure in comparison with a single hairpin; (ii) each pair was equipped with a small, mostly hydrophobic amino acid that caused a better affinity to the proto-aaRS.

The proto tRNAs with 9-base loop and 10-base loop hairpins give a clue about the emergence of the modern-type tRNA. D-loop and D-stem of the modern tRNA probably descended from 10-base loop proto tRNAs, and T-loop and T-stem from 9-base loop proto tRNAs.

3.2. Stop Codons, Noncanonical Amino Acids, and Deviations from SGC in Mitochondria

The combinatorial fusion rules establish a strong correlation between the stop codons, non-canonical amino acids, and deviation from the SGC in mitochondria. For example, stop codons UAG and UGA code the non-canonical amino acids Pyl and Sec. The list of the deviations from the SGC in the mitochondria [35] exactly matches the codon reassignments during the fusion (Table 5).

Table 5. Deviations from the SGC in mitochondria and the protocode fusion involving start and stop codons.

Occurrence	Codon	SGC	Deviation	Protocode Fusion
Mitochondria by all studied organisms	UGA	stop	Trp	stop GGG → Trp UGG + stop UGA (fusion rule 3)
Vertebrate mitochondria, *Drosophila*, and protozoa	AUA	Ile	Met	Ile AUA → Ile AUA + Met AUG (fusion rule 2)
Invertebrate mitochondria	AGG, AGA	Arg	Ser	Arg GGG → Ser AGC + Ser AGU (fusion rule 3)
Vertebrate mitochondria	AGG AGA,	Arg	stop	Arg GGG → Arg AGG + Arg AGA (fusion rule 3)
Drosophila	AGA	Arg	stop	Arg GGG → Arg AGG + Arg AGA (fusion rule 3)

These experimental results allow for the following evolutionary scenario of the SGC around the four-code fusion (Figure 2). The fusion might be considered as the origin of the LUCA. Initially, LUCA should additionally include X1–X4 amino acids. Their exclusion resulted in generating stop codons, which significantly reduced the stochastic translation of the amino acid sequences.

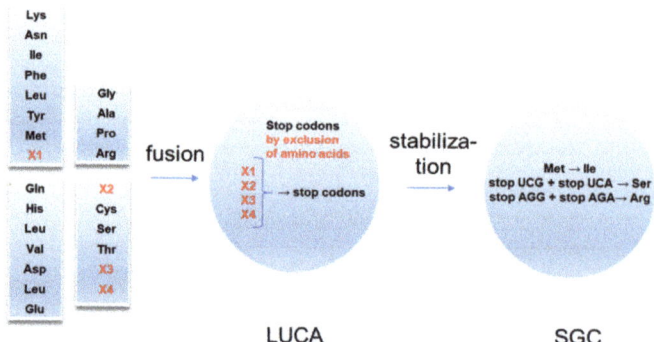

Figure 2. Origin of the SGC from the four-code fusion. Four primordial codes included four additional amino acids. After fusion, LUCA appeared. X1-X4 amino acids were excluded in favor of stop codons. In SGC, the part of stop codons was substituted by Ser and Arg.

This conclusion correlates with the "ambush" hypothesis [36,37]. Along with this hypothesis, an adaptive mechanism mitigates the effects of slippage prone ribosomes by increasing the density of off-frame stop codons. Such a mechanism is reasonable to compensate for reduced translation efficiency in the case of unstable rRNAs. The loss of amino acid X1 from the dominant AU protocode resulted in the two stop codons UAA and UAG (fusion rule 2 for the dominant code). The stop codon UAG was adapted by prokaryotes for the non-canonical amino acid Pyl under evolutionary pressures to develop the methane metabolism [38,39]. The recessive GC code had lost the most amino acids after the fusion. Referring to the deviation from the SGC in mitochondria, Trp was very probably the amino acid X2. After the fusion, one of its triplets UGA became a stop codon. This free codon became available for Sec during evolution [40]. Although Sec is found in the three domains of life, it is not universal in all organisms [41]. The origin of X3 and X4 is unknown. Probably, the primordial amino acid X3 had properties similar to Ser. X4 was probably similar to the positively charged amino acid Arg. An X3-candidate can be one of the extraterrestrial serine derivatives (isoserine, homoserine, and β-homoserine) recently found in significant amounts in the Murchison meteorite [42]. The extremophilic prokaryotes are characterized by a significant content of AGC, AGU, AGG, or AGA codons. In particular, thermophiles and barophiles have high AGG content (X4), although dominant Arg codons are not used to increase the content of the protein stabilizing arginine [43].

The recessive GC code has lost the largest number of amino acids. The lack of hydrophobic amino acids in its cluster **GCā** probably caused their loss.

The reduction in the stop- and start-codons towards the SGC indicates the development of a less error-prone translation system. The stop codons of X3 and X4 were replaced by Ser and Arg, while Met by Ile. These changes lead to the maximal number of codons assigned to Arg and Ser and explain the exceptional odd number of codons for Ile, Met, and Trp (3, 1, 1 correspondingly). This is in line with observations that the evolution of the mitochondrial genetic codes seems best reconstructed when assuming the insertion of amino acids at stop codons [44].

3.3. Protocodes and Modern-Type Aminoacyl-tRNA Classes

The partition AU/GC affected the formation of the modern-type aaRSs (Figure 3). We use the definition of aaRS classes and subclasses and the corresponding amino acid assignments as presented in the review of Kim Y. et al. [45]. Amino acids from the recessive clusters **AUā** and **GCā** are catalyzed by the same aaRS subclasses IA and IID, respectively. Amino acids from the dominant clusters **AUā** and **GCā** show slight inhomogeneity in aaRS classes and subclasses. Charged and polar amino acids from the complementary clusters **a** exhibit significant inhomogeneity: IE, IC, IB, ID, IIB, and IIA.

Interestingly, the subsequent distribution of charged and polar amino acids over aaRS classes is associated with the initial codons from the protocodes. For example, Asn and Asp shared the same codon AAU in the protocodes, and both are catalyzed with the same aaRS subclass IIB. Arg and Cys shared the same codon CGC, and both are catalyzed with the same aaRS class I. By analogy, Arg and Trp (codon CGG) belong to the aaRS class I.

All amino acids from clusters **AUā** (Phe is an exception) belong to aaRS class I, and all amino acids from the **GCā** clusters belong to aaRS class II. Recall that these small, mostly hydrophobic amino acids may play a primary role in charging the proto-aaRS with amino acids from clusters **AUa** and **GCa**. Remarkable is the feature of the Phe-aaRS. Although Phe changed to aaRS class II, Phe retained its feature from the aaRS class I. Phe is coupled to the 2′OH of the ribose of the tRNA terminal adenosine [46]. In contrast, all aaRSs from class II attach amino acids to the 3′OH [47].

Note that some observations suggest that class I and class II tRNA synthetases originate from complementary strands of a single ancestral gene [48,49]. This gene would have originated from tRNA gene pairs coded by complementary strands of a given sequence [50].

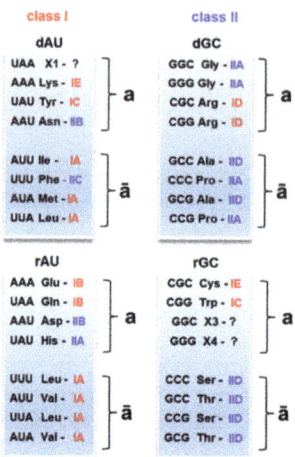

Figure 3. Distribution of the amino acids from the protocodes over two modern-type aminoacyl-tRNA classes and subclasses. The red color stands for aaRS class I, the blue for aaRS class II. The amino acids within clusters ā belong to the same aaRS class except for Phe. dAU and dGC indicate the dominant AU and GC protocodes, respectively. rAU and rGC indicate the recessive AU and GC protocodes.

3.4. Primordial Partition of the Genetic Code

As mentioned above, no assumptions about the evolutionary inclusion of the canonical amino acids into the genetic code are necessary to construct the SGC from the protocodes. Thus, the question about the evolution of the genetic code shifts to the question about the validity of the AU/GC partition. Is this just an unexpectedly simple mathematical trick or an indication of really coexisting ancient protocodes where two amino acids from different protocodes could share the same base triplet?

Besides the proposed AU/GC combinatorial partition, two additional nucleotide partitions exist: the purine/pyrimidine partition AG/CU (Table 6) and the keto/amino partition GU/AC (Table 7). Fusion rules specific to each of these partitions can be derived, to consider alternative fusion processes with exact mathematical descriptions of codon assignments. However, the alternative partitions AG/CU and GU/AC differ principally from the partition AU/GC. For AG/CU and GU/AC partitions, a significant number of new canonical amino acids should be assigned to the new codons after the fusion. This occurs in the case of amino acids presented in the SGC with two codons: Lys, Asn, Asp, Glu, Gln, His, and Phe. For example, two initial codons AAA and AAG of Lys (Table 7, first row) after fusion will be transformed to the codons AAU and AAC of Asn. Such fusions would require many additional assumptions that seem to be a significant disadvantage in comparison with the AU/GC partition. AU/GC partition includes most amino acids before the fusion.

Table 6. AG/CU partition before the fusion. Codons are followed by their codon/amino acid affinity [51] according to the dominant/recessive protocodes. The affinities are dimensionless (Sections 3.8 and 4.2).

	AG Dominant		AG Recessive			CU Dominant		CU Recessive	
Shared Codon	Amino Acid	Affinity	Amino Acid	Affinity	Shared Codon	Amino Acid	Affinity	Amino Acid	Affinity
AAA	Lys	−27	Stop	n.d.	CCC	Pro	−27	Ala	−51
AAG	Lys	−7	Stop	n.d.	CCU	Pro	6	Ala	−1
AGA	Arg	−7	Trp	−17	CUC	Leu	−19	Val	−1
AGG	Arg	4	Stop	n.d.	CUU	Leu	7	Val	22

Table 6. Cont.

GAA	Glu	−54	Arg	−18	UCC	Ser	−51	Ile	−24
GAG	Glu	−23	Gln	29	UCU	Ser	−40	Met	−18
GGA	Gly	−20	Arg	−7	UUC	Phe	−40	Thr	−36
GGG	Gly	−55	Gln	1	UUU	Phe	28	Thr	36

Table 7. AC/GU partition before the fusion. Codons are followed by their codon/amino acid affinity according to the dominant/recessive protocols. The affinities are dimensionless (Sections 3.8 and 4.2).

	AC Dominant		AC Recessive			GU Dominant		GU Recessive	
Shared Codon	Amino Acid	Affinity	Amino Acid	Affinity	Shared Codon	Amino Acid	Affinity	Amino Acid	Affinity
AAA	Lys	−27	Asp	−18	GGG	Gly	−55	Ser	−51
AAC	Asn	86	Glu	52	GGU	Gly	−24	Arg	−4
ACA	Thr	−68	Ala	−37	GUG	Val	14	Ile	26
ACC	Thr	−88	Ala	−44	GUU	Val	−5	Ile	2
CAA	Gln	35	Ser	−50	UGG	Trp	27	Arg	−4
CAC	His	13	Ser	−79	UGU	Cys	−1	Arg	−4
CCA	Pro	−37	Tyr	6	UUG	Leu	−12	Leu	−12
CCC	Pro	−27	Stop	n.d.	UUU	Phe	28	Leu	−47

3.5. Primordial Partition and Hypotheses on Amino Acid Inclusion Ranks in the Genetic Code

The partition AU/GC (as well the other AG/CU and AC/GU) imply the existence of the dominant and recessive protocols. In this respect, we calculated the mean of the amino acid inclusion ranks in the genetic code for amino acids assigned to the dominant versus recessive protocols assuming that one of the protocols would be older than the other. Therefore, we used genetic code origin hypotheses from [52,53]. We also included in analyses some more recent, rather complete hypotheses, the self-referential model [27], and Rogers's hypothesis [54]. Note that these hypotheses are congruent with the mean positions of amino acids in proteins [55,56] and with tRNA and ribosomal RNA secondary structures [57,58].

There was no difference between the mean genetic code inclusion ranks of amino acids coded by the dominant protocol pair vs. the remaining amino acids for fusion hypotheses based on the AU/GC and the AC/GU partitions. However, we found that amino acids coded by the dominant AG/CU protocols are on average significantly more ancient than the remaining twelve amino acids, for most genetic code origin hypotheses, besides 11 among the 40 hypotheses reviewed by Trifonov [53]. The greatest congruence was with Harada and Fox experimental amino acid yields at high temperatures [59] with a statistical p-value of 6.4×10^{-8}, followed by Miller's experiment [60] with $p = 3.3 \times 10^{-6}$, and Wong's nucleotide/amino acid metabolism coevolution hypothesis [61] with $p = 4.3 \times 10^{-6}$. Notable in this list are also hypotheses based on the amino acid contents of Murchison's meteorite ($p = 2.1 \times 10^{-5}$) [62], the hypothesis by Rogers ($p = 8.6 \times 10^{-5}$) [54], the self-referential hypothesis ($p = 0.0013$), and the tRNA Urgen hypothesis of Eigen and Winkler-Oswatitsch ($p = 0.0041$) [63].

Thus, the averaging over the hypotheses, which are based on the step-by-step inclusion of amino acids into the code, do not identify any temporal relation between the dominant and recessive protocols AU/GC.

The genetic code origin hypotheses reviewed by Trifonov [53] are not independent of each other and overall might have been selected for matching results of Miller's experiment.

In the next sections, we examine the relation of the primordial partition to other hypotheses that were not included in [53].

3.6. Primordial Partition and Self-Correcting Properties of the Natural Circular Code

The natural circular code consists of 20 codons that are overrepresented in the coding frame of genes as opposed to the remaining non-coding frames [64–66]. As a group, they have mathematical properties that enable the detection of the coding frame, a self-correcting property of genes, and of the genetic code. It is, hence, hypothesized that the natural circular code is somehow used by the ribosome to detect the coding frame. This assumption is strengthened by observations that specifically those ribosomal RNA regions that are in contact with mRNAs during translation are enriched in nucleotide triplets belonging to the natural circular code [67,68]. The natural circular code presumably arose as a result of selection for non-redundant coding in very short oligonucleotide chains [69,70].

The hypothesis that the natural circular code could have been an initial protocode is also strengthened by the observation that all amino acids coded by these 20 codons are listed as the most likely most ancient amino acids according to Miller's experiment and related hypotheses. Hence, one would predict an overrepresentation of these circular code codons in at least one of the protocodes assumed by the fusion hypothesis. However, all these protocodes include exactly two codons belonging to the natural circular code, which is less than a third expected by chance. None of the dominant codes predicted by the fusion hypothesis converges with the natural circular code observed in natural genes and theoretical minimal RNA rings [71,72].

It is worth noting that the transition from the natural circular code to the SGC remains unexplored, while the transition from the coexisting protocodes to the SGC occurs automatically by the use of the universal and simple fusion rules. Very probably, the natural circular code was selected from the SGC in translation systems that are prone to frameshift errors under unstable environmental conditions. The natural circular code probably played a role in the genetic code evolution. However, the circular code does not explain the codon assignments in the SGC.

3.7. Primordial Partition and Ribosomal Structure

The three-dimensional structure of ribosomes may also include information about the genetic code and its origins. Nucleotide triplets in rRNA in direct contact with ribosomal proteins are biased in such a way that eight amino acids are selectively enriched near their respective codons and eleven amino acids are selectively enriched near their respective anticodons [73]. These observations suggest that anticodons and translation by tRNAs arose in a second phase of the evolution of the genetic code and the ribosome, while direct codon/amino acid contacts ruled the earliest translation mechanisms [51,74].

Thus, the fusion hypothesis would expect a distribution of amino acids within the protocodes according to these observations. For example, the earliest amino acids, coded by dominant protocodes, would have negative values if the bias for contacts with their codons is subtracted from the bias for contacts with their anticodons. The average of these differences was indeed negative for dominant AU/GC protocodes, and the average was positive for recessive AU/GC protocodes, but the difference was not statistically significant (one tailed t test, $p = 0.147$). No pattern was detected for the two remaining partition scenarios.

3.8. Primordial Partition and Codon/Amino Acid Affinities

The stereochemical hypothesis on genetic code origins derived from amino acid/nucleotide contacts in ribosomes is based on the stereochemical affinities between codons and amino acids [75–77]. This hypothesis is in line with affinities observed between mRNAs and the peptides they encode [78–80]. Observations indicate that triplet/amino acid affinities are highest for amino acids that presumably integrated earliest the genetic code. Presumed "more recent" amino acids have low affinities for their assigned nucleotide triplets [81]. We compare the affinities for the three primordial partition AU/GC, AG/CU, and the AC/GU, using the values as reported previously [51]. The only statistically relevant scenario was obtained for the AG/CU partition. In this case, dominant code assignments have greater codon/amino acid affinities than recessive code assignments in ten among thirteen cases (excluding stop codons, one-tailed sign test, $p = 0.023$, Table 6). Hence, the dominant/recessive code division according to the AU/GC partition does not match the rationale of high/low affinities. However, the recent review on the stereochemical hypothesis taking into account high-throughput screens with aptamers leaves reasonable doubts that the weak specificity of amino acid interactions with RNA could play a central role in the code evolution [1].

3.9. How Could Protocodes Coexist?

AU/GC primordial partition distinguishes between only AU protocodes and only GC protocodes. However, chemical changes in A->G and G->A, as well as C->U and U->C are the most spontaneously occurring mutation types [82,83]. This implies that if one of the purines or one of the pyrimidines is available, the other purine, or the other pyrimidine, will spontaneously arise.

We believe that the four nucleotides and most of the canonical amino acids were available as building blocks before the formation of the protocodes and the SGC. The prerequisite of an existing protocode is the self-assembling of its building blocks to an ancient translation apparatus. Thus, if a building block does not involve interactions with such apparatus, its coexistence does not deliver the evidence that the protocode is not possible.

The protocode fusion can be divided into two stages. The first stage included the integration of G/C or A/U bases into the respective AU and GC protocodes. After this inclusion, the dominant and recessive protocodes could still exist as orthogonal translation systems, because new bases conserved the geometry of the kissing proto tRNAs. At this stage, the modern codon assignment of the most canonical amino acids was completed.

In the second stage, modern tRNAs and the aaRS classes emerged. According to the different hypotheses, the modern tRNA was formed by a fusion of two [84,85] or three hairpins [86,87]. Assuming the random nature of this fusion and the equal number of complementary proto tRNAs in the respective protocodes, we evaluated the relationship between the proto tRNA concentrations in dominant and recessive codes. Figure 4 shows the probabilities of the loop sets within the cloverleaf geometry versus the ratio of the 10-base loop concentration to the 9-base loop concentration $\nu = n_{10L}/n_{9L}$. As the cloverleaf has three positions and only two types of loops (9-base- and 10-base loop), these probabilities are described with known combinatorial formulas (Section 4.3).

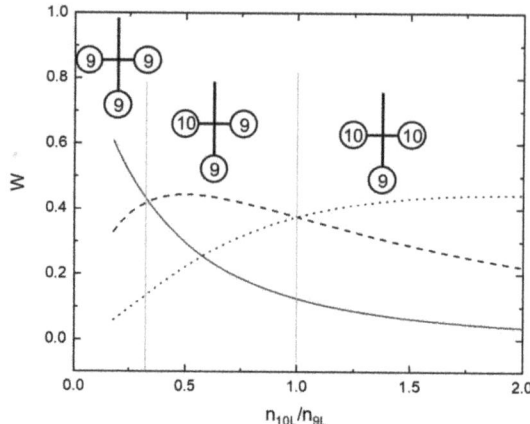

Figure 4. Probabilities W of three different 9-base and 10-base loop sets for the cloverleaf tRNA geometry versus the ratio of the 10-base loop concentration n_{10L} to the 9-base loop concentration n_{9L}. The solid line corresponds to the loop set (9;9;9), the dashed line to (10;9;9), and the dotted line to (10;10;9). The probability of modern-type set (10;9;9) is the largest by n_{10L}/n_{9L} in an interval (0.3:1).

The maximum of the probability for the modern type cloverleaf (10;9;9–D-loop, anti-codon loop, T- loop) is achieved by ν = 0.5. As the ratio of the 10-base loop to 9-base loop hairpins in the protocodes was 1:3 (see Section 3.1), this value means that the concentration of proto tRNAs in the dominant codes was twice as high as in the recessive ones. If tRNA was formed according to the two-hairpin-fusion models, then this ratio should be even higher.

This estimation delivers a simple explanation for the orthogonality in translation between the dominant and recessive protocodes. The hairpins from the dominant codes just inhibited the recessive translation via specific binding to the complementary recessive hairpins according to fusion rule 2. Thus, the translation of the recessive protocodes could work only with leftover codons according to fusion rule 3.

The conclusion in Figure 4 can be used to support the statement that the D-loop of real tRNAs originated from one of the 10-base loop hairpins from the dominant protocodes that was outnumbered inside coexisting protocodes.

4. Materials and Methods

4.1. Amino Acid Inclusion Ranks, Chou-Fasman Conformational Indices, Protocodes, and Aminoacyl-tRNA Classes

Amino acid inclusion ranks in the genetic code were as reviewed by Trifonov [53]. Additional hypotheses were also considered by Guimarães et al. [28] and Rogers et al. [54]. Chou-Fasman conformational indices and other amino acid properties are from ProtScale [78,88].

4.2. Trinucleotide/Amino Acid Affinities

We used the calculated affinities of all 64 trinucleotides with all 20 amino acids summing single-nucleotide affinities for amino acids from [51]. Single nucleotide/amino acid affinity scores were calculated based on contact frequencies between nucleotides and amino acids in crystal structures of interacting RNA–protein complexes [88].

We did not use affinities in solution, only affinities as determined for surfaces. Affinities follow the Gibbs equation $\Delta G = -RT\ln K_d$, with R is the gas constant, T temperature, and K_d the binding constant [88].

Polyansky and Zagrovic [41] estimated affinities as the negative of the log-transformed ratio between all observed contacts between an amino acid and a nucleotide in its assigned codons/anticodons (N^{ij}_{obs}) and the expected contact number assuming random contacts

(N^{ij}_{exp}). Expected random contact frequencies are the product of the frequency of that amino acid in the protein(s) forming a complex with that RNA and the frequencies of nucleotides in that amino acid's cognate codons/anticodons in that RNA:

$$\varepsilon^{ij} = -\ln(N^{ij}_{obs}/N^{ij}_{exp}), \quad (1)$$

where i = 1, ... ,20 for amino acids, and j = 1, ... ,4 for nucleotides. This estimates biases for these contacts in the 3D structure of the RNA–protein complex. This bias corresponds to the binding constant Kd in the Gibbs formulation of affinities: the binding constant is proportional to the bias for observed vs. expected contacts. Note that these ratios are dimensionless and have no unit. Standard quasi-chemical approximations estimate amino acid/amino acid contact energies to predict protein structures and their stabilities [89,90]. The same principles are applied in the context of nucleotide/amino acid contacts.

4.3. Probability of the Emergence of Different Loop Sets for the Cloverleaf Geometry as a Result of a Random Fusion of 9-Base Loop and 10-Base Loop Hairpins

We assume that

$$\nu = n_{10L}/n_{9L}, \quad (2)$$

where n_{10L} is the concentration of the hairpins with the 10-base loop, and n_{9L} the concentration of the hairpins with the 10-base loop. This implies that the probabilities of the emergence of the cloverleaf sets: Three 9-base loops W(9,9,9), two 9-base loops with one 10-base loop W(10,9,9), and two 10-base loops with one 9-base loop W(10,10,9):

$$W(9,9,9) = \left(\frac{1}{1+\nu}\right)^3, \quad (3)$$

$$W(10,9,9) = 3\frac{\nu}{1+\nu}\left(\frac{1}{1+\nu}\right)^2, \quad (4)$$

$$W(10,10,9) = 3\left(\frac{\nu}{1+\nu}\right)^2\frac{1}{1+\nu} \quad (5)$$

Here, W(9,9,9), W(10,9,9), and W(10,10,9) are probabilities of the emergence of the cloverleaf sets: Three 9-base loops, two 9-base loops with one 10-base loop, and two 10-base loops with one 9-base loop, respectively.

5. Conclusions

The fusion rules are no hypothesis, but the mathematical reality of the SGC. According to our knowledge, the AU/GC partition and the fusion rules are the simplest analytical way to describe the emergence of the SGC. Why has this solution not been noticed for more than half a century since the discovery of the code table? Firstly, the fusion rules contradict the postulates of the gradual expansion of the genetic code. This postulate has long dominated the science of the origin of the genetic code. Its cognitive potential is currently being questioned. Secondly, fusion rules imply coexisting protocodes. Interest in orthogonal translation systems has grown only in recent years. The question of why evolution did not use the same orthogonal approach seems no longer to be abstract.

Using the fusion rules, we propose a fusion hypothesis of the origin of the genetic code. The fusion hypothesis states that the SGC originated from the four real protocodes. Their biochemical meaning consists of retaining the complementarity of the codons of the "small" amino acids from the clusters ā to the codons of the "large" amino acids from the clusters **a**. Before the fusion, most of the canonical amino acids were already involved in the coexisting translational apparatuses of the protocodes. Our hypothesis proposes the existence of kissing proto tRNAs responsible for the emergence of the SGC code. The combinatorial fusion rules established the connections between the stop codons, non-canonical amino acids, and the deviation from the standard genetic codes in mitochondria.

Two alternative partitions of the genetic code AG/CU and AC/GU were also examined. The AC/GU partition would reflect keto-amino groupings of nucleotides, which imply relatively rare isoforms of nucleotides. The AG/CU partition reflects a purine/pyrimidin grouping of nucleotides. This partition, unlike the two other partitions, is statistically compatible with most historical hypotheses of integration ranks of amino acids in the genetic code. However, both alternative partitions require additional assumptions for expanding genetic codes after the fusion.

The large diversity of code origin hypotheses including the stereochemical hypothesis produce partially congruent predictions about the historical integration of amino acids into the code. This means that the genetic code, as we know, is compatible with a large number of evolutionary scenarios. Hence, the healthiest approach to this problem is no "natural selection" between hypotheses, because they probably reflect more or less different independent periods/conditions of the code's development. In contrast to most hypotheses, the fusion hypothesis exactly generates the SGC at its last stage.

The fusion hypothesis raises new questions: How did the protocodes appear? What amino acids are missing after the protocode fusion? How was the transition from protopeptide-synthetases to the modern-type aaRSs? Answering these questions requires experimental research. Many powerful methods are available today for screening peptide interactions with various targets, including phage display and peptide arrays. From the experimental point of view, the fusion hypothesis has an advantage. It allows for the study of primordial translation mechanisms with a reduced number of amino acids within single protocodes.

Author Contributions: A.N.-M. conceived the fusion of four protocodes to explain the standard genetic codes. A.N.-M. and H.S. worked out the relation of the fusion to the state-of-the-art SGC hypotheses. A.N.-M., H.S., and R.P. wrote the manuscript and discussed the results. All authors have read and agreed to the published version of the manuscript.

Funding: This research was funded by DFG, grant number AOBJ655892.

Institutional Review Board Statement: Not applicable.

Informed Consent Statement: Not applicable.

Acknowledgments: The authors thank Eugene V. Koonin for valuable and constructive criticism, William F. Martin for valuable remarks on the first version of the fusion concept. This work was supported by the International Excellence Fellowship of Karlsruhe Institute of Technology (KIT) funded within the Framework of the University of Excellence Concept "The Research University in the Helmholtz Association I Living the Change". We acknowledge support by the KIT-Publication Fund of the Karlsruhe Institute of Technology.

Conflicts of Interest: The authors declare no conflict of interest.

Abbreviations

SGC	standard genetic code
aaRS	aminoacyl-tRNA synthetase
tRNA	transfer RNA
stop	stop codon
Lys	lysine
Asn	asparagine
Asp	aspartic acid
Ile	isoleucine
Phe	phenylalanine
Leu	leucine
Tyr	tyrosine
Met	methionine
Pyl	pyrrolysine
Sec	selenocysteine
Gly	glycine
Ser	serine
Ala	alanine
Pro	proline
His	histidine
Val	valine
Arg	arginine
Cys	cysteine
Gln	glutamine

References

1. Koonin, E.V.; Novozhilov, A.S. Origin and Evolution of the Universal Genetic Code. *Annu. Rev. Genet.* **2017**, *51*, 45–62. [CrossRef] [PubMed]
2. Chatterjee, S.; Yadav, S. The Origin of Prebiotic Information System in the Peptide/RNA World: A Simulation Model of the Evolution of Translation and the Genetic Code. *Life* **2019**, *9*, 25. [CrossRef] [PubMed]
3. Higgs, P.G.; Pudritz, R.E. A Thermodynamic Basis for Prebiotic Amino Acid Synthesis and the Nature of the First Genetic Code. *Astrobiology* **2009**, *9*, 483–490. [CrossRef] [PubMed]
4. Gospodinov, A.; Kunnev, D. Universal Codons with Enrichment from GC to AU Nucleotide Composition Reveal a Chronological Assignment from Early to Late Along with LUCA Formation. *Life* **2020**, *10*, 81. [CrossRef]
5. Shore, J.A.; Holland, B.R.; Sumner, J.G.; Nieselt, K.; Wills, P.R. The Ancient Operational Code is Embedded in the Amino Acid Substitution Matrix and aaRS Phylogenies. *J. Mol. Evol.* **2020**, *88*, 136–150. [CrossRef]
6. Knight, R.D.; Freeland, S.J.; Landweber, L.F. Selection, history and chemistry: The three faces of the genetic code. *Trends Biochem. Sci.* **1999**, *24*, 241–247. [CrossRef]
7. Higgs, P.G. A four-column theory for the origin of the genetic code: Tracing the evolutionary pathways that gave rise to an optimized code. *Biol. Direct* **2009**, *4*, 16. [CrossRef]
8. Koonin, E.V.; Novozhilov, A.S. Origin and Evolution of the Genetic Code: The Universal Enigma. *IUBMB Life* **2009**, *61*, 99–111. [CrossRef]
9. Philip, G.K.; Freeland, S.J. Did Evolution Select a Nonrandom "Alphabet" of Amino Acids? *Astrobiology* **2011**, *11*, 235–240. [CrossRef]
10. Antoneli, F.; Forger, M. Symmetry breaking in the genetic code: Finite groups. *Math. Comput. Model.* **2011**, *53*, 1469–1488. [CrossRef]
11. Hornos, J.E.M.; Hornos, Y.M.M. Algebraic Model for the Evolution of the Genetic-Code. *Phys. Rev. Lett.* **1993**, *71*, 4401–4404. [CrossRef] [PubMed]
12. Lenstra, R. Evolution of the genetic code through progressive symmetry breaking. *J. Theor. Biol.* **2014**, *347*, 95–108. [CrossRef] [PubMed]
13. Gonzalez, D.L.; Giannerini, S.; Rosa, R. On the origin of degeneracy in the genetic code. *Interface Focus* **2019**, *9*, 20190038. [CrossRef] [PubMed]
14. Baranov, P.V.; Venin, M.; Provan, G. Codon Size Reduction as the Origin of the Triplet Genetic Code. *PLoS ONE* **2009**, *4*, e5708. [CrossRef] [PubMed]
15. Seligmann, H.; Labra, A. Tetracoding increases with body temperature in Lepidosauria. *Biosystems* **2013**, *114*, 155–163. [CrossRef] [PubMed]
16. Seligmann, H. Putative anticodons in mitochondrial tRNA sidearm loops: Pocketknife tRNAs? *J. Theor. Biol.* **2014**, *340*, 155–163. [CrossRef] [PubMed]

17. Riddle, D.L.; Carbon, J. Frameshift suppression: A nucleotide addition in the anticodon of a glycine transfer RNA. *Nat. New Biol.* **1973**, *242*, 230–234. [CrossRef]
18. Atkins, J.F.; Bjork, G.R. A gripping tale of ribosomal frameshifting: Extragenic suppressors of frameshift mutations spotlight P-site realignment. *Microbiol. Mol. Biol. Rev.* **2009**, *73*, 178–210. [CrossRef]
19. Atkins, J.F. Culmination of a half-century quest reveals insight into mutant tRNA-mediated frameshifting after tRNA departure from the decoding site. *Proc. Natl. Acad. Sci. USA* **2018**, *115*, 11121–11123. [CrossRef]
20. Seligmann, H. Natural mitochondrial proteolysis confirms transcription systematically exchanging/deleting nucleotides, peptides coded by expanded codons. *J. Theor. Biol.* **2017**, *414*, 76–90. [CrossRef]
21. Seligmann, H. Codon expansion and systematic transcriptional deletions produce tetra-, pentacoded mitochondrial peptides. *J. Theor. Biol.* **2015**, *387*, 154–165. [CrossRef] [PubMed]
22. Jukes, T.H. Possibilities for the evolution of the genetic code from a preceding form. *Nature* **1973**, *246*, 22–26. [CrossRef] [PubMed]
23. Lei, L.; Burton, Z.F. Evolution of Life on Earth: tRNA, Aminoacyl-tRNA Synthetases and the Genetic Code. *Life* **2020**, *10*, 21. [CrossRef] [PubMed]
24. Rodin, S.N.; Ohno, S. Four primordial modes of tRNA-synthetase recognition, determined by the (G,C) operational code. *Proc. Natl. Acad. Sci. USA* **1997**, *94*, 5183–5188. [CrossRef] [PubMed]
25. Salim, N.; Lamichhane, R.; Zhao, R.; Banerjee, T.; Philip, J.; Rueda, D.; Feig, A.L. Thermodynamic and Kinetic Analysis of an RNA Kissing Interaction and Its Resolution into an Extended Duplex. *Biophys. J.* **2012**, *102*, 1097–1107. [CrossRef] [PubMed]
26. Paillart, J.C.; Skripkin, E.; Ehresmann, B.; Ehresmann, C.; Marquet, R. A loop-loop "kissing" complex is the essential part of the dimer linkage of genomic HIV-1 RNA. *Proc. Natl. Acad. Sci. USA* **1996**, *93*, 5572–5577. [CrossRef]
27. Guimaraes, R.C. Self-Referential Encoding on Modules of Anticodon Pairs-Roots of the Biological Flow System. *Life* **2017**, *7*, 16. [CrossRef]
28. Guimaraes, R.C.; Moreira, C.H.C.; de Farias, S.T. A self-referential model for the formation of the genetic code. *Theor. Biosci.* **2008**, *127*, 249–270. [CrossRef]
29. Agmon, I. Hypothesis: Spontaneous Advent of the Prebiotic Translation System via the Accumulation of L-Shaped RNA Elements. *Int. J. Mol. Sci.* **2018**, *19*, 4021. [CrossRef]
30. Durand, G.; Dausse, E.; Goux, E.; Fiore, E.; Peyrin, E.; Ravelet, C.; Toulme, J.J. A combinatorial approach to the repertoire of RNA kissing motifs; towards multiplex detection by switching hairpin aptamers. *Nucleic Acids Res.* **2016**, *44*, 4450–4459. [CrossRef]
31. Windbichler, N.; Werner, M.; Schroeder, R. Kissing complex-mediated dimerisation of HIV-1 RNA: Coupling extended duplex formation to ribozyme cleavage. *Nucleic Acids Res.* **2003**, *31*, 6419–6427. [CrossRef] [PubMed]
32. Wallace, M.I.; Ying, L.M.; Balasubramanian, S.; Klenerman, D. Non-Arrhenius kinetics for the loop closure of a DNA hairpin. *Proc. Natl. Acad. Sci. USA* **2001**, *98*, 5584–5589. [CrossRef]
33. Kushiro, T.; Schimmel, P. Trbp111 selectively binds a noncovalently assembled tRNA-like structure. *Proc. Natl. Acad. Sci. USA* **2002**, *99*, 16631–16635. [CrossRef] [PubMed]
34. Weiss, M.C.; Preiner, M.; Xavier, J.C.; Zimorski, V.; Martin, W.F. The last universal common ancestor between ancient Earth chemistry and the onset of genetics. *PLoS Genet.* **2018**, *14*, e1007518. [CrossRef] [PubMed]
35. Jukes, T.H.; Osawa, S. The Genetic-Code in Mitochondria and Chloroplasts. *Experientia* **1990**, *46*, 1117–1126. [CrossRef]
36. Seligmann, H.; Pollock, D.D. The ambush hypothesis: Hidden stop codons prevent off-frame gene reading. *DNA Cell Biol.* **2004**, *23*, 701–705. [CrossRef]
37. Seligmann, H. Localized Context-Dependent Effects of the "Ambush" Hypothesis: More Off-Frame Stop Codons Downstream of Shifty Codons. *DNA Cell Biol.* **2019**, *38*, 786–795. [CrossRef]
38. Srinivasan, G.; James, C.M.; Krzycki, J.A. Pyrrolysine encoded by UAG in Archaea: Charging of a UAG-decoding specialized tRNA. *Science* **2002**, *296*, 1459–1462. [CrossRef]
39. Hao, B.; Gong, W.M.; Ferguson, T.K.; James, C.M.; Krzycki, J.A.; Chan, M.K. A new UAG-encoded residue in the structure of a methanogen methyltransferase. *Science* **2002**, *296*, 1462–1466. [CrossRef]
40. Donovan, J.; Copeland, P.R. The Efficiency of Selenocysteine Incorporation Is Regulated by Translation Initiation Factors. *J. Mol. Biol.* **2010**, *400*, 659–664. [CrossRef]
41. Longtin, R. A forgotten debate: Is selenocysteine the 21st amino acid? *J. Natl. Cancer Inst.* **2004**, *96*, 504–505. [CrossRef]
42. Koga, T.; Naraoka, H. A new family of extraterrestrial amino acids in the Murchison meteorite. *Sci. Rep.* **2017**, *7*, 636. [CrossRef] [PubMed]
43. Khan, M.F.; Patra, S. Deciphering the rationale behind specific codon usage pattern in extremophiles. *Sci. Rep.* **2018**, *8*, 15548. [CrossRef] [PubMed]
44. Seligmann, H. Phylogeny of genetic codes and punctuation codes within genetic codes. *Biosystems* **2015**, *129*, 36–43. [CrossRef] [PubMed]
45. Kim, Y.; Opron, K.; Burton, Z.F. A tRNA- and Anticodon-Centric View of the Evolution of Aminoacyl-tRNA Synthetases, tRNAomes, and the Genetic Code. *Life* **2019**, *9*, 37. [CrossRef] [PubMed]
46. Sprinzl, M.; Cramer, F. Site of Aminoacylation of Transfer-Rnas from Escherichia-Coli with Respect to 2'-Hydroxyl Group or 3'-Hydroxyl Group of Terminal Adenosine. *Proc. Natl. Acad. Sci. USA* **1975**, *72*, 3049–3053. [CrossRef]
47. Moras, D. Structural and Functional-Relationships between Aminoacyl-Transfer Rna-Synthetases. *Trends Biochem. Sci.* **1992**, *17*, 159–164. [CrossRef]

48. Martinez-Rodriguez, L.; Erdogan, O.; Jimenez-Rodriguez, M.; Gonzalez-Rivera, K.; Williams, T.; Li, L.; Weinreb, V.; Collier, M.; Chandrasekaran, S.N.; Ambroggio, X.; et al. Functional Class I and II Amino Acid-activating Enzymes Can Be Coded by Opposite Strands of the Same Gene. *J. Biol. Chem.* **2015**, *290*, 19710–19725. [CrossRef]
49. Carter, C.W.; Li, L.; Weinreb, V.; Collier, M.; Gonzalez-Rivera, K.; Jimenez-Rodriguez, M.; Erdogan, O.; Kuhlman, B.; Ambroggio, X.; Williams, T.; et al. The Rodin-Ohno hypothesis that two enzyme superfamilies descended from one ancestral gene: An unlikely scenario for the origins of translation that will not be dismissed. *Biol. Direct* **2014**, *9*, 11. [CrossRef]
50. Rodin, S.N.; Rodin, A.S. On the origin of the genetic code: Signatures of its primordial complementarity in tRNAs and aminoacyl-tRNA synthetases. *Heredity* **2008**, *100*, 341–355. [CrossRef]
51. Seligmann, H. First arrived, first served: Competition between codons for codon-amino acid stereochemical interactions determined early genetic code assignments. *Sci. Nat.-Heidelberg* **2020**, *107*, 20. [CrossRef] [PubMed]
52. Demongeot, J.; Seligmann, H. RNA Rings Strengthen Hairpin Accretion Hypotheses for tRNA Evolution: A Reply to Commentaries by ZF Burton and M. Di Giulio. *J. Mol. Evol.* **2020**, *88*, 243–252. [CrossRef] [PubMed]
53. Trifonov, E.N. Consensus temporal order of amino acids and evolution of the triplet code. *Gene* **2000**, *261*, 139–151. [CrossRef]
54. Rogers, S.O. Evolution of the genetic code based on conservative changes of codons, amino acids, and aminoacyl tRNA synthetases. *J. Theor. Biol.* **2019**, *466*, 1–10. [CrossRef]
55. Seligmann, H. Protein Sequences Recapitulate Genetic Code Evolution. *Comput. Struct. Biotechnol. J.* **2018**, *16*, 177–189. [CrossRef] [PubMed]
56. Demongeot, J.; Seligmann, H. Theoretical minimal RNA rings recapitulate the order of the genetic code's codon-amino acid assignments. *J. Theor. Biol.* **2019**, *471*, 108–116. [CrossRef] [PubMed]
57. Demongeot, J.; Seligmann, H. Accretion history of large ribosomal subunits deduced from theoretical minimal RNA rings is congruent with histories derived from phylogenetic and structural methods. *Gene* **2020**, *738*, 144436. [CrossRef] [PubMed]
58. Demongeot, J.; Seligmann, H. The Uroboros Theory of Life's Origin: 22-Nucleotide Theoretical Minimal RNA Rings Reflect Evolution of Genetic Code and tRNA-rRNA Translation Machineries. *Acta Biotheor.* **2019**, *67*, 273–297. [CrossRef]
59. Fox, S.W.; Harada, K. The Thermal Copolymerization of Amino Acids Common to Protein. *J. Am. Chem. Soc.* **1960**, *82*, 3745–3751. [CrossRef]
60. Miller, S.L. A Production of Amino Acids under Possible Primitive Earth Conditions. *Science* **1953**, *117*, 528–529. [CrossRef]
61. Wong, J.T. A co-evolution theory of the genetic code. *Proc. Natl. Acad. Sci. USA* **1975**, *72*, 1909–1912. [CrossRef] [PubMed]
62. Kvenvolden, K.; Lawless, J.; Pering, K.; Peterson, E.; Flores, J.; Ponnamperuma, C.; Kaplan, I.R.; Moore, C. Evidence for extraterrestrial amino-acids and hydrocarbons in the Murchison meteorite. *Nature* **1970**, *228*, 923–926. [CrossRef]
63. Eigen, M.; Winkler-Oswatitsch, R. Transfer-RNA, an early gene? *Naturwissenschaften* **1981**, *68*, 282–292. [CrossRef] [PubMed]
64. Michel, C.J. The Maximal C(3) Self-Complementary Trinucleotide Circular Code X in Genes of Bacteria, Archaea, Eukaryotes, Plasmids and Viruses. *Life* **2017**, *7*, 20. [CrossRef]
65. Dila, G.; Michel, C.J.; Thompson, J.D. Optimality of circular codes versus the genetic code after frameshift errors. *Biosystems* **2020**, *195*, 104134. [CrossRef] [PubMed]
66. Michel, C.J. The maximality of circular codes in genes statistically verified. *Biosystems* **2020**, *197*, 104201. [CrossRef] [PubMed]
67. Dila, G.; Ripp, R.; Mayer, C.; Poch, O.; Michel, C.J.; Thompson, J.D. Circular code motifs in the ribosome: A missing link in the evolution of translation? *RNA* **2019**, *25*, 1714–1730. [CrossRef] [PubMed]
68. Michel, C.J.; Thompson, J.D. Identification of a circular code periodicity in the bacterial ribosome: Origin of codon periodicity in genes? *RNA Biol.* **2020**, *17*, 571–583. [CrossRef]
69. Demongeot, J.; Seligmann, H. Spontaneous evolution of circular codes in theoretical minimal RNA rings. *Gene* **2019**, *705*, 95–102. [CrossRef]
70. Demongeot, J.; Seligmann, H. Pentamers with Non-redundant Frames: Bias for Natural Circular Code Codons. *J. Mol. Evol.* **2020**, *88*, 194–201. [CrossRef]
71. Demongeot, J.; Besson, J. The genetic code and cyclic codes. *C. R. Acad. Sci. III* **1996**, *319*, 443–451. [PubMed]
72. Demongeot, J.; Seligmann, H. Why Is AUG the Start Codon? Theoretical Minimal RNA Rings: Maximizing Coded Information Biases 1st Codon for the Universal Initiation Codon AUG. *Bioessays* **2020**, *42*, e1900201. [CrossRef] [PubMed]
73. Johnson, D.B.; Wang, L. Imprints of the genetic code in the ribosome. *Proc. Natl. Acad. Sci. USA* **2010**, *107*, 8298–8303. [CrossRef] [PubMed]
74. Demongeot, J.; Seligmann, H. Theoretical minimal RNA rings mimick molecular evolution before tRNA-mediated translation: Codon-amino acid affinities increase from early to late RNA rings. *C. R. Biol.* **2020**, *343*, 111–122. [CrossRef] [PubMed]
75. Pelc, S.R.; Welton, M.G.E. Stereochemical Relationship between Coding Triplets and Amino-Acids. *Nature* **1966**, *209*, 868–870. [CrossRef] [PubMed]
76. Yarus, M. The Genetic Code and RNA-Amino Acid Affinities. *Life* **2017**, *7*, 13. [CrossRef]
77. Koonin, E.V. Frozen Accident Pushing 50: Stereochemistry, Expansion, and Chance in the Evolution of the Genetic Code. *Life* **2017**, *7*, 22. [CrossRef]
78. Polyansky, A.A.; Zagrovic, B. Evidence of direct complementary interactions between messenger RNAs and their cognate proteins. *Nucleic Acids Res.* **2013**, *41*, 8434–8443. [CrossRef]
79. Bartonek, L.; Zagrovic, B. mRNA/protein sequence complementarity and its determinants: The impact of affinity scales. *PLoS Comput. Biol.* **2017**, *13*, e1005648. [CrossRef]

80. Bartonek, L.; Braun, D.; Zagrovic, B. Frameshifting preserves key physicochemical properties of proteins. *Proc. Natl. Acad. Sci. USA* **2020**, *117*, 5907–5912. [CrossRef]
81. Kahana, A.; Lancet, D. Protobiotic Systems Chemistry Analyzed by Molecular Dynamics. *Life* **2019**, *9*, 38. [CrossRef]
82. Graur, D.; Li, W.-H.; Gojobori, T. Patterns of nucleotide substitution in pseudogenes and funcftional genes. *J. Mol. Evol.* **1982**, *18*, 360–369.
83. Francino, M.P.; Ochman, H. Deamination as the basis of strand-asymmetric evolution in transcribed Escherichia coli sequences. *Mol. Biol. Evol.* **2001**, *18*, 1147–1150. [CrossRef] [PubMed]
84. Tanaka, T.; Kikuchi, Y. Origin of cloverleaf of transfer RNA—The double-hairpin model: Implication for the role of tRNA intron and the long extra loop. *Viva Origino* **2001**, *29*, 134–142.
85. Di Giulio, M. The origin of the tRNA molecule: Implications for the origin of protein synthesis. *J. Theor. Biol.* **2004**, *226*, 89–93. [CrossRef] [PubMed]
86. Kanai, A. Disrupted tRNA Genes and tRNA Fragments: A Perspective on tRNA Gene Evolution. *Life* **2015**, *5*, 321–331. [CrossRef] [PubMed]
87. Burton, Z.F. The 3-Minihelix tRNA Evolution Theorem. *J. Mol. Evol.* **2020**, *88*, 234–242. [CrossRef]
88. Du, X.; Li, Y.; Xia, Y.L.; Ai, S.M.; Liang, J.; Sang, P.; Ji, X.L.; Liu, S.Q. Insights into Protein-Ligand Interactions: Mechanisms, Models, and Methods. *Int. J. Mol. Sci.* **2016**, *17*, 144. [CrossRef]
89. Miyazawa, S.; Jernigan, R.L. Residue-residue potentials with a favorable contact pair term and an unfavorable high packing density term, for simulation and threading. *J. Mol. Biol.* **1996**, *256*, 623–644. [CrossRef]
90. Miyazawa, S.; Jernigan, R.L. Estimation of Effective Interresidue Contact Energies from Protein Crystal-Structures—Quasi-Chemical Approximation. *Macromolecules* **1985**, *18*, 534–552. [CrossRef]

Article

Further Characterization of the Pseudo-Symmetrical Ribosomal Region

Mario Rivas and George E. Fox *

Department of Biology & Biochemistry, University of Houston, Houston, TX 77204, USA; mrivasme@central.uh.edu
* Correspondence: fox@uh.edu; Tel.: +1-713-818-1601

Received: 6 August 2020; Accepted: 11 September 2020; Published: 14 September 2020

Abstract: The peptidyl transferase center of the modern ribosome has been found to encompass an area of twofold pseudosymmetry (SymR). This observation strongly suggests that the very core of the ribosome arose from a dimerization event between two modest-sized RNAs. It was previously shown that at least four non-standard interactions exist between the two halves of SymR. Herein, we verify that the structure of the SymR is highly conserved with respect to both ribosome transition state and phylogenetic diversity. These comparisons also reveal two additional sites of interaction between the two halves of SymR and refine our understanding of the previously known interactions. In addition, the possible role that magnesium may have in the coordination, stabilization, association, and evolutionary history of the two halves (A-region and P-region) was examined. Together, the results identify a likely site where structural elements and Mg^{2+} ions may have facilitated the ligation of two aboriginal RNAs into a single unit.

Keywords: origin of life; peptidyl-transferase center; pseudo-symmetry; proto-ribosome; SymR; emergence of biological systems; RNA ligation; dimerization

1. Introduction

Ribosomes are responsible for the coded synthesis of proteins one residue at a time in all living systems. Atomic resolution structures have revolutionized our understanding of the ribosome. Peptide bond synthesis occurs in the peptidyl transferase center (PTC), of the large ribosomal subunit, which is comprised exclusively of RNA [1,2]. The synthesis itself is facilitated by proper positioning of substrates [3]. In order to make room for the next synthesis cycle, the growing peptide enters an RNA cavity or pore [4,5] that over evolutionary time has grown to be the exit tunnel, from which the product peptide emerges [6–8]. The most widely accepted hypothesis on the evolution of the ribosome states that the most ancient part of it is represented by the PTC and the ribonucleotides that delineate it [9–14]. Alternative views suggest that components of the small subunit evolved earlier than the PTC [15] or that the rRNA evolved by accretion from preexisting tRNA-like molecules [16].

An important advancement was presented at the 28th FEBS meeting in 2002 [3] when it was first reported that there was an area of twofold pseudo-symmetry (SymR) in and around the PTC which is comprised of ≈180 RNA residues. SymR has two halves, one associated with the ribosomal A site and the other with the P site [17]. It has been suggested to be the equivalent of the proto-ribosome, the ancestor of the current ribosome [18,19]. SymR is thought to have come into existence by a heterodimerization event in the prebiotic world. Recent studies have pointed out the possible role of tRNA fragments in this and similar events in the prebiotic world [20,21]. In order to facilitate dimerization, the two separate halves would have to have been able to associate. Following dimerization, the resulting SymR would likely utilize RNA/RNA interactions to form a stable structure. A number of potential standard base pairs exist within each half, but none of these connect the two halves. Thus, the interaction between the two halves must involve non-standard interactions.

An early examination of atomic resolution ribosome structures from the bacterium *Deinococcus radiodurans*, the archaea *Haloarcula marismortui*, and the low-resolution structure of the thermophile *Thermus thermophilus* was undertaken [17,18]. This comparison allowed the authors to conclude that the structure of the SymR was in fact likely stabilized by four non-standard interactions between the two halves [17]. Subsequently, SymR was experimentally examined using the quantum kernel energy method [22]. This study confirmed that the full SymR was in fact likely to form a stable structure when alone or when interacting with substrates.

Although it was not explored in earlier studies, another likely stabilizing element is metal ions. Mg^{2+} ions, in particular, are well-known to be essential for the folding and stability of large RNAs that form compact structures, such as the ribosome [23]. On their own, Mg^{2+} ions seem to be crucial for ribosome stabilization and function [24]. Although less frequent, Mg^{2+} ions also coordinate RNA/protein interactions, such as the association of ribosomal protein uL2 with the 23S rRNA [25].

In the present study, we present evidence for the persistence of SymR and its RNA/RNA interactions with respect to both ribosome transition states and phylogenetic diversity. The possible role that some of these interactions and the magnesium ion contacts may have in the stabilization of the A-region and P-region is also discussed. The results suggest a region where RNA/RNA interactions and Mg^{2+} ion coordination may have facilitated the ligation of two aboriginal RNAs.

2. Materials and Methods

The pseudo-symmetric region (SymR) proposed by Agmon and coworkers [17] was superimposed over various ribosomal large subunit structures using the cealign algorithm that is integrated into PyMOL (The PyMOL Molecular Graphics System, Version 2.3.3, Schrödinger, New York, NY, USA). This allowed identification of equivalent SymR positions in each structure. SymR structures were initially obtained from crystallographic structures of the ribosome's large subunit from *Escherichia coli* (*E. coli*) and *Thermus thermophilus* (*T. thermophilus*). These structures capture the ribosomes in four transition states: classical PRE translocation (PDB ID 4V9D, 4WOI, 3JBU, 4WPO, and 5F8K), chimeric hybrid (PDB ID 4V7B, 4W29 and 4V9L), hybrid (PDB ID 4V9D, 4WOI, 4V9H, and 4V90), and classical POST translocation (PDB ID 3JCD, 4V67, and 4V51). Next, each corresponding SymR structure was analyzed using the Dissecting the Spatial Structure of RNA (DSSR software, v1.8.9) [26] to search for common structural features with the purpose of identifying those features that may vary as the ribosome transitions across each state. The exact position of the bases that form SymR in the context of the secondary structure of the large subunit (LSU) is shown in Figures S1 and S2.

The same procedures were repeated using crystallographic structures of the ribosome's large subunit from distantly related organisms belonging to the three domains of life [27] to obtain the corresponding SymR ribonucleotides. In particular, the crystallographic structures, (PDB IDs 1NKW and 1JJ2 1NJP, 1VQN, 1FFK, and 4V4Q) that were originally utilized by Agmon [17,18] to define SymR were revisited. These structures belong to *Deinococcus radiodurans (D. radiodurans)*, *Haloarcula marismortui (H. marismortui)*, and *E. coli*. Other *D. radiodurans*, *Methanocaldococcus jannaschii (M. jannaschii)*, *Pyrococcus furiosus (P. furiosus)*, *Saccharomyces cerevisiae (S. cerevisiae)*, and *Homo sapiens (H. sapiens)* crystallographic structures (PDB IDs: 5DM7, 4V4N, 4V6U, 4V88, and 6EK0), as well as previous SymR structures from *E. coli* and *T. thermophilus* in classical PRE and hybrid transition states (PDB IDs 4V9D, 4WOI, 3JBU, 4WPO, 5F8K, 4V9H, and 4V90) were also included in this analysis.

All corresponding SymR structures were analyzed using the DSSR software v1.8.9 [26] in the search for common elements that would indicate universal A- to P-region interactions. The DSSR software incorporates the Leontis-Westhof classification of base pairs [28] to identify non-standard base–base interactions. Phylogenetic conservation of bases described in the text was based on the RiboVision suite [29]. Visual inspection over all SymR structures was also performed to avoid discarding possible interactions that might be ignored by the DSSR software due to threshold values, local low-resolution elements, or nonstandard predefined interactions. Visualization of structures and distance measurements of potentially relevant hydrogen bonds was performed over each SymR crystallographic structure with the PyMOL software, and superimpositions were done with the cealign algorithm that is integrated into PyMOL.

Analysis of the conservation of magnesium ions associated with SymR was facilitated by using the Mg^{2+} ions catalogue proposed by Klein and coworkers [23]. This catalogue was originally generated using the *H. marismortui* large subunit crystallographic structure (PDB ID 1S72). Then, the original SymR [17] alone was superimposed over this *H. marismortui* LSU structure to obtain the equivalent ribonucleotides involved in the SymR. Mg^{2+} ions that have contact with at least one ribonucleotide within the *H. marismortui* SymR were selected for analysis. Next, SymR from *H. marismortui* was superimposed over crystallographic structures from phylogenetically distant organisms for which Mg^{2+} ion data was reported and included within each crystallographic structure. This included a different structure from *H. marismortui* (PDB ID 1JJ2), as well as crystal structures from *D. radiodurans* (PDB ID 5DM7), *T. thermophilus* (PDB ID 4WPO), *E. coli* (PDB ID 4V9D), and *S. cerevisiae* (PDB ID 4V88). The LSUs from these structures were aligned against the SymR derived from *H. marismortui* (PDB ID 1S72) using the cealign algorithm that is integrated into PyMOL. This allowed identification of SymR ribonucleotides that may coordinate equivalent Mg^{2+} ions within their respective crystallographic structures. Then, each SymR was displayed with its respective Mg^{2+} data over it to corroborate whether these contacts were in fact conserved on each structure.

3. Results

3.1. On the Stability of the Structural Features of SymR During Translation

SymR seems to be a stable structure that is well-conserved at the core of the LSU [17]. It is comprised, on average, of 178 nucleotides that form six helices, 13 stems, and eight single-stranded segments. In order to fully assess the extent to which the structure of SymR is conserved during protein synthesis, it was initially cut out from the large subunit crystallographic structures of *E. coli* and *T. thermophilus*. These structures represented four different transition states of the ribosome, including classical PRE translocation, chimeric hybrid, hybrid, and the classical POST translocation state, as defined by Mohan and Noller [30]. Detailed examination using the DSSR software [26] allowed identification of structural elements that are consistently seen as the ribosome transitions between the four states. SymR remains a consistent structure from one state to the next. Minor changes are seen that can be attributed to an intrinsic capacity of its phosphorous backbone to absorb tension. There are also some changes in the orientation of sugar and bases, such as bases that are splayed apart. The latter may result in changes at the nucleotide association level, which immediately impact the number of base pairs, the association of nucleotides into multiplets, and even the formation of long-range interactions, like the formation or dissociation of an extended A-minor. Nevertheless, the impact on the main elements of the structure, such as the number of helices, stems, and single-stranded segments appears to be minimal (Table 1). This is despite the fact that some expected variations appeared when comparing structures from different organisms or structures that were solved by distinct high-resolution methods at different resolutions (Tables S1 and S2).

Table 1. Summary of the structural elements detected by DSSR software on SymR derived from eight *T. thermophilus* LSU crystallographic structures captured in four transition states. Base–base interactions include standard and non-standard base pairs. Each SymR structure is comprised of 178 nucleotides.

Transition State	Classic PRE		Chimeric Hybrid		Hybrid		Classical POST	
PDB ID (Chain)	4WPO (CA)	5F8K (2A)	4W29 (BA)	4V9L (BA)	4V9H (BA)	4V90 (BA)	4V67 (BA)	4V51 (BA)
Base-Base interactions	89	86	85	79	86	87	83	85
Multiplets	17	15	13	14	14	16	15	14
Helices	6	6	6	6	6	6	6	6
Stems	13	13	13	13	13	13	13	13
Isolated WC/wobble pairs	5	6	5	7	5	5	5	5
Atom-base-capping interactions	14	13	13	14	12	11	10	15
Splayed-apart dinucleotides	33	31	33	32	36	34	31	35
Hairpin loops	3	3	3	3	3	3	3	3
Bulges	7	8	7	8	8	8	8	8
Internal loops	3	3	3	4	2	2	2	2
Junctions	1	1	1	1	1	1	1	1
Non-loop single-stranded segments	8	8	8	8	8	8	8	8
A-minor (types I and II) motifs	4	4	4	4	4	4	4	4
eXtended A-minor (type X) motifs	1	1	0	1	1	1	0	0
Ribose zippers	3	3	2	3	3	3	3	3

3.2. The Conservation of SymR over the Three Domains of Life

In order to assess the conservation of SymR in the three domains of life [27], the corresponding nucleotides were obtained from the large subunit crystallographic structures of phylogenetically distant organisms. These were I) bacteria (*E. coli*, *D. radiodurans*, and *T. thermophilus*), II) archaea (*H. marismortui*, *M. jannaschii*, and *P. furiosus*), and III) eukaryotes (*S. cerevisiae* and *H. sapiens*). This again included crystallographic structures of ribosomes in classic PRE translocation and hybrid states, as previously defined [30]. Using these structures, the structure of SymR [17] was verified. In particular, it was found that most of the A- to P-region stabilizing interactions were conserved among organisms from all three domains of life, as well as between ribosomal states.

Based on a limited sample of crystallographic structures then available, four major points of interaction between the two halves were previously suggested [17], although not fully described. In the present study, the original four points of interaction were verified and two additional regions of nucleotide interaction between the A-region and the P-region were consolidated from previous suggestions [17]. All six regions have now been characterized in detail, as described below.

The first region (Figure 1A, Figure A1, and Table S3) involves multiple interactions between a GNRA tetraloop motif [31] that is formed at the tip of H93 and the helical region of H74. This motif was previously reported as mediating RNA tertiary interactions, thereby stabilizing the proposed symmetrical dimers [17,18]. It is represented by the almost universal sequence GUGA from H93 that is conserved in all the structures that were used in the analysis. Of the four bases, the triplet UGA represents those nucleotides that are splayed-apart from H93 (A-region) and are exposed into the solvent, thereby allowing interaction with the accepting helical section of H74 from the P-region. Because of its position with respect to H74, U2596 of the triplet is mainly involved in base-stacking with succeeding bases G2597, A2598, and in some cases with U2076. The latter is a non-paired base located at the intersection of H74 and H75. The second base of the triplet, G2597, established a non-standard cSS (Sugar-Edge/Sugar-Edge Anti-parallel) base pair with U2074 from helix 74, which in some cases, by extension, forms part of a multiplet of four bases into which U2244 and A2435 from H74 are included. The third base of the triplet is a universally conserved adenine (A2598) that sometimes establishes a tSH (Sugar-Edge/Hoogsteen Parallel) base pair with G2595. But by far, the most persistent interaction of this adenine is the creation of a Type I A-minor interaction with a conserved C–G base pair (C2073:G2436) from H74. In addition, it sometimes forms part of a multiplet of five bases in which U2245 from the H75-H80 region is also included.

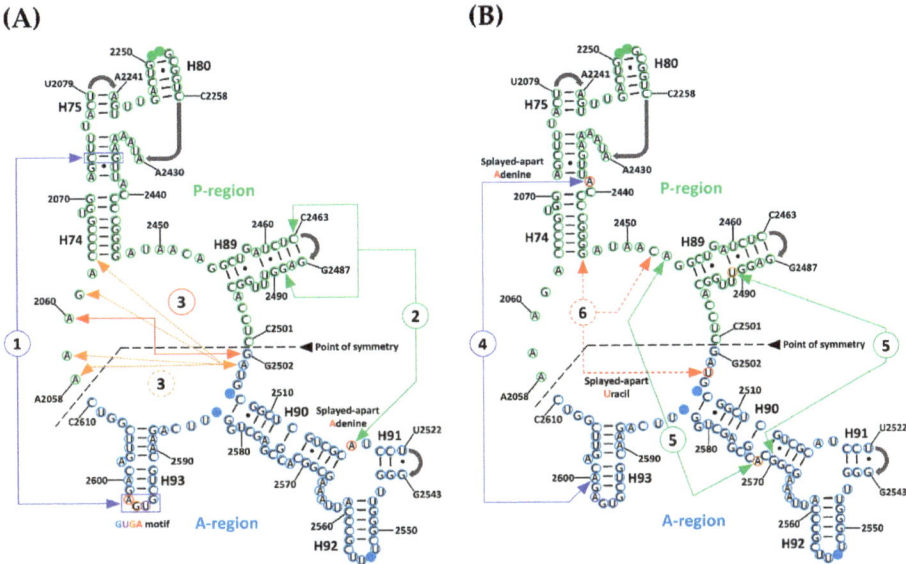

Figure 1. A- to P-region interactions in the symmetric region (SymR) proposed by Agmon [17] pictured over the *T. thermophilus* secondary structure. Bases in the A-region and P-region from the SymR are circled in blue and green, respectively. Bases that interact with the CCA end of the A-site tRNA are marked as solid blue circles (U2506 and G2583 from H90 and G2553 from H92) and bases that interact with the CCA end of the P-site tRNA are marked as solid green circles (G2251 and G2252 from H80). Gray arrows at the end of H75, H80, H89, and H91 represent points of connection that were not detailed by the proponents of the SymR [17]. Six highly conserved regions of long-range RNA interactions have been identified either by the DSSR software or by visual inspection. In order to avoid confusion, the first three of these interactions are illustrated on Panel (**A**) (1, 2, and 3), and the other three are illustrated on Panel (**B**) (4, 5, and 6). Splayed-apart A2519, A2439, U2491, and U2504 are circled in red. As described in the text, the third and fifth interactions both consist of two elements.

The second interaction region (Figure 1A, Figure A1, and Table S4) that likely contributes to the stabilization of SymR, as previously suggested [17] is represented by a universally conserved adenine (A2518). Due to its location, within the three-way junction of helices 90, 91, and 92, A2518 is never base-paired and is forced to be splayed apart, thereby enhancing the chances of establishing hydrogen bonds (HBs) with distant parts of the structure. Visual analysis of the SymR structures revealed that A2518 mostly established HBs with the available ribose hydroxyl groups (O2') of the nearest nucleotides from H89, C2463, and G2489 in the case of *T. thermophilus*. The observed hydrogen bond (HB) distances ranged from moderate (2.5–3.2 Å) to weak (3.2–4.0 Å), according to Jeffrey's HB classification [32]. In the case of two *D. radiodurans* crystallographic structures (PDB IDs 1NKW and 1NJP), distance measurements seemed to be out of HB range. This can be a discrepancy that arises from crystal resolution issues or because the ribosome crystal was captured in the middle of an association–dissociation event. In any case, there are plenty of structures, even from *D. radiodurans* (PDB ID 5DM7), that showed consistent HB distances suggesting that the interaction of splayed-apart A2518 and helix 89 is beyond doubt (Table S4).

The third region of interaction, (Figure 1A, Figure A1, and Table S5), between the two halves of SymR is divided into two consecutive base-stacking elements. The first is integrated by two bases, the highly conserved splayed-apart G2502 located between H89 and H90 at the beginning of the A-region and the universally conserved splayed-apart A2060 located between H73 and H74. The second stacking element is a more robust one, integrated by the highly conserved, splayed-apart

A2503 from the H89–H90 region and A2058, A2059, G2061 from the H73–H74 region, as well as C2063 from H74. This second stacking element varies in the number of bases that are part of it, but it remained constant over all crystallographic structures that were analyzed. For example, the incorporation of A2577 from H90 into the second base-stacking element is a variation that seemed to be exclusive to the *T. thermophilus* crystal structures (Table S5).

Suggested by Agmon [17] and confirmed to be a persistent element by visual inspection of SymR structures, the fourth point of interaction, (Figure 1B, Figure A1, and Table S6), is characterized by another universally conserved splayed-apart adenine (A2439). It is located in the middle of H74. This adenine typically creates a single hydrogen bond that extends from its sugar to the closest phosphate, usually with A2600 (from *T. thermophilus*), which is located in H93. These potential HB distances range from moderate (2.5–3.2 Å) to weak (3.2–4.0 Å). A2439 can also create other HBs with bases or sugars from nucleotides that belong to the H90–H93 region, like U2585 or U2586 from *E. coli* crystal structures. However, these last HBs seemed to vary depending on the crystallographic structure (Table S6).

The fifth interaction region, (Figure 1B, Figure A1, and Table S7), is represented by two hydrogen bonds (HBs) that can be established at two different points where H89 and H90 meet when the structure is completely folded. Distances associated with these HBs again seem to be in a range that goes from moderate (2.5–3.2 Å) to weak (3.2–4.0 Å). The most persistent of these HBs is established between two bases that are not base-paired. The highly conserved splayed-apart A2572 located in H90 established a hydrogen bond (HB) with hydroxyl (O2') of the ribose from the highly conserved A2453, which is localized in the H74–H89 region. The less persistent of these HBs is created between the sugar of the highly conserved U2491 from H89 and the phosphate of the highly conserved G2570 from H90, although the latter HB seems to vary depending on the crystallographic structure (Table S7).

Finally, the sixth point of interaction (Figure 1B, Figure A1, and Table S8) is characterized by a highly conserved splayed-apart uracil (U2504) from the H89–H90 region that usually creates a cHH (Hoogsteen/Hoogsteen anti-parallel) base pair with the highly conserved G2447 localized at H74 and a cWW (Watson-Crick/Watson-Crick Anti-parallel) base pair with a highly conserved C2452 from the H74–H89 region. Although archaeal crystallographic structures show a different plane of insertion for this uracil, it still creates several hydrogen bonds with other nucleotides in the P-region with distances that also range from moderate to weak (Table S8).

3.3. Magnesium Role in Folding and Evolution of the SymR

Initially, Mg^{2+} ion contacts within SymR were assigned to one of three classes based on the bases they stablished contacts with. Class I is characterized by Mg^{2+} ions that are exclusively coordinated by RNA residues that are part of SymR, implying that these interactions could be among the oldest. Class II includes contacts with bases from Domain V outside the SymR. This suggests they arose as the proto-ribosome grew. Class III includes Mg^{2+} ions that contact bases from more distant parts of the LSU RNA, as well as contacts with ribosomal proteins, such as uL2 and uL3 (Table S9). This suggests a more recent acquisition. Conservation of these Mg^{2+} ions over crystal structures from distantly related organisms revealed that the most conserved ions belong to any of these three classes indistinctly (Table 2).

Table 2. Classification of the 16 highly conserved magnesium ions that contact SymR. These were derived from comparisons of crystallographic structures from phylogenetically distant organisms. The numbering system used here was proposed by Klein and coworkers [23]. Magnesium ions that were found present in all structures are represented in bold numbers, while Mg^{2+} ions that were found to be absent only in one crystal structure among those analyzed are represented in italic numbers.

Symmetrical Region Contacts	Mg^{2+} Ion Number	Class
P-region	**1**	I
	90	
	101	
	38	II
	2	III
	32	
	54	
P-region and A-region	**14**	I
	13	III
A-region	**23**	I
	26	
	9	II
	37	
	3	III
	30	
	33	

As can be appreciated in Figure 2, there are two naturally occurring types of Mg^{2+} ions. The first type facilitates the interaction between distant residues within the SymR secondary structure, thereby assisting in its folding. The second group are coordinated locally by nearby residues, most likely neutralizing electrostatic forces, which is a different mechanism of enhancing the folding.

A total of 28 magnesium ion contacts were observed within SymR (Figure 2). Of the 28 Mg^{2+} ions contacting SymR nucleotides, only Mg13 and Mg14 established contacts between the A- and P-regions of the SymR (Figure 2, Table 2, and Table S9). However, they are not equally conserved. Of the 28, only 10 are present in all the crystallographic structures of the SymR (Table S10). This last number increases to 16 if we consider those ions that lack an equivalent ion in no more than one structure under analysis (Table 2). The latter speculation rests on the assumption that some of these interactions have an equivalency. For example, in *S. cerevisiae*, Mg14 is absent. However, its coordination has been replaced by the combination of Mg3964 and Mg4328 (Figure 3).

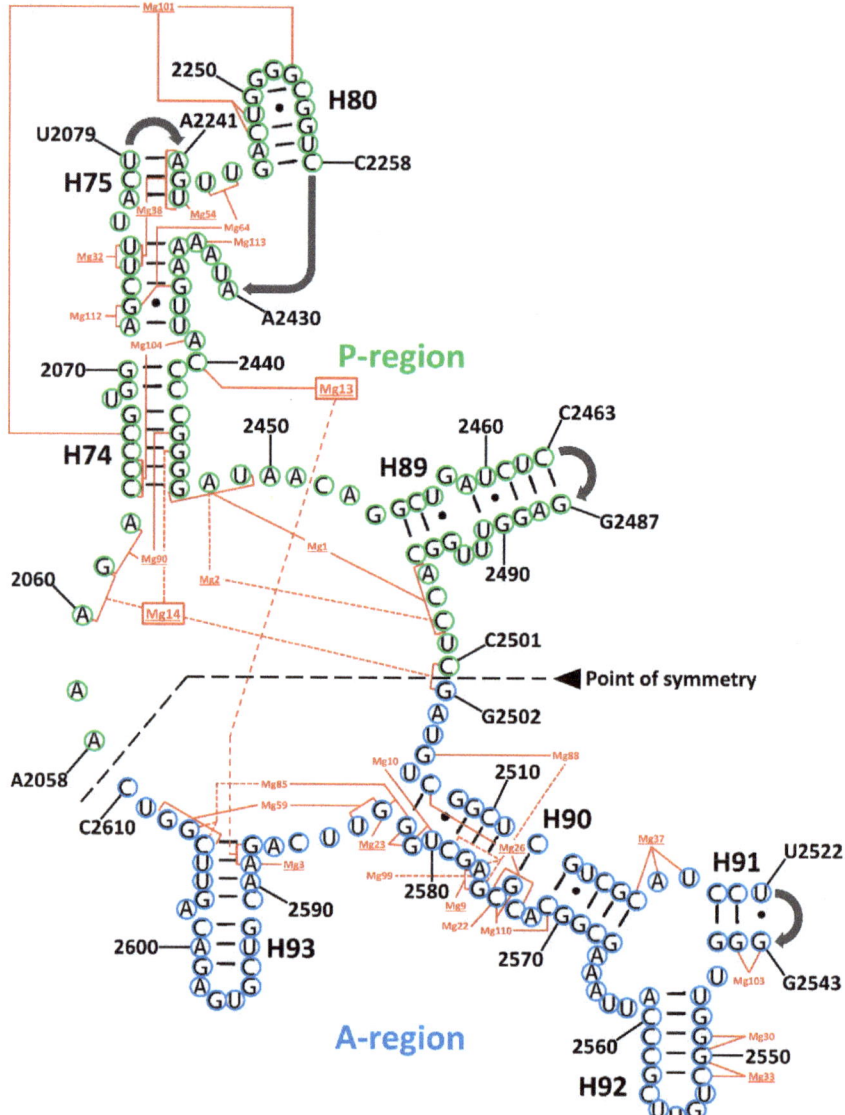

Figure 2. Magnesium ion contacts in the symmetrical region (SymR) modeled over the *T. thermophilus* secondary structure. Bases in the A-region and P-region from the SymR are circled in blue and green, respectively. Mg^{2+} ions are numbered after Klein and coworkers [23]. Underlined Mg^{2+} numbers represent conserved ions that were found in most crystallographic structures (Table S9). Dotted red lines, connecting the bases that coordinate a particular Mg^{2+} ion, were used exclusively to differentiate among lines that crossed with others at some point to avoid confusion. Mg13 and Mg14 are the only metallic ions that established contacts with bases from both the A-region and P-region (red boxes). Full data are presented in Supplementary Tables S9–S11.

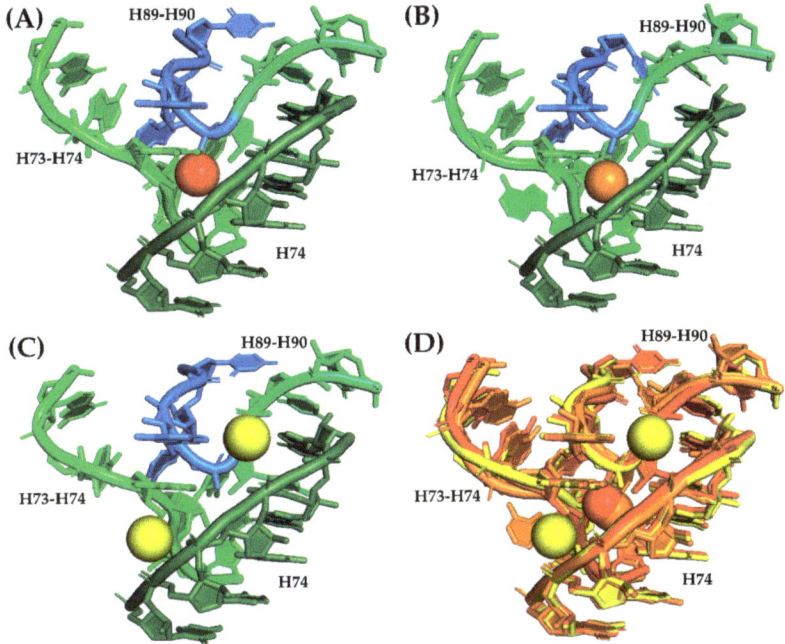

Figure 3. Replacement of conserved Mg14 in *S. cerevisiae*. Mg14 is coordinated in the crystallographic structure from *H. marismortui* (PDB ID 1S72) by A2101, G2102 from the H73–H74 region, G2480 from H74, and C2536 and G2537 from the H89–H90 region, as previously stated by Klein and coworkers [23]. This magnesium ion is conserved among crystallographic structures of *E. coli*, *T. thermophilus*, *H. marismortui*, and *D. radiodurans* (Table S10). (**A**) The coordination region of Mg14 (red sphere) in a different *H. marismortui* structure (PDB ID 1JJ2). Ribonucleotides from the P-region are colored in three tones of green to denote the fact that they belong to distant areas of the P-region, while nucleotides from the A-region are colored in blue. (**B**) Coordination region of Mg14 (orange sphere) in *T. thermophilus* (PDB ID 4WPO). The color code of the bases is the same as in panel A. (**C**) Mg14 coordination region on *S. cerevisiae* (PDB ID 4V88) with two Mg^{2+} ions (yellow spheres, Mg3964 and Mg4328) that seemed to have replaced Mg14 by establishing contacts with the phosphate groups of most of the bases that were supposed to be coordinating Mg14. The color code of the bases is the same as in Panel A. (**D**) Structural alignment of the Mg14 coordination region from *H. marismortui* (red), *T. thermophilus* (orange), and *S. cerevisiae* (yellow) that shows the relative positions of the corresponding Mg^{2+} ions in this region.

4. Discussion and Conclusions

The GUGA motif seems to be among the strongest forces that hold the two RNA pieces together, as previously suggested [17,18]. The two splayed-apart adenines (A2439 and A2518) and the contacts between H89 and H90 were also first observed by Agmon and coworkers [17]. These contacts between H89 and H90 and one splayed-apart adenine (A2518) were reported as features that alternated in different crystal structures [17], most likely due to the lack of available high-resolution structures at that time. We have confirmed that the splayed-apart adenines and the H89 to H90 contacts are indeed consistent features of the SymR that were extracted from extant crystallographic structures. They seem to be among the easiest to form and dissociate, because in most cases, they solely depend on a single HB. We speculate that this feature may have facilitated an early polymerization cycle to take place. Extant ribosomes exhibit regions where tension must be absorbed by its structure, allowing the ribosome to transition from one state to another [3,30]. This may have been the case in the structure of the earliest ribosome as well.

The base-stacking region that forms between two consecutive sets of nucleotides (G2502 and A2060, as well as A2503, A2058, A2059, G2061, and C2063) was previously mentioned as separate interactions that differed in the number of nucleobases that integrate in what we call the second element [17]. The third region of interaction is of special interest because it includes bases in the nearby area surrounding the PTC cavity, that may have served as a scaffold that holds in place the first base of the A-region (G2502). This would facilitate the interaction with the last base of the P-region (C2501). This is similar to what occurs in the active site of many enzymes where a substrate is held before catalysis. In the case of the SymR, it may have eventually resulted in the formation of a covalent bond between the two regions.

Finally, the splayed-apart uracil (U2504) was previously identified as creating a single contact with the ribose of A2572 in the *H. marismortui* crystal structure [17]. This uracil has been observed to be creating non-standard base pairs with a couple of bases over several structures that belong to a considerable number of phylogenetically distant organisms. We speculate that because U2504 is right next to the base-stacking region, it also contributed to the A- to P-region stability and the ligation of both regions.

In summary, the six A- to P- interaction regions are widely conserved in the SymR structures of distantly related organisms from the three domains of life. They also remain present in ribosomes captured in different transition states. This is a good indication of their relevance to the stability of the extant peptidyl transferase center (PTC) and its possible contribution to the assembly of the earliest PTC version, now called the proto-ribosome.

On the other hand, despite discordant visions [15,16] there is a widely accepted hypothesis that assumes that the ribosome is a molecular fossil whose structure recapitulates its history and states that the most ancient part of it is represented by the PTC and the ribonucleotides that delineate it [9–14]. By assuming this scenario, we classified Mg^{2+} ions that interact with the SymR in three classes that might say something about the relative time in which these interactions were established. A more classic view, derived from comparative biology, assumes the conservation level over phylogenetically distant organisms as a different way of dating biologically relevant features. With this in mind, we also performed a search of the magnesium ion coordination over several crystallographic structures from distantly related organisms from the three domains of life. The results derived from both approaches do not completely agree on the relative age of the 28 magnesium ions that have contacts with SymR. These contrasting results may reflect the structural dependency of the highly conserved Mg^{2+} ions that has been long-established over evolutionary time within the extant ribosome. Nevertheless, the overlapping results immediately point to relevant Mg^{2+} ion interactions, like the previously described magnesium ion microcluster formed by Mg1 and Mg2 within the P-region (Table 2), that has been recognized as a primeval coordination motif relevant for RNA folding, function, and evolution [24]. Other well-conserved Mg^{2+} ions within classes that suggest early SymR interactions are also of special interest.

If the vision that the proto-ribosome arose by fusion of two similar structures is correct [17,18], then to understand how this occurred is a central question. Based on our results, we also suggest that Mg14 coordination could represent the force that acted over the two pseudo-symmetrical units, thereby facilitating a ligation process between the last nucleotide from the P-region and the first nucleotide from the A-region. This interaction is now represented by C2501 and G2502 (Figures 2 and 3), from which the proto-ribosome could have arisen as a single RNA chain. It would be of value to experimentally determine if two separate RNA fragments would anneal to one another if they exhibit similar magnesium coordination. After all, the ligation of two ribonucleotides aided by magnesium ions has been observed within the extant RNA polymerase II, whose active site coordinates two Mg^{2+} cations that are considered essential for its activity [33].

The results presented here suggest that metallic cations like magnesium might be relevant not only for the folding and catalytic capabilities of the proto-ribosome, but they also play a role in the evolutionary pathway this structure might have followed.

It is well-known that several unsuccessful attempts to synthetize a functional proto-ribosome have been made in the past. The current results may contribute to the understanding of the relevant interactions that must be observed if a proto-ribosome is expected to exhibit structural integrity and catalytic function. Moreover, the well-established requirement of magnesium should not be overlooked if experimental approaches are to be taken, at least during early stages.

Supplementary Materials: The following are available online at http://www.mdpi.com/2075-1729/10/9/201/s1, Figure S1: SymR contrasted against the secondary structure of *Thermus thermophilus* ribosome's large subunit. Figure S2: *Thermus thermophilus* structure (PDB ID 4WPO) of entire large subunit that indicates the area occupied by the SymR and the RNA/RNA interactions within SymR, Table S1: Summary of the structural elements as reported by the DSSR software [26] when applied to SymR derived from crystallographic structures from *E. coli* at four transition states; Table S2: Summary of the structural elements as reported by the DSSR software [26] when applied to SymR derived from crystallographic structures from *T. thermophilus* at four transition states, Table S3: Conserved elements of the GUGA motif from H93 in the A-region that stablished interactions with bases from H74 in the P-region as reported by DSSR software [26], Table S4: Conserved splayed-apart adenine in the H90-H91 zone from A-region that creates hydrogen bonding with bases in H89 from P-region, Table S5: Base-stacking region that is integrated by two consecutive base staking elements as reported by DSSR software [26], Table S6: Conserved splayed-apart adenine in H74 from P-region that creates hydrogen bonding with bases in the H90-H93 zone and H93 from A-region, Table S7: Points of contact between the H90 from A-region and H74-H89 zone and H89 from P-region, Table S8: Conserved splayed-apart uracil (SA-Uracil) from the H89-H90 zone in the A-region that base pair with bases (A-Bases) from H74 and the H74-H89 zone in the P-region, Table S9: Magnesium ion contacts in the symmetrical region that was modeled over the *Haloarcula marismortui* crystallographic structure (PDB ID 1S72), Table S10: Magnesium ion contacts associated to the symmetrical region were verified to persists on other crystallographic structures from *H. marismortui* (PDB ID 1JJ2), *D. radiodurans* (PDB ID 5DM7), *T. thermophilus* (PDB ID 4WPO, Chain CA), *E. coli* (PDB ID 4V9D, Chain DA) and *S. cerevisiae* (PDB ID 4V88, Chain A1), Table S11: Number of the bases that are involved in the Mg^{2+} ion coordination as they appear in the different numbering systems for the *H. marismortui* (PDB ID 1JJ2), *D. radiodurans* (PDB ID 5DM7), *T. thermophilus* (PDB ID 4WPO), *E. coli* (PDB ID 4V9D) and *S. cerevisiae* (PDB ID 4V88) crystallographic structures.

Author Contributions: M.R. and G.E.F. conceived the work. M.R. conducted all the computational work. Results were discussed with G.E.F. Both authors contributed equally to the writing process and preparation of the manuscript. Both authors have read and agreed to the published version of the manuscript.

Funding: Mario Rivas Medrano's research was supported by an appointment to the NASA Postdoctoral Program at the NASA Astrobiology Institute, administered by Universities Space Research Association under contract with NASA. This work was also supported in part by NASA Grant NNX14AK16G to George E. Fox and NASA Contract 80NSSC18K1139 under the Center for Origin of Life, at the Georgia Institute of Technology.

Acknowledgments: The authors would like to thank Quyen Tran for help in implementing and maintaining computational tools.

Conflicts of Interest: The authors declare no conflict of interest. The funders had no role in the design of the study; in the collection, analyses, or interpretation of data; in the writing of the manuscript, or in the decision to publish the results.

Appendix A

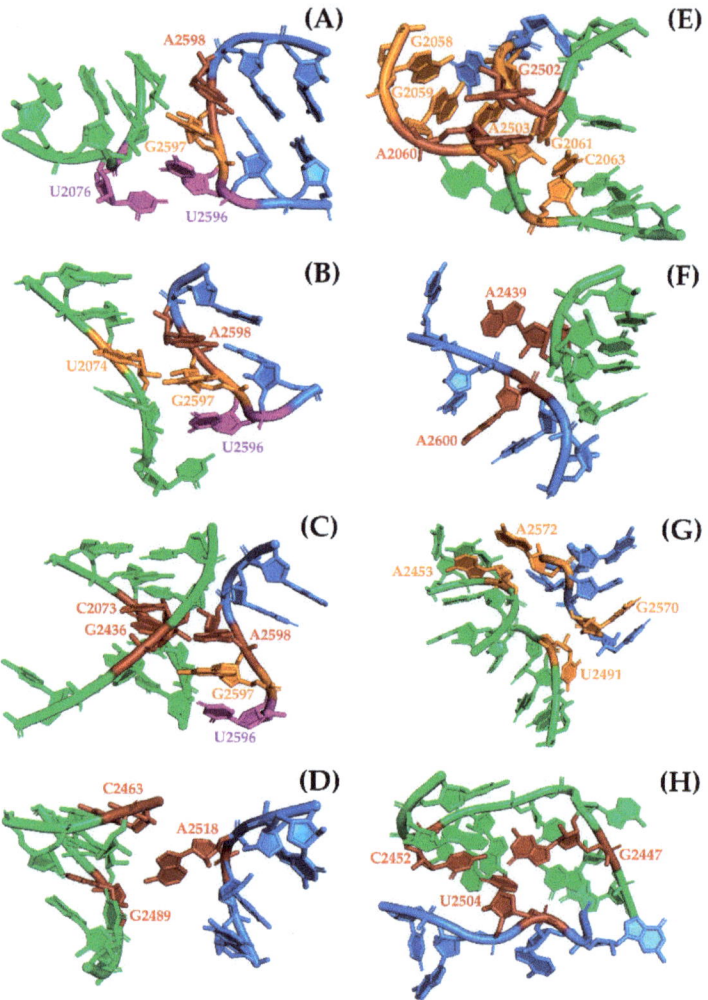

Figure A1. Close view of the A- to P-region interactions that contribute to the association and stabilization of SymR. Bases that are not directly involved in interactions are visualized in cartoon mode and colored blue and green to denote that they belong to either the A-region or the P-region respectively (same color code as Figure 1). (**A**) U2596 (purple) from the GUGA motif base stacks with G2597 (orange), A2598 (red) and U2076 (purple). (**B**) G2597 (orange) from the same motif base pairs with U2074 (orange). (**C**) A2598 (red) from the GUGA motif creates a type I A-minor with C2073 and G2436 (red). (**D**) Splayed-apart A2518 establishes hydrogen bonds with C2463 and G2489 (red). (**E**) Base-stacking region with two elements, the first involves G2502 and A2060 (red) and the second involves A2503, A2058, A2059, G2061, C2063 and A2577 (orange). (**F**) Splayed apart A2439 that created hydrogen bond with A2600 (red). (**G**) Two H89 to H90 contacts represented by A2572 and A2453, and U2491 and G2570 (orange). (**H**) Splayed-apart U2504 that base pairs with G2447 and C2452 (red). A detailed description of each interaction is provided in the text.

References

1. Bashan, A.; Zarivach, R.; Schluenzen, F.; Agmon, I.; Harms, J.; Auerbach, T.; Baram, D.; Berisio, R.; Bartels, H.; Hansen, H.A.S.; et al. Ribosomal crystallography: Peptide bond formation and its inhibition. *Biopolymers* **2003**, *70*, 19–41. [CrossRef] [PubMed]
2. Polacek, N.; Mankin, A.S. The Ribosomal Peptidyl Transferase Center: Structure, Function, Evolution, Inhibition. *Crit. Rev. Biochem. Mol. Boil.* **2005**, *40*, 285–311. [CrossRef] [PubMed]
3. Agmon, I.; Auerbach, T.; Baram, D.; Bartels, H.; Bashan, A.; Berisio, R.; Fucini, P.; Hansen, H.A.S.; Harms, J.; Kessler, M.; et al. On Peptide Bond Formation, Translocation, Nascent Protein Progression and the Regulatory Properties of Ribosomes: Delivered on 20 October 2002 at the 28th FEBS Meeting in Istanbul. *Eur. J. Biochem.* **2003**, *270*, 2543–2556. [CrossRef]
4. Fox, G.E.; Tran, Q.; Yonath, A. An Exit Cavity Was Crucial to the Polymerase Activity of the Early Ribosome. *Astrobiology* **2012**, *12*, 57–60. [CrossRef] [PubMed]
5. Rivas, M.; Tran, Q.; E Fox, G. Nanometer scale pores similar in size to the entrance of the ribosomal exit cavity are a common feature of large RNAs. *RNA* **2013**, *19*, 1349–1354. [CrossRef]
6. Milligan, R.A.; Unwin, P.N.T. Location of exit channel for nascent protein in 80S ribosome. *Nature* **1986**, *319*, 693–695. [CrossRef]
7. Yonath, A.; Leonard, K.R.; Wittmann, H.G. A tunnel in the large ribosomal subunit revealed by three-dimensional image reconstruction. *Science* **1987**, *236*, 813–816. [CrossRef]
8. Voss, N.R.; Gerstein, M.; Steitz, T.; Moore, P.B. The Geometry of the Ribosomal Polypeptide Exit Tunnel. *J. Mol. Boil.* **2006**, *360*, 893–906. [CrossRef]
9. Fox, G.E.; Naik, A.K. The Evolutionary History of the Translations Machinery. In *The Genetic Code and the Origin of Life*; Springer: Boston, MA, USA, 2004; pp. 92–105.
10. Hury, J.; Nagaswamy, U.; Larios-Sanz, M.; Fox, G.E. Ribosome origins: The relative age of 23S rRNA domains. *Orig. Life Evol. Biosphere* **2006**, *36*, 421–429. [CrossRef]
11. Fox, G.E. Origin and Evolution of the Ribosome. *Cold Spring Harb. Perspect. Boil.* **2010**, *2*, a003483. [CrossRef]
12. Hsiao, C.; Mohan, S.; Kalahar, B.K.; Williams, L.D. Peeling the Onion: Ribosomes are Ancient Molecular Fossils. *Mol. Boil. Evol.* **2009**, *26*, 2415–2425. [CrossRef] [PubMed]
13. Petrov, A.S.; Bernier, C.R.; Hsiao, C.; Norris, A.M.; Kovacs, N.A.; Waterbury, C.C.; Stepanov, V.G.; Harvey, S.C.; Fox, G.E.; Wartell, R.M.; et al. Evolution of the ribosome at atomic resolution. *Proc. Natl. Acad. Sci. USA* **2014**, *111*, 10251–10256. [CrossRef] [PubMed]
14. Petrov, A.S.; Gulen, B.; Norris, A.M.; Kovacs, N.A.; Bernier, C.R.; Lanier, K.A.; Fox, G.E.; Harvey, S.C.; Wartell, R.M.; Hud, N.V.; et al. History of the ribosome and the origin of translation. *Proc. Natl. Acad. Sci. USA* **2015**, *112*, 15396–15401. [CrossRef] [PubMed]
15. Harish, A.; Caetano-Anollés, G. Ribosomal History Reveals Origins of Modern Protein Synthesis. *PLoS ONE* **2012**, *7*, e32776. [CrossRef]
16. Demongeot, J.; Seligmann, H. Comparisons between small ribosomal RNA and theoretical minimal RNA ring secondary structures confirm phylogenetic and structural accretion histories. *Sci. Rep.* **2020**, *10*, 1–14. [CrossRef]
17. Agmon, I.; Bashan, A.; Zarivach, R.; Yonath, A. Symmetry at the active site of the ribosome: Structural and functional implications. *Boil. Chem.* **2005**, *386*, 833–844. [CrossRef]
18. Agmon, I. The Dimeric Proto-Ribosome: Structural Details and Possible Implications on the Origin of Life. *Int. J. Mol. Sci.* **2009**, *10*, 2921–2934. [CrossRef]
19. Davidovich, C.; Belousoff, M.; Wekselman, I.; Shapira, T.; Krupkin, M.; Zimmerman, E.; Bashan, A.; Yonath, A. The Proto-Ribosome: An Ancient Nano-Machine for Peptide Bond Formation. *Isr. J. Chem.* **2010**, *50*, 29–35. [CrossRef]
20. Root-Bernstein, R.; Root-Bernstein, M. The Ribosome as a Missing Link in Prebiotic Evolution III: Over-Representation of tRNA- and rRNA-Like Sequences and Plieofunctionality of Ribosome-Related Molecules Argues for the Evolution of Primitive Genomes from Ribosomal RNA Modules. *Int. J. Mol. Sci.* **2019**, *20*, 140. [CrossRef]
21. Prosdocimi, F.; Zamudio, G.S.; Palacios-Pérez, M.; De Farias, S.T.; José, M.V. The Ancient History of Peptidyl Transferase Center Formation as Told by Conservation and Information Analyses. *Life* **2020**, *10*, 134. [CrossRef]
22. Huang, L.; Krupkin, M.; Bashan, A.; Yonath, A.; Massa, L. Protoribosome by quantum kernel energy method. *Proc. Natl. Acad. Sci. USA* **2013**, *110*, 14900–14905. [CrossRef] [PubMed]

23. Klein, D.J.; Moore, P.B.; Steitz, T.A. The contribution of metal ions to the structural stability of the large ribosomal subunit. *RNA* **2004**, *10*, 1366–1379. [CrossRef] [PubMed]
24. Hsiao, C.; Williams, L.D. A recurrent magnesium-binding motif provides a framework for the ribosomal peptidyl transferase center. *Nucleic Acids Res.* **2009**, *37*, 3134–3142. [CrossRef] [PubMed]
25. Petrov, A.S.; Bernier, C.R.; Hsiao, C.; Okafor, C.D.; Tannenbaum, E.; Stern, J.; Gaucher, E.; Schneider, D.; Hud, N.V.; Harvey, S.C.; et al. RNA–Magnesium–Protein Interactions in Large Ribosomal Subunit. *J. Phys. Chem. B* **2012**, *116*, 8113–8120. [CrossRef] [PubMed]
26. Lu, X.-J.; Bussemaker, H.J.; Olson, W.K. DSSR: An integrated software tool for dissecting the spatial structure of RNA. *Nucleic Acids Res.* **2015**, *43*, e142. [CrossRef]
27. Woese, C.R.; Fox, G.E. Phylogenetic structure of the prokaryotic domain: The primary kingdoms. *Proc. Natl. Acad. Sci. USA* **1977**, *74*, 5088–5090. [CrossRef]
28. Leontis, N.B.; Westhof, E. Geometric nomenclature and classification of RNA base pairs. *RNA* **2001**, *7*, 499–512. [CrossRef]
29. Bernier, C.R.; Petrov, A.S.; Waterbury, C.C.; Jett, J.; Li, F.; Freil, L.E.; Xiong, X.; Wang, L.; Migliozzi, B.; Hershkovits, E.; et al. RiboVision suite for visualization and analysis of ribosomes. *Faraday Discuss.* **2014**, *169*, 195–207. [CrossRef]
30. Mohan, S.; Noller, H.F. Recurring RNA structural motifs underlie the mechanics of L1 stalk movement. *Nat. Commun.* **2017**, *8*, 14285. [CrossRef]
31. Jaeger, L.; Michel, F.; Westhof, E. Involvement of a GNRA tetraloop in long-range RNA tertiary interactions. *J. Mol. Boil.* **1994**, *236*, 1271–1276. [CrossRef]
32. Jeffrey, G.A. *An Introduction to Hydrogen Bonding*; Oxford University Press: New York, NY, USA, 1997.
33. Svetlov, V.; Nudler, E. Basic mechanism of transcription by RNA polymerase II. *Biochim. Biophys. Acta* **2013**, *1829*, 20–28. [CrossRef] [PubMed]

© 2020 by the authors. Licensee MDPI, Basel, Switzerland. This article is an open access article distributed under the terms and conditions of the Creative Commons Attribution (CC BY) license (http://creativecommons.org/licenses/by/4.0/).

Concept Paper

Viroids and Viroid-like Circular RNAs: Do They Descend from Primordial Replicators?

Benjamin D. Lee [1,2] and Eugene V. Koonin [1,*]

1. National Center for Biotechnology Information, National Library of Medicine, National Institutes of Health, Bethesda, MD 20894, USA; benjamin.lee@nih.gov
2. Nuffield Department of Medicine, University of Oxford, Oxford OX3 7BN, UK
* Correspondence: koonin@ncbi.nlm.nih.gov

Abstract: Viroids are a unique class of plant pathogens that consist of small circular RNA molecules, between 220 and 450 nucleotides in size. Viroids encode no proteins and are the smallest known infectious agents. Viroids replicate via the rolling circle mechanism, producing multimeric intermediates which are cleaved to unit length either by ribozymes formed from both polarities of the viroid genomic RNA or by coopted host RNAses. Many viroid-like small circular RNAs are satellites of plant RNA viruses. Ribozyviruses, represented by human hepatitis delta virus, are larger viroid-like circular RNAs that additionally encode the viral nucleocapsid protein. It has been proposed that viroids are direct descendants of primordial RNA replicons that were present in the hypothetical RNA world. We argue, however, that much later origin of viroids, possibly, from recently discovered mobile genetic elements known as retrozymes, is a far more parsimonious evolutionary scenario. Nevertheless, viroids and viroid-like circular RNAs are minimal replicators that are likely to be close to the theoretical lower limit of replicator size and arguably comprise the paradigm for replicator emergence. Thus, although viroid-like replicators are unlikely to be direct descendants of primordial RNA replicators, the study of the diversity and evolution of these ultimate genetic parasites can yield insights into the earliest stages of the evolution of life.

Keywords: viroids; ribozyviruses; primordial replicators; ribozymes; origin of life

1. Introduction

The leading current scenario for the origin of life involves the stage of a primordial RNA world, in which RNA molecules are postulated to have doubled in the roles of genomes (replicators) and catalysts (enzymes, or in this case, ribozymes) [1–3]. The key ribozymes in the RNA world, obviously, would have been RNA-dependent RNA polymerases (RdRP) that would catalyze their own replication as well as replication of other RNA molecules [4,5]. The major properties of RNA world replicators are easy to surmise: these would be small, stable RNA molecules endowed with the RdRP or capable of recruiting ribozyme RdRP in trans. Obviously, such replicators would not code for any proteins. The phenotype of these primitive replicators would be chemically identical to their genotype.

Can such a gene-free genome exist? Viroids, small circular satellite RNAs (satRNAs) and several other groups of similar RNA elements prove that such genomes can and do exist (Figure 1) [6–10]. However, there are no known ribozyme RdRPs. Instead, to replicate without relying on encoded proteins, these agents coopt host enzymes although some of them catalyze certain steps of their own replication using ribozymes formed by the genomic RNA itself. All viroid-like genomes (Table 1) are small, covalently closed circular RNAs (cccRNAs), or at least, go through a cccRNA phase in their replication cycle, and replicate when a polymerase repeatedly rolls around the RNA circle to produce multiple copies, in a process known as rolling circle replication (RCR) [11,12]. The known diversity of these autonomous circular RNAs is growing, with representatives spanning a broad range of genetic elements, from viruses to retrotransposons [13–15]. This viroid-like "brotherhood" [14] attracts considerable attention, both because of the importance of some viroid-like agents for human health and agriculture [16], but also because their origin(s) remains enigmatic. Due to their unusual properties, viroids have long been suspected to be relics of the RNA world [17–19].

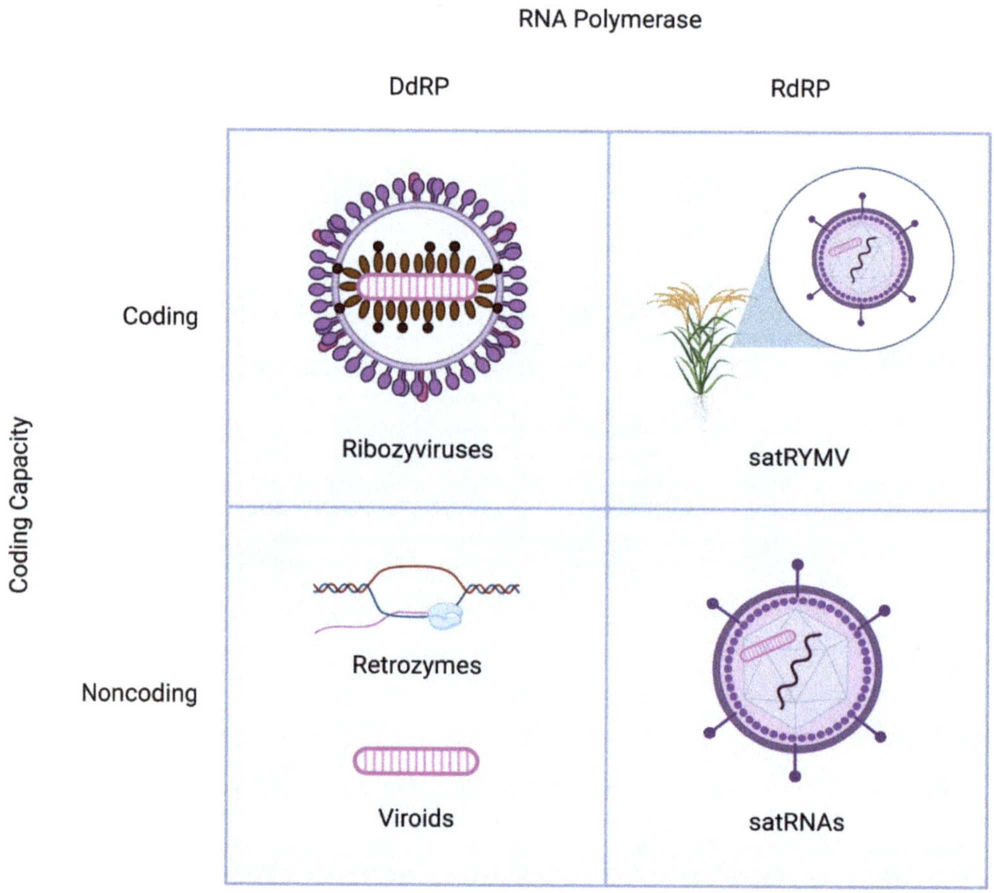

Figure 1. Schematic structures of distinct classes of viroid-like RNAs. All viroid-like agents are divided into four groups depending on whether they encode proteins and are replicated by DNA-dependent or RNA-dependent RNA polymerase. The satRNAs (lower right quadrant) are encapsidated by the helper virus capsid proteins. In one case (top right quadrant), a satRNA of rice yellow mottle virus (satRYMV) appears to encode a protein.

Viroids were the first subviral agents to be discovered and remain the most diverse group in the brotherhood. First described by Diener in 1969 [20,21], these agents are unencapsidated cccRNAs 200–400 nt in length that can cause diseases in plants, some of these fatal. To date, no viroid has been conclusively demonstrated to cause infection in any organisms outside of plants although some can replicate in yeast [22] and may even infect phytopathogenic fungi [23]. Viroid RNAs exhibit extensive self-complementarity resulting in compact, robust, rod-like structures [24,25]. Viroids as well as ribozyviruses replicate via the RCR mechanism, in which the circular RNA is reiteratively transcribed by hijacked host DNA-dependent RNA polymerase (DdRP) into multimeric intermediates of the opposite polarity [12,26–30]. These intermediates are then transcribed themselves and cleaved to unit length, after which the original circular positive-polarity sequence is regenerated via ligation by a co-opted host ligase. The cleavage step varies between the two families of viroids. In members of the *Pospiviroidae* family, a host RNAse cleaves the intermediates, whereas in members of the family *Avsunviroidae*, cleavage of the intermediate is catalyzed by an autocatalytic RNA ribozyme formed by the viroid RNA itself [11,12,29]. Viroids of the *Avsunviroidae* engage in a symmetric process, in which intermediates are ligated into circular forms of the opposite polarity before undergoing the same RCR process again, to produce circular RNAs of the original polarity [31,32]. This process requires viroids to contain two ribozymes (one per RNA strand of each polarity). In members of the *Pospiviroidae* family, only one RCR cycle occurs to create the negative strand [26]. The linear, multimeric negative strand is transcribed, and the positive polarity transcript is cleaved to unit length by a host RNAse. The two families of viroids also vary in their choice of the ligase: members of *Pospiviroidae* use host DNA ligase 1 that they repurpose as RNA ligase [33], whereas members of *Avsunviroidae*, which replicate within plastids, use chloroplast tRNA ligase [34].

Table 1. The major types of viroid-like cccRNAs.

Viroid-like cccRNAs	Size	Host	Ribozymes	Known Coding Capacity
Viroids	246–450 nt	Plants	HHR when present	None
Ribozyviruses	1547–1735 nt	Metazoans	HDVR or HHR	One conserved protein
Retrozymes	300–1116 nt	Eukaryotic genomes	HHR	None
satRNAs	220–457 nt	Plants	HHR or hairpin	None (except satRYMV)
Hypothetical primordial replicator	~200 nt	None (RNA world)	HHR	None

Broadly similar to viroids in terms of structure, replication, and range, satRNAs differ in that they are encapsidated, albeit by a helper virus rather than by proteins they encode themselves [10,35]. Furthermore, satRNAs do not rely on host DdRP for replication, but instead employ the RNA-dependent RNA polymerase (RdRP) of the helper virus [12,36]. Additionally, of note, satRNAs vary in their replication mechanisms, beyond using a different polymerase, in that they do not always contain one ribozyme per strand polarity and, when they do, each strand can contain a distinct type of ribozyme [36]. Given that viroids and satRNAs share so many features and are present in the same types of hosts, these agents are thought to share a common ancestor. Phylogenetic reconstructions have shown evidence of such an evolutionary relationship although the exact nature of the common ancestor remains unclear [37].

Neither is the range of viroid-like RNA agents limited to plants nor are they all strictly non-coding. Members of the realm *Ribozyviria*, of which the only well-characterized one is human hepatitis delta virus (HDV), infect animals and encode a single protein that undergoes post-translational modification, producing two distinct forms that perform various functions in virus reproduction including the role of nucleocapsid inside the virions [38,39]. Apart from this distinction, ribozyviruses exhibit the key features of viroids and satRNAs. They too replicate via the rolling circle mechanism catalyzed by host DdRP and employ a virus-encoded ribozyme to cleave the replication intermediates [40–42]. HDV, a human pathogen that uses hepatitis B virus as its helper, was the first ribozyvirus to be discovered, but recent metagenomics studies resulted in the identification of multiple ribozyviruses that apparently infect diverse animals [43–46]. Similarly to satRNAs, ribozyviruses use different types of ribozymes, one of which is unique to members of *Ribozyviria* [47].

Where did the diminutive viroid-like replicators come from? While conclusive evidence is lacking, hypotheses abound, more or less, mirroring the main scenarios considered for the origin of viruses: origin from the pre-cellular replicator pool, regression from existing replicators, and escape from existing genomes [48]. However, the pre-cellular origin hypothesis has a twist in the case of viroids. Unlike the virus-first pre-cellular origin scenario, which postulates that viruses evolved from pre-cellular replicators, the viroid version holds that viroids are not only direct descendants of pre-cellular replicators but also holdovers from the pre-protein RNA world as evidenced by their use of ribozymes for (some stages of) replication [17]. Later, ribozymes would again provide the engine for the transition to the RNA-protein world by serving as the catalytic core of the ribosome. In some of these "viroid early" scenarios, viroids or viroid-like RNAs come across not only as relics from the RNA world but also the first RNA world replicators [19]. Hence, potential special interest of viroid-like agents as a unique window into the hypothetical pre-cellular RNA world. In this article, we discuss the evidence pro and contra the primordial status of viroids, coming to the conclusion that these agents are likely to have much more recent origins, but nevertheless, can offer valuable clues to the features of minimal replicators, even including pre-cellular ones.

2. Viroids: Ancient Relics or Recently Emerged Parasites?

Whether all viroid-like agents are monophyletic remains an open question although parsimony suggests a common ancestry [49]. The attributes of the putative common ancestor are hard to pin down in detail but, given that viroids are both the simplest and most autonomous agents within the brotherhood, it seems likely that this ancestor was either viroid-like or satRNA-like [50]. Indeed, the hypothesis on the emergence of these agents within the RNA world is predicated on that assumption. The alternative possibility, that the ancestor was a ribozy-like virus that gave rise to viroids through reductive evolution, is clearly less parsimonious given the higher complexity and apparent narrow spread of ribozyvirus among hosts compared to viroids.

A broad variety of possible scenarios for viroid emergence have been proposed (Figure 2). The most radical idea is that viroids emerged de novo [17]. Hard evidence for this mechanism is scant and difficult to obtain although both computational modeling [51] and in vitro experiments [52] indicate that RNAs with properties similar to those of viroids might be produced de novo. Specifically, simulations suggest that rod-shaped and branched circular structures can form when short random RNA sequences fold into hairpins, which then catalyze ligation with other random hairpins, finally producing circular structures. These cccRNAs could then grow in length by both recombination and random insertions. More importantly, such RNAs have the potential to evolve such that sequences with greater stability, robustness, and polymerase affinity being selected for. Thus, Catalan and colleagues [51] conclude that de novo emergence of viroids through a stepwise evolutionary process seeded by any of the diverse small RNAs present in cells, such as microRNAs, is plausible.

Experimental evidence bears a similar picture, albeit with a different replication mechanism. Inspired by the role of DdRPs in viroid and ribozyvirus replication, Jain and colleagues [52] demonstrated that high concentrations of T7 phage DdRP can generate populations of replicating structured linear RNAs from DNA seeds in vitro. Unlike the standard model of viroid replication, which is based on RCR, with monomers ligating to form antigenomic cccRNAs, the mechanism these replicators used bypasses ligation altogether: instead, the polymerase jumps from the 5' end of the template back to the 3' end. Recent research also demonstrates that viroid-like self-complementary cccRNAs are readily formed in autocatalytic RNA reaction networks [53]. These lines of research suggest that replicators with the structural hallmarks of viroids could potentially emerge de novo, at least in the presence of a suitable polymerase.

Other proposed mechanisms for the emergence of viroids are also of considerable interest when viewed in the broader context of replicator evolution. The genome escape hypothesis posits that viroids are genomic sequences that managed to achieve semi-autonomous replication [54]. Although arguably less exotic than the de novo scenario, this model nonetheless provides an example of the emergence of a replicator from a non-replicator. In this case, potential ancestors of viroids could be, for example, group I self-splicing introns, the simplest of which form viroid-sized cccRNAs after splicing catalyzed by a ribozyme present in the intron itself [55,56].

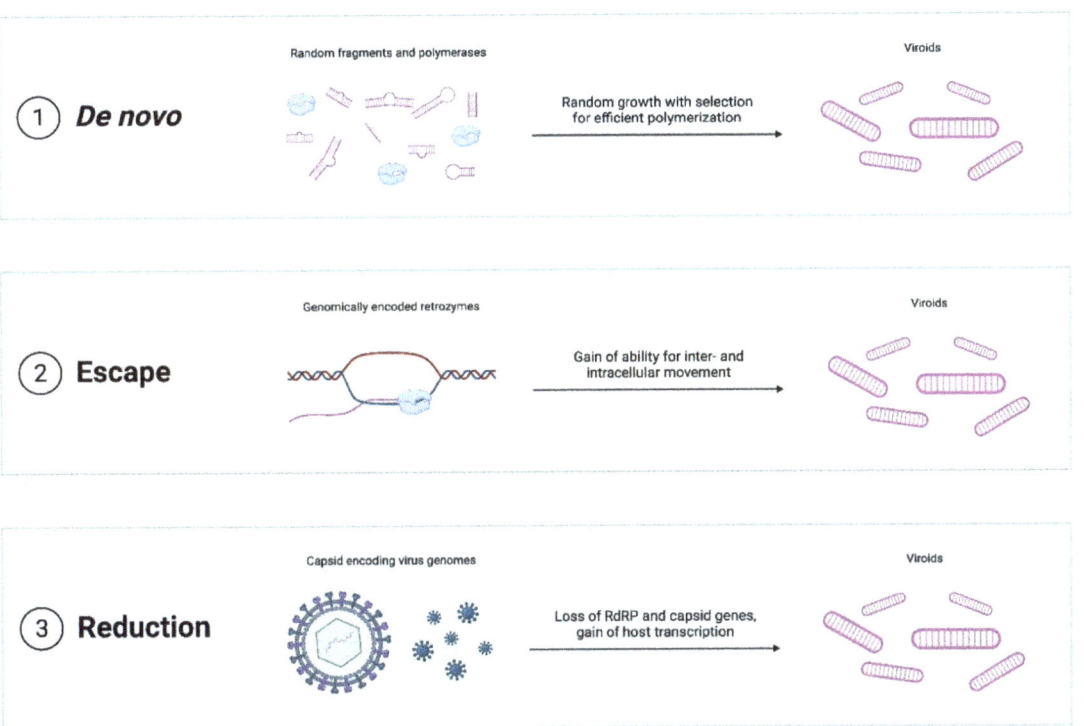

Figure 2. Evolutionary scenarios for the emergence of viroid-like replicators.

In the genome reduction model, viroids emerge from pre-existing replicators, such as viruses, by shedding all coding regions and relying entirely on host enzymes. The classic in vitro evolution experiments of Spiegelman and colleagues [57–59] clearly demonstrate the phenomenon of viral genome reduction to viroid size by completely eliminating coding capacity. In these paradigmatic experiments, the 3800 nt genome of RNA bacteriophage Qβ ultimately shrank to a 218 nt non-coding RNA (known as "Spiegelman monster") under selection for replication speed in the presence of an excess of phage RdRP [60]. A similar process could have conceivably occurred in vivo. Genome reduction is a common phenomenon in obligate parasites of all kinds ranging from bacteria and archaea to worms to arthropods [61–63]. In some such cases, the genome reduction can be extreme, with the genomes of some obligate parasitic and symbiotic bacteria shrinking more than an order of magnitude, and mitochondrial genomes more than two orders of magnitude [64,65]. Reductive evolution also readily occurs in viruses, in particular, giving rise to defective interfering particles, and to dramatic contraction of virus genomes upon passages in cell cultures [66–68]. Thus, it appears entirely plausible that viroids are ultimately reduced viruses. Identification of sequences homologous to viroids within viral or cellular genomes could lend credence to either of these hypotheses.

Of the three principal models of viroid emergence, two implicitly answer the question of when viroids emerged: both the escape and genome reduction hypotheses, by design, require viroids to have emerged in a cellular context. However, what type of cells? Viroids and satRNAs are limited in range to higher plants, whereas ribozyviruses are limited to animals. Assuming that viroids and viroid-like agents share a common ancestry, horizontal transfer between plants and animals could be an explanation [49]. Hard evidence for the direction of the transfer is lacking, but the greater complexity of ribozyviruses compared to viroids and their limited spread in animals, taken together, seem to point towards a likely origin in plants, with acquisition of the protein-coding gene via recombination in an animal host.

Evidence of emergence of viroids in planta seems to be mounting. Recently, retrotransposons containing hammerhead ribozymes (HHR) similar to those present in viroids, satRNAs, and some ribozyviruses were discovered in plant genomes [69–71]. These ribozyme-containing retrotransposons, named retrozymes, are not autonomous and apparently depend on the reverse transcriptase activity of Ty3-like retrotransposons. The retrozymes consist of a 300–700 nt non-coding sequence flanked by long terminal repeats (LTRs) and, apart from the presence of the HHR, resemble other small, non-autonomous retrotransposons that are abundant in plants, such as terminal-repeat retrotransposons in miniature (TRIMs) [72] or small LTR retrotransposons (SMARTs) [73]. The retrozymes are actively transcribed, the transcripts are self-cleaved by HHR and form abundant cccRNAs of both polarities [74,75]. Notably, the circularization is catalyzed by chloroplast tRNA ligase as is the case also in avsunviroids. The RNA sequences of retrozymes are predicted to fold into stable, branched structures resembling those of avsunviroids. Even smaller, non-LTR retrozymes of 170–400 nt have been discovered in genomes of diverse invertebrate and invertebrate animals and also shown to form abundant cccRNAs [76]. The broad representation of retrozymes in plant genomes is compatible with the origin of viroids via retrozyme escape in plants. Given their limited (currently known) host range, such escape seems to be the most likely scenario for the emergence of viroids, in accord with the prescient early hypothesis of Diener [53]. Conversely, the animal retrozymes potentially might have independently given rise to ribozyviruses, as an alternative to the horizontal transfer of viroids discussed above. The evolution of viroid-like replicators from retrozymes seems to combine elements of the escape and reduction scenarios, assuming the retrozymes themselves are products of reductive evolution of autonomous retrotransposons.

In contrast, the viroid-first hypothesis seems to face multiple difficulties. Probably the most damning for this scenario is the limited host range of viroid-like agents, especially their apparent absence in bacteria or archaea. Like spliceosomal introns and many other varieties of non-coding RNAs, viroid-like cccRNAs belong to the expanded "new RNA world" of eukaryotic cells [77,78], and thus, their origin most likely postdates the origin of eukaryotes. Beyond this major problem, replication of viroid-like RNAs in the hypothetical RNA world also seems problematic. Indeed, the extensive, robust secondary structures of these cccRNAs are adapted to mimic DNA and so fool DdRPs and, possibly, also to avoid RNAi host response [9]. Neither of these factors that apparently shape the evolution of viroid-like RNAs is relevant in the primordial setting. On the contrary, replication of such highly structured RNAs by the hypothetical ribozyme polymerases would be hampered, even beyond the more generic problems faced by such polymerases despite extensive experimental attempts to evolve accurate and processive ones [4].

Thus, the retrozyme escape hypothesis appears to be the most parsimonious scenario for the origin of viroid-like cccRNAs. Refutation of this hypothesis would imply that retrozymes either are integrated viroids or a case of convergence. Although HHRs appear to have evolved on multiple occasions independently [79], the similarities between retrozymes and viroids are so pronounced that the most likely explanation remains that viroids are escaped retrozymes.

3. Viroids as Paragons of Minimal Replicator Emergence

As discussed, the possibility that viroids and other similar cccRNAs are direct descendants of primordial replicators that inhabited the hypothetical ancient RNA world appears to be remote. However, the likely relatively recent origins of viroid-like RNAs cannot deprive them of the status of minimal replicators. It does not appear to be a chance coincidence that the Spiegelman monster, the minimal parasite artificially evolved under conditions when the sole selective factor was the speed of replication, is the same size as the smallest viroids and satRNAs [60]. This size, approximately 200 nt, is likely to be close to the ultimate low bound for a replicator.

Replicators are genetic elements that encode information that is necessary and sufficient for their replication but lack the full complement of genes required for providing energy and building blocks for replication [80–84]. For energy and building blocks, replicators depend on reproducers, biological entities whose reproduction involves physical rather than only informational continuity as in the case of replicators [81,84]. The paradigmatic reproducers are, obviously, cells—*omnis cellula e cellula*, according to Virchow. More precisely, all cellular life forms can be conceived as symbioses of a reproducer and a replicator, the genome [85]. Cellular genomes are ultimate cooperative replicators, but generally, replicators span wide ranges of replicative autonomy and aggressiveness/cooperativity with respect to the host reproducer [83,86].

Some replicators, such as many large viruses or self-synthesizing transposons, encode a (nearly) complete replication and sometimes transcription machinery, and thus possess a high degree of replicative autonomy. Others rely mostly or completely on the host replication and expression apparatus. Viroids are the extreme manifestation of the latter strategy—arguably, the ultimate parasites. Generally, the information content of replicators splits into "replicase expression signal" (RES), which can consist of one or more genes encoding replicative enzyme (s) and possibly various accessory proteins, and "replicase recognition signal" (RRS), non-coding elements that are necessary and sufficient to hijack the host replication (or expression) machinery [87]. Viroids have adopted the ultimate parasitic strategy, namely, selection for maximum efficiency of RRS accompanied by complete elimination of RES, under the reduction scenario or by never acquiring RES, under the de novo scenario. More precisely, this is the evolutionary strategy of pospiviroids, whereas avsunviroids add a notable twist by engaging viroid-embedded ribozymes in some stages of replication. The HHRs appear easy to evolve [79] and could be the simplest evolutionary step towards increased replicative autonomy.

Evolution towards ultimate parasitism occurs already at the stage of retrozymes as well as other non-autonomous transposons that hijack the replication and transposition machinery of autonomous transposons with closely similar RRS. However, retrozymes have only limited replication capacity restricted to copy-paste transposition within the host genome. Viroids likely escaped from the host genomes by evolving the ability to redirect the host DdRP towards viroid RNA replication. Identification of an intermediate on the proposed evolutionary path from retrozymes to viroid-like agents—namely, a retrozyme capable of replicating autonomously within a cell—would be a major step in demonstrating the emergence of viroid-like replicators as an ongoing process. There are some indications that retrozymes might possess that capacity, in particular because multimeric intermediates of both polarities have been detected and the high error rate of retrozymes suggests polymerization by DdRP, but definitive experimental evidence is still lacking [69,76].

Although viroids are unlikely to be relics of the pre-cellular stage of life evolution, they provide a unique window into the origin of parasitic replicators, which most likely occurred on multiple, independent occasions, via retrozyme escape and, possibly, de novo. Similar events could have transpired during the pre-cellular stage of evolution where ultimate parasites similar to viroids with respect to size and reproduction cycle, but not necessarily secondary structure, could have evolved from autonomous replicators.

4. Viroid-like RNAs and Emergence of Protein Coding

If emergence of viroid-like replicators is an ongoing process, there seems to be no reason why some of these replicators would not evolve to encode proteins, and indeed, although neither viroids nor retrozymes encompass protein-coding genes, other members of the viroid brotherhood do. The best studied case, obviously, are the ribozyviruses that encode a single protein of unclear provenance, which is employed for a typical virus function, encapsidation of the genome [38,39]. An unrelated example is a satRNA of rice yellow mottle virus (satRYMV), the smallest known satRNA at 220 nt, that has been shown to encode proteins in overlapping open reading frames, covering most of the genome, and reiteratively translated, in parallel to the replication of satRYMV via reiterative transcription [88]. Translation of satRYMV proceeds through multiple rounds by switching frames and thus yielding proteins of supergenome size. Notably, satRYMV contains ribozymes in both polarities within the protein-coding region. Although too far reaching conclusions, perhaps, should not be drawn from this single case, satRYMV seems to present a remarkable example of de novo emergence of proteins. Search for other similar cases might shed light on the evolution of protein coding in new replicators and, possibly, even at processes that occurred in the primordial RNA world.

More generally, it appears likely that the diversity of protein-coding viroid-like agents is substantially underappreciated as suggested, in particular, by the recent discovery of an expanded range of ribozyviruses [43–47]. Furthermore, it seems likely that evolution of coding and non-coding viroid-like RNAs is a two-way street, with protein-coding genes both gained and lost on multiple occasions. Comprehensive screening of the rapidly growing metagenomic data for different types of such agents should reveal their actual diversity and evolutionary relationships.

5. Conclusions

Viroids and viroid-like RNAs are minimal replicators, the only known ones that encode no proteins and probably do not far exceed the theoretical minimum replicator size. Therefore, the idea that these RNA replicators are relics of the hypothetical primordial RNA world appears both straightforward and highly attractive. Yet, this scenario seems to be poorly compatible with the available evidence including both the host range of these agents that currently appears to be limited to multicellular eukaryotes and the properties of the viroid-like RNAs that are clearly attuned to replication by sophisticated protein polymerases. In contrast, a direct line of descent appears to be traceable from retrozymes, non-autonomous retrotransposons, to viroids. It appears likely that viroid-like replicators evolved from retrozymes, possibly, on several independent occasions. However, this likely relatively recent origin hardly diminishes the fundamental interest of viroid-like RNAs as the paradigm for the emergence of new replicons as ultimate parasites, at different stages of life's evolution, including the pre-cellular stage. The existence of protein-coding viroid-like genomes, those of ribozyviruses and the single so far characterized protein-coding viroid-like satRNA, indicates that there is no impassable gulf between the non-coding, minimal RNA replicators and protein-coding replicators including bon fide viruses. Conceivably, evolutionary transitions between these two types of replicators occurred repeatedly during evolution. The recent major expansion of the diversity of ribozyviruses that until then have been represented solely by human HDV suggests that we are presently unaware of the full range of viroid-like replicators, and even might be missing most of their diversity. Comprehensive search of transcriptomes and metatranscriptomes for viroid-like RNAs using dedicated computational pipelines can be expected to substantially expand their diversity and shed light on their origins and evolutionary relationships.

Author Contributions: Conceptualization, E.V.K., writing—original draft preparation, B.D.L., writing—review and editing, B.D.L., E.V.K. All authors have read and agreed to the published version of the manuscript.

Funding: B.D.L. was supported by a fellowship from the National Institutes of Health Oxford-Cambridge Scholars Program. E.V.K. is supported by the Intramural Research Program of the National Institutes of Health of the USA (National Library of Medicine).

Acknowledgments: Figures 1 and 2 were created with BioRender.com (accessed on 15 December 2021). This article is dedicated to Theodor Otto Diener, the discoverer of viroids, on the occasion of his 100th anniversary.

Conflicts of Interest: The authors declare no conflict of interest.

References

1. Gilbert, W. The RNA World. *Nature* **1986**, *319*, 618. [CrossRef]
2. Joyce, G.F. The antiquity of RNA-based evolution. *Nature* **2002**, *418*, 214–221. [CrossRef] [PubMed]
3. Joyce, G.F.; Szostak, J.W. Protocells and RNA Self-Replication. *Cold Spring Harb. Perspect. Biol.* **2018**, *10*, a034801. [CrossRef]
4. Tjhung, K.F.; Shokhirev, M.N.; Horning, D.P.; Joyce, G.F. An RNA polymerase ribozyme that synthesizes its own ancestor. *Proc. Natl. Acad. Sci. USA* **2020**, *117*, 2906–2913. [CrossRef] [PubMed]
5. Horning, D.P.; Joyce, G.F. Amplification of RNA by an RNA polymerase ribozyme. *Proc Natl. Acad. Sci. USA* **2016**, *113*, 9786–9791. [CrossRef]
6. Diener, T.O. The viroid: Biological oddity or evolutionary fossil? *Adv. Virus Res.* **2001**, *57*, 137–184.
7. Diener, T.O. Discovering viroids—A personal perspective. *Nat. Rev. Microbiol.* **2003**, *1*, 75–80. [CrossRef] [PubMed]
8. Diener, T.O. Viroids. *Adv. Virus Res.* **1972**, *17*, 295–313.
9. Navarro, B.; Flores, R.; Di Serio, F. Advances in Viroid-Host Interactions. *Annu. Rev. Virol.* **2021**, *8*, 305–325. [CrossRef]
10. Badar, U.; Venkataraman, S.; AbouHaidar, M.; Hefferon, K. Molecular interactions of plant viral satellites. *Virus Genes* **2021**, *57*, 1–22. [CrossRef]
11. Wang, Y. Current view and perspectives in viroid replication. *Curr. Opin. Virol.* **2021**, *47*, 32–37. [CrossRef] [PubMed]
12. Flores, R.; Grubb, D.; Elleuch, A.; Nohales, M.A.; Delgado, S.; Gago, S. Rolling-circle replication of viroids, viroid-like satellite RNAs and hepatitis delta virus: Variations on a theme. *RNA Biol.* **2011**, *8*, 200–206. [CrossRef] [PubMed]
13. Lasda, E.; Parker, R. Circular RNAs: Diversity of form and function. *RNA Microbiol.* **2014**, *20*, 1829–1842. [CrossRef]
14. Rocheleau, L.; Pelchat, M. The Subviral RNA Database: A toolbox for viroids, the hepatitis delta virus and satellite RNAs research. *BMC Microbiol.* **2006**, *6*, 24. [CrossRef]
15. Lee, B.D.; Neri, U.; Oh, C.J.; Simmonds, P.; Koonin, E.V. ViroidDB: A database of viroids and viroid-like circular RNAs. *Nucleic Acids Res.* **2021**, *50*, D432–D438. [CrossRef]
16. Flores, R.; Owens, R.A.; Taylor, J. Pathogenesis by subviral agents: Viroids and hepatitis delta virus. *Curr. Opin. Virol.* **2016**, *17*, 87–94. [CrossRef]
17. Diener, T.O. Circular RNAs: Relics of precellular evolution? *Proc. Natl. Acad. Sci. USA* **1989**, *86*, 9370–9374. [CrossRef] [PubMed]
18. Flores, R.; Gago-Zachert, S.; Serra, P.; Sanjuan, R.; Elena, S.F. Viroids: Survivors from the RNA world? *Annu. Rev. Microbiol.* **2014**, *68*, 395–414. [CrossRef]
19. Moelling, K.; Broecker, F. Viroids and the Origin of Life. *Int. J. Mol. Sci.* **2021**, *22*, 3476. [CrossRef]
20. Raymer, W.B.; Diener, T.O. Potato spindle tuber virus: A plant virus with properties of a free nucleic acid. I. Assay, extraction, and concentration. *Virology* **1969**, *37*, 343–350. [CrossRef]
21. Stollar, B.D.; Diener, T.O. Potato spindle tuber viroid. V. Failure of immunological tests to disclose double-stranded RNA or RNA-DNA hybrids. *Virology* **1971**, *46*, 168–170. [CrossRef]
22. Delan-Forino, C.; Maurel, M.-C.; Torchet, C. Replication of Avocado Sunblotch Viroid in the Yeast Saccharomyces Cerevisiae. *J. Virol.* **2011**, *85*, 3229–3238. [CrossRef] [PubMed]
23. Wei, S.; Bian, R.; Andika, I.B.; Niu, E.; Liu, Q.; Kondo, H.; Yang, L.; Zhou, H.; Pang, T.; Lian, Z.; et al. Symptomatic plant viroid infections in phytopathogenic fungi. *Proc. Natl. Acad. Sci. USA* **2019**, *116*, 13042–13050. [CrossRef] [PubMed]
24. Giguere, T.; Adkar-Purushothama, C.R.; Perreault, J.P. Comprehensive secondary structure elucidation of four genera of the family Pospiviroidae. *PLoS ONE* **2014**, *9*, e98655.
25. Giguere, T.; Adkar-Purushothama, C.R.; Bolduc, F.; Perreault, J.P. Elucidation of the structures of all members of the Avsunviroidae family. *Mol. Plant Pathol.* **2014**, *15*, 767–779. [CrossRef]
26. Branch, A.D.; Benenfeld, B.J.; Robertson, H.D. Evidence for a single rolling circle in the replication of potato spindle tuber viroid. *Proc. Natl. Acad. Sci. USA* **1988**, *85*, 9128–9132. [CrossRef] [PubMed]
27. Daros, J.A.; Marcos, J.F.; Hernandez, C.; Flores, R. Replication of avocado sunblotch viroid: Evidence for a symmetric pathway with two rolling circles and hammerhead ribozyme processing. *Proc. Natl. Acad. Sci. USA* **1994**, *91*, 12813–12817. [CrossRef] [PubMed]
28. Macnaughton, T.B.; Shi, S.T.; Modahl, L.E.; Lai, M.M. Rolling circle replication of hepatitis delta virus RNA is carried out by two different cellular RNA polymerases. *J. Virol.* **2002**, *76*, 3920–3927. [CrossRef] [PubMed]
29. Flores, R.; Gas, M.E.; Molina-Serrano, D.; Nohales, M.A.; Carbonell, A.; Gago, S.; De la Pena, M.; Daros, J.A. Viroid replication: Rolling-circles, enzymes and ribozymes. *Viruses* **2009**, *1*, 317–334. [CrossRef] [PubMed]
30. Flores, R.; Minoia, S.; Carbonell, A.; Gisel, A.; Delgado, S.; Lopez-Carrasco, A.; Navarro, B.; Di Serio, F. Viroids, the simplest RNA replicons: How they manipulate their hosts for being propagated and how their hosts react for containing the infection. *Virus Res.* **2015**, *209*, 136–145. [CrossRef] [PubMed]
31. Daros, J.A.; Elena, S.F.; Flores, R. Viroids: An Ariadne's thread into the RNA labyrinth. *EMBO Rep.* **2006**, *7*, 593–598. [CrossRef] [PubMed]
32. Molina-Serrano, D.; Suay, L.; Salvador, M.L.; Flores, R.; Daros, J.A. Processing of RNAs of the family Avsunviroidae in Chlamydomonas reinhardtii chloroplasts. *J. Virol.* **2007**, *81*, 4363–4366. [CrossRef] [PubMed]
33. Nohales, M.A.; Flores, R.; Daros, J.A. Viroid RNA redirects host DNA ligase 1 to act as an RNA ligase. *Proc. Natl. Acad. Sci. USA* **2012**, *109*, 13805–13810. [CrossRef] [PubMed]

34. Nohales, M.A.; Molina-Serrano, D.; Flores, R.; Daros, J.A. Involvement of the chloroplastic isoform of tRNA ligase in the replication of viroids belonging to the family Avsunviroidae. *J. Virol.* **2012**, *86*, 8269–8276. [CrossRef] [PubMed]
35. Gnanasekaran, P.; Chakraborty, S. Biology of viral satellites and their role in pathogenesis. *Curr. Opin. Virol.* **2018**, *33*, 96–105 [CrossRef] [PubMed]
36. Navarro, B.; Rubino, L.; Di Serio, F. Small circular satellite RNAs. In *Viroids and Satellites*; Academic Press: San Diego, CA, USA, 2017; pp. 659–669.
37. Elena, S.F.; Dopazo, J.; de la Pena, M.; Flores, R.; Diener, T.O.; Moya, A. Phylogenetic analysis of viroid and viroid-like satellite RNAs from plants: A reassessment. *J. Mol. Evol.* **2001**, *53*, 155–159. [CrossRef] [PubMed]
38. Taylor, J.M. Structure and replication of hepatitis delta virus RNA. In *Hepatitis Delta Virus. Medical Intelligence Unit*; Springer: Boston, MA, USA, 2006.
39. Sureau, C.; Negro, F. The hepatitis delta virus: Replication and pathogenesis. *J. Hepatol.* **2016**, *64*, S102–S116. [CrossRef]
40. Modahl, L.E.; Macnaughton, T.B.; Zhu, N.; Johnson, D.L.; Lai, M.M. RNA-Dependent replication and transcription of hepatitis delta virus RNA involve distinct cellular RNA polymerases. *Mol. Cell Biol.* **2000**, *20*, 6030–6039. [CrossRef]
41. Macnaughton, T.B.; Lai, M.M. HDV RNA replication: Ancient relic or primer? *Curr. Top. Microbiol. Immunol.* **2006**, *307*, 25–45. [PubMed]
42. Tseng, C.H.; Lai, M.M. Hepatitis delta virus RNA replication. *Viruses* **2009**, *1*, 818–831. [CrossRef] [PubMed]
43. Wille, M.; Netter, H.J.; Littlejohn, M.; Yuen, L.; Shi, M.; Eden, J.S.; Klaassen, M.; Holmes, E.C.; Hurt, A.C. A Divergent Hepatitis D-Like Agent in Birds. *Viruses* **2018**, *10*, 720. [CrossRef] [PubMed]
44. Chang, W.S.; Pettersson, J.H.; Le Lay, C.; Shi, M.; Lo, N.; Wille, M.; Eden, J.S.; Holmes, E.C. Novel hepatitis D-like agents in vertebrates and invertebrates. *Virus Evol.* **2019**, *5*, vez021. [CrossRef] [PubMed]
45. Hetzel, U.; Szirovicza, L.; Smura, T.; Prahauser, B.; Vapalahti, O.; Kipar, A.; Hepojoki, J. Identification of a Novel Deltavirus in Boa Constrictors. *mBio* **2019**, *10*, e00014-19. [CrossRef] [PubMed]
46. Szirovicza, L.; Hetzel, U.; Kipar, A.; Martinez-Sobrido, L.; Vapalahti, O.; Hepojoki, J. Snake Deltavirus Utilizes Envelope Proteins of Different Viruses to Generate Infectious Particles. *mBio* **2020**, *11*, e03250-19. [CrossRef]
47. De la Pena, M.; Ceprian, R.; Casey, J.L.; Cervera, A. Hepatitis delta virus-like circular RNAs from diverse metazoans encode conserved hammerhead ribozymes. *Virus Evol.* **2021**, *7*, veab016. [CrossRef]
48. Krupovic, M.; Dolja, V.V.; Koonin, E.V. Origin of viruses: Primordial replicators recruiting capsids from hosts. *Nat. Rev. Microbiol.* **2019**, *17*, 449–458. [CrossRef] [PubMed]
49. Diener, T.O. Viroids: "living fossils" of primordial RNAs? *Biol. Direct* **2016**, *11*, 15. [CrossRef] [PubMed]
50. Koonin, E.V. The origins of cellular life. *Antonie Van Leeuwenhoek* **2014**, *106*, 27–41. [CrossRef]
51. Catalan, P.; Elena, S.F.; Cuesta, J.A.; Manrubia, S. Parsimonious Scenario for the Emergence of Viroid-Like Replicons De Novo. *Viruses* **2019**, *11*, 425. [CrossRef] [PubMed]
52. Jain, N.; Blauch, L.R.; Szymanski, M.R.; Das, R.; Tang, S.K.Y.; Yin, Y.W.; Fire, A.Z. Transcription polymerase-catalyzed emergence of novel RNA replicons. *Science* **2020**, *368*, eaay0688. [CrossRef] [PubMed]
53. Jeancolas, C.; Matsubara, Y.J.; Vybornyi, M.; Lambert, C.N.; Blokhuis, A.; Alline, T.; Griffiths, A.D.; Ameta, S.; Krishna, S.; Nghe, P. RNA Diversification by a Self-Reproducing Ribozyme Revealed by Deep Sequencing and Kinetic Modelling. *Chem. Commun.* **2021**, *61*, 7517–7520. [CrossRef]
54. Diener, T.O. Are viroids escaped introns? *Proc. Natl. Acad. Sci. USA* **1981**, *78*, 5014–5015. [CrossRef] [PubMed]
55. Grabowski, P.J.; Zaug, A.J.; Cech, T.R. The intervening sequence of the ribosomal RNA precursor is converted to a circular RNA in isolated nuclei of Tetrahymena. *Cell* **1981**, *23*, 467–476. [CrossRef]
56. Kruger, K.; Grabowski, P.J.; Zaug, A.J.; Sands, J.; Gottschling, D.E.; Cech, T.R. Self-splicing RNA: Autoexcision and autocyclization of the ribosomal RNA intervening sequence of Tetrahymena. *Cell* **1982**, *31*, 147–157. [CrossRef]
57. Haruna, I.; Spiegelman, S. Autocatalytic synthesis of a viral RNA in vitro. *Science* **1965**, *150*, 884–886. [CrossRef] [PubMed]
58. Spiegelman, S.; Haruna, I.; Pace, N.R.; Mills, D.R.; Bishop, D.H.; Claybrook, J.R.; Peterson, R. Studies in the replication of viral RNA. *J. Cell Physiol.* **1967**, *70*, 35–64. [CrossRef]
59. Spiegelman, S. An approach to the experimental analysis of precellular evolution. *Q. Rev. Biophys.* **1971**, *4*, 213–253. [CrossRef]
60. Mills, D.R.; Kramer, F.R.; Spiegelman, S. Complete nucleotide sequence of a replicating RNA molecule. *Science* **1973**, *180*, 916–927. [CrossRef]
61. Wernegreen, J.J. For better or worse: Genomic consequences of intracellular mutualism and parasitism. *Curr. Opin. Genet. Dev.* **2005**, *15*, 572–583. [CrossRef]
62. Koonin, E.V.; Wolf, Y.I. Genomics of bacteria and archaea: The emerging dynamic view of the prokaryotic world. *Nucleic Acids Res.* **2008**, *36*, 6688–6719. [CrossRef] [PubMed]
63. Wolf, Y.I.; Koonin, E.V. Genome reduction as the dominant mode of evolution. *Bioessays* **2013**, *35*, 829–837. [CrossRef] [PubMed]
64. Moran, N.A. Microbial minimalism: Genome reduction in bacterial pathogens. *Cell* **2002**, *108*, 583–586. [CrossRef]
65. McCutcheon, J.P.; Moran, N.A. Extreme genome reduction in symbiotic bacteria. *Nat. Rev. Microbiol.* **2011**, *10*, 13–26. [CrossRef] [PubMed]
66. Boyer, M.; Azza, S.; Barrassi, L.; Klose, T.; Campocasso, A.; Pagnier, I.; Fournous, G.; Borg, A.; Robert, C.; Zhang, X.; et al. Mimivirus shows dramatic genome reduction after intraamoebal culture. *Proc. Natl. Acad. Sci. USA* **2011**, *108*, 10296–10301. [CrossRef] [PubMed]

67. Sanjuan, R. The Social Life of Viruses. *Annu. Rev. Virol.* **2021**, *8*, 183–199. [CrossRef]
68. Senkevich, T.G.; Yutin, N.; Wolf, Y.I.; Koonin, E.V.; Moss, B. Ancient Gene Capture and Recent Gene Loss Shape the Evolution of Orthopoxvirus-Host Interaction Genes. *mBio* **2021**, *12*, e0149521. [CrossRef]
69. Cervera, A.; Urbina, D.; de la Pena, M. Retrozymes are a unique family of non-autonomous retrotransposons with hammerhead ribozymes that propagate in plants through circular RNAs. *Genome Biol.* **2016**, *17*, 135. [CrossRef]
70. De la Pena, M. Circular RNAs Biogenesis in Eukaryotes Through Self-Cleaving Hammerhead Ribozymes. *Adv. Exp. Med. Biol.* **2018**, *1087*, 53–63. [PubMed]
71. Cervera, A.; de la Pena, M. Cloning and Detection of Genomic Retrozymes and Their circRNA Intermediates. *Methods Mol. Biol.* **2021**, *2167*, 27–44.
72. Witte, C.P.; Le, Q.H.; Bureau, T.; Kumar, A. Terminal-repeat retrotransposons in miniature (TRIM) are involved in restructuring plant genomes. *Proc. Natl. Acad. Sci. USA* **2001**, *98*, 13778–13783. [CrossRef] [PubMed]
73. Gao, D.; Chen, J.; Chen, M.; Meyers, B.C.; Jackson, S. A highly conserved, small LTR retrotransposon that preferentially targets genes in grass genomes. *PLoS ONE* **2012**, *7*, e32010. [CrossRef] [PubMed]
74. De la Pena, M.; Cervera, A. Circular RNAs with hammerhead ribozymes encoded in eukaryotic genomes: The enemy at home. *RNA Biol.* **2017**, *14*, 985–991. [CrossRef] [PubMed]
75. De la Pena, M.; Ceprian, R.; Cervera, A. A Singular and Widespread Group of Mobile Genetic Elements: RNA Circles with Autocatalytic Ribozymes. *Cells* **2020**, *9*, 2555. [CrossRef]
76. Cervera, A.; de la Pena, M. Small circRNAs with self-cleaving ribozymes are highly expressed in diverse metazoan transcriptomes. *Nucleic Acids Res.* **2020**, *48*, 5054–5064. [CrossRef]
77. Jandura, A.; Krause, H.M. The New RNA World: Growing Evidence for Long Noncoding RNA Functionality. *Trends Genet.* **2017**, *33*, 665–676. [CrossRef] [PubMed]
78. Quake, S.R. The cell as a bag of RNA. *Trends Genet.* **2021**, *37*, 1064–1068. [CrossRef] [PubMed]
79. Salehi-Ashtiani, K.; Szostak, J.W. In vitro evolution suggests multiple origins for the hammerhead ribozyme. *Nature* **2001**, *414*, 82–84. [CrossRef]
80. Dawkins, R. *The Selfish Gene*; Oxford University Press: Oxford, UK, 1976.
81. Szathmary, E.; Maynard Smith, J. From replicators to reproducers: The first major transitions leading to life. *J. Theor. Biol.* **1997**, *187*, 555–571. [CrossRef]
82. Szathmary, E. The evolution of replicators. *Philos. Trans. R. Soc. Lond. B Biol. Sci.* **2000**, *355*, 1669–1676. [CrossRef]
83. Koonin, E.V.; Starokadomskyy, P. Are viruses alive? The replicator paradigm sheds a decisive light on an old but misguided question. *Stud. Hist. Philos. Biol. Biomed. Sci.* **2016**, *59*, 125–134. [CrossRef]
84. Koonin, E.V.; Dolja, V.V.; Krupovic, M.; Kuhn, J.H. Viruses Defined by the Position of the Virosphere within the Replicator Space. *Microbiol. Mol. Biol. Rev.* **2021**, *85*, e0019320. [CrossRef]
85. Copley, S.D.; Smith, E.; Morowitz, H.J. The origin of the RNA world: Co-evolution of genes and metabolism. *Bioorg. Chem.* **2007**, *35*, 430–443. [CrossRef] [PubMed]
86. Jalasvuori, M.; Koonin, E.V. Classification of prokaryotic genetic replicators: Between selfishness and altruism. *Ann. N. Y. Acad. Sci.* **2015**, *1341*, 96–105. [CrossRef] [PubMed]
87. Koonin, E.V.; Wolf, Y.I.; Katsnelson, M.I. Inevitability of the emergence and persistence of genetic parasites caused by evolutionary instability of parasite-free states. *Biol. Direct* **2017**, *12*, 31. [CrossRef] [PubMed]
88. AbouHaidar, M.G.; Venkataraman, S.; Golshani, A.; Liu, B.; Ahmad, T. Novel coding, translation, and gene expression of a replicating covalently closed circular RNA of 220 nt. *Proc. Natl. Acad. Sci. USA* **2014**, *111*, 14542–14547. [CrossRef] [PubMed]

MDPI
St. Alban-Anlage 66
4052 Basel
Switzerland
Tel. +41 61 683 77 34
Fax +41 61 302 89 18
www.mdpi.com

Life Editorial Office
E-mail: life@mdpi.com
www.mdpi.com/journal/life

www.ingramcontent.com/pod-product-compliance
Lightning Source LLC
LaVergne TN
LVHW070208100526
838202LV00015B/2017